Tales of the Telegraph

Tales of the Telegraph

The early development of British Army communications, and their deployment on operations 1850 to 1902

Lieutenant Colonel David Mullineaux

ROYAL SIGNALS INSTITUTION
BLANDFORD CAMP
BLANDFORD FORUM
DORSET DT11 8RH

https://royalsignals.org/royal-signals-institution

First published 2021

Royal Signals Institution
Blandford Camp
Blandford Forum
DORSET DT11 8RH
https://royalsignals.org/royal-signals-institution

© Royal Signals Trustee Limited, 2021

The right of Lt Col David Mullineaux to be identified as the author of this work has been asserted in accordance with the Copyright, Design and Patents Act 1988.

All rights reserved. No part of this book may be reprinted or reproduced or utilised in any form or by any electronic, mechanical or other means, now known or hereafter invented, including photocopying and recording, or in any information storage or retrieval system, without the permission in writing from the Royal Signals Trustee Limited.

British Library Cataloguing in Publication Data.
A catalogue record for this book is available from the British Library.

ISBN-13: 978-1-9162643-2-8 Paperback

Typesetting and origination by Adam Forty, from an orginal artwork by Lt Col D Mullineaux.
Printed in the UK by Holbrooks Printers Ltd, Hilsea, Portsmouth, PO3 5HX.

This book was commissioned by the Royal Signals Trustee Ltd, it includes two original books by Lt Col D Mullineaux and was reworked to this single edition by Adam Forty.

Foreword

The latter part of the 19th century saw what in those days were significant advances in the development of communication methods. They offered much to the progress of army signalling but these opportunities were largely ignored at senior levels in the moribund Victorian army, and what was achieved was mainly due to the initiative of more junior officers

With "Tales of the Telegraph" Lieutenant Colonel David Mullineaux has fully addressed this failing. Having visited battlefields in South Africa, and found that the guides had little or no knowledge of the communications involved - and realising it was a subject of little interest to military historians - he decided to investigate. Harnessing his training and knowledge as a military communications engineer to great energy in tracking down primary sources, visiting battlefield sites and collating personal detail, he has produced an authoritative, entertaining and readable account of the communications involvement in the most significant engagements over a period of 80 years up to and including the Boer Wars.

Reading through these stories, one is inevitably struck by the relative youth of those intrepid young officers of the Royal Engineers who planned, mounted and led the communications support for operations in the Middle East, India and Africa. With little or no senior officer level direction, they performed astonishing deeds of courage, resilience and initiative to realise the ambitions of those at senior levels in Whitehall and elsewhere in the growing British Empire. Warring tribesmen, Treasury reluctance, professional resentment, equipment scarcity, heat, jungle and disease were all obstacles to be overcome in achieving their mission. Their exploits are an example to all, and a testament to the power of the human will in overcoming adversity in all its forms.
To communications practitioners, as well as anyone with an interest in the history of these times, this book is an indispensable reference. I thoroughly recommend it to everyone.

Major General SPM Nesmith
Master of Signals
May 2021

The Author

David Mullineaux was commissioned from Sandhurst into the Royal Signals in December 1957. After the young officers qualifying course he was posted to Germany to the Signal Troop attached to 47 Guided Weapons Regiment, Royal Artillery, which had been recently deployed there with Corporal guided missiles. His next posting was to Kenya in 1961 for nearly five years, firstly with a British Brigade Signal Squadron during which time he climbed Mount Kilimanjaro and saw active service for three months in Kuwait, and then with the King's African Rifles Signal Squadron

On completion of this tour he returned to Catterick for the two-year Telecommunications Engineering course, during which he qualified as a Chartered Engineer and as a Member of the Institution of Electrical Engineers. This was followed by another tour in Germany as second-in-command of an Armoured Brigade Signal Squadron.

On promotion to Major he was posted to a technical staff appointment at the Signals Research and Development Establishment where he was involved with the introduction of the new Clansman radios system. After two years this was followed by a tour at Sandhurst as the Chief Instructor of the Signals Wing. Then it was back to Germany, this time to the Corps Signal Regiment, and subsequently to a staff appointment in the Ministry of Defence. After this, in 1978, he attended a course at the National Defence College.

Now a Lieutenant Colonel, his next appointment was as Commander Royal Signals in Cyprus for a pleasant three years during which he organised a five-day visit by the recently appointed Colonel-in-Chief of the Corps, HRH Princess Anne. This was followed by his last tour in the Army, in the Ministry of Defence planning the introduction of the Ptarmigan communication system.

He retired in 1985 after 27 years of service and took up appointments in the civil sector, firstly with Racal Communications, and then with the overseas division of British Telecom, retiring finally in 1994. Since then he has enjoyed many holidays of worldwide travel.

Contents

Foreword

Part 1

The early development of British Army communications
and their deployment on operations 1850 to 1898

Pages 1 - 198

Part 2

The Telegraph Battalion in the Anglo-Boer War, 1899 to 1902

Pages 1 - 124

Tales of the Telegraph

Part 1

The early development
of
British Army communications
and their deployment on operations
In the late Victorian period, 1850 to 1898

by

Lieutenant Colonel David Mullineaux

Tales of the Telegraph

Contents

		Page
	Introduction	- ii -

Chapter

1	Early Telegraphs	1
2	The Crimean War, 1854-56	9
3	Telegraph Developments in India	18
4	The Telegraph Route to India, 1858-70	26
5	Army Signalling Developments in the 1860s	40
6	The Abyssinian Expedition, 1867-68	48
7	The Formation of Army Telegraph Units, 1870-72	61
8	The Ashanti Expedition, 1873-74	72
9	The Heliograph	79
10	The 2nd Afghan War, 1878-80	84
11	The Anglo-Zulu War, 1879	96
12	The Campaign in Egypt, 1882	120
13	The Bechuanaland Expedition, 1884-85	131
14	The Nile Expedition, 1884-85	137
15	Suakin, 1885	159
16	The Second Ashanti Expedition, 1895-96	171
17	Reconquest of the Sudan, 1896-98	179
	Bibliography	196

Introduction

The story of army signalling in any meaningful way begins in the second half of the 19th century - the mid-Victorian era.

Visual signalling of various sorts using sun, fire, and smoke, goes back to ancient history but these haphazard methods form no part of the narrative. Effective visual telegraphs, aided by the invention of the telescope, had been introduced in France in 1792, and soon afterwards in England. They used signalling codes so that coherent messages could be sent and acknowledged. But they could only be established in friendly territory, along pre-determined routes, and were slow to construct; the army needed something more mobile and less conspicuous.

By the late 1820s optical developments had produced the heliostat, and developments in chemistry had produced the limelight, a brilliant source of light. These inventions were initially used for other purposes and it was to take an inordinately long time, until the mid-1860s, for the army to exploit their visual signalling potential.

Following a series of scientific discoveries the electric telegraph became a practical means of communication, and generally came into civil use in the 1840s; with it came the Morse signalling code, which could be applied also to visual signalling. From these two separate threads, visual signalling and electric telegraphy, each with their strengths and weaknesses, army signalling began to make practical advances from about 1850 onwards.

But its progress was slow, for reasons that will be explained as the narrative unfolds. The main contributory factors to this sluggish development were general retrenchment of the army in the post-Waterloo period, and lack of awareness and disinterest in the higher levels of the moribund mid-Victorian army. When new methods were eventually introduced they were often not used to best advantage; there were organisational defects in both the wider aspect of army staff organisation, and more relevant to this story, the organisation needed to establish and integrate efficiently the two separate signalling methods - visual and electrical.

All of that is set against the changing course of history over the period, moving away from operations focused on Europe towards Britain's growing empire, principally in India and Africa, and the many army operations that accompanied this expansion. Such a scenario, far from home territory, into comparatively undeveloped places where distances were great, transport capacity very limited, and topography and climate quite different, all contrasted with the European scene and presented numerous challenges to the provision of army communications.

Tales of the Telegraph follows the course of these events chronologically. It describes the signalling developments and their application, with varying degrees of success, to the wars and expeditions in the second half of the 19th century, stopping just before the Boer War of 1899-1902. Signalling in that war, which included an unsuccessful trial of newly developed wireless communications, was extensive, and to do it justice is another story. *Tales of the Telegraph* is not just about signalling methods and communications in the numerous operations of the developing British Empire; it is also about some of the personalities who were involved, many of whom wrote of their experiences.

I have written this book because, although the history of the period is thoroughly well-known and well-documented, the signalling story - a specialised aspect that neither attracts general interest nor profitable publishing - does not exist in one entity. What has been published is fragmented and unfortunately sometimes inaccurate. My research into primary sources has included contemporary accounts by those directly involved, either in written documents or delivered in presentations to organisations such as the Royal United Services Institution and the Society of Telegraph Engineers. It has been conducted mainly in the archives of the Royal Engineers and Royal Signals, and in the Prince Consort's Library, Aldershot, where many contemporary books and reference material have been made freely available to me. There are other sources, referred to in the narrative.

I hope I have put together in one document an accurate account of how army signalling developed in the second half of the 19th century, how it was used in the principal army operations of the period, how relatively junior officers carried great responsibility, and what a struggle it was in comparison with the pace of modern technology and the relative luxury of communication methods now available.

David Mullineaux

Chapter 1

Early Telegraphs

Introduction

By way of introduction, this first chapter reviews the early stages of telegraphy. It summarises the development and use of visual telegraph systems starting in 1792 and, succeeding them, the early developments in the electric telegraph up to about 1850. These preceded the main period to be described in this book - the second half of the 19th century.

The Chappe Semaphore System in France

Although he was not the first to dabble with the idea, the first practical telegraph system able to send and receive messages intelligently was developed in France by Claude Chappe, and introduced in the unsettled post-revolutionary period of the early 1790s when, apart from its internal problems, France was also at war with Austria. As we shall see later, in several contexts, the threat of war proved a great stimulus to improvements in military communications.

Chappe had considered electrical methods for his system but at the end of the 18th century the scientific developments were insufficiently advanced. Meanwhile, with his country at war, the need for better communication was pressing. He chose instead a visual telegraph system using a semaphore signalling code. Without becoming too involved in semantics, 'telegraph' means literally *writing at a distance*, and it is immaterial whether that is by visual or electrical methods, and 'semaphore' means *bearing a sign,* both being derived from Greek. Thus, semaphore is simply one method of telegraphy, using visual signals.

Chappe's method of producing a semaphore code consisted of a bar of wood about four-and-a-half metres long, called the 'regulator', placed at the top of a vertical mast. The regulator could be turned at forty-five degrees on each side of the mast, as well as horizontally and vertically. At each end of the regulator was another arm, the 'indicator', about two metres long. All these parts were moveable, using a system of pulleys, and by arranging them at different angles according to his semaphore code, 196 different characters could be produced. Chappe had researched his subject well. Within the limitations of any visual system, it was regarded during its lifetime of nearly sixty years as the most effective of the various other contemporary visual methods that came into use.

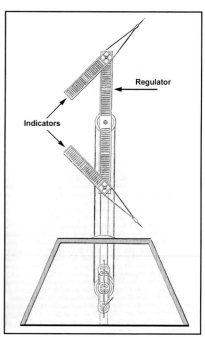

After some initial experiments, the first stretch of the Chappe system was constructed between Paris and Lille, about 135 miles with sixteen intermediate stations, and it was brought into service in mid-1794. The French and Austrian armies were at that time confronting each other in the area of Lille, so the new telegraph system brought the seat of government in Paris in rapid communication with the war front. Messages could be exchanged accurately in both directions at a speed and over a distance never before achieved. Meanwhile, as the French Revolution continued, gruesome executions took place in Paris - but that is not part of this story.

Ten years later, in 1804, after a coup d'état, Napoleon came to power in post-revolutionary France. Tragically, and for somewhat obscure reasons, Claude Chappe committed suicide in January the following year, but his three brothers continued to develop the telegraph network.

Chappe's semaphore.

Napoleon's militant activities gave further impetus to Chappe's highly successful telegraph system which, as well as being a strategic asset for military purposes, was turning out to be commercially profitable , and the network continued to expand. At its peak there were 556 Chappe semaphore stations stretching over a total distance of 4,800 kilometres in France and beyond. They radiated from Paris to Boulogne and Brussels in the north, Brest in the west, south through Lyons and over the Mont Cenis Pass into Milan and

Venice, and west through Strasbourg, the deployment beyond French borders influenced mainly for military reasons.

After nearly sixty years service, Chappe's hugely successful system was eventually overtaken by the electric telegraph, and was phased out in the early 1850s. [1]

The Admiralty's Shutter Telegraph System

The success of Chappe's system in France had soon been noted by other countries, and chains of visual telegraph stations were subsequently developed elsewhere in Europe and in the United States of America. The first of these, prompted because of the outbreak of war with France in 1793 and the consequent threat of invasion, was between the Admiralty in London and the naval bases at Portsmouth and Deal.

In developing the British system the Admiralty asked the Reverend John Gamble to investigate signalling methods. Gamble was a former mathematics don at Cambridge, and interested in scientific developments. He subsequently became Chaplain General to the Forces and chaplain to the Duke of York, the Commander-in-Chief of the army - not appointments that one might usually associate with the development of communication systems. His proposed solution was a shutter telegraph in which the individual shutters in an array could each be in either the vertical (visible) or horizontal (not visible) position and, depending on the form of signalling code and the arrangement of open and closed shutters, could spell out the characters - either letters or figures. The Gamble code that emerged was a five-shutter telegraph, limiting the number of characters to thirty-one – scarcely enough. The prime reason for adopting a shutter telegraph system was that, with rapid implementation demanded, it was the simplest, quickest and cheapest method available.

A slightly different proposal for the Admiralty system came from Lord George Murray, a son of the Duke of Atholl and later Bishop of St David's, who also became involved in the development. With Gamble and Murray both involved, it would appear, even in those early days, that a strong connection with the Almighty was considered beneficial to good communications for military purposes. Chappe, too, it might be observed, had been trained for and initially entered the church, but the anti-religious influence of the French Revolution had induced him to change course.

Murray used a frame of six shutters, thus enabling a code of sixty-three different characters, and after trials this swayed the Admiralty towards his method for their planned new telegraph chains. A £2,000 award went to Murray! Gamble, who seemed unaware that such money was at stake, was understandably peeved, saying that he only developed a five-shutter system because the Admiralty said they wanted something simple, and he could easily have designed a six-shutter system. Definition of operational requirements was clearly somewhat lax. (Lord George Murray, incidentally, should not be confused with Donald Murray, a New Zealander, who later invented the five-bit Murray code and other devices used in 20th century electric machine telegraphy.)

With some urgency, because of the impending French threat, construction of the Admiralty's shutter telegraph system was ordered in September 1795, and the route to Deal (70 miles) was opened four months later, in January 1796. The route to Portsmouth (65 miles) was completed later in 1796. The speed of construction was excellent, although the buildings housing the telegraph stations were temporary structures, economy as well as speed being a factor. With naval wartime activity

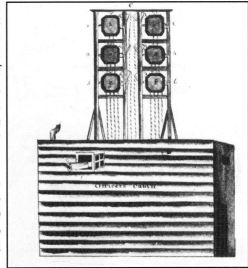

Murray's Shutter Telegraph.

high, the telegraph was much in use. In October 1805, coinciding with the Battle of Trafalgar, extension was authorised to Plymouth. Interestingly, one of the intermediate telegraph stations on the Plymouth route was located at Telegraph Clump, 1.2 miles east of Pimperne, close to the site of the present day army Royal School of Signals at Blandford Camp, complete with its modern equivalent, a microwave tower. In 1807 a further extension to Yarmouth on the east coast was authorised, and completed in 1808.

In 1814, Napoleon having been defeated at Leipzig and sent to Elba, and war apparently over, the Admiralty's shutter telegraph network was quickly, and rather prematurely, dismantled - to save money! Even by the time Napoleon escaped from Elba, ultimately to be defeated at Waterloo in 1815, the network was no longer workable. Little thought was given to the commercial benefits that might have been exploited. The shutters, quite literally, were closed. Nevertheless, many in England in those days had been amazed that messages could be sent over such distances so quickly.

The Admiralty's Semaphore Telegraph System

However, after the 'Hundred Days' following Napoleon's escape from Elba, and subsequent events, the shock of what might have happened was enough to reconsider that over-hasty decision. Only eleven days after Waterloo, on 29 June 1815, a new Act was issued by parliament for re-establishing a chain of telegraph stations. It included powers to take over private land as sites for telegraph stations - something that was to prove unpopular with many landowners, and impeding construction.

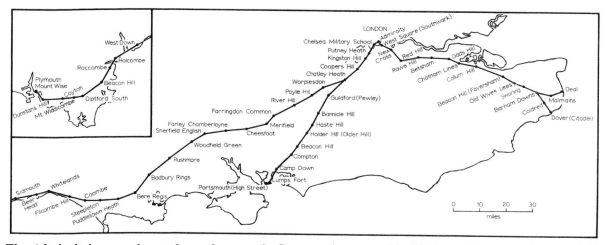

The Admiralty's semaphore telegraph network. Construction started in 1816, and it was used up to 1847.

This time, with the war over, Napoleon distanced to St Helena and thus no immediate threat, a more measured approach was taken. With visual telegraph still the only practical option, the principal terminal stations around the south coast were still much the same, but Yarmouth on the east coast was no longer needed. However, the exact location of telegraph stations along the south coast routes were in many cases modified, and the telegraph station buildings were more permanent structures rather than the earlier wooden shacks. The network came into service in 1822.

As part of this redesign, it was also decided, after trials, that the new telegraph system would no longer be a six-shutter system but would use a semaphore code - in this case the Popham code. Who was Popham, and what were the advantages of his signalling code?

Rear Admiral Sir Home Riggs Popham was a colourful character, and had even been court-martialled during his career, but he made major contributions to ship and land-based signalling at the time. [2] The feature of his signalling code that made it attractive was that it was a vocabulary code; in other words it allowed a conversational style of communication rather than a rigid code of pre-arranged phrases that covered the more common situations. [3] Others were involved in the development of comparable signalling codes at the time and quite who invented what remains a little obscure, but it was Popham to whom it was eventually attributed. Amongst those involved was Lieutenant Colonel Sir Charles William Pasley KCB, an officer in the Royal Engineers, who felt that Popham's semaphore greatly copied the one he had invented some years previously. [4] As with other inventions to follow, both in the electric telegraph and in other methods of visual signalling yet to be developed, there were numerous contributors with competing claims.

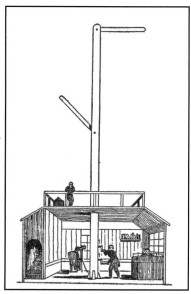

The Popham code telegraph station at the Admiralty.

The Popham Telegraph Station and Semaphore Code

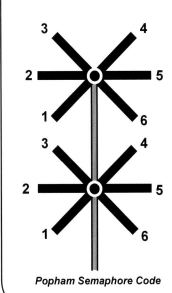

Popham Semaphore Code

The hollow hexagonal mast was formed of fir boards 10 inches wide and was 1 foot 8 inches in diameter, and stood 30 feet high above the roof of the building. The signalling arms were pivoted 12 feet up, and at the top, with 16½ feet between the pivots. The arms themselves were 8 feet long and 15 inches wide, and when at rest were housed within the mast. The whole apparatus could be rotated around its collar in the roof of the station, and could be fixed at 16 points around the compass. The arms were actuated by bevel wheels and rods leading from winch handles in the operating room below.

Signalling was achieved by the coordinated positioning of the two arms on the mast, and they could signal 48 different characters. The upper and lower arms could be used separately, and when individually positioned as shown in the sketch were able to signal 6 characters each (the arm not in use being in the vertical position against the mast). The combined arms when used together could signal 6 x 6, or 36 characters. Thus the two arms could signal a total of 48 characters; these were the 26 letters of the alphabets, 10 single digit numerals, and the remainder were a set of operating signals - preparative, received, closing down, etc. The operator at the receiving station would read the signals using a telescope.

One interesting sideline is that in 1826, after deliberations by an Admiralty committee about signalling between ships, Pasley's visual telegraph system was found to perform better than Popham's, and replaced it on navy ships, and so the navy continued to use a signalling system developed by an army officer for much of the Victorian era! Perhaps as compensation for this inter-service rivalry, the Popham code was used to provide emergency communications for the army in 1857 – from the besieged Lucknow Residency during the Indian Mutiny, but more about that later (chapter 3).

The Admiralty chain of semaphore telegraph stations using the Popham code, authorised to be built in 1816 and introduced into service in 1822, lasted until 1847 around when, as in France and elsewhere, it was superseded by the electric telegraph. [5] It existed during a period when Britain was not at war, so it was never used operationally - perhaps an example of planning defence strategy and budgets based on 'fighting the last war'.

Within the scientific limitations of their time, these visual telegraph systems were effective and had enabled communications at speeds and distances not previously possible. Their biggest failing was poor visibility - caused not just by weather, but also due to smoke pollution and fog ('smog') around London. Communication with the Admiralty signal station in central London was interrupted for an average of 100 days a year for this reason, whereas in Portsmouth it was around twenty-five days a year.

Army Use

What may have struck the reader by now is that the army, in contrast to the navy, seems to have made little use of the developments described so far. Despite their relative success, telegraph systems such as these could not meet most army operational requirements: planning and construction took a long time; the telegraph stations were naturally conspicuous; they could only work in daylight; and above all, they lacked mobility. How was the army going to develop their field communications?

Gamble, despite losing out rather unfairly on the £2,000 award that went to Murray for the shutter telegraph, had not given up in disgust at his shabby treatment and continued to work on the subject. In 1796 he developed what was called a 'radiated telegraph'. Designed purposely for mobile military use in the field, it was a semaphore system, consisting of a single mast and five arms. Each arm was five feet long,

Gamble's mobile telegraph cart.

with a disc at the end to improve its visibility to the distant observer. It had five positions, as shown in the diagram, enabling thirty-one different characters to be signalled. It could be mounted on a cart drawn by a horse. The arms were painted white on one side and black on the reverse, and depending on the background, the cart could be oriented so that the better contrast between signalling arm and background was presented to the distant station. A demonstration was given to senior officers between Woolwich and Shooters Hill over a distance of about five miles, and the performance of the system was praised.

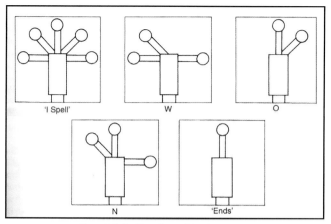

Gamble's Radiated Telegraph.

Later, in 1798, the army constructed a Gamble radiated telegraph station on a prominent point on Hampstead Heath and communicated with another station on top of the 225-foot north-west tower of Westminster Abbey. That seems, at a slight stretch of the imagination, to have been the first use of telegraphy by the army, but there are no details available, and any subsequent use of the Gamble radiated telegraph is obscure and of no operational significance.

During the Peninsular War the Duke of Wellington, with help from the Royal Navy, made some attempts to provide army visual telegraph communication. A small number of mobile telegraph stations using the Popham semaphore code were deployed by the army in 1810-11 around Torres Vedras in Portugal. This probably led to the introduction of telegraph sections attached to Divisional HQs in 1814, equipped with poles, flags, and codebooks evolved from the naval flag dictionary. Wellington probably intended them as a permanent part of Divisional HQs, but after Napoleon's final defeat the British army lapsed and many such ideas were lost.

In South Africa a chain of visual telegraph stations was built by the Royal Engineers between forts along the eastern frontier of the Cape Colony during the Kaffir Wars in the 1840s. Based on Pasley's semaphore system (it would, after all, have been unthinkable for the Royal Engineers to use any other!) it suffered various defects: siting of some of the stations proved to be imperfect, the telescopes provided from England were of inadequate power, and on hot days refraction distorted visibility over a distance and made the semaphore signals unreadable. It was used with only limited success for less than ten years.

But that was it for the time being. After Waterloo in 1815, and after the visual systems that have so far been described were brought into use, peace prevailed as far as Britain was concerned for some forty years. Meanwhile the industrial revolution gathered pace. Scientific developments started to present other opportunities for army signalling. New visual discoveries could easily have been adapted - but they weren't. Indeed, the adaptation of heliography and the limelight to army signalling, simply for lack of operational urgency and practical application by anybody with sufficient initiative, languished for some three decades after their invention. Their development into army signalling at a tactical level, along with flag signalling, will be explained in later chapters.

The Electric Telegraph

The other new opportunity was the development of the electric telegraph. It was for some time foreseen, but lacked certain elements that prevented it being a practical method of communication until the late 1830s. But there was an inordinate time after its introduction - some thirty years - until the electric telegraph was properly absorbed into army communications systems. Along with the electric telegraph came the Morse code - yet another signalling code, but this time adaptable to both visual signalling and the electric telegraph, with obvious advantages to signallers.

From the early 1840s, in the more developed countries such as in Europe and America, the electric telegraph took over rapidly. There is no need to trace the development of the electric telegraph in any detail for it is contained in numerous books (see bibliography).

EARLY TELEGRAPHS

The development of electric telegraphy began with a number of scientific discoveries from 1770, summarised in the chronology below.

1770.	The Italians Volta and Galvani developed what we would today call a battery, giving a steady source of electric current. It was a primary cell (that is, it could not be recharged), and it produced electricity from a chemical reaction involving a zinc plate, a copper plate, and dilute sulphuric acid. Chemical action was the only practical source of electricity for telegraphs until almost the end of the nineteenth century.
1820.	Some earlier experimenters noted that an electric current affected a magnetic needle although it was not until Oersted published his detailed observations in 1820 that anybody took much notice. Ampère then quickly worked out the magnetic action of a current, and Schweigger, in the same year, showed how to intensify its effect by winding the conductors in coils around the needle. It was then realised that it was possible to detect the reception of a signal, and thus the needle, which came to be called a galvanometer, became an early receiving device.
1825.	The electromagnet was discovered by William Sturgeon. He wound the wire around a soft magnetic core and discovered that, while electric current flowed, it could exert force on an iron armature. Its potential to act as a telegraph 'relay', thus regenerating a signal, was not realised until about 1835.
1827.	Ohm's study of the 'Galvanic circuit' led to Ohm's Law – the relationship between electric voltage, resistance, and current. Schoolboy stuff today, but few people understood electricity then. It was invisible, and to most, incomprehensible.
1832.	In the USA Professor Samuel Morse and his partner Alfred Vail (who seemed to do most of the work while Morse got the credit) were hard at it. His experiments were conducted in isolation of the work being done in Europe but he filed his patent for a telegraph system in the same year as Cooke and Wheatstone, 1837.
1835.	The next step was the invention of the 'repeating relay', in about 1835. Up to then there was a limit, determined by electrical parameters, which limited the distance over which a signal could be received – about ten miles. But of course the 'signal' was only 'on' or 'off', (in today's vocabulary it was 'digital'), so the use of an electromagnetic relay could automatically 'repeat' it without operator intervention.
1836.	A needle telegraph had been developed by Baron Pawel Schilling. This was seen by the young William Cooke who immediately realised the possibilities of a telegraphic system linking the major towns in the country. But he was unknown to investors and to get commercial support for his scheme he turned to Professor Charles Wheatstone as partner.
1837.	Steinheil, an Austrian, unlike many of the names in this list, does not often figure in the development of the electric telegraph, yet he made many contributions to early practical systems. The most effective, certainly as far as army field telegraph was concerned, was that he discovered the 'earth return' circuit. In other words, between two places, a single wire connected at each end to earth, would enable the electric current to flow. This was crucial to early army telegraph circuits and was invariably used - until World War I, when the density of telegraph circuits in France using 'earth return' circuitry led to 'cross talk', with consequent confusion and security problems, and two-wire circuits had to be used .
1837.	The partnership of Cooke and Wheatstone introduced the first railway telegraph; it worked between Euston and Camden, about two miles. It was an alphabetical system with five needles, but was clumsy to operate and, because it needed five wires, was expensive in cable. Nevertheless, the man walking ahead of the train with a red flag was already facing redundancy! It was not long before it was replaced by the double-needle system, with only two wires, using a code to indicate the letters. In turn the double-needle soon gave way to the single-needle system, using only one wire. It used the Morse code, transmitted from a device known as a commutator. The commutator had two keys, which made current flow in opposite directions in the line, corresponding to a dot and a dash. The receiving operator had to watch the swing of the needle but this was slow and a second operator was needed to write it down.

At the time there was great competition as to who had been first to invent what, and patents were contested fiercely, with the usual accompanying enrichment for lawyers.

Further Improvements

The next improvement in electric telegraphy was the development of better methods of receiving the signal. Up to now transmission had been what could be called 'transitory' – if the operator missed any part of the incoming signal, it was lost and had to be repeated. This led to the invention of 'recorders' (called 'registers' in America). 'Recording' was the term used to provide some permanent record of the received signal, such as some form of printing, and was preferred because it provided a permanent record and did not need a skilled operator. Morse had invented the 'embosser' as part of his system in 1837, but it was a rather feeble affair, difficult to read. In 1846 an improved method of sending and recording was invented by Alexander Bain, a Scotsman. Bain's inventions were an automatic punched paper tape sender which replaced the Morse key, and a new form of receiver which, using a stylus and an electromagnet, printed on to special chemical paper.

This was followed in 1854 by the Morse inker, invented by Thomas John of Vienna. Here an inked wheel in contact with the paper tape produced easily readable characters. Inkers in one form or another were to be around for many years, and were used by the army.

Vail, Morse's partner, then developed the 'sounder', consisting simply of an electromagnet which emitted clicks that could be recognised by skilled operators as resulting from a short or a long period of excitation, i.e. the dots or dashes of the Morse code. This enabled faster receiving, as the single operator wrote down what he heard. It was a basic instrument for hand speed operation which, together with the basic Morse key, could be used in case of the breakdown of more sophisticated equipment. The Morse key and sounder, due to their simplicity and ruggedness, were widely used by the military telegraph operators.

The sounder was further developed, in two ways. Firstly by Charles Bright, to activate a two-tone bell system so that the dots and dashes could be more easily distinguished, the apparatus being known as 'Bright's bells'. This was used in commercial telegraph offices but was not suitable for military applications. Secondly, after the telephone was invented in 1876 (an analogue instrument, able to transmit and receive the complex waveform of human speech), the telephone receiver was used together with a sounder in an instrument known as a vibrating sounder. The improved sensitivity enabled telegraph circuits to work greater distances over poor quality lines, and thus had considerable military application although it suffered the disadvantage of interfering with other circuits on the same line.

The inker, the sounder, and the vibrator were the three principal instruments used by the army. Many other instruments came to be used in civilian telegraph systems, but they were mostly unsuited for one reason or another to the rather rough army environment.

Early Users

Railways were the first users, but the scope for commercial telegraph traffic expanded rapidly. During the 1840s telegraph systems were developed in the principal European countries and in the United States of America. In Europe the telegraph was controlled by the governments of the countries, Britain being the exception. In Britain numerous private telegraph companies were formed to meet this growing demand; their subsequent 'nationalisation' under the Post Office in 1868 was to have an effect on army signalling organisation, as will be explained later (chapter 7).

Submarine Cable

The next major hurdle was international telegraphy. Commercial demand was increasing, and the fact that the electric telegraph could not cross the English Channel or, more ambitiously, the Atlantic Ocean, led to the next area of development – the submarine cable. Again, the development of submarine cable has been described in numerous books and need not be elaborated here. Where appropriate in the narrative to follow, the relevant development of submarine cable will be explained.

Britain led the world in this technology as it developed rapidly from about 1850, once waterproof methods of insulation had been discovered, although not without some famous disasters. The repercussions of one of these, a failed attempt to connect India by submarine cable telegraph, had a direct effect on the army, which as a consequence became heavily involved with the implementation of an overland alternative, to be described in chapter 4. Generally though, the army had no control of the development and operation of submarine cable links, these mostly being decided on commercial grounds by entrepreneurs and private funding, yet these strategic rear links were quite crucial to forthcoming military operations. The Indian Mutiny (1857-58) and the Zulu War (1879) are two examples where the absence of international telegraph communications had severe consequences, again to be described in later chapters.

Protracted Introduction to the British Army

Against that background summary of early developments in visual and electric telegraphy, each with their strengths and weaknesses, advances in army signalling of any significance began from about 1850 onwards. But progress was slow. Suffice to repeat at this point that the main contributory factors were general retrenchment of the army in the post-Waterloo period, and lack of awareness and lack of professional interest in the subject in the higher levels of the decadent mid-Victorian army. Coupled with this were organisational defects, both the wider aspect of army staff organisation, and more relevant to this story, the organisation needed to integrate visual and electrical methods.

Endnotes

1. An excellent description of Chappe's system is given in *Send it by Semaphore* by Howard Mallinson. See bibliography.
2. A biography has been written by one of his descendants, *A Damned Cunning Fellow* by Hugh Popham. See bibliography.
3. Those who were involved in army communications from the mid-1940s to the mid-1960s will recall the Slidex system, used at the lower levels of command. Slidex, although it was only used for short term security reasons at tactical level, was a system similar in principle to Popham's. There were a variety of cards, different cards being used by different users such as infantry, artillery, etc, with a set of frequently used phrases on them set out in a grid. There were also letters and figures. Thus a message could be sent partly in clear and partly in Slidex code; the secure part of the message could use the code either for pre-arranged phrases, or to spell out words letter by letter.
4. Called the 'Polygrammatic Telegraph' by Pasley, it would work by day and by night using six fixed lights. Pasley produced a descriptive pamphlet in 1823, the *Description of the Universal Telegraph for Day and Night Signals*.
5. These Admiralty telegraph routes and codes are described in detail in *The Old Telegraphs* by Geoffrey Wilson. See bibliography.

Chapter 2

The Crimean War, 1854-56

The British Army's first use of the electric telegraph

The advances in electric telegraphy up to the early 1850s had no impact on army communications. The moribund early Victorian army, incapable in many things, as the Crimean War was about to demonstrate, was simply not organised for such matters, nor were the vast majority of its officers able to grasp the issues, even had they been interested. (Great efforts were still needed to get many of them to use the field telephone during operations at the start of World War I in 1914!) But in the Crimean War the British and their allies, the French, had a strategic need to be in communication between London and Paris with their troops in the Crimea, and now there was a way of doing it. With much help from civilian sources a small detachment of the British Army was about to use the electric telegraph for the first time, and this chapter will describe it.

The War

In Britain, the Crimean War is principally remembered for three reasons: maladministration, the inept Charge of the Light Brigade, and Florence Nightingale. That it was the first time the army used electric telegraph in support of its operations comes well down the list of memorabilia and generally gets little mention.

The underlying cause of maladministration is not difficult to see. At the top there was a very inefficient system of control. The Army was the responsibility of the Secretary of State for War and the Colonies. He was assisted by the Secretary-at-War, who conveyed the government's wishes to the Commander-in-Chief of the Army, a general whose headquarters was the Horse Guards. This organization was completely divorced from that of the Master General of the Ordnance, who at that time was not a soldier but a Member of Parliament. He controlled the entire production of all forms of military equipment, including weapons, and the supply of food to troops stationed at home. Half-way through the war, in May 1855, the appointment was abolished and its functions transferred to the Commander-in-Chief and Minister for War. Another peculiar department was the Commissariat which was responsible for feeding troops abroad, and which came under the Treasury. The structure, with such divided responsibilities, was completely dysfunctional.

The novel method of communicating with the battlefield over such a distance introduced some new dimensions. It brought the field commander into rapid communication with his superiors in London, but with poor staff organisation at both ends, this led to much time-consuming trivia which the commander (Lord Raglan, and his successor, General Simpson) found a somewhat irksome distraction. It also enabled war correspondents to send despatches that were published in their newspapers within a day or two of the event, such topical reports from afar causing much interest back at home. "The press and the telegraph are enemies we had not taken into account but as they are invincible there is no use in complaining to them", lamented the British foreign secretary, the Earl of Clarendon.

Before describing how the telegraph communications were achieved, a brief background to the war. Many wars have been fought on the grounds of the strategic importance of a region, and many wars have been fought over religious differences. The Crimean War arose from both factors, and the reasons for Britain getting involved with it are still questionable. During the years leading up to the war, France, Russia and Britain were all competing for influence in the Middle East, particularly with Turkey. Religious differences were also a catalyst, and control of access to religious sites in the Holy Land - hardly something Britain should become involved with, one might think - had become a cause of tension between Catholic France and Orthodox Russia. Britain and France were primarily concerned with Russian expansion into the weakening Ottoman Empire.

In 1853 matters came to a head. In July that year, to pressure the Turks, Russia occupied the Danubian Principalities of Moldavia and Wallachia (both now part of Romania) which were nominally under Ottoman (Turkish) control. Diplomatic efforts to resolve the situation failed. Turkey declared war in

THE CRIMEAN WAR, 1853-56

October 1853 and attacked the Russians. As events unfolded, Britain and France, fearful of Russian expansion in the region, were drawn in. They demanded that Russia evacuate the Danubian Principalities, and issued an ultimatum. It expired in late March 1854 without response, and so Britain was at war.

The war was fought mainly on the Crimean Peninsula between the Russians and the alliance of British, French, and Ottoman Turks, with later support from the army of Sardinia. Britain, it may be observed, after many years of hostility, was now fighting with France! A joint invasion force of British, French and Turkish troops, over 60,000 strong, landed at Kalamatia Bay, north of Sevastopol, in mid-September 1854. The war lasted until February 1856.

The detailed course of the war, and the lesser known naval aspects of it in the Baltic, are not relevant, but the outline from the date of British participation is summarised in the chronological table below.

Chronology of Principal Events in the Crimean War

1854

22 September	The Battle of Alma. Attacked by the newly landed British and French forces, the Russians were defeated and forced to withdraw to Sevastopol, having suffered heavy casualties.
23 September	The Siege of Sevastopol begins. British and French bases established at Balaklava and Kamiesch Bay respectively. Russians maintain line of communication north and east of Sevastopol.
25 October	The Battle of Balaklava. Charge of the Light Brigade. Heavy British casualties, but British and Russian positions remained unchanged.
5 November	The Battle of Inkerman. Russians make surprise attack. Fierce fighting, spirited defence, thousands of casualties, mostly Russian. British held their positions.
	Over the coming severe winter, the British and French continue to besiege Sevastopol.

1855

Spring	The allies, with reinforcements and improved logistics, begin to batter their way into Sevastopol. British gunboats cut Russian supply lines across the Sea of Azov.
9 September	Russsians evacuate Sevastopol after 12-month siege. Allies destroy docks.

1856

30 March	Treaty of Paris. Russia returned southern Bessarabia (an area that included Moldavia, Wallachia, and surrounding territories) and the mouth of the Danube to Turkey; Moldavia and Wallachia were placed under international guarantee; the Sultan promised to respect the rights of his Christian subjects; and the Russians were forbidden from maintaining a navy in the Black Sea.

Civilian Support

In Britain, the country at that time leading the Industrial Revolution, there was a remarkable series of loyal private initiatives to support the national war effort. The railway contractor Morton Peto and his partner Thomas Brassey created a Railway Construction Corps from their own labourers and built a railway line from the base at Balaklava to the front line, and the mining industry in Leeds contributed two locomotives to work it. Joseph Paxton, the former head gardener at Chatsworth, who had turned from horticulture to architecture and constructed the Crystal Palace Exhibition building in 1851, organised an Army Works Corps to erect a township of wooden huts to protect the troops in the bitter winter. Isambard Brunel, the railway engineer, designed and had built a huge hospital from prefabricated components. William Fairbairn, the ironmaster and shipbuilder, constructed a pair of floating workshops to undertake all manner of repair and maintenance tasks for the besieging army. One of the recently formed private telegraph companies, with their telegraph engineering experts and suppliers also joined in, and it is to the telegraph story that we now turn.

THE CRIMEAN WAR, 1853-56

The Telegraph - an Overview

In contrast to the development of the electric telegraph in other European countries, where the matter was under governmental control, the telegraph network in Britain had developed in a rather uncoordinated manner under privately owned companies. (This eventually led to problems which were to have an impact on the army, as will be described in chapter 7.) One of these private companies, the Electric Telegraph Company, had been formed in 1845 by a group of London city merchants, who saw commercial potential from this newly developed science and bought in the patents of Cooke and Wheatstone, the inventors of numerous telegraph devices.

Some nine years later, when the Crimean War started, the Electric Telegraph Company had become the largest private telegraph company in Britain. It had invested heavily, and now connected many towns across Britain, provided telegraph for the country's railways, and had become involved in international telegraphy, having laid submarine cables to Holland. The brothers Edwin and Latimer Clark, leading telegraph engineers at the time, held senior appointments in the Company in the 1850s and 1860s. In support of the war the Company trained a cadre of soldiers in telegraphy and provided the army with a wagon-train carrying all the apparatus – instruments, batteries and underground cable – for a field electric telegraph, the first in the world.

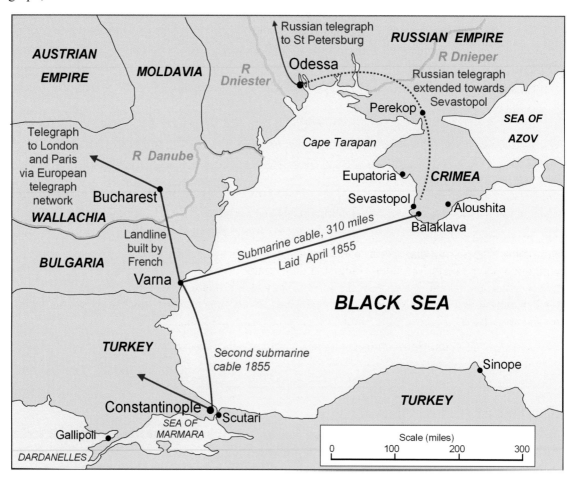

The telegraph that eventually ran between the Crimean peninsula and the seats of government in London and Paris was a combined effort using a range of resources.

- The European telegraph network that had been developed in the preceding decade, including the cross-Channel submarine cable between Dover and Calais built in 1851, provided telegraph facilities from London and Paris as far as Bucharest and Constantinople. Numerous national telegraph agencies were involved.
- From Bucharest, the civil telegraph system was extended by constructing an overland line to Varna on the Black Sea coast. This was built by the French Army.

- From Varna, the British laid a submarine cable under the Black Sea to the Crimean peninsula. Another submarine cable between Varna and Constantinople (Istanbul) was laid later.
- On the Crimean peninsula, a small, newly-formed telegraph detachment of the British Army, with the help of the Electric Telegraph Company, constructed and operated a line from the submarine cable terminal to the British headquarters at Balaklava and a number of other tactically important sites around the area.

It is these last two aspects that will be described - the Black Sea submarine cable and the Army detachment on the Crimean peninsula.

The Army Telegraph Detachment

In November 1854 the government in London authorised the formation of a military Telegraph Detachment, to be commanded by an officer of the Royal Engineers. This was hardly a rapid response. Britain had been at war since March, and the allied troops had been assembled, had landed on the Crimean peninsula, and had already fought three major battles (see chronology above), before the provision of communications was addressed. It was a failing of poor staff organisation and planning, and was to be repeated in other army campaigns to follow, even after regular army telegraph units were formed.

The Army Telegraph Detachment comprised twenty-five men from the Royal Corps of Sappers & Miners, the army's artisan corps. [1] As mentioned above, they were equipped, and a small cadre of them were trained, by the Electric Telegraph Company. The *History of the Corps of Royal Sappers & Miners* records that:

> The two sappers were specially instructed in the electric telegraph establishment at Lothbury [the Electric Telegraph Company's main telegraph office in the city of London] in the mode of working the instruments, laying the wire, and in the ingenious manipulation required to give effect to the process. [2]

The detachment was commanded from November 1854 by Lieutenant George Montagu Stopford, then from April 1855 by Captain Edmund Frederick du Cane, and finally, on his illness, from September 1855 by a Lieutenant Fisher. The greatest burden fell on Stopford who supervised the entire construction of the field telegraph, making preparations during the bitter winter months and entrenching the cables in the spring of 1855 between the base at Balaklava and Sebastopol. [3]

The two sappers in charge of the field telegraph arrived at Balaklava on 7 December 1854 with their equipment - the instruments, batteries, twenty-four miles of copper wire on twelve coils, the wire insulated with gutta percha resin for underground and underwater use, and other tools and equipment, all packed in two wagons, each wagon drawn by six horses. They also brought a plough, to be hauled by six men, used for burying the wire. All this equipment was designed by the engineer Latimer Clark and had been made for the army by the Electric Telegraph Company at their workshops.

The wire was uncoiled from a drum revolving horizontally and was laid in a shallow trough made by the plough, which served the double purpose of cutting the furrow and depositing the line. The trough was intended to be just deep enough to protect the wire from ordinary accidents. The plough often failed in the heavy, water-logged earth around the British positions, and eighteen-inch deep trenches then had to be dug and filled by hand. Unfortunately the gutta percha insulated line, better at insulating cables under the sea than under the ground, was frequently broken - by troops digging, by traffic, by burials, by exploding artillery shells, by ignorant soldiers who used the insulation to make

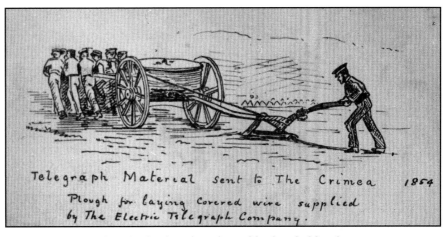

The plough, for burying the cable, hauled by six men.

pipe stems, and in one instance, by a family of mice. Fault finding and repair of a buried cable by the inexperienced soldiers must have been slow. The field telegraph service was not wholly reliable!

The other soldiers who formed the Telegraph Detachment were sappers and non-commissioned officers drawn from the ordinary field companies of the Sappers & Miners in the Crimea, and were taught in the field by Corporal Peter Fraser, one of those who had been taught to use the single-needle instrument by the Electric Telegraph Company in London. Two of his trainees were soon able to read code at a very effective 16½ words a minute. Sappers each received one shilling (5p in modern currency) a day extra allowance for their proficiency and extra duties, and the two sergeants, Anderson and Montgomery, received five shillings a day.

The telegraph wagon. On top is a folding boat. The wagon was normally drawn by six horses.

Apart from the time taken to train the men, the winter weather was severe, with deep snow on the ground, so laying the cable was delayed. After the snow melted the ground remained extremely wet, causing yet more delay.

On 7 March 1855, with the arrival of more favourable springtime weather, the first telegraph was opened from the headquarters at Balaklava to Kadikoi, three miles distant.

Each field station had two sappers to work a single-needle instrument. They were assisted by two orderlies from infantry regiments to carry messages.

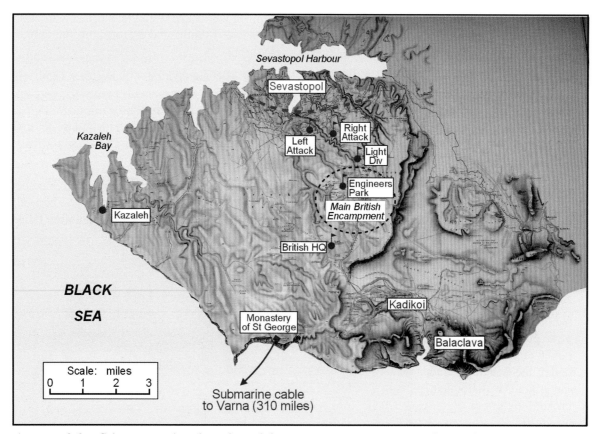

A map of the Crimean peninsula, adapted from a contemporary map drawn by RE officers. The locations of the telegraph stations are shown. The Engineer and Light Division stations were in bell tents, the Monastery in a ruined inn, four others in wooden huts and one, the Left Attack, in a cave. [4]

The single-needle instrument manufactured by W T Henley was adopted for field service. It was a miniature single-needle galvanic telegraph instrument in a box-like mahogany case. The needle was calibrated so that it could also be used as a galvanometer; and it had two button keys let into the base. It was easily put into circuit with small butterfly nuts on either side of the case, and was carried by a brass ring on its flat top. This very neat instrument, unlike Henley's magneto-telegraphs widely used in public circuits in Britain, required portable batteries; it could be worked with just two sulphate cells, and used either light iron wire or a resin-insulated field cable for its operation.

S J G Calthorpe, a staff officer at the Headquarters, recorded in his diary on 29 March 1855:

The single-needle telegraph instrument

> ... that a field telegraph, which was sent out here near two months ago, is now in use. Lines have been laid down from Headquarters to Balaklava, to each of our Attacks, as well as to a station between the 3rd and 4th Division camps, and another between those of the 2nd and light Divisions. Lord Raglan [the commander-in-chief] can therefore now communicate in a few minutes with any of his generals at any time, day or night. It is also a great advantage to have it in the trenches, as in the event of any sortie by the enemy, reinforcements can be sent for and instructions asked by the commanding officers in either Attack. [5]

Interestingly, there was no telegraph link established to the French HQ for liaison purposes, even though it was only about five miles north-west of the British HQ.

The Black Sea Submarine Cable

While the Telegraph Detachment was getting itself established on the Crimean peninsula, work was in progress to procure and lay the submarine cable to be laid across the Black Sea from Varna. The cable was manufactured and laid by R S Newall & Co, who had laid the first successful submarine telegraph across the English Channel in 1851. As part of the national contribution to the war, Newall made his proposal to lay a cable at cost price to the War Department in London on 9 December 1854; it was accepted three days later! By 16 January 1855, 400 miles of cable had been insulated by the Gutta Percha Company in London and shipped to Newall's Gateshead works for finishing. A construction gang of sixty men was assembled and a transport ship chartered. But things then went wrong; the new ship proved unseaworthy and the winter weather was terrible. The cable and equipment had to be transferred to a stronger vessel and only left England on 25 February. The replacement ship, *Argus*, carrying Newall's business partner Charles Liddell, his engineer Henry Woodhouse and Captain Edmund du Cane RE, arrived at Varna on 30 March 1855, to be joined soon afterwards by its navy escorts, HMS *Spitfire* and HMS *Terrible*.

It was originally intended to lay the cable from Cape Kaliakria, thirty miles north of Varna, to Monastery Bay in the Crimea. The connection at the mainland end was then to be by a landline between Kaliakria and Varna. But in the event there was

> ### Gutta Percha Insulation
> Insulation for telegraph cable had first been made from gutta percha. This was a product of the isonandra gutta tree found in the Malay peninsula, and examples were brought to Europe and exhibited at the Royal Society of Arts in London in 1843. In 1845 S W Silver & Co of Stratford, East London, invented a means of extruding it to cover wire. The discovery of gutta percha and the method of extruding it were keys to the development of submarine cable, and thus 19th century international telegraphy. However, there were disadvantages to gutta percha in other situations – it was not very flexible, and contact with air and movement tended to dry it out and cause it to crack. This made it unsuitable for land line. The only recorded British army use of gutta percha insulated cable was in the Crimean War. The buried cable proved unreliable. Other improved methods of insulating cable for land use were brought into service in 1868.

enough cable to run the line directly between Varna and Monastery Bay, so that plan was adopted instead. The little fleet set out on 1 April and completed laying the cable on 13 April 1855.

The cable consisted, throughout most of its length, simply of one No. 16 copper wire insulated with gutta percha but otherwise wholly unprotected. The shore ends had an iron sheathing, extending to a distance of ten miles from the Varna shore, and six miles from the Crimean coast. It remained in perfect working order for nearly twelve months, notwithstanding the many violent storms to which it was exposed in the Black Sea, until during a storm of more than usual severity it was broken on 5 December 1855. It was the longest submarine cable which up to that date had been operated successfully anywhere in the world. [6]

S J C Calthorpe, the staff officer at the HQ, noted in his diary on 13 April that:

> The end of the wire of the submarine electric telegraph was landed about midday near the monastery of St George. ..... The telegraph is now laid down from England to the Crimea, with the exception of from Giurgevo to Varna; but that will be finished in less than a month. As it is, Lord Raglan can communicate with the government in London in about 30 hours.

Meanwhile, Varna was connected with Bucharest by a line constructed by the French Army, and thence to London and Paris through European telegraph circuits. Using the Black Sea cable, the first message from Balaklava to London was sent on 28 April 1855.

Also in 1855 Newall laid another submarine cable for the British government, 171 miles from Varna direct to Constantinople, the Turkish capital, where another overland circuit existed to Vienna and the European network.

Communications on the Peninsula

On the Crimean peninsula the sappers worked the telegraph. Sergeant Anderson, the senior non-commissioned officer, was stationed at the Monastery, on the Black Sea coast, where he sent and received the messages from England by way of the submarine cable from Varna, and relayed them to Headquarters. The Monastery station later also handled the telegraph messages for the Sardinian Army.

At the British Headquarters, where Corporal Peter Fraser was the chief telegraphist, there were three corporals and three buglers, with three infantrymen acting as message orderlies. Traffic in August 1855 was 464 messages received and 402 messages sent - an average of 15 and 13 despatches a day respectively. The messages to the Commander-in-Chief were sent and received in numeric cipher; all other despatches were sent in plain English.

In August 1855 Samuel Varley, on loan from the Electric Telegraph Company, and younger brother of the company's chief engineer, C F Varley, was appointed civil superintendent of the field telegraph and, with ten civil clerks (as telegraph operators were then called), was sent out to work the equipment under Captain du Cane. It was intended to have one sapper and one civilian clerk to each field telegraph station. This proved unnecessary as the sappers worked the line perfectly well, without complaint, and the remaining civilians were soon posted to work the new submarine circuit on the other side of the Black Sea between Varna and Constantinople.

The Electric Telegraph Company arranged that any officer engaged in the campaign could send a message by the continental telegraph system to its station at The Hague and it would be forwarded to their relatives or friends in Britain without further charge.

The British also equipped the Turkish element of the force in 1856 with another telegraph detachment. Unlike its own unit this was provided with ten miles of very light No 10 gauge galvanised wire and lightweight porcelain insulators for attaching to trees and fencing, with only a few miles of gutta percha insulated underground cable.

The neighbouring French Army relied on a rather crude version of the Chappe semaphore telegraph system for its field signalling between the front line near Sevastopol and its base at Kameisch throughout the war. [7]

The Russians used a mixture of electric telegraph, semaphore telegraph and, some ten years before the British army got round to developing the instrument for signalling purposes, the heliostat.

THE CRIMEAN WAR, 1853-56

The electric telegraph system in Russia at the time had been built by the German company Siemens and Halske between 1853-55. The network was approximately 10,000 km in length connecting St Petersburg, Finland, the Baltic provinces, Warsaw, Moscow, Kiev and Odessa. During the war the line from Odessa was extended towards Sevastopol where their force was besieged (see map page 13). It became operational just in time to inform Moscow that the city was about to surrender!

It seems that the Russians semaphore telegraph was similar to the Chappe system in France. There was a signalling tower on the top of 'Telegraph Hill', and during the initial allied landing at Kalamatia Bay, north of Sevastopol, in mid-September 1854 the Russian General, Todleben, noted in his record of the *Defence of Sevastopol* that:

> The telegraph threw up its [semaphore] arms in despair when later in the day it announced that the enemy's flotilla swarmed by hundreds, and at last a panting cossack arrived with the news that the number was so great it was impossible to count them.

The Russians used the heliostat for their local communications around Sevastopol. In *The Times* on 11 July 1853 their correspondent describes how "a long train of provisions came into Sevastopol today, and the mirror telegraph which works by flashes from a mound over the Belbeck was exceedingly busy all the forenoon". And towards the end of the fighting, on 1 February 1856, Lieutenant-General Ewart noted in a letter written on the heights above Sevastopol that: "The Russians opened a regular bombardment from all their batteries, but happily we had not a single casualty. On the heights we were at a loss to understand the cause of such heavy firing, but the Russians were observed to be signalling with lights. I telegraphed the information to Headquarters ..."

Conclusions

That an electric telegraph had been set up and worked by joint army and civilian resources was a step forward, and that it could work both locally on the Crimean peninsula by hastily trained troops, and between the Crimea and London, was a definite achievement, even if there were shortcomings in organisation and equipment. It catered for an essentially static situation and could not of course cope with mobile warfare, and there was still no signalling below HQ level, but it was a start.

After the war the Royal Engineers established a small telegraph wing at Chatham, and began to adopt the electric telegraph for internal communication in fortresses; firstly at Malta and then at Gibraltar, soon to be joined by those at Portsmouth, Gosport, Chatham and Plymouth in the late 1850s. They also despatched telegraph detachments to China in 1859, where field telegraph followed the headquarters of the advancing expedition many miles inland from the port of Canton.

At higher levels, though, post-war British Army interest in military communications continued to languish. Any further advance to establish permanent army telegraph units in support of fighting formations was restricted for another fifteen years. And available methods of visual signalling, which were already waiting in the wings and could provide much-needed tactical signalling capability had anybody applied themselves to it with any imagination and vigour, were disregarded until the following decade.

Endnotes

1. The Corps of Sappers & Miners, which effectively worked the electric telegraph in the Crimea, in addition to its main function of creating the siege works about Sevastopol, was re-titled the 'Royal Engineers' in October 1856, merging with the officer-only corps of that name.

2. The *History of the Corps of Royal Sappers & Miners* by T W J Conolly, pub 1855.

3. Captain du Cane (1830-1903) was later to become General Sir Edmund du Cane and was the last military Surveyor General of Prisons before the appointment was civilianised. In this appointment he was involved with many well known prisons, Wormwood Scrubs being completed in 1883 during his tenure.

4. Entitled *Plan of the Attack and Positions of the Allied Armies around Sebastopol in 1884-5,* it was compiled by RE officers on the ground (they are named on the map), and engraved in the Ordnance Survey office, Southampton in 1856. Unfortunately it does not show the telegraph line routes. A copy is held in the Prince Consort Library, Aldershot.

THE CRIMEAN WAR, 1853-56

5. *Letters from Head-Quarters* by S J C Calthorpe, pub John Murray, London, 1859.

6. A first attempt at a cross-Channel submarine cable in 1850 had failed, but was completed successfully between Dover and Calais in 1851. Another cable was laid in 1854 between Corsica and La Spezia. The first, and unsuccessful transatlantic submarine cable was laid in 1858, failing only a few weeks after completion. The transatlantic link was finally established successfully in 1866.

7. Some fifty years previously the British had copied the successful French Chappe system, and established semaphore telegraph stations from the Admiralty to their principal naval bases; after the Crimean war the French were to copy the English! For their brief and bloody campaign into northern Italy against Austria in June and July 1859 the French organised a 'service télégraphique'. This was formed of civilian staff engaged to follow the army in two 'brigades' of ten wagons. To maintain communication with France, they successfully laid 400 kilometres of line and created thirty-five telegraph stations along the route of advance, using light 2mm gauge wire on 6 metre poles.

THE WORSHIPFUL COMPANY OF INFORMATION TECHNOLOGISTS

The Worshipful Company of Information Technologists, the 100th Livery Company of the City of London, is proud of its long association with the Royal Signals. We are delighted to support the publication of 'Tales of the Telegraph' and wish the Corps every success for the future.

Chapter 3

Telegraph Developments in India

Introduction

Surprisingly perhaps, but placing it in chronological order, the scene now shifts to India. In the Victorian era India was the principal asset in the expanding British Empire, and much of the British army served there. The telegraph was introduced at an early stage.

The early British influence in India stemmed from the East India Company, which extended its power from simply trading in the early 1600s to involvement in political and military matters. At the beginning of the eighteenth century, British possessions in India were divided into the three 'Presidencies' of Bengal, Bombay, and Madras, each independent, each with its own military and civil officials, and each under a 'Governor and President-in-Council' who was directly responsible to the Company's Court of Directors in London. Bengal was the largest, and always the most dominant.

In the period 1770-90, the power of the East India Company was progressively brought under British government control. By 1822, after the defeat of the Maratha Confederacy, which had controlled central and western India, Britain was established as the paramount power.

Whilst the south was mostly docile, wars against tribes and neighbouring countries in the north of India were to follow, making the Bengal Presidency the most active and the most important in military terms. These days it should be remembered that, by 1850, British India was composed of what are now the separate and independent countries of India, Pakistan, Bangladesh, and Burma (Myanmar); it was a huge territory. The north-west frontier, bordering Afghanistan, was the most turbulent - a potent concoction of conflicting tribal rivalries, local politics, and religion - *plus ça change* ... .

With that brief historical reminder, back now to telegraph matters. Following the pattern of early semaphore telegraphs in France and Britain, India also attempted its own version in the 1820s. Then, in the early 1850s, under its innovative Governor-General, Lord Dalhousie, newly developed European inventions were introduced, amongst them the electric telegraph, and the army became involved to some extent with its introduction. Soon afterwards the telegraph was to play its part in overcoming the Mutiny.

This chapter will describe the events from 1815 up to the end of the Mutiny and its immediate aftermath. After that, connecting London and India by international telegraph became important for strategic and commercial reasons, and the army was to take a leading part; that will be described in chapter 4. [1]

Visual Telegraphs

Having seen the benefits of the extensive Chappe system in France and the British Admiralty's telegraph system, similar systems in India were mooted. By 1815, the year of the battle of Waterloo and the end of the Napoleonic wars, the Bengal and Bombay Presidencies were in discussion about what in retrospect can be seen as grossly over-ambitious plans for visual telegraph systems. They were advised by William Boyce, a surveyor who had previous experience of the construction and working of such telegraph systems in Ireland. The novel idea of east and west India being in rapid communication between Calcutta and Bombay was attractive but the practicalities over such a distance, it seems, were not fully understood. The initial proposal was a telegraph link between those two places, a chain of 75 stations over a distance of some 1,200 miles. The dream continued, and the idea was then enlarged to include extensions southwards to Mangalore and Madras in a chain of 133 telegraph stations. The Surveyor-General of Bengal was pessimistic, questioning Boyce's experience, not in telegraphy but in Indian matters - such things as hostile tribes, the security and defence of remote telegraph stations, and the logistics of resupply.

Discussions rumbled on for a few years, and surveys were carried out. One of these surveys, in 1818 along the 400-mile route between Calcutta and Chunar, about sixteen miles south-west of Benares (now Varanasi), involved Captain George Everest of the Bengal Artillery, later to be the Surveyor-General of India whose name was given to Mount Everest. Decisions were reached, and this route, which could later be extended from Chunar on to Bombay via Nagpur (a further 800 miles, through much more difficult terrain), was built, and opened in about 1821. Construction costs were high. The line consisted of forty-five stations, some with towers 100-feet tall. It used a semaphore system of four balls, each 6-feet in diameter, which singly or combined could form only nine characters. Why this inferior method of balls - aptly named - was chosen, when other methods such as shutters or semaphore arms with better signalling codes able to signal many more characters already existed, is obscure. A Director of Telegraph Communications was appointed from the Bengal Infantry; Indians were instructed in using it, and manned the stations. Then running costs caused concern, outweighing any apparent benefit; adverse criticisms mounted. By 1829 Sir Charles Metcalfe, a senior administrative officer, in a letter on the subject observed that:

> ... the whole expense incurred on account of the Telegraph is to be regretted, its services not being likely to counterbalance the outlay. ... as yet, for about ten years, it seems to be nothing better than an expensive plaything. We were of this opinion from the beginning ... and the sooner the Establishment is abolished the better.

The pessimists, or perhaps one should say the realists, having been proved right, the Calcutta-Chunar telegraph line was closed in 1830, although some of the towers proved useful in conducting the great trigonometrical survey of India being carried out at that time. [2]

Another chain of fifteen semaphore telegraph stations was built around Calcutta and the nearby Saugor Island, and seems to have operated between about 1833 and 1850, presumably serving local communication requirements.

The Electric Telegraph

Following that, the advances being made in electric telegraphy offered a much more practical alternative in a large country such as India. The first electric telegraph line in India ran from Calcutta to Diamond Harbour, 30 miles to the south, and was opened in December 1851. It was the culmination of experiments

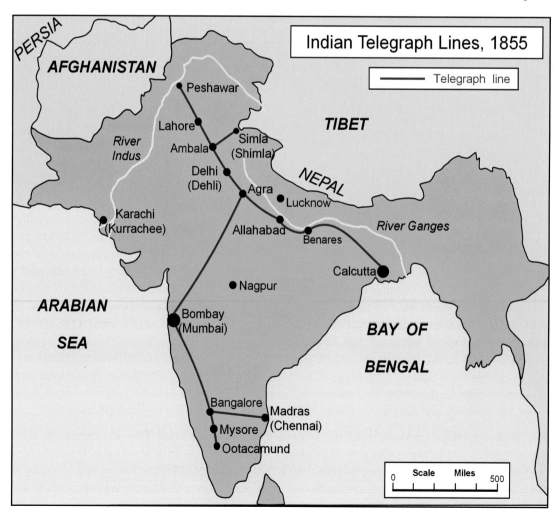

carried out since 1839 by Dr (later Sir) William O'Shaughnessy of the Bengal Army. O'Shaughnessy, a doctor of medicine who had graduated from Edinburgh University in 1829 when he was twenty-one, had pursued medical research and had published his findings in *The Lancet*. He then joined the Bengal Army in 1833 as an assistant surgeon, and became a professor of medicine at the Calcutta medical college. One would have thought his career was set in that direction, but for some reason, and while continuing his career in medicine, he turned to the electric telegraph and conducted trials with a primitive experimental line in Calcutta. In 1850 he managed to bring his work to the attention of the Governor-General, Lord Dalhousie. A former President of the Board of Trade in the British government in the mid-1840s, Dalhousie was well-known for his keenness to introduce the inventions of Victorian times to India, and authorised the line from Calcutta to nearby Diamond Harbour. Its success prompted Dalhousie to send O'Shaughnessy to England to win the approval of the Court of Directors of the East India Company for a telegraph network in India.

As a result, in 1852, the Court sanctioned the construction of telegraph lines from Calcutta to Peshawar, Bombay and Madras. Further investigation caused some modification to the original plan. It was decided that a northern line should follow the valley of the Ganges to Allahabad, running thence through Agra and Delhi to Lahore, and finally to Peshawar, with a branch line from Ambala to Simla. Connection between Calcutta and Bombay was to be established by a line taking off from the northern line at Agra and running southwards through Indore and Nasik; while between Bombay and Madras there was to be a direct line across the Deccan. Lord Dalhousie appointed O'Shaughnessy as Superintendent of the new Indian electric telegraph, and work commenced in 1853.

O'Shaughnessy began to organize the work in Bengal before sailing back to England to arrange for the procurement and shipment of materials and stores. By May 1854, the telegraph was established between Bombay and Calcutta, and in the following year between Bombay and Madras. By 1856 telegraph lines stretched from Calcutta to Peshawar (1,600 miles), from Agra to Bombay, and from Bombay to Madras, Mysore, and the hill-station of Ootacamund in the extreme south.

Army Participation

In May 1853, when O'Shaughnessy went to England to make these arrangements, matters in India were left in the hands of a young officer of the Bengal Engineers, Lieutenant Patrick Stewart. Born in Kircudbrightshire, Scotland, his career perhaps influenced by his maternal grandfather who had served in the East India Company, Stewart attended the East India Company's military college at Addiscombe, near Croydon, in the period 1848-1850. He passed out top of his term, winning the Pollock medal, and in June 1850 was commissioned as a 2nd Lieutenant in the Bengal Engineers. He attended the young officer's engineering course at Chatham, and arrived in India in October 1852, a few months before his twenty-first birthday. It was perhaps an accident of timing that a year later this young man found himself standing in for O'Shaughnessy, involved with something that was to set the course for his future career.

On appointment, the young Stewart toured his novel fiefdom and found much to surprise him:

> The lightning has come into the offices several times lately, destroying parts of the instruments completely and sounding all the alarms. On the lower parts of the line, near the Bay, there are immense numbers of wild buffaloes that continually come to rub themselves on the posts, and not infrequently knock them down. Great numbers of birds have been killed, while sitting on the iron rods, by flashes of lightning passing along them. [3]

The construction of the Indian telegraph lines should perhaps be described, as they differed greatly from those in Britain. O'Shaughnessy had been influenced by American designs, and to begin with had copied their methods. British practice favoured No. 8 gauge wire, 1/6 inch in diameter and weighing about 1/5 ton per mile, but in India the wire was so thick that it might almost be termed a rod, being No. 1 gauge and weighing half a ton per mile. Most of the heavy Indian lines were supported on stout bamboos instead of metal poles – easily and cheaply procured, but with limited lifespan.

Besides making tours of inspection, Stewart had to arrange for the delivery of stores and instruments for the whole line from Calcutta to Lahore, and for the Bombay line from Agra to Indore, where his men would connect with the parties under another Superintendent. He had also to send materials for a line from Prome, on the Irrawaddy in Burma, to the Arakan Coast at Sandoway.

It was no sinecure for the young officer, but he thrived in the situation. The line from Calcutta to Lahore,

> ### Addiscombe Military College
>
> With the increasing need in India for professional training of officers in the technical arms (artillery and engineers), or ordnance as it was then called, and the difficulty of getting them trained within the British army establishments at Woolwich and Sandhurst, the Court of Directors of the East India Company decided to form its own officer training establishment. In 1809 they acquired Addiscombe Place, near Croydon, south of London, a large and stately mansion built in English Baroque style.
>
> Cadets were taken in at around the age of fifteen and commissioned after a two year training course into the appropriate arm of the Indian Army. Their training at Addiscombe closely followed the organisation and syllabus of the Royal Military Academy, Woolwich, where their counterparts destined for the British army were trained Military officers provided the command and staff, and shared with civilian professors the duties of instructing in such subjects as fortification, mathematics, landscape and military drawing, chemistry, geology, French, and Hindustani. As well as these academic subjects the cadets were taught drill, both gun and musketry, and the use of the sword. Academic and domestic posts were filled by the patronage of the Directors.
>
> As Addiscombe could house more than the number of ordnance cadets required each year it was decided that the Seminary should be filled up, and that the final order of merit should decide whether the cadets were appointed to the engineers, artillery or infantry. Young men wishing to join the infantry as cadets could do so by going directly to India, and no cavalry cadets at all went to Addiscombe. After their training at Addiscombe young officers destined for the engineers underwent two years further training at the School of Military Engineering at Chatham.
>
> Addiscombe closed in 1860, after the Indian Mutiny and the consequent transfer of power to the British Government. Cadets for the artillery and engineers were subsequently all trained at the RMA Woolwich and granted Royal commissions, enabling them to serve in Britain or India.

a distance of 1,280 miles, required the establishment of depots at seven different places. Forty artificers arrived, and then came tons of wire, hundreds of thousands of posts, and machines for straightening the lines. This all had to be done in a country where there was only one road and no railways (they, too, were being built as part of Lord Dalhousie's plans), where any delay that may occur was accepted complacently as normal practice, and by a young officer with no authority over anyone except his immediate staff. Nevertheless, the artificers were despatched by river steamer to the depots, and the wire was carried by steamer and bullock carts to thirty-two different places between Calcutta and Benares alone, all without delay. When Dr. O'Shaughnessy returned all had been done, and he was able to continue without any delay to the project.

Lieutenant Patrick Stewart was then appointed Aide-de-Camp to the Lieutenant-Governor of the North-West Provinces, no doubt as part of some career plan, but one that in a few years was going to change its intended course in a rather unexpected way. Whilst in his appointment as ADC, Stewart very nearly lost his life. In November 1853 he had left the Lieutenant-Governor's camp at Mirzapore for a day's shooting with three others, and succeeded in wounding a tigress. While tracking the wounded animal, it suddenly and unexpectedly attacked him, and having severely bitten him in the calf and thigh, left him for dead. He was rescued by his companions, and after treatment and a period of recuperation was able to return to work. Then, in late 1854, he suffered a severe fall from his horse when at full gallop, which left him severely concussed. In addition to this, at the beginning of 1856, a racquet ball struck him near the left eye, knocked him down, and cut an artery, causing him the loss of much blood. These incidents probably contributed to later health problems.

Having completed his tour as ADC, and probably somewhat to the relief of the Lieutenant-Governor who could well have regarded him as an accident-prone young tearaway, he was in 1856 reappointed to work under O'Shaughnessy in the Indian Telegraph Department. His previous work on the telegraph project had been very competent, and it had not passed unnoticed.

TELEGRAPH DEVELOPMENTS IN INDIA

The Indian Mutiny – 'The Electric Telegraph has saved us'

But momentous events were soon to happen – the Indian Mutiny, to give it its historically accepted British name, although it has been described otherwise since. The causes of the mutiny need not concern this narrative. Suffice to say that they were much wider than the lame old excuse offered for some time afterwards in school history books, of greasing new cartridges with pig and cow fat – abhorrent both to Hindus and Muslims. Various political decisions made by the British had caused unrest amongst wide swathes of the Indian population. Radical reforms introduced by Lord Dalhousie, Governor-General from 1847 to 1856, together with unfounded Hindu and Moslem suspicions about enforced conversion to Christianity, were seen by the population as signs of 'westernisation'. The rapid spread of the telegraph, as part of the modernisation programme, was seen in similar vein and became a minor catalyst in inciting the mutiny. It was also to become a vital factor in initially containing it and then defeating it.

The Mutiny first broke out at Meerut, about 40 miles north-east of Delhi, on Sunday 10 May 1857. Meerut was by then connected to the main telegraph line at Delhi. In Delhi the telegraph office was following its normal Sunday routine. It had opened at daybreak and, being the sabbath, it would close during the heat of the day from 9.00 am to 4.00 pm, before a late afternoon shift for a few hours. The European telegraphist in charge, Charles Todd, and his two young Eurasian assistants, were about to close the office at the end of the early morning shift when one of the assistants noticed the needle of the telegraph instrument had started to move. It was an unofficial message from the telegraph office in Meerut, describing some disturbing events that were happening there. Despite this, the true nature of things was not yet realised, and the Delhi office closed. On reopening at 4.00 pm it was discovered that the line to Meerut was 'dead'. Tests showed that the line was broken beyond the River Jumna, but by then darkness was falling. The following morning, unaware of the nature of the disturbance, Todd set off to repair the line; he was never seen again, killed by mutineers.

Meanwhile fugitives escaped from Meerut, and raised the alarm in Delhi. After evacuating the Delhi telegraph office, one of the telegraph assistants returned with an army officer and escort and sent a message down the line to other stations, giving a brief description and alerting them. The then Captain, later Field-Marshal, Fred Roberts, in his autobiography *Forty-One Years in India*, describes how the news was received:

> ..... we were quite thunderstruck when on the evening of 11th May, as we were sitting in the mess [at Peshawar], the telegraph signaller rushed in breathless with excitement, a telegram in his hand, which proved to be a message from Delhi 'to all stations in the Punjab', conveying the startling intelligence that a very serious outbreak had occurred at Meerut the previous evening, that some of the troopers from there had already reached Delhi, that the native soldiers at the latter place had joined the mutineers, and that many officers and residents at both stations had been killed. [4]

It was the last transmission from Delhi before the telegraph office there was captured. In Lahore, where the majority of British troops were stationed, precautions were quickly taken and sepoys' arms and ammunition there were confiscated and secured. A message was also sent from Ambala to the C-in-C of India, then at Simla.

The telegraph messages had enabled quick reactions to take place, preventing rapid escalation of the Mutiny, and perhaps even the preservation of British India. "The electric telegraph has saved us", wrote Donald MacLeod, the Financial Commissioner of the Punjab. The fact that there was no international telegraph from India to transmit the news quickly and arrange reinforcements from outside India was to prove an enormous hindrance to the conduct of the campaign to suppress the uprising.

As the British in the Punjab organised themselves to form a military force to take on the mutineers the telegraph was much in use. Roberts himself became a staff officer in what became known as 'the Movable Column' and describes how, as they advanced towards Delhi, and "... away from telegraph stations, which were few and far between in 1857, a signaller accompanied us, and travelled with his instruments on a second cart, and wherever we halted for the day he attached his wire to the main line". [5] In that way, they kept in communication as they advanced.

Lucknow

After Delhi the mutineers went on to seize Lucknow, which was to see some of the bloodiest fighting. In May, when the mutiny broke, out Lieutenant Patrick Stewart was on reconnaissance for the construction of a telegraph line in Ceylon (now Sri Lanka). When he heard the news he returned at once to Bengal, and after leaving instructions at Madras for the commencement of a line towards Calcutta, he spent the next four months in rapid journeys between Calcutta and Allahabad, organizing the telegraph arrangements in the theatre of war to keep the seat of government in Calcutta in communication with the British forces as they engaged the mutineers.

On 3 November he began to lay a line from Cawnpore (mid-way between Allahabad and Agra, and served by the main telegraph line shown on the map above) towards Lucknow for Sir Colin Campbell's first advance to relieve the garrison there. Stewart brought the line into the Alambagh (Garden of the World), a large house with a walled enclosure to the south of the city, which had been captured by the British before the attack, but the line was soon cut by the rebels. The Alambagh remained a defended base for the troops. It was to be some months before Lucknow was recaptured, but during that time Stewart managed to maintain communications with Calcutta.

Contemporay extracts of telegrams between high ranking officers at Lucknow and the Government at Fort William, Calcutta, are occasionally interspersed with telegrams from Lieutenant Stewart, mostly to the Secretary of the Government, C Beadon Esq, to keep him updated. For example, on 2 December 1857 from Cawnpore:

> The line of the electric telegraph to the old offices in Cawnpore passes through part of the station still in the enemy's hands and was destroyed by them. A branch line has been carried into camp, and an office opened there. Lines to Alum Bagh almost entirely destroyed. ... *etc* . [6]

Meanwhile, the remnants of the Lucknow garrison were holding out in the Residency. The Residency at Lucknow, on a site slightly higher than the surrounding city, had been besieged since 1 July 1857. An interesting piece of improvisation enabled local communications between the Residency and the relief force's base at the Alambagh, about three-and-a-half miles from the Residency. Some in the besieged Residency constructed a Popham semaphore, details of which they had found in a copy of the *Penny Cyclopaedia* in the Residency, and Eurasian youths from the Martinière College who were sheltering in the Residency were taught how to use it. A plan for opening communication was made, and smuggled to the Alambagh by means of a runner. Stewart, at the Alambagh, constructed a Popham semaphore station there, and visual telegraph communication between the two began. But at first it suffered a hitch; those at the Residency could not sensibly decode the semaphore signals they were receiving. After some investigation it was realised that the Alambagh signallers were standing on the wrong side of their semaphore, thus the signals received at the Residency showed the semaphore arms reversed, left to right. Once this problem was resolved, communications were properly established. Plans were exchanged, and the besieged Residency was successfully evacuated soon afterwards, after a siege lasting some three-and-a-half months. The mutineers quickly occupied the Residency and smashed the temporary semaphore system there.

The Alambagh, with Popham code semaphore arms on the roof, to communicate with the besieged Residency at Lucknow.

On 15 January 1858, after the rescue of the Residency garrison, Stewart handed over charge of the civil telegraph system to Dr O'Shaughnessy, and was then free to devote all his attention to field telegraph lines for use in the final assault on Lucknow. He re-opened an office at the Alambagh on 19 February and extended his line to the Dilkusha on 6 March during Sir Colin Campbell's attack.

The correspondent of *The Times* described it:

> Never since its discovery has the electric telegraph played so important a role. It has served the

> Commander-in-Chief better than his right arm. In this war, for the first time, a telegraph wire has been carried along under fire and through the midst of a hostile country. The telegraph was brought into connection with the Governor-General at Allahabad, with Outram at the Alam Bagh with Calcutta, Madras, Bombay and the most remote districts. It is mainly to the zeal, energy and ability of a young officer of the Bengal Engineers, Lieutenant Patrick Stewart, that these advantages are due. At one time his men are chased for miles by the enemy's cavalry; at another they are attacked by Sowars [cavalry troopers], and they and the wires are cut to pieces. Again, their electric batteries are smashed by the fire of a gun, or their cart knocked to pieces by a round shot; but still they work on, creep over plains, cross water courses, span rivers, and pierce jungles, till one after another the rude poles raise aloft their slender burden, and the quick needle vibrates with its silent tongue amid the thunder of the artillery. As Sir Colin Campbell advanced towards Lucknow, the line was carried with or soon after him; a tent was pitched near his, a hole was dug in the ground and filled with water, and down dropped the wire from a pole stuck up in haste, dived into the water otter-like, the simple magnet was arranged, the battery set in play, and at once the steel moved, responsive to every touch. The wire is thick and is not protected by non-conductive coatings of any kind; it is twisted round the top of a rude pole, and is found to answer perfectly. [7]

Patrick Stewart received much approbation for his outstanding work during the Mutiny in bringing the telegraph from the seat of government at Calcutta right to the commanders at the scene of action. He was Mentioned in Despatches on a number of occasions. In Sir Colin Campbell's first despatch from Lucknow, on 18 November 1857:

> I would especially draw attention to ..... Lieutenant P. Stewart, Bengal Engineers, Superintendent of the Electric Telegraph, who accompanied the force and made himself particularly useful throughout.

Then on 10 December 1857, the following General Order from the Governor-General in Council (Lord Canning):

> His Lordship would like to thank ... [list of names]...as well as Lieutenant P. Stewart, Bengal Engineers, Superintendent of the Electric Telegraph, whom the Commander-in-Chief [Sir Colin Campbell] mentions with much praise.

Further Mentions in Despatches in similar vein were accorded by the C-in C and the Governor–General on 22 March 1858 and 5 April 1858, and need not be quoted. Suffice to say that Patrick Stewart had excelled himself.

The Aftermath

So much for the Indian Mutiny. It was eventually quelled in 1858, and big changes were to follow. Principally, the East India Company's powers were transferred to the British Crown, the Governor-General in India (called the Viceroy) taking his instructions from the Secretary of State for India in London. In 1877 Queen Victoria became Empress of India.

Within that major change of policy there were also changes to army organisation. After the Mutiny it was necessary to create a new Bengal Army, that army having been at the heart of the rebellion, and to make changes to the other two Presidencies. As part of the army reorganisation, and relevant to military telegraph, the Bengal, Madras, and Bombay Engineers were amalgamated with the Royal Engineers in April 1862, the engineer officers at the time in those Presidencies being granted royal commissions, though remaining on their separate lists for promotion and seniority purposes. British Royal Engineer officers could spend their whole career in India or could, after a period of time, choose to return to Britain and elsewhere. Three years later, in 1865, a scheme was introduced to improve the quality and supply of British NCOs to the Royal Engineers in India and companies were formed, one in each Presidency, kept supplied by a training arrangement at the School of Military Engineering, Chatham. [8]

As a result of these arrangements, there was a fairly free exchange of postings between Britain and India for engineer officers and NCOs, and with this interchange went up-to-date training and ideas. It occurred just at the time that developments in British army communications methods, both visual signalling and electric telegraph, led by work at the School of Military Engineering at Chatham, were being introduced.

The Mutiny had been a demonstration of the value of the electric telegraph in war, but the experience gained was put to no useful military purpose. Proposals to form Indian army telegraph units were rejected. Understandably the available resources were instead devoted to the extension of civil telegraphs within India, and the strategically important linking of India with England by telegraph. The result was that electric telegraphy within the army in India languished, as it had done in Britain after the Crimean War, and following the familiar theme, was to continue so until stimulated by the 2nd Afghan War in 1878, to be described in chapter 10.

The development of the Indian civil telegraph system continued apace. By 1872 every important place was connected by telegraph, while additional lines also followed the 5,373 miles of railway then opened. Until 1884 some military engineers participated in the operation of the Indian Telegraph Department, work which was outside the normal range of duty at that time. A notable incumbent was Colonel D G Robinson CB of the Royal (Bengal) Engineers, who was appointed Director-General of Indian Telegraphs in 1865. During Daniel Robinson's tenure of twelve years, the telegraphs continued to develop in India, and his technical and administrative ability placed the Indian Telegraph Department on a thoroughly sound footing. He took a leading part in international conferences on telegraphic communication which were held in Berne, Rome, and St. Petersburg. He died suddenly in 1877, while on his way to England. Gradually the Indian Telegraph Department became the civil, government-controlled organisation that it was in every other country, the army relinquishing all responsibilities within it by 1884.

Endnotes

1. A very comprehensive description of the Indian Army at the time is to be found in *The Indian Army, The Garrison of British Imperial India 1822-1922*. See bibliography.
2. Further detail will be found in *The Old Telegraphs*, Chapter XXIV, pp 195-98. See bibliography.
3. *The Military Engineer in India, Vol 2*, Chap XV, by Lieutenant Colonel E W C Sandes, published by the Institution of Royal Engineers, 1935.
4. *Forty-One Years in India* by Field-Marshal Lord Roberts of Kandahar, p 34.
5. ibid p 63.
6. *Selections from Despatches* by G Forrest.
7. Extract from *The Times* quoted in *Telegraph and Travel* by Colonel Sir F. J. Goldsmid, pp 45-46.
8. A more detailed description is given in *The Military Engineer in India*, by Lieutenant Colonel E W C Sandes, Vol 1, pp 268–372.

Chapter 4

The Telegraph Route to India

Although the electric telegraph had been successfully introduced in India, and continued to expand, one big requirement remained after the Indian Mutiny - an international telegraph link between Britain and India. The inability of the two governments to intercommunicate rapidly at that time had been a serious problem. In those days the fastest route for communication was mail by steamer between Britain and Alexandria on the Mediterrranean coast of Egypt, overland to Suez (the canal was not opened until 1869), and then by steamer from Suez through the Red Sea to India. It took about two months to get answers to letters passing between the seats of government in London and Calcutta. When the Mutiny was quelled in 1858, the urgent need to be in telegraphic communication between London and India required no further justification. As events turned out, the army became closely involved in the venture.

The Options

To establish the link, two options were considered initially. One was a submarine cable from Suez through the Red Sea to Karachi (Kurrachee). A submarine cable was being laid across the Mediterranean by a privately owned company to connect Marseilles, Malta, and Alexandria, and Marseilles was connected by telegraph to London. The construction of an overland section between Alexandria and Suez presented no technical problem, but Suez needed to be connected with India by another submarine cable. The second option was an overland line between Constantinople (Istanbul) and Karachi. At that time Constantinople was the limit of the European telegraph network, and Karachi was served by the newly developed Indian telegraph. With this second option, the intervening route between Constantinople and Karachi, devoid of any telegraph, was the problem to be solved - and it turned out to be a big one, as will be described.

The Red Sea and India Line to Karachi

It was decided in 1858, possibly with speed in mind, to choose the submarine cable option from Suez to India. At this time submarine cable laying and maintenance was still in its infancy. The first international submarine cable had been laid between Dover and Calais in 1851, after an unsuccessful attempt the previous year. But the first attempt at laying a trans-Atlantic ocean submarine cable in that same year of 1858 was to prove a failure, costing private investors a huge financial loss, and they were no longer willing to speculate on the still immature technology. Consequently the £800,000 capital sum needed for the Red Sea and India line, as the proposed route between Suez and Karachi was called, was funded in equal shares by the British and Indian governments under the terms of an unconditional guarantee of a 4½ per cent loan for fifty years. It was somewhat speculative.

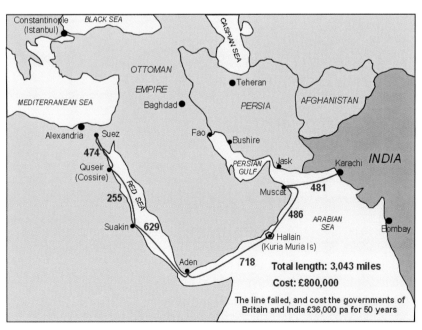

The cable was manufactured by Messrs R S Newall and Company, and it was laid in sections during 1859 and 1860, as shown in the map, the total length being 3,043 miles. It proved to be a disastrous failure. The design of the cable itself was faulty, and it was badly laid, with insufficient slack. Numerous problems arose, and it was never able to be brought into service. But it had satisfied

The Red Sea and India Line, 1859-60. A failure.

the terms of its guarantee and the two governments were committed to pay an annual bill of £36,000 for fifty years for a line that never passed any telegraph traffic. [1]

The Line between Constantinople and Karachi

Following this costly fiasco it became necessary to adopt the second option, the overland route between Constantinople and Karachi by way of the countries bordering the Persian Gulf and Arabian Sea - the present day Iraq, Iran, and Pakistan. It should be remembered that in the 1860s, when the telegraph link was being planned, what is now Iraq was part of the Ottoman (Turkish) Empire, and so Baghdad and Mesopotamia were under the hand of Constantinople – 'controlled' might be too strong a word.

Many diplomatic difficulties had to be surmounted before a convention could be concluded with Turkey for the construction of the section of line across their territory, but agreement was reached in 1860. A leading figure in the negotiations was Lieutenant Colonel Frederic Goldsmid, about whom more shortly.

The new plan developed as follows:

- The line from Constantinople would run through Baghdad and along the Tigris river valley through Mesopotamia to the top of the Persian Gulf at Fao, this entire section to be constructed by the British but then controlled and operated by Turkey.

- From Fao a submarine cable would be laid through the Persian Gulf with landings at Bushire and Jask to Gwadur, a fishing village then under the control of the Sultan of Muscat. Despite the earlier submarine cable failure it was considered feasible to lay a cable through the Persian Gulf as the sea was not deep and the intermediate landings not far apart.

- The route would be completed by a landline in Baluchistan along the Makran coast between Gwadur and Karachi, where it would connect with the existing Indian telegraph system and on to the seat of government in Calcutta. The plan was subsequently modified slightly, so that the submarine cable ran all the way from Fao to Karachi.

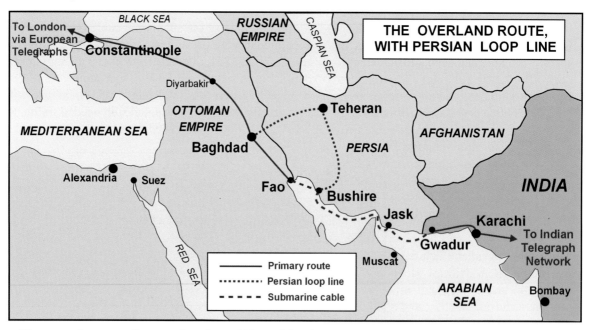

The route between Constantinople and Karachi. The Persian loop line was planned as a backup.

The principal problems to be faced in developing this route were going to be extended and devious diplomacy in middle-eastern countries, poor security arising from local antipathy to disliked foreigners and the purpose of their misunderstood inventions, and the unknown practicalities of long distance international telegraphy without agreed procedures in a number of languages and different scripts. In retrospect it appears a brave decision, influenced no doubt by the failure of the submarine cable through the Red Sea in 1860. As might have been foreseen, these matters were going to take years to resolve, but it was the only option open at the time.

Moreover, it was now going to be a 'boots on the ground' job. The planning and management of this ambitious project thus fell to a large extent on the army, in particular a small team of relatively junior officers of the Royal Engineers whose dedication, diplomacy, and commitment went far beyond a simple knowledge of telegraph engineering. And interestingly, the solution, reached in 1870, turned out to be something never envisaged at the outset, as will be explained.

The Team is formed

Put in charge was none other than Patrick Stewart who had featured prominently in the development of the telegraph in India and communications during the Indian Mutiny, described in the previous chapter. With a brilliant career predicted, the substantive Captain, acting Major, local Lieutenant Colonel Stewart, now thirty years old, took charge of the challenging Indo-European telegraph venture – not the sort of thing one of his age or rank would expect these days.

Patrick Stewart

Appointed as his assistant in February 1862 was another Royal (Bengal) Engineers officer, Lieutenant John Champain, then aged 26. Educated at Cheltenham College, Champain followed Stewart through Addiscombe (1851-53), and like Stewart passsed out top of his intake and was awarded the Pollock medal. His competence at mathematics and Latin were proof of a logical mind, and he was also an amateur artist. After his course at Chatham he arrived in India in 1854 as a 2nd Lieutenant in the Bengal Engineers, and saw active service in the Mutiny, being present at the siege and capture of Delhi, and later at Agra, Cawnpore, Lucknow, and during final pursuit operations in 1858. He subsequently held engineering appointments in the public works department (many engineer officers in India held such appointments, which were not 'civilianised' until much later) then he volunteered as Stewart's assistant.

Prospecting a Loop Line through Persia

From the start one particular problem worried those who planned this optimistic telegraph venture – the section of the line from Baghdad to Fao ran through disturbed tribal territory in Mesopotamia, and its security was judged to be at high risk. It was decided to try to run a loop line through Persia, from Bushire through Tehran to Baghdad, intended as a back-up route to circumvent this foreseen problem. Whatever the difficulties in Mesopotamia might have been, they were certainly not reduced by the Persian scheme. It was an outrageously ambitious project, and proved to be a considerable undertaking.

In the spring of 1862, while preparation for the Baghdad to Karachi section through the Persian Gulf was getting under way, Stewart and Champain left India by boat for Bushire to undertake a reconnaissance for the loop line through Persia. They travelled along a proposed route from Bushire to Tehran, about 800 miles, traversing much rugged and virtually unknown country. The western part of the country is very mountainous, with passes up to 8,000 feet high, and was completely undeveloped. There were no roads, only mule tracks - and bandits roamed. Travelling through Persia involved using the local *chapar* sytem. This meant travelling between *caravanserai* about twenty miles apart. These 'rest houses' provided crude accommodation facilities and stables, where it was possible to hire riding horses and baggage mules to the next station along the route, where they were handed in and fresh ones obtained.

Their route took them through the principal cities of Shiraz and Ispahan (nowadays Isfahan, the site of the presently contentious nuclear reactor). It was hoped that the offer to build a telegraph line connecting these places, as well as suiting their own intended purpose, might prove attractive to the Persian authorities, there being at that time no telegraph communications anywhere in the country. On reaching Tehran they engaged in discussion with Persian officials who, as it might be imagined, were unimpressed by this strategy, viewed their presence as an intrusion, and the whole preposterous idea with great suspicion. Numerous obstructions to the plan were fabricated.

After these preliminary soundings, prospects for the Persian loop line looked distinctly unencouraging, so Stewart, who was suffering health problems, went to England to get on with arranging the manufacture and supply of materials for the planned line through Mesopotamia and the submarine cable through the Persian

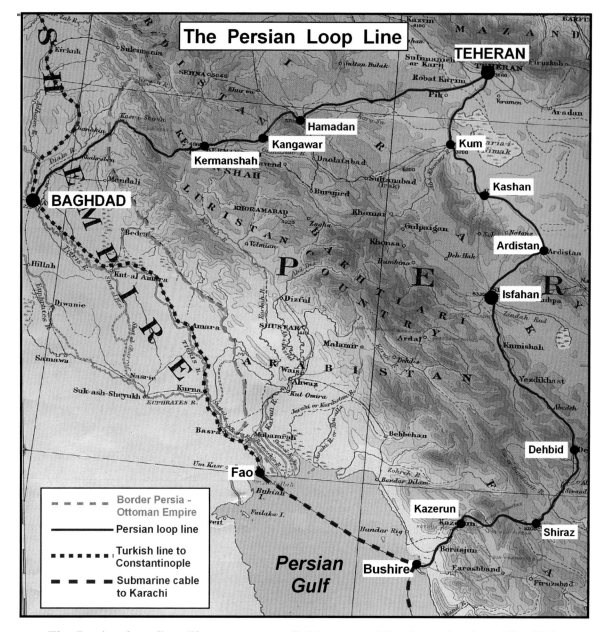

The Persian loop line. There were no reliable maps of Persia at the time. The 1862 reconnaissance route, subsequently used for the line itself, is shown here overlaid on a relief map published by the War Office in 1906. Although an anachronism, it accurately indicates the mountainous terrain, place names, and territorial boundaries at the time.

Gulf. Champain was left to reconnoitre the country between Tehran and Baghdad, about 500 miles, again going through principal cities *en route*, Hamadan and Kermanshah, so that the telegraph line, if ever agreed, could serve Persian as well as British and Indian interests. The total length of their proposed Persian route from Bushire through Tehran to Baghdad was about 1,250 miles, an enormous distance through unknown and unfriendly territory, rugged and in parts mountainous terrain, and much longer than the dubious 300 mile section between Baghdad and Fao that it was circumventing.

Development of the Main Route

While negotiations with Persia for the loop line continued, work started on the main route. The line from Constantinople through Anatolia to Baghdad was completed by the end of 1861. Royal Engineer non-commissioned officers were in charge of the construction of sections of the line, and it appears to have been completed quite quickly and without any special difficulty. At the other end of the route, construction of the landline from Karachi to Gwadur along the dreary Makran coast began in June 1862 under a civilian

Superintendent of the Indian Telegraph Department, Mr H V Walton, reaching Gwadur in April 1863. It was an unenviable task for Henry Walton and his men.

In anticipation of laying the submarine cable, Lieutenant Stiffe of the Bombay Navy carried out a route reconnaissance of the seabed along the Persian Gulf, taking soundings to determine the cable's exact course. A new design of submarine cable was constructed by Sir Charles Bright and Mr Latimer Clark, leading submarine cable engineers, who had been employed by the government. It consisted of four copper wires of high conductivity passing through a copper cylinder, the whole drawn out under pressure to form a solid wire. This conductor was covered with four coats of gutta percha insulation and four coats of Chatterton's insulating compound. On top of this was a layer of hemp, twelve steel guard wires spiralled around to give physical protection, more hemp and a final coating of two layers of a tar and silica compound. The finished cable was one and a quarter inches in diameter, weighed four tons per mile, at a cost of £200 per mile. After manufacture it was taken from England round the Cape of Good Hope to India, and laid towards Fao.

Laying the cable began in February 1864, with landings being made at Gwadur, Jask, Mussendam, and Bushire, but bringing it ashore in the mud flats at Fao was a problem. The principal cable laying ships had to anchor some six miles offshore, a flat-bottomed boat, the Comet, was requisitioned from the Bombay Marine Service. Even so, the men eventually had to crawl on their stomachs over the deep mud, and the cable had to be cut into mile-and-a-half lengths, each section being carried by several hundred Arabs to its position before being joined up again. The operation was successfully completed on 8 April 1864, and Fao was now in direct telegraph communication with Karachi.

The Baghdad to Fao section was still under construction so the connection with Europe was not yet complete. This line, along the Tigris valley, was under the overall direction of the British Resident in Baghdad, Lieutenant Colonel (later Sir Arnold) Kemball, Bombay Artillery, and its construction was under the detailed superintendence of a telegraph engineer, Mr Greener, supported by a team of Royal Engineer non-commissioned officers from Chatham and a local labour force. It was completed in February 1865.

Landing the submarine cable at Fao. A sketch originally published in the Illustrated London News.

The Persians Agree a Convention

While construction of the main route was under way, the wrangling in Tehran continued. Quite unexpectedly, after a year spent in protracted negotiations, agreement was reached and a telegraph convention with Persia was signed, although some impracticable concessions had to be made to get any agreement at all. In Persia, the idea that others might construct the telegraph line was unacceptable, even if electric telegraphy was something entirely novel in the country and the technicalities not in any way understood. The convention dictated that the line was to be built by Persian labour, the cost of the telegraph equipment from Britain was to be paid by Persia in instalments, and that Persia was to work the line but British and Indian traffic could be passed for a tariff to be decided later. The only intervention by Britain to be permitted was the presence of *one* British officer for a limited period in a purely advisory role. As that was the best that could be obtained, and in order to get things moving, it was agreed. When it came to the implementation of this farcical arrangement, hatched by petty bureaucrats in Tehran, a Nelsonian blind eye was turned to the impracticable parts of it.

Constructing the Route through Persia

The construction of the Persian loop line now agreed, more people were needed, and volunteers were called for. Patrick Stewart was still in overall charge. The recently promoted Major John Champain was placed in charge of the Persian portion of the undertaking, to be based in Tehran, and although the arrangement with the Persian government limited the British assistance to one person this was discreetly ignored, fortunately by both sides, and three other Royal Engineers officers, twelve Royal Engineers non-commissioned officers, and two civilian telegraph superintendents were selected to assist him. The officers were Lieutenants Robert

Murdoch Smith, Oliver St John, and William Pierson. These were to be three outstanding officers. The two civilian Superintendents from the Bengal Telegraph Department were Mr H Man and Mr Henry Walton, the latter having already built the line from Karachi to Gwadur.

Robert Murdoch Smith, Oliver St John, and William Pierson

Robert Murdoch Smith was the son of a GP in Kilmarnock, Scotland. He was educated at Kilmarnock Academy and Glasgow University, and in 1855 was commissioned into the Royal Engineers. The following year he was selected to command a party of engineers in support of an archaeological expedition to Halicarnassus, and as a result became interested in archaeology. Several years later, after a further archaeological expedition to North Africa, and then based in London, he read in *The Times* about the agreement with Persia for the Indo-European telegraph, and volunteered.

Oliver St John was born in Ryde, Isle of Wight, in 1837, the eldest son of a Captain in the Madras infantry. A great-grandson of the tenth Baron St John of Bletso, he was educated at Norwich grammar school and attended Addiscombe, the East India Company's military college near Croydon, was commissioned into the Bengal Engineers in 1858, and went to India in 1859. In October 1863 he volunteered for the Indo-European telegraph expedition.

William Pierson, also aged twenty-five, was educated at Cheltenham College and obtained a cadetship to Addiscombe military college, winning the Pollock medal awarded to the best cadet of his intake. He joined the Bengal Engineers in December 1858. At the Chatham School of Military Engineering he was also considered an outstanding student. At Chatham he studied German, was a high-class chess-player, went deeply into music, and taught himself the cornet and concertina, having long played the piano. On arrival in India he was sent on active service in 1861 with the Sikkim Field Force, carrying out a number of bridging operations for which he was three times mentioned in despatches. After a short spell in Oude he was selected to work on the Indo-European telegraph.

Work began in the autumn of 1863. The line through Persia followed the route previously envisaged from the reconnaissance by Stewart and Champain in 1862 (see map above). The plan now was to divide it into five sections, each about 200-250 miles long. Baghdad to Kangawar under Pierson, Kangawar to Tehran under Walton, Tehran to Kohrud under Murdoch Smith, Kohrud to Murghab under Man, and Murghab to Bushire under St John.

Lieutenant Murdoch Smith landed at Bushire on 17 November 1863 with twelve NCOs of the Royal Engineers and a mass of telegraph stores, and travelled to Tehran with Champain. Leaving Tehran on 24 February 1864, Murdoch Smith soon reached his headquarters at the holy city of Kum, the burial place of Fatima, and began work on his 200-mile section.

In accordance with the agreed terms the Persian authorities were involved in the construction work, which proved to be exceedingly frustrating. Administration of such things as transport and stores was poor. Much time was wasted by inefficiency, as well as the local concept of what constituted a day's work. Things could, however, be speeded up by *mudakhil*, literally translated as income, but otherwise known as perquisites!

Murdoch Smith found that neither wire, insulators nor tools had been sent from Ispahan by the Persians. In March he received the wire and insulators, but no tools; and when at last he had got his tools and applied to the Persian officials for workmen, he was given a few old men and boys. These were followed by thirty labourers, but the latter decamped because they had received no pay. Having tried unsuccessfully to obtain some money from the Governor of Kum, he appealed to Tehran and in the end was given about three dozen labourers, half of whom, for want of mules, had always to be employed in carrying poles.

A rigid disciplinarian, Murdoch Smith abhorred the corrupt dealings of the Persian officials and contractors. "I found", he wrote, "that all were actuated by the same principles. When they thought it possible to make *mudakhil* they were all activity, but when their income was interfered with by the system I adopted of seeing everything paid in my presence, and warning the villagers on no account to give 'presents' to anyone, they relapsed into their usual state of obstinate indifference." Nevertheless he completed his section during the autumn of 1864.

To begin with, the people were often hostile. Local governors regarded the new-fangled *Feringhi* (English) invention as something that was going to diminish their powers. Before the line was completed many miles of it were swept away by the nomadic tribes at the secret instigation of the local authorities. Many years later Murdoch Smith gave an account of the difficulties encountered by the construction parties:

> ....for the presence of none of whom had any provision whatever been made in the convention, and whose arrival in the country was not unnaturally viewed with the very strongest suspicions. The situation was altogether false and unsatisfactory. A line of 1,250 miles through an extremely difficult and troublesome country had by hook or by crook to be made with Persian materials, at Persian expense, by a handful of foreigners, whom every man in the kingdom, from the Shah downwards, then regarded as pestilent interlopers. Looking back with the knowledge of subsequent experience, the writer is astounded at the cool impudence of the whole undertaking. The marvel is that our throats were not promptly cut by patriotic brigands. The work, however, advanced somehow until the erection of the line was nearly completed, when matters came to a deadlock. For two or three months we withdrew altogether from the telegraph, and our departure from the country seemed all but certain. But Champain never despaired. Negotiations were renewed as a sort of forlorn hope, and, somewhat to our surprise, they resulted in a working arrangement for five months, after which, in the words of the convention, we should 'cease to have any connection with the telegraph, and at once quit the country.

At one time Oliver St John had fever and had to go to the coast for a few weeks to recuperate. He returned to find twenty miles of his line destroyed but managed to trace the offenders, one a local clan chief. The authorities in Tehran were much annoyed, and the Shah issued instructions for punishment. He despatched the royal executioner from Tehran, who would cut off any heads that St John indicated. The offending clan chief was sent in chains to Tehran. St John, exercising well-judged diplomacy, waited two months, keeping them in suspense, then reprieved all. His action secured the friendship of the local tribes, and ensured no more wanton damage in the area, but it did cause a three-month delay.

William Pierson was in charge of the construction of the section between Baghdad and Kangawar. "His work was of the most arduous possible nature," wrote Champain. "Innumerable obstacles were put in his way, and many vexatious political difficulties had to be overcome. We owed our eventual success chiefly to Pierson's indefatigable exertions, and to his personal influence with the Persian authorities and with the Kurdish chiefs of the neighbourhood." [2] It is recorded that, on three occasions, Pierson rode from Kermanshah to Tehran, a distance of over 300 miles, in four days.

Despite the difficulties, the construction of the Persian loop line from Bushire via Tehran to Baghdad, a distance of about 1,350 miles, was completed in just over eleven months, on 13 October 1864.

Kotal Dukhtar.

Kotal Pir Zan.

Sketches drawn by Oliver St John of two kotals (passes) on his section of the mountainous route between Kazerun and Shiraz. Reproduced from Telegraph and Travel.

The Death of Lieutenant Colonel Patrick Stewart

Thus by early 1865 the line between Karachi and Constantinople existed, including both the Fao to Baghdad section and the Persian loop line. Electric signals may have passed, but turning it into an efficient telegraph system that passed traffic reliably was another matter. The working of the line was dire. Patrick Stewart went to Constantinople in July 1864, to try both to induce the Ottoman authorities to improve the working of their section of the line and to improve the arrangements for traffic from Constantinople onwards through the European system where numerous uninterested national telegraph departments with their own operators speaking different languages, and no agreed international telegraph procedures, passed this not particularly profitable traffic in a desultory fashion.

Stewart was awarded the CB for his work, but was suffering from the vigorous life he had led for over ten years in countries and climates not beneficial to health. In Constantinople he was as energetic as ever, but he had pushed himself too far. In December he contracted fever, complications followed, and he died on 16 January 1865, aged only 32, to be buried along with the dead of the Crimean War in the military cemetery at Scutari. He was to be the first of the dedicated team who, for health reasons, mostly came to a premature end.

Reorganisation and Further Progress

Reorganisation of the senior posts followed Patrick Stewart's demise. Lieutenant Colonel F J Goldsmid, an officer of the Madras staff corps, was appointed as the new Director-in-Chief of the Indo-European Department, and it fell to him to organise the working of the line, a task requiring organisational and diplomatic skills rather than telegraph engineering expertise.

Frederic Goldsmid had a very wide experience in the region before becoming involved in the Indo-European Telegraph project, including a period in the Indian civil service when, with interpretership qualifications in Hindustani, Persian, and Arabic, he held appointments as a Deputy Collector and Assistant Commissioner for lands in the recently annexed province of Sind. He had volunteered for service in the Crimean War, was attached to the Turkish contingent, and with his linguistic talent, quickly learnt to speak Turkish.

He first became connected with the Indo-European telegraph project in 1861 when he negotiated and got agreement with the tribal chiefs in Baluchistan and Makran for the construction of the line through their territory to Gwadur. He then negotiated with the Turkish authorities for the construction of the telegraph line from Constantinople across Anatolia and Mesopotamia. Further negotiation with the Persians clearly being needed at the time of Stewart's death, he was ideally suited to the appointment.

Goldsmid's first task was to go to Tehran to negotiate a new and more comprehensive treaty with the Persian government, and he set off on this mission in June 1865. Major John Champain was appointed as his deputy, and left Tehran for London to carry out duties in the London office in the absence of its new Chief. For two years their efforts were concentrated in Turkey, where Champain spent the greater part of 1866 in frustratingly trying to get the Constantinople-Baghdad line into some semblance of operating efficiency. Then in 1867 Champain was sent to St. Petersburg, at that time the capital of Russia, to carry on some difficult negotiations, about which more shortly. He had handed over his duties in Persia to the newly-promoted Major Robert Murdoch Smith who, as things turned out, was going to be the representative in Tehran for twenty-two years. In 1866 his staff was increased by the arrival of Lieutenant B Lovett R(B)E, who joined as an assistant to St John and worked on the Persian telegraphs for four years.

Operating and Maintaining the Line through Persia

Although construction of the Persian loop line was completed in October 1864, it could not be left to local resources to maintain and operate it, as the Persians themselves now realised. The team of officers stayed there and were reinforced by non-commissioned officers from the Royal Engineers and staff from the Indian Telegraph Department. It was difficult to cope with the maintenance and improvement of the lines. "Much of our time between 1864 and 1870," said Murdoch Smith, "was taken up in special employments, such as Captain St. John's service with the Army in Abyssinia (described in chapter 6) and with the Eastern Persia Frontier Commission; Captain Pierson's detachment on special duty to Mazanderan, to the Caucasus, to Vienna for the International Telegraph Conference, to Bombay, to London, and finally to Tehran to superintend the erection of the new British Legation; Lieutenant Lovett's service with the squadron on the

Arabian Coast and on the Frontier Commission; and my own absence on special duty in Baluchistan, Arabia and India. During this period, however, every opportunity of exploring and surveying the country was eagerly seized. To the skill and perseverance of Captain St. John we are indebted for the admirable map which must form the groundwork of all future surveys of Persia. From 1871 to 1873 [when Murdoch Smith was on extended leave] Captain Pierson was the only officer in Persia, and I was the only one from his departure in 1873 until the arrival of Captain H. L. Wells RE, in the beginning of 1881".

The life of those charged with operating and maintaining the line was far from easy. The long distances and rough terrain, especially the mountainous parts where snow lay in winter, made maintenance and repair very difficult. When snow was on the ground a traveller could hope for no better shelter than a damp and dirty *caravanserai*. Working parties on the passes often laboured up to the waist in snow, and took refuge in a cave for the night. In the early days there was frequent wilful damage to line and poles about which the local authorities did nothing. Every Persian man seemed to own a gun, and it became a popular new national sport to shoot the insulators on the telegraph poles - they shattered quite spectacularly.

The wives and children of the British telegraph employees, who followed their husbands and fathers into the wilds, travelled on horseback or rode, as the Persian women did, in *kajawahs* or panniers slung on each side of a mule, or in palanquins carried between two mules. Wheeled conveyances were almost unknown, for the roads were execrable. The conditions in which they lived were primitive.

Mrs Smith, the wife of an Inspector at one of the remote telegraph stations, having learnt, as many such wives did, to receive and send messages, kept the telegraph office open and worked it alone for six weeks while nursing her husband, who was ill with enteric fever.

Robbers lay in wait for unguarded travellers. Corporal Collins RE was attacked and killed after accounting for three of his assailants. Three others of the gang were caught by the local Persian Governor, who buried them up to their waists and then built gypsum around them up to their necks so that they suffocated slowly as the gypsum solidified and contracted.

In March 1867, Oliver St John was riding along his section of the line near Shiraz and had just entered an oak forest when a lioness appeared some thirty yards in front of him. His horse stopped, so did the lioness, and for a few seconds they stood looking at each other. Then St John cracked his whip and shouted; but instead of sneaking back into the forest as he expected, the lioness charged and sprang at the horse's throat. Having no weapon except a small revolver, he tried to spur his horse forward, but the animal refused to move, and in another moment the lioness had sprung on to its hindquarters. St John jumped off. The horse plunged violently, knocking him in one direction and throwing the lioness in another, and then, trotting away, was overtaken by the lioness who sprang on to its hindquarters once more and both thus vanished into the forest. After a time, St John followed, only to discover that although the lioness had vanished, the horse was so frantic with terror that it could not be approached, so he left it till next morning, when he found it badly mauled but grazing quietly. Its wounds were sewn up, and within a week it was as well as ever.

When Corporal Blackman RE, a Telegraph Inspector in the same area, arrived one winter's afternoon at Dehbid, he learnt that his servant had been bitten by a mad wolf. This beast, after biting many other people, all of whom are reported to have died, next seized a man in a dark passage leading into the telegraph station; whereupon Blackman ran out and shot it dead with his revolver.

The telegraph offices in Persia were usually considered as sanctuaries by the law-abiding population because the wires were supposed to end 'at the foot of the throne', in other words it was imagined that they had direct access to the Shah at Tehran. But they were sometimes attacked by rioters or brigands who recognized no authority. It was said that these attacks were the result of rumours that the telegraph instruments were of pure gold - a tribute to the care with which the brass polished.

The isolated telegraph stations along the Persian Gulf, although not suffering the problems of those in Persia, were anything but desirable havens. Jask was a depressing spot. According to one contemporary description, "No human ingenuity could make it an acceptable residence for an ordinary mortal. The only materials for a landscape are a glaring, glittering, glabrous sea for a foreground; a glaring, shiny, sandy plain for the mid-distance; and bare volcanic hills for the background. Add to this the heat which must make a nervous sinner most apprehensive, and the fact that you never see a soul from the outside world, and only get letters twice a month." The chief difficulty at all stations was water supply. At a few feet below ground-level, a kind of crust existed above which pure water collected and could be tapped in small quantities; but if the crust was perforated, brackish water was encountered, and the well, and all adjacent wells, were ruined for ever. At Gwadur the supply was so bad that all drinking water had to be condensed from the sea.

During this period the officers took well-deserved leave. Oliver St John happened to be in England during his leave when he got diverted to the Abyssinian Expedition, to be described in chapter 6. In 1866 William Pierson was sent to the Caucasus to help the Persians build a telegraph line from Tehran north into Russia. (By now attitudes had changed and the Persians had got used to the idea of the telegraph, the influential factor being the realisation that it was a good source of revenue.) On his return journey Pierson was attacked by a band of robbers, who killed his attendant, but he beat them off, saved his baggage, and brought in the body of his servant.

In 1870, after five years at the helm, Lieutenant Colonel Frederic Goldsmid resigned to undertake other duties and Lieutenant Colonel John Champain, as he had become, succeeded him as the Director-in-Chief. Major Robert Murdoch Smith remained in charge in Tehran. St John and Pierson had moved on to new appointments elsewhere. [3] [4]

A New Idea - the Route though Russia

Shortly before the Crimean War (1854-56) a telegraph system had been constructed in Russia by the German firm of Siemens and Halske (see chapter 2). This had continued to be developed in the subsequent period, and in 1867 the Russian system had been extended to the Persian frontier to connect with a rickety line which the Persians had built through Tabriz to Tehran, and a telegraph convention had been concluded between Russia and Persia. Before Champain went to Russia he had received an offer from Messrs Siemens and Co to build a double line from London to Tehran, and his visit to St Petersburg (then the capital) in 1867 was largely in connection with this. Although he secured no definite agreement from the Russian Government he established, in his eminently successful and engaging style, cordial relations with the Russian Director-General of Telegraphs. It paved the way for the concession subsequently granted to a British concern called the Indo-European Telegraph Company, a new commercial entity, not to be confused with the Indo-European Telegraph Department which managed the telegraph line between Karachi and Tehran.

This concession provided for the erection of a line through Russia to connect London with Tehran, thus avoiding the inefficient Turkish system and completing a chain of direct and special communication between England and India. The new Siemens route through Germany and Russia was opened on 31 January 1870. The rickety Persian section from the Russian frontier to Tehran had been renewed, and two wires, specially devoted to Indian traffic, were available over the whole distance from London to Tehran.

From Tehran, through Bushire, to Karachi, the line remained under the direct control of the Indo-European Telegraph Department and a programme of upgrading was put in hand. The single submarine cable from Bushire, through Jask, to Gwadur was deemed insufficient, so it was decided that the Makran Coast line should be extended from Gwadur to Jask, and that a second submarine cable should be laid from Jask to Bushire. A third wire was to be attached to the Bushire-Tehran land line, so that one wire might be reserved for local traffic and the other two devoted entirely to international business. The wooden poles were replaced by metal ones, and other aspects generally improved. The extension by land from Gwadur to Jask was completed in August 1869, and the new cable from Jask to Bushire in November. Champain came from England to supervise the latter undertaking, and nearly lost his life on the voyage to India, when the the ship in which he was travelling, the P and O line's *SS Carnatic,* was wrecked in the Gulf of Suez.

Siemens telegraph equipment used on the line.

The Russian route quickly took the lead. At a stroke many of the operating problems had been resolved and the line was worked with much greater efficiency, leading to an increased volume of traffic and greater revenue. In 1870, when a message between England and India usually took 30 hours to get through by the Turkish route, it took only six hours by the Russian route. Never part of the original scheme, its inception owed much of its success to John Champain's persistence and quiet diplomacy.

New Technology - Submarine Cable from England to Bombay

Then in 1870 this ten year epic of endurance and perseverance in the Middle East to connect India with London was overtaken by developing technology, bringing a rapid solution. Submarine cable design, laying, and maintenance techniques, had taken enormous strides forward since the early fiasco of the Red Sea and India line a decade before. Britain led the world in this technology. Coupled with this, new entrepreneurs were prepared to invest, foremost amongst them Sir John Pender who had made his first fortune in cotton in Lancashire. Confident of the new methods, he decided to get India connected by submarine cable.

The Telegraph Construction and Maintenance Company provided a fleet of five ships comprising the *Great Eastern* under the command of Captain Halpin, the *Hibernia* on charter from the Anchor Line, the *William Cory* (known in the trade as *Dirty Billy*), the *Chiltern*, and the small shore-end steamer *Hawk*. The *Great Eastern* and the *Chiltern* left Portland on 6 November 1869, carrying between them 2,600 miles of cable, the *Hibernia* left in December with 600 miles of cable and went direct to Aden, and the smaller *William Cory* and the *Hawk* with some 400 miles of cable went through the newly opened Suez Canal and awaited the others at Suez.

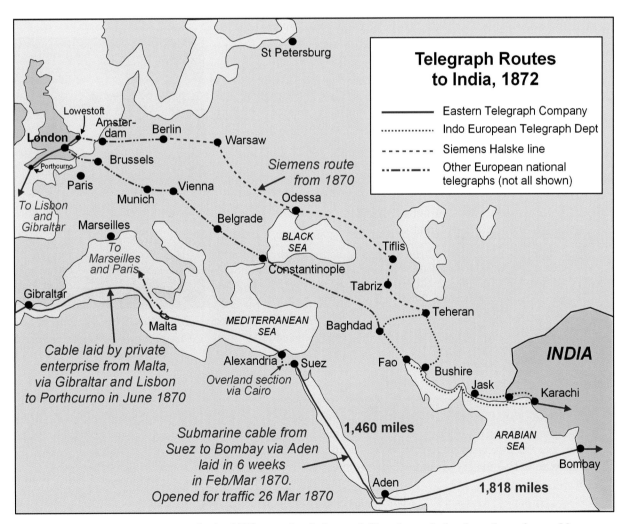

The telegraph routes to India in 1870, overland through Russia and also by submarine cable

After their long voyage round the Cape of Good Hope the *Great Eastern* and the *Chiltern* arrived at Bombay at the end of January and began laying towards Aden on 7 February 1870. In Aden they joined up with the *Hibernia* and continued on through the Red Sea towards Suez, where they met up with the *William Cory* and the *Hawk* and joined the cables. The final splice of the cables was made on 22 March, the whole distance of some 3,000 miles having taken just over six weeks to lay. More to the point, it worked straight away.

The route from India was now Bombay to Suez by the new cable, overland to Alexandria on the Mediterranean coast, and then to Malta. From Malta it had then to continue via foreign landlines through Italy or France to Great Britain. This unsatisfactory situation was quickly remedied, and the section from Malta to Porthcurno, near Lands End in Cornwall was laid. This cable was owned by the Falmouth Gibraltar and Malta Company, and laying was again contracted to the Telegraph Construction and Maintenance Company. Laying began in mid-May 1870, in three sections. Malta to Gibraltar was laid by the cableships *Scanderia* and *Edinburgh*, the second section, Gibraltar to Carcavelos, near Lisbon, was laid by the *Scanderia*, and the third section, Carcavelos to Porthcurno was laid by the *Hibernian*, straight from her earlier work in the Red Sea. The final splice was made on 8 June 1870, and later that month the line opened for service – the first long distance chain of direct submarine telegraph in the British Empire.

Further Improvements

Now there were two reliable routes to India – the new submarine cable route, and the Russian route. Submarine cables were, of course, not immune to problems, so it was essential to keep both routes. At a stroke many of the operating problems had been resolved and communications achieved much greater efficiency. It was correctly anticipated that increased traffic would follow this much improved service. The troublesome route from Fao to Constantinople fell into disuse. The original submarine cable between Karachi and Bushire was duplicated in October 1876, and again in March 1877, as a safeguard against faults or damage, and to cope with increasing traffic demands. [5] [6]

In 1885 the cables laid between Jask and Bushire were becoming unreliable and were replaced in November that year, again under the personal supervision of Colonel Bateman-Champain (he had added the additional name Bateman in 1871), and since 1870 the Director, who had come from England for that purpose. This was to be his last visit to the East. On the completion of the work he proceeded to Calcutta to confer with the Government on telegraphic matters, and afterwards went to Delhi to visit his old friend, General Sir Frederick Roberts. While at Delhi he learnt that his services had been recognized by the award of the KCMG, and this, following on the bestowal of a sword of honour by the Shah, made his visit to India a memorable one for him.

Unfortunately, he had long been subject to asthma, complicated at times by bronchitis. He was frequently urged to take some months of complete rest in a more genial climate than that of England, but always refused to do so. Back in England he continued at his task, but in an attempt to restore his health he went to the south of France in January 1887, and died at San Remo on 1 February, aged fifty-one. [7]

Colonel Sir John Bateman-Champain. This portrait presently hangs in the Royal Signals HQ Officers Mess.

Life in Tehran

Robert Murdoch Smith remained in Tehran to oversee the Indo-European Telegraph in its new form. He was one of a small European community comprising British, French and Russians, mostly the staff of their Legations, about fifty in all in the early days, and he enjoyed a life style that apparently suited him. He could converse fluently with the Persians, was well versed in their customs and diplomatic niceties, and gained their confidence. Many years later he wrote:

> Gradually I managed to get rid of the jealousy and opposition of the Persians, and at last to gain their active goodwill and friendship to get wilful damage stopped, and the lines to work perfectly. ..... With our help and advice Persia has been covered with lines in all directions which have brought the distant provinces directly under the control of the Central Government at Tehran, and thereby got rid of a vast amount of local misgovernment and oppression.

One amusement in Tehran in the early days was amateur theatricals, and during these he met his future wife. Eleanor Baker, the daughter of a RN Captain, had come to Tehran to visit her brother, who was attached to the telegraph staff, and she had a part in one of the plays. Robert and Eleanor duly got married in 1869, but not without difficulty as there was no official in Tehran authorised to marry Christians. After some delay, the Foreign Office somehow managed to deal with the necessary legalities.

Their marriage was happy but tragic. Their first child, a son, lived only a day, and Eleanor developed a partial paralysis and was to remain an invalid for the rest of her life. They went to England on long leave from 1871 to 1873, during which time William Pierson managed things in Tehran. In England a second son was born to the Murdoch Smiths, and on returning to Tehran five more children were born, three sons and two daughter. Disease was prevalent and medical services in Tehran were scant. Another son died in January 1877, and a daughter died in 1881, a further daughter was born in 1883 but Eleanor died soon afterwards from post-natal complications. Five surviving children were left, two boys and three girls. It was arranged that these children would be taken back to Scotland to grow up with Robert's widowed sister in Edinburgh. Murdoch Smith set off with the children along the caravan route from Tehran to Bushire and thence home by boat. At Kashan, *en route* to Bushire, three of the children caught diptheria and died there within forty-eight hours. Eventually two surviving daughters reached Scotland, and grew up to lead normal lives. Robert Murdoch Smith never remarried.

He retired to Edinburgh in 1885, handing over his duties in Tehran after twenty-three years there, but in 1887 Colonel Murdoch Smith agreed as an interim measure to fill the gap left by Champain's death, and later that year he was sent out by the Foreign Office on special duty, partly departmental and partly diplomatic, to Baluchistan, the Persian Gulf, and Tehran, and in the course of this he succeeded in obtaining a renewal of the treaties for working the telegraph lines throughout the Shah's dominions, which had expanded considerably in the last twenty years. Whilst at Tehran the Shah presented him with a diamond snuff-box as a personal souvenir. He received the KCMG for his services on this mission. On final retirement he was given the honorary rank of Major-General. Sir Robert Murdoch Smith died in Scotland on 3 July 1900, at the age of sixty-five. [8]

Maj Gen Sir Robert Murdoch Smith in 1898

The Eventual Demise of the Indo-European Telegraph

With the departure of the last of the original team it is time to leave the story of the telegraph route to India, although much more could be said. Army officers had played a dominant part in establishing it, and running it in its formative years. The Indo-European Telegraph Company existed until 1931 when two things changed the situation. Firstly, the Persians wanted to run it themselves, and secondly, wireless and other commercial competition had resulted in the merger of cable companies. In 1931 the Indo-European Telegraph company was wound up and the line and all equipment was given to the Persians. The company Cable and Wireless had been formed, one of its constituents being the former Eastern Telegraph Company, and was more than capable of handling the Indo-European work.

Endnotes

1. A more detailed description of the Red Sea and India line is given in *Submarine Telegraphs* by Sir Charles Bright, pp 57-58, pub 1898.

2. *Royal Engineers Journal*, 1 Aug 1881, p 166.

3. After resigning as Director of the Indo-European telegraph in 1870, Goldsmid was appointed the following year as a Commissioner for boundary agreements between Persia and Baluchistan. Other similar appointments followed, and he was created KCSI in 1871. He retired from the army with a special pension and the honorary rank of Major-General in 1875. He took up writing, and continued in public life with further diplomatic work on various Commissions in the Middle East and Africa. He settled in London in 1883, and became involved with

work in the Royal Asiatic Society, the Royal Geographical Society, and various other religious and philanthropic institutions. He died on 12 January 1908, aged ninety - a remarkable age in those days for one who had for so long served in intemperate climates. Goldsmid wrote a book describing his experiences with the Indo-European telegraph project, *Telegraph and Travel,* published in 1874.

4. Pierson was detached for a period from duties in Persia and acted as Secretary to the British representative at the International Telegraph Conference in Vienna. He then returned to Persia, charged with the design and construction of a new building for the British Legation in Teheran. Between 1871-73 he acted as Director of the Persian Telegraph while Murdoch Smith took a long leave in Britain. He went to England for various duties in 1874, returning to India in 1876. In 1877 he was appointed Secretary to the Indian Defence Committee, and in 1880 he was appointed Military Secretary to the Viceroy, Lord Ripon, a prestigious appointment. In March 1881 he was appointed C.R.E. to the field-force involved in operations against the Mahsoud Waziri tribe in the North-West Frontier area. In this campaign he was greatly exposed to fearful heat, contracted dysentery, and this talented man died suddenly at Bannu on 2 June 1881, aged only forty-one. A memoir of Major Pierson was published in the *Royal Engineers Journal* 1 August 1881, pp 165-66.

5. A series of articles in the *Royal Engineers Journal* give a detailed technical description of the construction and operation of the routes up to 1878, covering various aspects not described in this chapter, and based on a paper read to the Society of Arts by Major Bateman-Champain in May 1878. See *REJ* 1 Jun to 2 Sep 1878 pp 62 to 96.

6. A further detailed description of *'The Telegraph Line from Bushire to Teheran'*, was given to the Society of Telegraph Engineers by Lieutenant Colonel Bateman-Champain, then the President of the Society. *See Journal of the Society of Telegraph Engineers*, Vol VIII, 1879 pp 400 to 441.

7. Champain married in 1865 Harriet Sophia, daughter of Sir Frederick Currie of the Indian civil service, and they had six sons and two daughters; three sons followed their father into the army, and one went into the navy. He was a member of the councils of the Royal Geographic Society and the Society of Telegraph Engineers, being President of the latter in 1879. He was also a talented draughtsman and amateur artist, and several of the illustrations in Sir Frederic Goldsmid's book about the Indo-European Telegraph project, *Telegraph and Travel*, are from Champain's watercolour sketches. After his death a message of sympathy was sent by the Shah of Persia to his widow. Memoirs of Colonel Sir John Underwood Bateman-Champain KCMG will be found in *The Royal Engineers Journal,* Vol 17, pp 56-60; in the *History of the Corps of Royal Engineers,* by Major-General W Porter, Vol II, pp 530-533; and in *Addiscombe - Its Heroes and Men of Note,* by Colonel H M Vibart, pp 603-608. He also contributed a number of articles on the Indo-European telegraph published in the *Journal of the Royal Geographic Society*. A very good description of contemporary life in Persia, entitled *On the Various Means of Communication between Central Persia and the Sea*, was given to the Royal Geographical Society in January 1883 by the then Colonel Bateman-Champain. See Proceedings of the RGS, March 1883.

8. Murdoch Smith's life in Persia is well described in *The Life of Major-General Sir Robert Murdoch Smith*, written in 1901 by his son in law, William Dickson, the husband of one of the two surviving daughters. The quotations attributable to him in this chapter are taken from it. An obituary was published in the *Royal Engineers Journal*, 1 Sep 1900, pp 192-194.

Biographies of Major-General Sir Frederic Goldsmid KCSI CB, Colonel Sir Oliver St John KCSI, and Lieutenant-Colonel Patrick Stewart CB can be found in *The Dictionary of National Biography*.

Chapter 5

Army Signalling Developments in the 1860s

Introduction

The scientific discoveries that made electric telegraphy a practical form of communication from around 1840 were described in chapter 1, and the British army's first use of it in the Crimean War, 1854-56, was described in chapter 2. The army's involvement in the introduction of the electric telegraph to India from 1853 and, following the Indian Mutiny, the construction of the Indo-European telegraph route connecting London and India, were described in chapters 3 and 4. It is now time to return to Britain and resume the story of the sluggish development of signalling in the British army.

It was not a subject that aroused much interest in the mid-Victorian army, and it was not until the early 1860s that the army started to pull itself out of its lethargy and address the subject. Such movement, it would appear, came more from the initiatives of a small number of individuals rather than any progressive policy originated in stultified enclaves such as the Horse Guards. As will become apparent in this chapter, certain developments over the previous three decades could, with a modicum of thought, have been applied to providing some solutions to military signalling requirements, in particular the need for tactical signalling where mobility was a crucial factor. But the army's response lacked enthusiasm and direction. By 1860 tactical signalling in the British army still depended on nothing more imaginative than a galloper on his horse carrying a message.

Unified Communication

The contemporary operational threat influenced signalling developments. In the 1860s, it may be recalled, other European countries were still variously in conflict, and the prime threat to England was seen as an invasion from across the English Channel. In this perhaps forgotten scenario the two services, the navy and the army, needed to be able to work together and intercommunicate - 'interoperability' it would later be called, but at that time 'unified' communication was the term used.

After the Crimean war, the Royal Engineers had formed a small telegraph wing at their headquarters at Chatham and, although there was not yet any established telegraph unit, they were following advances in electric telegraphy - more about their work later in the chapter. The Royal Engineers and the Royal Navy were jointly responsible for coastal defence communications and needed to be able to signal between coastal fortresses and ships offshore. The fortresses could be interconnected by electric telegraph, which was not a problem, but the telegraph of course only worked on land and could not communicate with ships at sea. Also, it had little application to the army's need for more mobile tactical land communications.

The ship-to-shore communications, and between ships at sea, had to be by visual signalling. Naval signalling up to then had developed as an uncoordinated mixture of often rather impractical ideas and inconsistent procedures with colourful arrays of flags strung between masts (perhaps most famously - 'England expects that every man will do his duty') and lamps, all of these developed independently by individual enthusiasts. During this period, as the Royal Navy entered the age of steam, the increased speed and manoeuvreability of ships had on occasions led to ships in a fleet colliding because of confused signalling instructions. The navy had their own reasons for needing improvement, but their methods were mostly unsuitable for army purposes. Nevertheless, unified communications were sought.

Despite the lack of higher level interest, two individuals were principally involved in the early 1860s in moving things forward - Captain Frank Bolton of the 12th (later Suffolk) Regiment, and Lieutenant Philip Colomb of the Royal Navy. They were hardly a high-ranking duo, but it is indicative of the low level of importance apparently attached to the subject. Bolton, originally a non-commissioned officer in the Royal Artillery, had subsequently been commissioned into the Gold Coast Artillery and served in West Africa for a few years before, in the strange ways of the Victorian army, transferring to the 12th Regiment (presumably by purchase of a vacant Captain's commission). With no apparent qualification or relevant experience in the subject, he then addressed the problems of army tactical signalling. Why he did so, either as an organised project or on his own initiative, is obscure, but it was probably the latter, for he was known to possess energy and inventiveness. At much the same time Colomb was seeking to improve the confused methods of naval signalling. There was a tendency, Colomb noted, for people to invent what they thought to be the signalling solution before really studying the problem and understanding the requirement. He approached the problem rigorously, having apparently been tasked to do so by the Navy.

In an attempt to sort things out and get the unified communications that were needed, principally by developing visual communication systems, Bolton and Colomb worked alongside the telegraph wing of the Royal Engineers establishment at Chatham. They each explained their work in detail in lectures to the Royal United Service Institution in 1863 and 1864 that were later published in its Journal. [1]

Captain Bolton's Work

Bolton's 1864 lecture to the RUSI described his approach to army tactical signalling, starting from the basic premise that *any* system is better than *no* system, and he had freely considered communications by electricity, light, signs, or sound. His lecture indicates that he was somewhat obsessed by an exaggerated idea of the prevalence of colour blindness in the population (probably as a result of alarmist papers then being produced by some in the medical profession), and for that reason, contrary to naval practice, he refused to consider colour as an element of visual communication - probably the correct conclusion, but not for the right reason.

He then considered signalling codes, and concluded that a modified form of Samuel Morse's code, where the most frequently used letters were given the shorter codes thus enabling faster transmission, was the most suitable, and it could be used for both the electric telegraph and lamp signalling. He introduced a few minor variations to Morse's code, but we can ignore that detail now, and he added a few 'particular signals' for repetitive operating terms such as 'understood', 'repeat', 'erase', 'alarm', etc. Next, he turned to the means of signalling, and considered signalling by electric telegraphy and visual methods.

The Portable Telegraph

Bolton produced a portable telegraph. He had presumably been instructed in basic telegraphy by the Royal Engineers at Chatham, with whom he was working, and developed his tactical version from this. The telephone not yet invented, it was intended for such roles as communicating with observation posts, or outposts as they were called at the time, to enable them to report back to their command post. Outpost telegraphy was something that was recognised as important, and a simple telegraph circuit between the two posts that could be laid quickly over short distances, did not depend on line of sight, could pass simple messages during day or night, and could be picked up again quickly and moved, was seen to have advantages.

Bolton's described it as follows:

> The electric telegraph instruments of the day already provide most of the means best adapted for use in the field. The single-needle instrument for rapid interchange of messages; the Morse printing instrument for recording important and cypher communications; and a small modified form of the Morse instrument (where the signals will be read by sound instead of sight, by means of a magnet striking between a stop and the helix) for obtaining secret information, or working in the enemy's position. This last description of instrument should be carried by one man, in the form of a set of accoutrements, the indicator on the shoulder close to the left ear, the battery in his pouch, and the finger-key, or contact-maker, attached conveniently to the waistband. Thus, with a proper supply of covered wire, each man would represent a complete telegraphic station in himself, being able either to send or receive messages.

Portable Field Electric Telegraph.

Visual Signalling by Lime Light

Despite its advantages, Bolton acknowledged the vulnerability of telegraph lines and turned to the need for visual signalling. He postulated that visual signalling would need to be "equally visible by day as by night". He described the means of producing the light. It could be either electric or the lime light, and the design of the lamp would incorporate lenses, reflectors, and other optical arrangements to concentrate and direct the beam towards the distant station.

Electricity of the requisite power would not be available in tactical situations, so he elaborated on the characteristics of the lime light (or oxy-hydrogen light, as it was sometimes known). The light is obtained from the ignition of a piece of lime, which is submitted to the intense heat of a flame of oxygen and hydrogen mixed in appropriate proportions. The lime does not burn but emits a brilliant and constant light, visible even in daylight, and its intensity may be increased by using more gas. Its disadvantage in the military environment was the supply of oxygen and hydrogen gas containers on faraway battlefields.

> **The Lime Light**
>
> The lime light was invented by Mr (later Sir) Goldsworthy Gurney in 1823, and for that achievement he was presented with a gold medal at the Society of Arts. He attended Bolton's RUSI lecture in 1864, although by then rather infirm, and explained how he had developed the light in 1823 and written a book on the subject. He related how, a few years later, in 1826, he had been approached by Lieutenant Drummond RE, who had attended the presentation of the medal to Gurney, and now wished to use the lime light on survey work. For this purpose Gurney developed a parabolic reflector to focus the light into a beam, and Drummond successfully put it to work. The light had been visible for great distances, enabling Drummond to carry out very accurate triangulation work at long range. It was first tried out in Ireland, and he quoted instances where the light on Knock Larg in Northern Ireland had been visible on Ben Lomond in Scotland, a distance of ninety-five miles. As a result of this application to survey work, it became known, somewhat erroneously, as 'The Drummond Light'. [3]
>
> It was tried in various other applications, notably in lighthouses, sometimes in theatres, from which its modern connotation derives, and once to illuminate the House of Commons, but generally its application was limited. Captain Bolton resurrected it in the early 1860s, and applied it to army signalling.

Bolton adapted it for signalling - something that could have been done many years before. The dots and dashes of the Morse code would be signalled by means of a shutter in front of the lime light - a form of Venetian blind similar in style to that used much later on the Aldis lamp - to cover and uncover the light for short or long flashes. In his 1864 lecture he described his experiments with the lime light signalling lamp over the previous two years. He had used it, he said, to signal between numerous places, including Aldershot and Crystal Palace (35 miles), and across the English Channel between Dover and Calais (22 miles). With the best operators, a speed of seventeen words per minute had been obtained. He described the system of working:

> ... The lantern is turned in the direction of the distant station, and the light exposed until seen or acknowledged by them. When the call-signal is sent, and replied to, the message is forwarded, each word being acknowledged at the distant station, by one flash if understood, and by a particular signal if not understood, when the word is repeated until it is. ...

> ***Bolton's Signalling Lamp***
>
> Bolton envisaged the signaller as a 'complete signalling station', with the lamp strapped to his chest and the gases for the lime light carried in compressed form in two small containers which fitted into his 'knapsack'. Accompanying the commander, he would flash orders or intelligence reports as required. There is no evidence that this ever happened in practice quite as Bolton imagined, but the lime light, mounted on a tripod, did come into service.

Bolton's work, good though it was, seems to have been restricted to the portable telegraph and the lime light - no mention by him of flag signalling or heliography which were to become the main tools of tactical signalling. And there was some diversity between his concepts and Colomb's. Colomb had studied his

> ### Colonel Sir Francis John Bolton
>
> Frank Bolton was the son of Dr Thomas W Bolton, a surgeon. He was born in 1830 and enlisted in the ranks of the Royal Artillery, becoming a non-commissioned officer. He was commissioned into the Gold Coast artillery corps in September 1857 and served in the expedition against the Crobboes in 1858, being promoted to Lieutenant in that year. In 1859 he was adjutant in an expedition against the Dounquah rebels.
>
>
>
> On his return to England he was transferred to the 12th (East Suffolk) Regiment, and promoted Captain in September 1860. Soon afterwards he began his work with Lieutenant Philip Colomb RN. From 1868 to 1869 Bolton was DAQMG and Assistant Instructor in visual signalling at the School of Military Engineering Chatham, under Captain Richard Stothard, the Instructor in Telegraphy. He was promoted to Major in July 1868.
>
> Bolton was instrumental in founding the Society of Telegraph Engineers in 1871, a few years later becoming its honorary secretary, and in 1885 its vice-president. In 1871 he was appointed to the Board of Trade and was involved in London's water supply system. He was promoted Lieutenant Colonel in June 1877, and retired from the army in the honorary rank of Colonel in July 1881. He was knighted in 1884. He died on 5 January 1887.

area extremely thoroughly, but the fundamental difference was that at that time flag signalling was the best basis for naval signalling, with an array of flags broadcasting its message to all around in the flotilla for as long as necessary; the lime light, with its narrow beam on an unstable ship's platform, was impractical in such circumstances. These matters all came out at some length in Colomb's second address to the RUSI on 1 June 1863 when their divergence of views is apparent. Thus, although progress was made, and they did jointly produce an Army and Navy Signal Book so that a degree of interoperability could be achieved, the differing requirements of the two services, and the limited methods then available to them, meant in practice that the effectiveness of unifying the two services' communications was somewhat limited.

Army Flag Signalling

The other method of visual signalling developed in this period for the army was flag signalling - again something that could have been done years before. It seems that the development of flag signalling was led by the telegraph wing of the Royal Engineers at Chatham, although Bolton, who was working with them, was almost certainly involved.

Most readers today might imagine that flag signalling was by semaphore (literally meaning *sign signal*), with two flags, one in each hand, the combined position of the two flags signalling a letter - latterly the more familiar method, which came into military use early in the 1900s. But in the 1860s, when flag signalling was introduced, a single flag was used, each letter was signalled in Morse code, the movement of the flag indicating the dots and dashes, as described below. This was sensible, because the signaller had to use the Morse code for other methods of signalling, either visual or electrical, so he was already trained in that skill.

The signaller may work from left to right or from right to left, or may turn his back upon the station to which he is signalling, according to the direction of the wind, so that the flag may be waved from the normal position

To make a Dot: Wave the flag from A to B and back to A without pause.

To make a Dash: Wave the flag from A to C, make a short pause in this position, then back to A.

against the wind. With a single flag, 3 feet square, on a pole 5 feet 6 inches long, held in both hands, signallers were expected to achieve speeds of nine words per minute. Using a telescope, and in good conditions, this was reckoned, perhaps rather optimistically, to have a range of twelve miles. (The reality, in later practice, was that telescopes, being relatively expensive, were in short supply, and were also of poor quality.) Sending a long message with a flag this size was a physically and mentally exhausting business, especially in wind - and being on a hill top for visual signalling reasons, quite apart from presenting oneself as a conspicuous target to the enemy, often meant that there was a wind. Brevity was a particularly desirable characteristic of messages that had to be signalled by flag! As an alternative, a small flag was used whenever possible (2 feet square, with a pole 3 feet 6 inches long); it was faster, but being less visible, the range was reduced. Signallers were expected to achieve speeds of twelve words per minute with small flags.

The colour chosen for the signalling flag was for some years a matter of debate. A dark coloured flag was more visible against a light background, and a light coloured flag was more visible against a dark background. The 'Pattern of Flags', involving colour, design, material, and procurement was discussed at some length by the Royal Engineers Committee during 1871 and 1872. In 1871 Captain Malcolm RE (a founder member of the Society of Telegraph Engineers, along with Bolton and Colomb), drew attention to the impropriety of using white flags "owing to their liability to be mistaken for flags of truce, and so discrediting the latter", and in consequence the Committee recommended a "white flag with a black centre" - a compromise if ever there was! Two months later, in December 1871, after further investigation into material, the Committee recommended replacing the black bunting with dark blue, as "it wears better". This was referred to higher authority, in this case the Deputy Adjutant General RE, and on 11 January 1872 the Committee's minutes record that "their proposal to substitute dark blue for black bunting is approved". [2] Thus the white-blue-white pattern was adopted for the signalling flag - exemplified by the crossed flags worn by regimental signallers for many years, and also the colour code that to this day has identified communication centres and indeed the Royal Signals.

Organisational Developments

Meanwhile, apart from Bolton's contributions, other developments were taking place in the 1860s. It was of course recognised at the time that, although better tactical mobile communications were needed, the electric telegraph was the way that the army's communications for higher level command purposes should develop. The American Civil War, 1861-65, had served as an example of how it could be used successfully in support of military operations, although in that war, conducted on 'home ground', existing civil telegraph lines were mainly used - a luxury that the British army was not going to enjoy in its forthcoming operations. Also, the telegraph could not reach everywhere, and beyond the telegraph line flag signalling was used. The American experience had not gone unnoticed by other European countries, which in the latter half of the decade set about establishing telegraph units in their armies.

The small telegraph wing that had been set up by the Royal Engineers at their headquarters at Chatham after the Crimean war was following telegraph developments and conducting equipment trials, as well as monitoring the organisation and roles of the telegraph units being established in the other European countries, but this was not enough.

In 1867, recognising the advances being made, the British army at last grasped the subject and held a review of signalling. A year later, in 1868, as a result of that review, the School of Military Engineering at Chatham formed a Signal Wing to combine visual and electrical methods under an 'Instructor in Army Signalling' (the first incumbent being Bolton, to be succeeded in due course by Captain Le Mesurier RE) and an 'Instructor in Telegraphy' respectively. Visual signalling was taught to soldiers in the cavalry, artillery, and infantry regiments, and detachments of trained visual signallers were formed in them.

The establishment of regimental signallers was twelve per cavalry regiment and eight per infantry battalion, with a small number of trained supernumeraries. However, there were no signallers for formation HQs, a reflection of the fact that at that time there were no properly established field formation HQs nor staff to go with them. These HQs were formed as necessary for operational requirements when they arose, to the detriment of good organisation and staff work, even up to the Boer War (1899-1902). The signalling upon which they depended was also formed on an *ad hoc* basis, by pooling the signallers from the regiments within the formation and appointing a suitably qualified signalling officer. This arrangement led to many unnecessary problems in the coming years, not just in signalling and associated staff work, but in administrative support, for such a new-found body of signallers were 'nobody's children' when it came to such basic matters as feeding them.

Electric telegraphy remained the province of the Royal Engineers, and Captain Richard Stotherd RE was appointed the Instructor in Telegraphy. Things at last were moving on to a more structured basis, but attempts to form a British regular army telegraph unit were unsuccessful for reasons that will be explained in chapter 7. Moreover, the policy of integrated training at Chatham only lasted for six years, and in 1875 the Instructor in Army Signalling, dealing only with visual signalling, moved to Aldershot, separating from the Royal Engineers who stayed at Chatham and retained responsibility for electric telegraphy.

Major W Hudson of the 21st Punjab Infantry, who attended one of the early signalling courses at Chatham before returning to India, described what he had learnt from Bolton's system of signalling. [3] "In doing so", he said, "it will be convenient to adhere as closely as possible to the language used in the Manual of Instruction, published for the use of the classes at the school of Military Engineering at Chatham."

> The system is known as the 'Flashing system', and the mode of signalling adopted is by a combination of long and short 'flashes', or 'appearances' of any given object with proper intervals or obscurations between them, which are made by the visual apparatus, such as revolving shutters or discs, collapsing cones, flags, bandroles, &c, by day, by lamps or lights by night; and by a combination of long and short sounds made by a fog horn or bugle in fogs, or when visible symbols are not available.

> Flashing signals are made by the motion of any single object. In some instances the object is made to appear and disappear; and in others it is made to change its position, so that one position shall represent the appearance and the other the disappearance of the object

> The symbols are determined by successive appearances and disappearances at regulated intervals, constantly recurring after a fixed pause, in a manner precisely similar to those of revolving or flashing lights in lighthouses.

He went on to describe flag signalling, and then "another very useful apparatus, the revolving shutter". This gets little mention anywhere, but was a tactical form of 'mini' shutter telegraph. Hudson described it as follows:

> It may be made of any size corresponding to the distance required to transmit signals. An apparatus exposing a surface of 72 square feet will give a range of about 12 miles in clear weather. It consists of a series of shutters, each working on a pivot, and all connected together in such a manneras to move simultaneously by the motion of the handle. When the shutters lie horizontal, representing the obscured state of the light, an observer sees nothing, ... but when the shutters lie vertically, representing an exposure of light, a very large surface comes into view. Signalling by this apparatus may be carried on with great rapidity, as this appearance and disappearance may be produced 100 times in a minute.

For all that, it does not seem to have been a device used in any real tactical situation. - cumbersome and impractical in the field.

Hudson continued by saying that: "For night signalling, the most effective apparatus is that known as 'Bolton's lime-light flashing apparatus', by means of which signalling may be carried on up to 25 miles. ... The flashes are made by means of a mobile disc worked by a lever and finger key for shutting off and displaying the light." He continued with a detailed description of the chemicals and gases involved in making the lime light work.

Another signalling lamp was also described, the 'Chatham light', which Hudson explained was:

> ... a simple oil lamp, on to the flame of which a jet of powder, called 'Chatham powder', is by means of an ingenious contrivance flashed by bellows which are attached to the apparatus. The powder gives intense brilliancy to the light for longer or shorter periods, as may be required to produce long or short flashes. ... This lamp gives an excellent light, and is very handy; it has a range of from 5 to 10 miles, according to the strength of the powder, which depends on the quantity of magnesium in the mixture.

Hudson concluded his description of the 'new system of Army Signalling' with this observation:

> Enough, it is hoped, has been said to show the advantages of the system, which recommends itself chiefly on account of its simplicity, the ease with which a knowledge of it may be acquired by men of ordinary intelligence; and last but by no means least in these days of financial pressure, the inexpensive nature of the apparatus.

The Field Telegraph Corps

By early 1869 the new organisation at Chatham had assembled a small group of people and equipment capable of conducting both visual signalling and simple electric telegraphy, and at the time rather grandiloquently referred to as 'The Field Telegraph Corps'. At Easter 1869 they took part in the Volunteer

Review at Dover, when a training exercise simulating an enemy attack on Dover was carried out. Their participation was reported in the *Illustrated London News*, the content of the article reproduced below. The pictures show the telegraph office and wire wagons that had been developed. The reference to 'magnetic office' refers to one of the private telegraph companies which at that time were providing civilian telegraph communications in Britain and Ireland. And it is interesting to note that signalling to the navy at sea still required a prominently displayed flagstaff.

THE VOLUNTEER REVIEW AT DOVER

The Field Telegraph Headquarters   *Paying Out the Telegraph Wire*

The two engravings represent the Headquarters of the Field Telegraph Corps, ordered from Chatham to take part in the operations of Easter Monday. They were engaged in the work of laying down a line of telegraph in the field. Their equipment consists of two travelling offices, in shape not unlike a small omnibus. Each office contains two telegraph instruments which print the messages on strips of paper. Underneath are the Voltaic batteries. It has also a writing desk, and miscellaneous stores are carried in a large cupboard at the front. Two large panniers serve as seats for the telegraph operators and contain flags and lamps for visual signals, which can be extended in advance of and are auxiliary to the field telegraph lines. These panniers, which are adapted for pack transport, also contain cooking vessels and a patrol tent so that a party despatched with same have the means to bivouac for the night. The office wagons normally have two horses, but the plough around Dover necessitated four.

Accompanying the office wagons are two wire wagons, each wagon carrying three miles of insulated wire on drums containing half a mile. Each wagon carries 18 tubular iron poles of telescopic pattern which when pulled out make one pole 18 feet long. These are used for road crossings; along other places the wire is laid on the ground. Wire is "payed out" direct from the drums on the wagon, and is picked up in similar manner by communicating the rotation of the wheels to the spindles of the drums by means of an endless band.

Along impassable places where the wagon cannot travel the drum is shifted to a small handbarrow which can be wheeled by one man or carried by two. Wire can thus be laid wherever a man can walk. The Wire wagon is drawn by four horses.

The Field Telegraph arrangements at the Review consisted of establishing a central station at a spot called Lone Tree, from here communication was kept up with magnetic office in the town (through courtesy of W. Walsh, Chief Engineer to the Company), also to easternmost point of the Castle where a flagstaff was erected for communication with the ships. A third line was carried from the Central Station to the Castle and along the cliffs to a point in rear of the invaders line.

As the invaders advanced a fourth line was laid out and followed them as quickly as the men marched and signals were thus kept up between the most advanced and original positions of the invading line. In all 6 miles were laid, of which 3 were laid and picked up as part of the operations of the day.

The telegraph was used during the review for the despatch of messages of importance between the several stations. The whole of the Field Telegraph and visual signals were in the charge of Captain Stotherd RE, and the Field Telegraph was in the immediate charge of Lieut. Anderson RE. A system of visual signals was established with the different divisions by Major Bolton, Lieuts. Armstrong, Blood, Courteny, RE, and Cornet Gough, 10th Hussars. Captain Maitland assisted. Many signals were passed.

So much for the signalling developments of the 1860s. Several other important developments were soon to take place. After the procrastination described above, and prompted by two quite different requirements, army telegraph units with separate roles were to be formed and their organisation and equipment will be explained later (chapter 7). And the most useful advance in tactical signalling for many years to come, heliography, was about to be introduced. 'Invented' would be the wrong word, for reflecting the sun had been around since ancient times, and the Morse code, on which heliography depended, had been developed in 1837, but turning those two things into a signalling apparatus had passed over the top of military heads - just another example of the lack of initiative in that period which was mentioned in the introduction to this chapter.

But next, and chronologically a short step backwards, the little-known Abyssinian expedition of 1867-68, in which visual signalling and the electric telegraph, using the methods so far developed, were to be used by a hastily assembled unit sent on a most unusual operation.

Endnotes

1. Bolton's description was published in the *Journal of the Royal United Services Institution*, Vol VII, 1864, pp 268-291, and Colomb's in the same volume, pp 349-393.

2. *Proceedings of the Royal Engineers Committee*, 1871, p 37. Pattern of Flags.

3. *Papers on Indian Engineering*, Vol VII, No CCLXXXIV, pp 291-300. (RE library.)

Chapter 6

The Abyssinian Expedition, 1867-68

The Abyssinian expedition, despite being little remembered these days, is of surprising interest in the evolution of army communications. Following the signalling developments at Chatham in the preceding few years, described in the previous chapter, it saw the effective use of both electric field telegraph and tactical visual signalling by the British army, and it successfully demonstrated their possibilities. This led in 1869 to the introduction of regimental signalling, albeit then only by visual methods, and contributed in the following year to the establishment of the first army telegraph unit.

A most explicit description of the expedition is given in a now rare book, *The Record of the Expedition to Abyssinia*, published in 1870 in two volumes and running to thirty-eight chapters (see bibliography). Its authors, Major T J Holland CB, Bombay Staff Corps, and Captain H H Hozier, 3rd Dragoon Guards, had both held appointments on the expedition's headquarters staff, and garnered their information from many contemporary sources, amongst them Lieutenant O B C St John RE, the expedition's Director of Telegraphs. The thirty-eight pages of Chapter XXII of the book are devoted to *Telegraph and Army Signals* and go into very fine detail, giving lengthy extracts from contemporary letters, planning reports, details of equipment lists, staffing arrangements, and Oliver St John's interesting and, as it turned out, prescient post-expedition report.

Present day readers might raise their eyebrows at the idea that the Director of Telegraphs for a major expedition involving 13,000 troops could be a Lieutenant, but in those formative days the junior signals officers, dealing with novel things that the majority simply did not understand, carried great independent responsibility.

Historical Background

What was this curious expedition all about? Abyssinia as it then was (now Ethiopia and Eritrea) was never anything to do with the British Empire or Britain's strategic interests, which were the usual reasons for despatching an expedition on one of Queen Victoria's little wars in the second half of the 19th century. Perhaps, too, not many readers have heard of King Theodore. A little explanation is necessary.

Theodore himself was of humble origin but had become a successful bandit and, by conquests over other tribes, had gained the throne. Unfortunately he became mentally unbalanced and unpredictable, not helped by an increasing fondness for alcohol. In January 1864 he imprisoned Captain Charles Cameron, Her Brittanic Majesty's Consul to Abyssinia, and then had him whipped and tortured. This undiplomatic behaviour was all about an unanswered letter from Theodore to Queen Victoria proposing to send a delegation to England to explain to her how Islam was oppressing good Christians such as the Abyssinians, and to suggest some form of alliance with Britain against his Muslim neighbours. Angered at the lack of response, which was due to unwillingness by the Foreign Office in London to become involved with this unstable man in such a remote country, Theodore imprisoned Cameron and a motley collection of other Europeans, mostly missionaries. Two representatives were sent to try and sort the problem out, and they too were incarcerated. The situation was bizarre, but in August 1867, three-and-a-half years after Cameron was imprisoned, the British government decided that it would not have its subjects treated in this way. An ultimatum was sent to Theodore, and an expedition was launched to rescue the prisoners.

The Expedition

It was decided that the expedition would be found mainly from the Indian army, but with some specialist help from the British army and the Royal Navy, and would be led by Lieutenant General Sir Robert Napier of the Royal Engineers, then Commander of the Bombay Army. The campaign plan that emerged was notable for its detail – not up to then a characteristic of the Victorian army. Napier rejected a mad dash strategy, and planned the logistics, transport, and communications with admirable care, as well as some questionable extravagance. It required 13,000 troops, an even larger retinue of servants and labourers, 291 ships, and 36,000 assorted transport animals - a menagerie of camels, horses, elephants, mules, and donkeys. It was quite excessive in numbers and cost for the mission. Despite the limitations of transport and logistics in those days, the idea of 'travelling light' never seemed to be acceptable although, as the expedition advanced, practical circumstances enforced severe reductions in 'baggage'.

The first problem facing the planners was how to get to Theodore's fortress, where the prisoners were held, located in an inaccessible mountain range deep inside Abyssinia at a place called Magdala. A reconnaissance party was sent from Bombay to select the best place on the Red Sea coast to form the expedition's base; they chose Zula, in Annesley Bay, now in Eritrea. The expedition was launched in November 1867, when the advance brigade landed at Zula and began to build the base camp. From there they would march southwards, through largely unknown and completely undeveloped territory, to Theodore's fortress, 380 miles from Zula and at an altitude of nearly 10,000 feet.

Planning the Communications

Planning for the communications started in late August 1867 when a joint Anglo-Indian contribution was agreed in principle. Initial advice was given by Major John Champain RE, the Deputy Director of the Indo-European telegraph project and at that time based in London after his work in Persia constructing the telegraph line there (see chapter 4). Champain had put together his own experience with advice he had received from Captain Richard Stotherd, the Instructor in Telegraphy at Chatham.

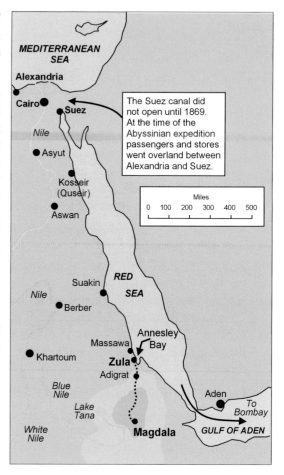

Champain proposed that the Abyssinian expedition's telegraph should consist of two components. He called these a 'semi-permanent aerial line' from the base at the port of disembarkation to follow the route taken by the force to the interior, and 'flying lines' (what might now be described as tactical telegraph communications), supported by visual signalling to work with the forward HQ and leading formations during the advance.

Noted for his logical mind and attention to detail, Champain then carried out some calculations. For the semi-permanent aerial line 350 miles of line would be needed, and for such a line 10,000 bamboo poles of eighteen-feet length should be procured in Bombay. 4,000 insulators would be needed, half to be of a design for fixing to the poles, the other half suitable for fixing to trees when they passed through the forestry belt. Copper wire should be used, and No 16 wire was recommended, weighing sixty-four pounds per mile (comparatively light, and with high conductivity, but not as strong as a mixture of galvanised iron and copper), and costing sixty shillings per mile. 350 miles (ie 10 tons) were to be ordered from Messrs T Bolton and Sons, Broad Street, Birmingham. He anticipated there would be eight telegraph offices along the line (an underestimate), and so eight Siemens and Halske Morse recording instruments would be needed. (These were often referred to as 'inkers', because they made a permanent record of the received signal in Morse code on paper tape.) He also suggested an inventory of the tools needed to erect the line and the supplementary equipment for the telegraph offices.

He then turned to the flying lines and proposed 50 miles of this, all carefully coiled on drums, the combined weight not to exceed 140 lbs per drum, so that they could be manhandled and two drums would make a good mule load. The line itself, or field cable as it would be called these days, was known as 'Hooper's core' and was a relatively new design. Records of the expedition show that 50 miles of Hooper's core, specified as 'three strands of copper wire, of not less than 85 conductivity, covered with india-rubber &c, including drums and packing, with 3% added for testing by Mr Latimer Clark', was to be obtained from 'Messrs. Hooper, 7 Pall Mall East', at £45 per mile, total £2,250.

It was then calculated that fifty mules would be needed to carry the flying lines and all their associated equipment. For instruments on the flying lines he recommended the acoustic field apparatus of Siemens and Halske, having used them very satisfactorily in Persia; they were portable, robust, and easy to adjust. (These items of recording equipment were more usually called Morse sounders, and they did of course depend on the telegraphist to receive accurately; there was no permanent record as with an 'inker'.) Nothing wrong with most of the calculations but come the day the practicalities were to prove otherwise, for many of the stores from India did not arrive, leaving much improvisation for those on the ground.

In addition to the field telegraph Major Champain proposed "apparatus and special staff for day and night signalling by means of flags and cones, heliotropes, and flashing light". (ie Bolton's and Colomb's recent visual signalling innovations.) This equipment was intended to be used in two roles: firstly, ship-to-shore signalling in Annesley Bay; and secondly, with the forward elements of the force, in conjunction with the

TABULAR Statement of Stores for the Equipment of the Field Telegraph Train.

Amount.	Specification.	From whom to be obtained.	Estimate of probable Cost. £ s. d.	Per	Probable Total Cost. £ s. d.
50 miles.	Hooper's core, 20 in diameter, consisting of three strands of copper wire, of not less than 85 conductivity, covered with india-rubber, &c., including drums and packing.	Messrs. Hooper, 7, Pall Mall East.	45 0 0	mile.	2,250 0 0
	Add three per cent. on the above, payable to Mr. Latimer Clark for testing the said core.	..	..	..	67 10 0
350 miles, at 64 lbs. per mile = 22,400 lbs.	No. 16 W. G. copper wire of 18° standard, including joining in lengths of two miles and winding.	Messrs. Thomas Bolton & Sons, Broad Street, Birmingham.	0 1 2	lb.	1,306 13 4
175	Drums for winding the same	,, ,,	1 0 0	each	175 0 0
4 cwt.	Iron wire, No. 8 gauge, in lengths of 30 feet, for stays.	Messrs. Siemens Brothers, Great George Street, Westminster.	2 0 0	cwt.	8 0 0
2,000	Insulators, each consisting of a cast-iron top, china insulator, and wrought-iron support, two hoop-iron bands, and 10 nails, packed in strong deal cases of 200 each, with compartment containing the following tools, viz., 2 hammers, 2 files, 2 pair cutting nippers, 2 gimblets, and 12 sheets emery paper.	,, ,,	0 2 6	each	250 0 0
2,000	Insulators, each consisting of one wrought-iron spike, one china insulator, with hook above and below, packed as above with similar tools.	,, ,,	0 2 6	,,	250 0 0
8	Portable recording relay instruments, in cases complete.	,, ,,	32 0 0	,,	256 0 0
12	Portable recording field instruments, in cases complete.	,, ,,	25 0 0	,,	300 0 0
12	Cases, each containing 12 "Marie Davy's" elements for field instruments.	,, ,,	0 15 0	cell	108 0 0
8	Cases each containing 24 ditto, ditto, for relay instruments.	,, ,,	0 15 0	,,	144 0 0
12	Portable cases, each containing one copper earth-plate, with connecting wires, one set of assorted repairing tools, 50 yards covered wire for connexions, six bottles of printing ink, 25 discs Morse paper.	,, ,,	8 0 0	,,	96 0 0
4	Cases, each containing one set of repairing instruments, one copper earth plate, 100 yards covered wire for connexions, 100 discs Morse paper, 12 bottles of printing ink.	,, ,,	10 0 0	,,	40 0 0
4	Magnetic field instruments, in cases complete.	Mr. Henley, 27, Leadenhall Street.	26 10 0	,,	106 0 0
8	Cases, each containing eight large size "Marie Davy's" elements for local batteries.	Messrs. Siemens Brothers, Great George Street, Westminster.	3 0 0	case	24 0 0
2	Galvanometers, in portable cases, complete for testing purposes.	,, ,,	8 0 0	each	16 0 0
100 lbs.	Pro-sulphate of mercury, packed in hermetically sealed tins of 5 lbs. each.	,, ,,	0 4 0	lb.	20 0 0
500 discs	Morse paper, for relay instruments	,, ,,	0 0 8½	each	17 14 2
500 disc	Morse paper, for field instruments	,, ,,	0 0 8½	,,	17 14 2
4	Lightning dischargers, with, each, two line plates	,, ,,	1 15 0	,,	7 0 0
100	Connecting screws	,, ,,	0 0 10	,,	4 3 4
72	Bottles of ink for instruments	,, ,,	0 1 6	,,	5 4 0
3 sets	Iron blocks and tackles, with 100 yards spare rope.	,, ,,	1 10 0	,,	4 10 0
50	Bill-hooks, in leather cases, for clearing thorns, &c.	,, ,,	0 10 0	,,	25 0 0
10	American axes, ditto, ditto	,, ,,	0 10 0	,,	5 0 0
20	Spare handles for the same..	,, ,,	0 1 0	,,	1 0 0
6	Earth borers, 3¼ inches diameter	,, ,,	0 12 6	,,	3 15 0
12	Earth scoops	,, ,,	0 2 6	,,	1 10 0
12	Hooks for fixing wire insulators	,, ,,	0 5 0	,,	3 0 0
56	Whistles	,, ,,	0 2 0	,,	5 12 0
25	Clasp knives, with lanyards	,, ,,	0 5 0	,,	6 5 0
6	Turnscrews	,, ,,	0 2 6	,,	0 15 0
12	Despatch bags, with straps..	,, ,,	1 0 0	,,	12 0 0
12	Two-wheeled barrows, for paying out and picking up wire, with tool box and grease pot attached.	,, ,,	12 0 0	,,	144 0 0

An extract from the stores list prepared for the expedition.

flying lines, so that tactical communications could be maintained with the leading troops. As it happened, the ships were able to anchor in Annesley Bay very close to the shore, so ship-to-shore signalling was never required and nothing more will be said about it.

Stores

Champain refined the stores list, showing items, supplier, quantities, weight, and cost. It was a model of detailed staff work of the time. It should be remembered that there were no military depots holding these stores, and everything had to be procured by the Royal Engineers from civilian suppliers. All the telegraph and signalling stores were to be supplied from England, and the transport animals and bamboo telegraph poles were to be supplied under Indian arrangements. Stores were further broken down into loading lists, suitable for transportation by mules.

The telegraph equipment was heavy, and for this expedition the stores sent from England weighed forty tons. A lesson to be learnt was that it was pointless to take so much, for on arrival it could not be transported. For many years to come, until mechanisation, the provision of field telegraph away from railways was dependent on animal transport, and that, together with the weight and bulk of the equipment, imposed many limitations.

```
No. 4 Mule carried—
2 panniers, with locks and keys.        48 rockets.
1 5-gallon spirit case.                 1 set poles for tents.
1 case Chatham powder.                  2 water decks.
1 2-gallon oil case.

No. 5 Mule carried—
1 set cooking utensils (spare).         2 tomahawks and case.
1 set tinman's tools (spare).           1 small size pack complete.
1 large size pack, complete.            1 powder case.
6 Chatham lamps.                        1 fog horn.
2 hand lamps                            3 flags.
1 spirit case.                          1 pair bellows.
1 oil case.                             2 pair field glasses.
2 cases, with wick and scissors.        24 message books.
4 bags.

No. 6 Mule carried—
18 sets of banderols and sticks.        2 files, square.
18 hand lamps.                          1 pliers.
18 cases for oil.                       1 quire blotting paper.
18 code books.                          1 case matches, large.
36 parchment codes.                     2 boxes nibs.
3 sets poles for flags and lamps.       1 dozen pencils.
1 lot brass solder.                     100 envelopes.
4 bags.                                 3 satchels for messages.
14 cotton wick balls.                   1 hand axe.
9 fog horns.                            1 felling axe.
36 fog tongues.                         1 pickaxe.
1 coil, brass wire.                     1 shovel.
```

An extract from the detailed loading list. Unfortunately, sufficient mules never arrived.

Manning

Champain then suggested manning arrangements, recommending that the control and superintendence of the entire telegraph train should be "in the hands of a Royal Engineers officer with experience in practical telegraphy as well as being accustomed to the direction of native working parties". For this post he recommended the employment of Lieutenant O B C St John RE, on account of "his recent experiences in a country not entirely dissimilar". Oliver St John, then thirty years old, had worked under John Champain in Persia on the construction of the Indo-European line (see chapter 4). Fortuitously, he happened to be in England at the time the planning for the Abyssinian expedition was undertaken, in the last week of three months leave, so he was duly selected. Lieutenant Arthur Puzey was appointed the Assistant-Superintendent of Telegraph, and Lieutenant Jeffrey Morgan was deputed to the flying lines, both Royal Engineers officers from Chatham.

The telegraphists for the semi-permanent line were to be a mixture of Europeans and Eurasians recruited in India. The manpower for the flying lines, which included the visual signallers, was to be provided by the Royal Engineers from Chatham. There was no established unit of trained telegraphists or signallers and so soldiers were selected and trained, twenty-five as telegraphists and ten as visual signallers, and formed into the 10th Company RE at Chatham, along with some other RE specialists, including a small photographic unit.

Movement

Lieutenant St John and the telegraphist and signalling party, with all their equipment, sailed from England in the SS *Mendoza* on 4 November 1867. Moving overland between Alexandria and Suez (the canal did not open until 1869), they arrived at Annesley Bay on 12 December 1867. It had taken about four months to plan, assemble, train, equip, and transport the troops – hardly a rapid reaction, but for this expedition it had to be implemented *ab initio*, and in those circumstances the time taken was creditable.

Command and Staff Arrangements

When they arrived in Abyssinia the 10th Company RE was placed under the command of the CRE for technical and regimental purposes, but under the staff of the Quartermaster-General's Department for

Telegraphists, flag signallers, well borers, and photographers of the 10th Company RE at Chatham.

operational direction. The principle of that arrangement, reached after due consideration but with no precedent, has lasted until this day, although of course staff responsibilities and nomenclature have changed over the years. The 'G' (*ie* operations) staff, under whose direction communications later fell, was formed in a restructuring of staff responsibilities in 1904. Until then, signalling remained a responsibility of the QMG's Department, even though Cardwell's Army Reforms in 1869 subordinated the Quartermaster-General to the Adjutant-General (prior to that, the two held equal status).

Strategic Communications

The strategic rear link for the expeditionary force was another aspect of the communications plan that was considered. Lieutenant Colonel Daniel Robinson RE, the Director of Telegraphs of the Indian army, was sent from Bombay to Egypt to arrange strategic communications, so that the expedition was in communication from its base at Zula. At that time there was no submarine cable through the Red Sea, and Robinson, helped by advice from Colonel Stanton, a Royal Engineers officer then on the staff of the British Consul in Cairo, and Mr Gisborne, the Director of the Sudan Telegraphs, quickly discovered he had three options: either to extend the line presently being built in the Sudan from Berber to Suakin, the port in the Eastern Sudan, along the coast to Zula – a distance of about 250 miles; or to extend existing lines from Kosseir in Egypt along the Red Sea littoral – a distance of about 850 miles; or to arrange for a submarine cable to be laid from Suez (see map). The line being built down the Nile valley to Berber in the Sudan, and thence to Suakin, was turning into a disastrous project. The Vice-Consul in Egypt reported that owing to negligence and general bad management on the part of the Egyptian authorities, and partly to unforeseen misfortunes, work on the line had been suspended. The two Englishmen who directed the project had become ill due to the effects of the climate, 1,500 camels had died because of starvation, wooden telegraph poles had been eaten by white ants, metal poles to replace them had not arrived, and so on. The line was still 120 miles from Suakin. Robinson's attempts to inject new momentum to the project were met with bureaucratic indifference and a lack of any sense of urgency, despite the availability in Egypt of the necessary stores. The Egyptian army and other officials were at that time, through maladministration, heavily in arrears of pay and were indisposed to any form of cooperation. Even if it were built, Robinson had no confidence that it would be kept in working order.

He turned next to the possibility of extending the existing line from the area of Kosseir along the Red Sea littoral through Suakin to Zula. However, it was a great distance; there was insufficient labour available to build it; it would require financial inducements to the tribal chiefs through whose areas the line would run to ensure its protection (such 'subsidies' to be paid direct by the British government, because if through the Egyptians it was doubtful if the money would reach them and in consequence the line would be cut); an airline would be vulnerable to the violent sandstorms in the region; and, as the final nail in this coffin, even if built, there were no telegraphists available to operate it. So, very sensibly, that option also was deemed impracticable.

Finally, the submarine cable option. It was certainly possible, as proved to be the case in 1870 when the submarine cable was laid between Suez and Bombay (described in chapter 4) but it was still uncertain, would take more time than was available, and would be very expensive. The British government weighed matters up, and decided the risks of cost, time, and technology were greater than any advantages that might accrue, so that last option was also rejected.

Robinson, frustrated, left Cairo on 24 January to return to Bombay to report. Time had not been on his side, and he had been sent on his mission too late. It was the first of numerous examples to come when communications planning was too low a priority, and there were no staff dedicated to it.

The best that could be done, it was decided, was a regular mailboat service, and so throughout the expedition mailboats plied a weekly schedule from the base at Zula to both Suez and Bombay. As things were to turn out, the lack of a strategic rear link had no impact on the campaign. Indeed one might imagine that Sir Robert Napier, in command, was probably quite glad to be left alone.

The Campaign

The campaign got under way in November 1867. After a slow start, due to lack of organisational ability on the part of the commissariat officers of the Indian Army, the base camp at Zula was established. The war correspondent H M Stanley, later more famous for finding Dr Livingstone in Africa, went to Abyssinia as a war correspondent, and he described the scene at Zula:

> Bewhiskered officers of the commissariat lounged in pyjamas in their luxurious tents, fanning themselves languidly; visitors were quizzed through glasses and, if found acceptable, were offered boundless liquid hospitality, before being ushered out with 'Ta-ta old fellow', nothing having been achieved.

The incumbents of the commissariat, never part of the 1st XI in any case, were quite inexperienced in operating outside their comfortable home territory. The arrival soon afterwards of regular staff and army units, notably the 33rd Regiment, induced a more energetic approach to administration.

Piers were constructed, wells sunk, a camp and roads built, sheds erected, and to meet the large demand for water by men and animals, a water chute on trestles was built to convey to storage tanks the water pumped from the sea and desalinated by a large condenser aboard HMS *Satellite*. The telegraph and signalling party from England arrived on 12 December 1867.

The route to Magdala, shown on the next page, crossed three areas of country: first, from Zula, a twelve-mile belt of low-lying land; then, starting at Kumayli, a tortuous fifty-mile defile, the Suru Pass, leading to the highlands; and lastly the high Abyssinian plateau, some 8,000 feet and more in altitude. The climate in the low coastal strip (now the independent country of Eritrea) was a health problem, particularly for Europeans; the defile to the highlands was to prove the most difficult stretch for constructing the telegraph line, and beyond Antalo the country became increasingly rugged and mountainous with deep ravines.

Despite all the careful communications planning in England, things went wrong from the start. Manpower from India did not arrive on time so alternative arrangements had to be made. The stores having been carefully planned into mule loads, there were insufficient mules and most of the heavy

Lieutenant General Sir Robert Napier (seated) and officers of the Royal Engineers.

cable of the flying line had to be left behind at the base at Zula. There were considerable problems with the expedition's transport train, due to a combination of insufficient numbers to begin with, poor transport conditions on ships, poor animal husbandry and management, and animal sickness. The bellowing, braying animal compound at Zula was a noisy and unhygienic shambles, littered with dead animals.

The 10,000 bamboo poles, on which to construct the airline, did not arrive from India until early February, by which time the telegraph line, on improvised poles, had reached Senafé, the first station in the highlands. There was then insufficient transport to move them forward. About 700 poles were eventually brought forward by local native carriage and used in places where no other timber of any sort was available. Otherwise, poles had to be procured locally from whatever was available, leading to a slow rate of construction and, through no fault of the construction party, a badly built line.

All this was symptomatic of the fact, as many must have thought at the time, that there was no permanent unit or organisation responsible for telegraph, and that everything was having to be planned on an *ad hoc* basis with executive responsibility and provision of resources shared between separate entities. The British-run Indian hierarchy appears to have been a bureaucracy, unable to cope with the logistics of an operation outside India for which they were not prepared, with lack of urgency, much shuffling of paper and delegation of responsibilities, leading as we might say these days to lack of 'ownership' and failure to deliver.

As a consequence of all this, the intended communication plan changed. There now being no flying line, Lieutenant Morgan was placed in sole charge of the visual signalling and was sent ahead with his party, accompanying the leading troops all the way to Magdala; their work will be described shortly. Lieutenants St John and Puzey gave their undivided attention to the construction of the semi-permanent telegraph line, something that had now become much more difficult than envisaged.

The construction of the semi-permanent line from Zula began on 24 December, and for this purpose the 2nd Company Bombay Sappers and 53 Lascars were placed at the disposal of Lieutenant St John. The first six miles were along the course of the railway it was proposed to build, and the telegraph line was supported on teak posts. Then mimosa poles were cut in the surrounding country, but they were unsatisfactory as they were green and warped

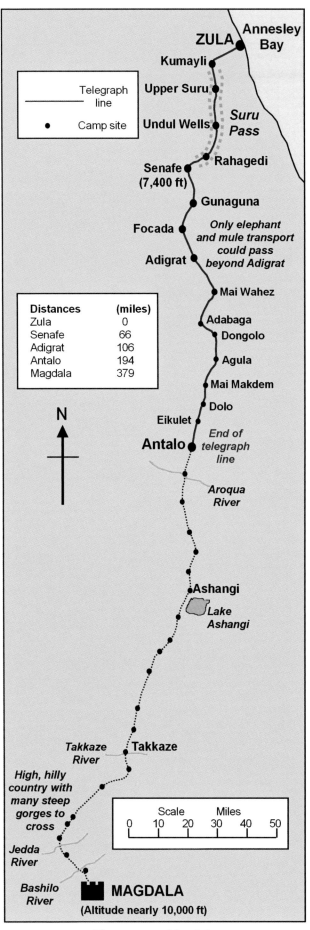

The route to Magdala.

badly when they dried out, and often were not long enough to raise the line to a satisfactory height. The mimosa poles were later replaced by more teak poles when they became available.

From Kumayli, at the start of the narrow Suru Pass leading to the highlands, where the line had to run right beside the mountainous track, the insulators were often fixed to the cliff itself. Sometimes it was possible to lay the wire in long spans between high points in the hilly terrain. Small sections of the line through the Suru Pass, where it was difficult to construct airline, were built with the cable intended for the flying lines but which mostly had to stay at Zula. Unintentional though that was, it appears to have been the first use of field cable in a military operation. Hooper's core, a recent innovation, laid through the remote Suru Pass in Abyssinia, and later to be developed into many more types of field cable, lays claim to this record.

Hooper's Core

Insulation for telegraph cable had first been made from gutta-percha. This was a product of the isonandra gutta tree found in the Malay peninsula, and examples were brought to Europe and exhibited at the Royal Society of Arts in London in 1843. In 1845 S W Silver & Co of Stratford, East London, invented a means of extruding it to cover wire. The discovery of gutta-percha and the method of extruding it were keys to the development of submarine cable, and thus 19th century international telegraphy. However, there were disadvantages to gutta-percha in other situations – it was not very flexible, and contact with air and movement tended to dry it out and cause it to crack. This made it quite unsuitable for land line. The only recorded British army use of gutta-percha insulated cable was in the Crimean War, 1854-56, when the army first used the electric telegraph. The buried cable proved unreliable (not helped by the fact that ignorant soldiers dug it up and used the gutta-percha insulation for pipe stems).

In 1849 Mr William Hooper, a chemist of some distinction, turned his attention from gutta-percha to rubber, and discovered that when vulcanised (a process involving heating the rubber in a sulphur solution to about 250 to 300 degrees Fahrenheit, causing the two to combine), and after curing, the treated rubber physically changed its characteristics. While retaining its flexibility, it was impervious to changes in temperature, did not oxidise in air, and absorbed less water. Until Hooper discovered this process, rubber had been no use as an insulator because of its unsuitable physical, mechanical, and durability characteristics.

Early tests on Hooper's core, by leading engineers of the day, Charles Bright, Latimer Clark, and Sir William Thomson (later Lord Kelvin), were all favourable. As well as its insulating qualities it was found to have lower electrostatic capacitance than gutta-percha, thus enabling a higher working speed for telegraphy over long submarine cables. When submarine cables had to be raised and repaired, it was found that Hooper's india-rubber compound was better than gutta-percha to restore the insulation around the cable. But its great advantage for military use was that it was more flexible, did not dry out and crack when exposed to air, and could take the rough handling it was going to get in army service.

The line emerged from the fifty-mile Pass into the highlands at Senafé. Once into the highlands it was possible to procure local juniper pine saplings, ten to fifteen feet in length but only about three inches in diameter, very flexible, and not really strong enough to support a tensioned airline. But for the last 130 miles of line this was all there was. Sometimes three poles were formed into a tripod for greater strength. The natives, quick to spot an opportunity, brought the saplings to the line as it was constructed, sometimes from considerable distances, and were paid cash (Marie Theresa dollars) according to quality. But the line could not be strongly constructed, and only at crossings was it possible to raise it to a safe height. For most of the way it was so low that a loaded camel could not pass beneath.

The line itself was No 16 gauge copper wire. Insulators were only used for the first sixty miles, and the line was found to work quite satisfactorily without them thereafter, even in rain. St John reported very favourably on that wire: "For facility of stretching and jointing, conductivity and portability, nothing could have been more satisfactory than this wire. An iron wire of even half the electrical advantage of the 16-copper wire would have been three times the weight, and apart from the larger amount of carriage, could not have been supported on poles of such tenuity as those necessarily used". He did find, however, that cattle used the poles as scratching posts, and tended to shake the wire vigorously and stretch it, so that it often had to be re-tensioned. It was found best to protect each pole by placing thorn brushwood around it.

St John explained how the wire used on the airline was jointed: "The joint used was the ordinary German or twisted joint, eight or ten turns being taken on each side. ... No solder was used. The wire was stretched by hand, on the ground where sufficiently level. Hedger's gloves were worn by the men stretching, as the thin wire often cut through the skin and caused painful sores. Where insulators were used, the wire was bound at every one. Much trouble arose from the insufficient strength of the drums on which the copper wire was wound to stand the terrible wear and tear of mule carriage". The drums had been specially made for the expedition, to carry the right weight of wire for mule loads, and St John made specific recommendations for future improvements.

It was found necessary to establish more telegraph offices along the semi-permanent line than had been anticipated, and it was possible to equip these with the Morse sounders that had been brought with the intention of being used on the flying line. The twelve sounders originally intended for the flying lines performed well, but were not equipped with relays.

The eight Siemens and Halske Morse recorders came equipped with relays. The use of the relay means that the on-off signal (*ie* in today's terminology, a digital signal) received at an intermediate station was detected by the sensitive relay, regenerated by the local battery, and retransmitted onwards. That, of course, meant that no manual operator intervention was needed, distance became irrelevant, and the signal continued on its way. So without this facility the sounders were somewhat limited and message traffic needed manual retransmission.

Another unusual instrument was used. Four magnetic instruments, obtained from Messrs Henley, were used on the short railway telegraph line. Their principal advantage was that they needed no batteries.

Contemporary photograph of the Devil's Staircase, part of the Suru Pass, through which the telegraph line ran.

All seemed to work well except the galvanometers, which were part of the Siemens and Halske recorders. They came in for severe criticism. (A galvanometer is a sensitive instrument that detects an electric current and the direction in which it is flowing, and was used in setting up the telegraph circuit.) The galvanometers proved to be easily damaged and were not robust enough for the rigours of military service, particularly transportation. St John's recommendation was that galvanometers should be detached from the equipment before transportation, carried separately under more benign conditions, and plenty of spares provided.

The travails with the poles and the transport meant that the line could not keep up with the forward headquarters and the advancing troops. It reached Antalo on 2 April 1868, 197 miles from Zula, by which time the leading troops were getting near Magdala, a further 180 miles ahead through increasingly difficult country. It was decided to terminate the telegraph at Antalo. The extension of the telegraph beyond Antalo would have been an ideal task for the heliograph, but the instrument had not yet been invented. A second line was also constructed from Zula along the line of a railway that had been constructed to Kumayli, and this line was continued through the pass to Undul, raising the total length of line to 234 miles.

The work of the telegraph department was then concentrated on keeping these lines in good working order. St John commented that "great as were the difficulties experienced in the erection of the line, they were trifling in comparison with the labour of maintaining it. Carts and camels, elephants and camp followers, apes and Abyssinians, all contributed in varying proportions to the interruptions". The "pilfering propensities of the natives" was another problem, and not infrequently they stole lengths of wire up to 1,000 yards. To maintain the line, it was divided into twelve sections and placed under the control of RE NCOs with a team of Lascar labourers, spare line, and mules to enable rapid repair. The line was never unbroken for long, but the repair parties were kept extremely busy.

The traffic was at first small, but by the time the line reached Adigrat it built up to such a point that the system became overloaded, and restrictions on its use were applied. St John arranged that only authorised officers were allowed to originate telegraph messages, and a simple method of two-level precedence was introduced - both novel procedures at that time, to be formalised and continued thereafter. By the end of the expedition, nearly 8,000 messages had been passed. In most expeditions over the years to come, the field telegraph never had enough capacity for the traffic demands placed on it.

> ### *Working Rules for the Field Telegraph, Abyssinian Expedition*
>
> Messages were given a precedence grading – SP for special precedence, to be originated only by specified senior officers, and PD for the others. When SP messages were to be sent, all other traffic ceased. No messages were accepted for transmission unless signed by an authorised officer.
>
> Each station was identified by a three-letter telegraphic code (the first three letters of its name). The station required was called by transmitting its code, then the letter 'V', and then the three-letter code of the calling station. The station that was called replied with its three-letter code and then the word 'Here'. The calling station then gave a message prefix, the date, and time.
>
> Each message was sent as follows: prefix; message number; number of words of text; date and time; name and address of sender; name and address of receiver; text. After transmission, the receiver counted the words and compared this with the number given. If these did not agree, the sender repeated the first letter of each word until the error was corrected. All messages were registered in a record book.
>
> The telegraph offices at Zula and Senafe were open continuously, day and night. The other offices worked a daytime-only schedule.

Visual Signalling

Lieutenant Morgan and his signallers, without the flying line and thus restricted to visual signalling with flags and lamps, were detached to support the advancing troops. In January, ten NCOs and men of the 33rd Regiment, one of the British units forming part of the expeditionary force, were attached to the 10th Company RE and trained to be flag signallers, presumably because the signalling service being provided was found to be so useful that more men were needed.

The 33rd (Duke of Wellington's) Regiment, was filled at that time, not with Yorkshiremen, as would later be the case, but with a motley team of hard-drinking Irishmen and about ninety Germans, most of whom spoke little or no English, and who had for some obscure reason ended up in the 33rd. Hardly fertile ground for the selection of signallers, one might suppose, but sufficient sober anglophone soldiers were found and trained by the signallers of 10th Company RE during January at Senafé. The 10th Company also practised signalling by night with their lamps, but these seem to have made little practical contribution to the communications.

One of the expedition's despatches notes how: "The signallers made themselves useful to the army the whole way from Senafé to Magdala, and their services were more especially valuable while the army was crossing the ravines of the Takkaze, the Jedda, and the Bashilo, and on the advance to Magdala".

It is also recorded that in March, as they approached Magdala, the flag signallers were used frequently to pass orders between brigades and other elements of the advancing force when they were distant from each other. A few days before the final assault, the expedition had to traverse a precipitous ravine with a deep river at the bottom. A reconnaissance party, including signallers, was sent ahead. After completing their task, a signaller sent the message 'passable for infantry', and soon afterwards, when further investigation was completed, 'passable for cavalry'. Today this sounds trivial, but at the time, in that rugged country, the alternative was a messenger who would have taken several exhausting hours to carry the message, had he survived. With the army on limited supplies, a very long and failing line of communication, and time not on their side, it was extremely useful.

Magdala Captured

The troops eventually reached Magdala on 13 April. For the last fifty miles it had been tough going. There had been a successful engagement at Arogi, a few miles short of Magdala. The men of the 33rd Regiment led the final assault on Theodore's stronghold (the last time the regiment carried its colours into battle, two Victoria Crosses were won, and the battle honour 'Magdala' was later awarded). The signallers were also

The capture of Magdala, 13 April 1868. A sketch by Lt Col R Baigrie, AQMG on the expedition.

involved. It was soon over, there was little serious resistance, casualties were few, and the prisoners were all safely released. Theodore shot himself with a pistol that had been given to him in happier times by Queen Victoria, and the fortress was destroyed, but not before the 33rd discovered Theodore's large stock of arrack (a form of distilled alcoholic spirit) and used it generously to celebrate their success. The troops did not linger - there was no need, and annual rains, which would flood rivers on their withdrawal route, were imminent. They began the long march back to Zula.

The news of the fall of Magdala was known in Zula a few minutes after the arrival of the messenger at the telegraph head at Antalo.

The Death of Lieutenant Morgan

The success of the expedition was marred, as far as the signalling operation was concerned, by the death of Lieutenant Jeffrey Morgan. He, it will be recalled, had been placed in sole charge of the signalling when it became evident that the flying lines, as originally envisaged, could not be implemented for want of mules to transport the cable. This had placed an additional burden on him. The visual signalling, crude though it may seem today, had been extremely useful and required energetic organisation - hence the addition and training of signallers from the 33rd Regiment. Suffering from health problems from the time of his arrival in the country, he had struggled on. He had been personally involved in the assault on Theodore's stronghold, using his revolver to clear defenders, and had been wounded in the shoulder and concussed. Like the others, he had been subjected to the very harsh conditions of the last few weeks, when the troops were advancing without proper supplies, on short rations, and with only what they could carry, without even tents for their protection, at altitudes approaching 10,000 feet. On the return march, suffering from exhaustion and exposure, compounded by his wound, he died at Takkaze on 26 April, and was buried there.

The grave of Lieutenant Morgan at Takkaze

THE ABYSSINIAN EXPEDITION

Sir Robert Napier, the commander of the expedition, wrote that: "he had received with great regret the report of the death of Lieutenant Morgan, R.E., in charge of the signals of the 10th Company R.E.", noting that he "had constant opportunities of observing the unflagging zeal and energy of this young officer, and the cheerful alacrity with which he embraced every opportunity to render his special work useful to the force". It is unfortunate that, due to his death, a more complete account of the signalling operations is not available.

Recommendations for Future Operations

The rather unlikely Abyssinian Expedition has largely faded into obscurity, although one can still wonder at the size and cost of it all in relation to the mission. But in the context of the development of army communications, it influenced a number of forthcoming changes.

After the campaign was over, Oliver St John reported on his experience. The telegraph line had been heavily used, at times overloaded, and he saw the need to increase traffic capacity. His suggestion was more than one line, but with weight and transport already a problem, this had obvious disadvantages. His immediate palliative had been to introduce precedence and restrict originators. In the years to come, the telegraph was always overloaded, although technical inventions enabling more than one circuit on one wire brought some relief.

Turning to organisational matters, he questioned: "Under what department of the army should the telegraph be placed?" In Abyssinia, the direction was subordinated to the QMG, but this arrangement, he suggested, would not work well in operations on a more extended scale. What he had hit upon, probably without realising it, was the wider question of staff organisation and responsibilities that were to be the subject of much debate and operational inefficiency over the ensuing decades.

He then turned to the training and experience of the telegraphists, commenting that the technical training at Chatham made them well-qualified in the manipulation and management of the instruments, but they lacked experience in the practical running of a telegraph office and traffic handling. He suggested that the only way they were going to get this sort of experience was that "some one or more of the public lines in England should be entirely managed and worked by the Royal Engineers. This would give a trained staff, which could be transferred, complete in organisation and material, to a seat of war". He went on to place this in the context of the impending transfer of private telegraph companies in England to the Post Office (ie 'nationalisation'). These were to be prophetic words in the evolution of British military telegraphy within the next few years, for just such an arrangement came into being.

Captain St John (he was promoted Local Captain during the expedition) and Lieutenant Puzey were mentioned in Sir Robert Napier's final despatch for their "intelligent and experienced direction of the telegraph ... which worked well and rendered important service". After the expedition Napier himself was ennobled as Lord Napier of Magdala, and a medal was struck and issued to those who had taken part.

Colonel Sir Oliver St John KCSI

After the Abyssinian expedition Oliver St John resumed his career in the Indian army, and in the early 1870s was involved in map-making in Persia and Baluchistan, his maps for a long time remaining the standard reference. Interestingly, the longitude of various points on the maps were calculated using time signals exchanged between the telegraph stations of the line, then operational between Karachi and London.

In January 1875, as a result of post-mutiny reforms, Mayo College ('the Eton of India', as it came to be known) was established at Ajmer, Rajasthan, and St John, then a Lieutenant Colonel, was appointed as its first Principal.

In 1878 he furthered his career by transferring to the political service, acting as a political officer in Afghanistan during the second Afghan war, 1878-80, and in Kandahar for a few years afterwards. Other political appointments followed, in Quetta, Kashmir (where he was the political Resident), Hyderabad, Quetta again, Baroda, and then as Resident and Chief Commissioner in Mysore, one of the top political jobs in the country but rather quiet for this man of action. By then Colonel Sir Oliver St John KCSI, he returned from Mysore at short notice to serve as the Governor-General's agent in Baluchistan, but the sudden change in climate proved fatal, for he contracted pneumonia and died at Quetta on 3 June 1891, aged 54.

THE ABYSSINIAN EXPEDITION

Influence of the Expedition on Army Signalling

The expedition influenced a number of forthcoming changes. Tactical visual signalling, with flags and lamps (but not yet heliographs), was seen to have had distinct advantages, and the detachment of the 33rd (Duke of Wellington's) Regiment had adapted easily to the role. A year later, in 1869, signalling sections were introduced into infantry and cavalry regiments – effectively the beginning of regimental signalling.

The electric telegraph had demonstrated its effectiveness in the field. Although there had been problems, one stood out - a permanent organisation, with properly trained soldiers and established equipment and procedures, did not exist. The deficiency was recognised and addressed, as will be described in the next chapter.

The Campaign Medal

The campaign medal shows on the obverse a bust of Queen Victoria, and on the reverse a laurel wreath and space for the recipient's name to be inscribed. It was awarded to all British and Indian Forces, army and navy, who took part. This led to questions about the award being given to non-combatants, for example those of the Indian Telegraph Department who took part. The Secretary of State for India in London disagreed with Lord Napier's recommendation that all should be awarded the medal, saying in May 1870 that: "this would, of course, include the followers of every description in the Indian Army" who, he continued, "equal in number, if they do not exceed, the combatant portion". So the Indian telegraphists, classified along with the servants and labourers, did not receive the medal.

Theodore's fortress at Magdala - a view sketched from the east.

Chapter 7

The Formation of Army Telegraph Units

In 1870 and 1871, and for different reasons, three army telegraph units were formed. The first, in 1870, was 'C' Telegraph Troop, intended to support the regular Army Corps. Later in 1870, and then in 1871, two Postal Telegraph Companies were formed from existing regular army units that were re-roled, the 22nd and 34th Postal Telegraph Companies. This chapter describes the evolution of these units.

'C' Telegraph Troop is formed

Procrastination and Politics

Despite the 1867 review of signalling, and despite plans and representations having been made, there was still, by early 1870, no permanently established army telegraph unit even though it had been clear for some years that it was needed. Why? The answer is not unusual – money. Considerable investment of resources in army 'engineering' had been made in about 1864, when the Royal Engineer Field Train was established. This was an organisation set up to provide field engineering support to the Army Corps – the fighting element of the British Army, stationed at Aldershot, and at the time principally intended for a role in European warfare. It consisted of 'A' and 'B' Troops, the former being responsible for pontoon equipment, and the latter for field engineering equipment. (In those days a Troop was equivalent to what is now a Squadron.) The funding for yet more 'engineer' troops and equipment was simply not forthcoming. The engineers had had their ration.

It was the time when a new Liberal government under Mr Gladstone came into power. From 1868 to 1874 Edward Cardwell was the Secretary of State for War, responsible for much-needed reforms of the organisation and administration of the moribund army, and perhaps remembered most for the abolition of the purchase of commissions. On entering office, Cardwell's immediate task was to prepare the Army Estimates, and significant reductions in military expenditure were being demanded. It was not an auspicious time to try and introduce a new unit.

'... We have no organisation except on paper ...'

In May 1870 the Instructor in Telegraphy at Chatham, Captain Richard Stotherd, addressed a meeting of the Royal United Service Institution on the subject of Military Telegraphy and Signalling. He reviewed the advances in signalling, particularly the work of Colomb and Bolton, described in chapter 5. The main thrust of their work had been 'unified' communications between ships and coastal fortresses, and the development of visual signalling. Their principal scenario was defence in the event of an invasion of the south coast of England – a far cry from what soon afterwards turned out to be the real operational requirement for army communications in the late Victorian era.

As he addressed the RUSI, Stotherd reviewed the advances in telegraphy and its method of use in other European armies, and outlined the proposal for a British unit. He ended up venting his frustration:

> ..... In this country we are undoubtedly slow in taking up any new question connected with the art of war; the sort of idea that active operations are a remote contingency, combined with the views of economy under which the Army Estimates are always drawn up, act prejudicially against the introduction and effectual development of any improvements, and the electric telegraph is no exception to this rule. Almost every European nation, except Great Britain, has now a properly organised field electric telegraph; even the small powers such as Bavaria, Belgium, Denmark, &c., have their properly constituted equipments. We have the most authentic information concerning the Prussian equipment; this consists of six complete units of field telegraph, or in other words 18 travelling offices and 180 miles of wire, these are in charge of the Engineer Corps at Berlin, and are ready to take the field at a few days' notice. ..... We have no organisation except on paper. [1]

Things weren't quite as bad as Stotherd made out - it wasn't all paper. The Telegraph Wing at Chatham had experimented, and did have a few wagons which took part in trials of field telegraph communications. Nevertheless, it was all quite outspoken stuff from a Captain to his senior audience in days when officers were generally more circumspect, but as it was a military audience they probably all agreed with him. He subsequently enjoyed a successful career, eventually to become Director General of Ordnance Survey, retiring in 1886 as Major General R H Stotherd CB.

Thanks to Bismarck

The stimulus for change came two months after Stotherd's little outburst, when in July 1870 the Franco-Prussian war broke out. Earlier that year the Prussian chancellor, Bismarck, put forward a candidate for the vacant Spanish throne, with the deliberate intention of provoking the French emperor, Napoleon III, into declaring war. With the help of the infamous Ems telegram (when Bismarck tampered with the contents during its transmission), that is just what Napoleon did. The Prussians defeated the French at Sedan, and then besieged Paris. The British government became alarmed about Bismarck's predatory intentions and their own military strength and defence capability. Orders were promptly issued to the naval and military authorities to place everything ready for immediate action, and purse strings were loosened.

The Prussians tapping French telegraph lines - an early example of SIGINT. The sketch was published in The Illustrated London News.

Here was the opportunity that had been awaited. Substantial funding was requested by the army and navy, and hastily obtained from a nervous Parliament. The details of the scheme for the addition of a Telegraph Troop to the Royal Engineer Field Train were already worked out, and a case was at once put forward for its establishment. Captain Robert Home, the Secretary of the Royal Engineers Committee, was responsible for its staffing. The plans were agreed, and the third Troop of the Royal Engineer Field Train was formed - 'C' Telegraph Troop. After much prevarication it had happened - thanks to Bismarck!

'The 'C' Troop is to be formed at Chatham from the 1st proximo ... '

The regimental order issued from the Horse Guards, shown below, is hardly the way that staff-trained officers today would write such a document, but in 1870 the staff college, set up in 1858, was also still languishing. It was regarded with disdain by leisurely officers who up to then purchased their commissions in infantry and cavalry regiments. Things were different in the Royal Engineers. Along with the Royal Artillery, then the two 'technical' arms, entry to and training at the Royal Military Academy at Woolwich was a much more competitive and professional matter.

As instructed, the new Telegraph Troop (a Squadron in latter day terms) was formed at Chatham on 1 September 1870. Its establishment was fixed at five officers, 245 WOs, NCOs and men, and 150 horses, although the strength authorised on formation was limited to two officers, 135 WOs, NCOs and men, and 55 horses. Over the years the numbers were to fluctuate slightly as changes were made. Captain Montague Lambert, Lieutenant George Tisdall, Troop Sergeant Major R Williams and Sergeant Dockrell, who was promoted to Quarter Master Sergeant, were all transferred from 'B' Troop.

> C. R. E., Aldershot.
>
> With reference to special Army circular, dated 24th inst., will you have the goodness to inform the Officer Commanding the R. E. Train that the Establishment of the Troops is to be as stated in the accompanying Return.
>
> The C Troop is to be formed at Chatham from the 1st proximo, inclusive, and it is intended at present only to raise it to the strength stated in Red Ink on the Return.
>
> The Officer Commanding R. E. Train is to be directed to submit the necessary promotions and transfers, and before doing so, to place himself in communication with the Commandant S. R. E., Chatham, in order to ascertain the names of the N. C. O.'s, and men, and the horses that Officer is desirous of having transferred to the C Troop.
>
> Lieutenant Tisdall's section of B Troop will form the nucleus of C Troop.
>
> Sd. H. FANE KEANE,
>
> D. A. General.

The Troop's Role and Equipment

A description of the Troop's role and equipment is given in the 'C' Troop Record, a manuscript journal of the Troop's activities containing a wealth of information, and is reproduced below. [2]

> The duties to be performed by the Troop consist in the carriage, charge, and working of 36 miles of insulated cable, with all the necessary stores; supplemented by visual signalling, for which purpose a body of the drivers, 20 in number, are equipped as mounted signallers. These men are equipped with telescopes and signal flags; their use is to prolong the line of communication and obtain the information which the regimental signallers may furnish, transmit it to the Office Waggon, from whence it is telegraphed to Head Quarters.
>
> The number of carriages is as follows:
>
> Wire 12, Office 4,, Pontoon 1, Forge 1, Store 6, Total, 24.
>
> The wire, office, and pontoon wagons are constructed with springs, a novelty in military carriages, and are lighter than those in general use at this date for military purposes.
>
> The pontoon waggon, which carries a bay of super-structure 15' and a pontoon boat of the new pattern for putting the cable across a river; the bay is for crossing a small opening.
>
> Each wire-wagon carries 3 miles of insulated wire, this wire is made up of 3 No 20 B.W.G. copper wires, tinned over: the diameter of the conductor so formed is .064 inch; and by the addition of the insulating material, which consists of vulcanized India-rubber, the diameter is increased to .206 inch; a layer of strong canvas is laid over the India-rubber, and the whole is bound round with two thicknesses of tape, primed with India-rubber, the cable thus formed has a diameter of .3 inch. The junctions of the wires are made with Mathieson's ebonite jointers
>
> Each wire-wagon has six drums, and each drum has ½ mile of cable wound on it. These drums revolve on pivots fixed in the sheers of the carriage. The wire is reeled up on the drums placed in the two rear pivots by a driving band worked by the revolution of the hind wheels of the carriage, each drum being moved to the rear in its turn. This method of reeling up the cable is the invention of Troop Sergeant Major Williams, who has been assisted in the mechanical details concerning the disconnecting gear by Corporal Wheeler Knight. As originally designed, the cable was reeled up by hand, and this method of using the wagin's own motion to do the work has very much increased the efficiency of the Troop and reflects great credit on its authors.
>
> The boxes of the wire-wagons carry the necessary small stores, such as jointers, pliers, knives, &c, and the arms and kit of the detachment who lay out and reel up the cable.
>
> The cable is generally laid out along roads; the place selected is the fence, or side of the road. When a road has to be crossed, iron telescopic poles are put up, and the cable is thus raised high enough to allow of carriages passing underneath. The cable is not easily cut by carts passing over it; indeed a battery of artillery has gone over it very often without doing harm.
>
> The office wagons carry the instruments, writing materials, &c, and their roofs and sides are made of Clarkson's patent material.

THE FORMATION OF ARMY TELEGRAPH UNITS

An elaboration of some extracts from this description given in the Troop Record might be interesting.

'... The wire wagon ...'

The 'wire wagon', and the arrangement for reeling up the cable by a belt connected to the 'hind wheel' is evident in the picture. There were twelve wire wagons carrying a total of thirty-six miles of cable.

'... The office wagons carry the instruments ...'

There were to be four telegraph office wagons in 'C' Troop. They were described by Stotherd in his RUSI lecture:

... It is simply a small omnibus mounted on springs, drawn by two horses, carrying two Morse recording telegraph instruments on a small table, always ready to commence work the moment the earth and line wires are attached. Two telegraph batteries, of a form designed by Quartermaster Sergeant J. Mathieson, R.E., a modification of Daniel's form, are always in position under the table carrying the instruments, and ready for work. Four spare Morse telegraph instruments, two recording and two sounding, and two spare batteries are carried in each travelling office. ... Besides these, a set of visual signalling apparatus, tools of various kinds, ...

The Morse recording instrument or 'inker' was based on a design invented in 1854 and used in civil telegraph offices, where the ability to receive messages by relatively unskilled (and therefore lower paid) operators was an advantage. A paper tape was moved through the instrument by a clockwork mechanism. As the Morse code signal was received an electromagnet responded to the current being keyed on and off by the sending telegraphist. When the current was 'on' it operated a lever which dipped the moving tape into an ink trough. When the current ceased the lever was restored to the 'rest' position by a spring. Thus the dots and dashes of the Morse code were inked, or recorded, on to the paper tape. Although it provided a permanent record of the message it was not generally popular with its military users due to weight, fragility, the need for copious amounts of paper tape and ink, and subsequent time-wasting transcription on to a message form.

The sounding instrument, or Morse sounder as it was generally called, was preferred by skilled telegraphists for its simplicity. It used an electromagnet which 'clicked' as the current was keyed on and attracted an armature, and distinctively 'clacked' when the current was off and the armature returned to its rest position by the action of a spring. It had developed from the recorder, which also used an electromagnet and clicked and clacked as it inked. Experienced telegraphists then realised, as they listened to messages being received by the recorder, that they could aurally unscramble the short or long intervals between clicks and clacks (a method quite alien to latter day Morse operators - if there are any left these days - who listen to the tones of oscillators or buzzers), translate them into the dots and dashes of the Morse code, and write the message on to a message form as they received it, doing away with all the inking rigmarole. The best telegraphists were those who learnt the skill while in their 'teens, and they could achieve speeds of up to about 25 words per minute or more with the sounder.

The batteries designed by QMS Mathieson, and consisting of zinc and copper plates in a copper sulphate solution, with a special design of cap to prevent spillage, had been trialled against the Prussian type and found superior. They were specially made by the India Rubber, Gutta Percha, and Telegraph Works Company of North Woolwich and, said Stotherd, "are extremely well finished."

'... The Wire, Office, and Pontoon wagons are constructed with springs, a novelty in military carriages ...'

A trial of the telegraph office wagons had been carried out and in his talk to the RUSI Stotherd revealed that "a travelling office complete marched from Chatham to Canterbury, a distance of 27 miles, in five-and-a-quarter hours, without distress to the pair of horses drawing it." It was the wagon's springs that did it, or words to that effect, he claimed at some length, envisaging a future wartime trot along European roads. Little did he know, as he spoke in 1870, that the telegraph office wagons, even though constructed with springs, would never leave Britain on operations – Chatham to Canterbury was nothing like Durban to Zululand or Cairo to the Sudan. In the real world yet to come, the telegraph instruments and associated equipment were usually dismounted and the ungainly office wagons left behind.

'C' Troop reaches full strength

The Troop, still forming at Chatham, continued to grow. On 12 November 1870 Lieutenant Bindon Blood, previously in charge of No 1 Section of 'B' Troop, was posted in. Later to be General Sir Bindon Blood GCB GCVO, Chief Royal Engineer, Col Comdt RE 1914-1940, he was born in November 1842, a member of an old, landed Irish family, one of whom, Colonel Thomas Blood, involved himself in much skulduggery and even attempted to steal the Crown Jewels in 1671! Bindon Blood had an outstanding career, much of it in India, and in his nineties wrote his autobiography *Four Score Years and Ten*. (The number of tigers he describes shooting would meet with some opprobrium today.) He died in 1940, aged ninety-seven.

On 1 February 1871 authority was received to complete the number of men in the Troop to the authorised establishment of 245 men and 115 horses. Then the first exercise took place, nothing too complicated, and was recorded in the 'C' Troop Record:

> *'... 29th May 1871. First extended exercise of the Troop is carried out. A Telegraph Line is laid from Brompton Barracks to Milton, near Sittingbourne, signal stations being thrown out in advance. ...'*

Aldershot

On 14 August 1871, just under a year since their formation had been ordered, the fully-formed Troop left Brompton Barracks, Chatham and, as the Troop Record tells us, ' *... proceeds by march route to Aldershot ...*', duly reached five days later on 19 August, to colocate in South Camp with the two other Royal Engineer Troops and the Army HQ they now served. It was to be their base for about thirty years.

1	Captain.
3	Lieutenants.
2	Staff Sergts.
9	Sergeants.
8	Corporals.
9	2nd Corporals.
9	Lance Corporals.
1	Farrier Sergt.
1	Artificer Sergt.
1	Telegrapher Sergt.
2	Artificer Corporals.
3	Telegrapher Corps.
12	Telegraphers.
6	Shoeing Smiths.
2	Collar Makers.
2	Wheelers.
2	Carpenters.
3	Trumpeters.
66	Sappers.
92	Drivers.
234	Total.

The 'C' Troop Establishment - copied from the 'C' Troop Record manuscript.

A *Guide to Aldershot*, written in 1885, described the scene:

> At the north side of South Camp – end of A and E lines – and abutting on the Basingstoke Canal, is the Royal Engineer Train Establishment, comprising stables for 44 officers' horses and 388 troop horses; an infirmary for sixteen horses, pharmacy, collar makers' and wheelers' shops, waggon sheds, forge, armourer's shop, &c., boat house and guard house. On the parade is the park train arranged with military precision and care.
>
> ... Quarters are provided in I, K, N, and O lines for the officers, non-commissioned officers, and men of two troops, pontoon and telegraph; two companies and detachment of postal telegraph company; .....

The description shows, as did the establishment reproduced earlier, what a disproportionate amount of the resources were needed for transport.

For those familiar with Aldershot, an 1888 map of the area shows their lines. It is interesting to muse that some ninety years later the Corps that grew from 'C' Troop won numerous Army Rugby Cup competitions in the stadium only a few hundred yards from their original lines.

Bindon Blood, when he arrived there, described Aldershot in those days as "a very pleasant place in spite of being rather in the rough". It is no longer in the rough. Aldershot has subsequently seen numerous phases of redevelopment. In 1871 the area of South Camp originally occupied by the Royal Engineers Train, including 'C' Troop, was a collection of huts originally built during the Crimean War. Aldershot continued to develop as a military camp and the huts were replaced by more substantial buildings in the 1880s. These buildings were in turn demolished and in about 1970 the former 'C' Troop area became known as Browning Barracks, occupied from then until about 2005 by the Airborne Forces. Further redevelopment of this 'brownfield' site can be expected.

But back to olden days. After their arrival at Aldershot in August 1871 'C' Troop was quickly into the swing of things. Some selected extracts from the Troop Record give the flavour:

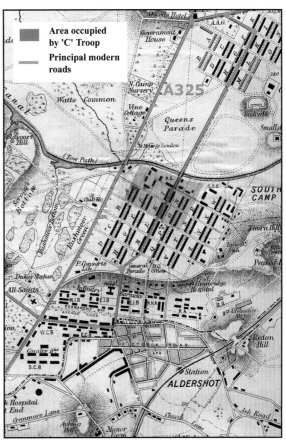

A map of Aldershot dated 1882, showing the 'C' Troop lines.

> *'... 9th to 21st September 1871. Autumn Manoeuvres. The Troop is employed keeping up communications between Head Qrs Aldershot Camp and the Camps of the Divisions in the Field. Lieut Tisdall is detached with the signallers and joins 1st Division. ... During these manoeuvres 80 miles of wire are laid and reeled up, and communication is maintained daily. ...'*

With a little whiff of things to come, the Troop went on exercise to Blandford, and in the Royal Signals archives there is a rather grainy photograph of this event (overleaf). Blandford, where the School of Signals has been located since 1967, like Aldershot, has changed considerably, but for those familiar with the area, it appears as though this photograph was taken near the site of the present car park on Mudros Road, looking towards the Roosevelt Memorial Garden and the Headquarters Officers Mess.

The Troop Record continues its chronicle:

> *'... 24th November 1871. Lieut Tisdall is struck off the strength of the Troop having been appointed Instructor in Signalling to Bengal Engineers. ...'*

The unfortunate George Tisdall, a founder member of the Troop, duly went to India where, like so many in that debilitating climate and without the benefits of modern medicine, he died of disease.

THE FORMATION OF ARMY TELEGRAPH UNITS

'C' Troop at Blandford in 1872. Just an exercise, or a far-sighted reconnaissance?

'... 8th December 1871. Telegraphers are ordered to London to assist in putting down strike of operators. 8 N.C.Offrs and men under Sergt. A Lewis proceed at once and are employed in Dublin until the end of strike. ...'

The privately owned telegraph companies had, under the Telegraph Act of 1868, been 'nationalised', and in 1870 the country's telegraph network formed part of the Post Office. It had taken only a few years for the new organisation, under government control, to decline into strike mode. Sergeant Lewis, clearly an outstanding NCO, later became the Troop Sergeant Major of 'C' Troop and saw active service in that appointment in Zululand and Egypt - operations to be described later.

The Troop Record maintains details of postings in and out. Other early comings and goings were:

'... 1st December 1872 Captn. Durnford took over command of the Troop from Captn. Lambert, who embarked for Barbadoes [sic] as Assistant Military Secretary on 17th December 1872. ...'

Arthur Durnford was the fourth generation of his family to serve in the Royal Engineers, and was the brother of Lieutenant Colonel Anthony Durnford RE, a somewhat controversial character, who was killed at the battle of Isandlwana during the Zulu War of 1879. Arthur later became Colonel Durnford, Deputy Inspector General of Fortifications. Montague Lambert, the first 'C' Troop commander, having returned from Barbados and taken up a major's appointment at Shoeburyness, died in 1880.

'23rd April 1873 Lieut Jelf RE is struck off the strength, on appointment as Adjutant, and is replaced by Lieut. Kitchener RE. ...'

Richard Jelf was later to command the Telegraph Battalion, as 'C' Troop became in 1884. Herbert Kitchener later became Field Marshal Earl Kitchener of Khartoum (and both will feature in later chapters).

Richard Stotherd
Instructor in telegraphy.
'We have no organisation except on paper ...'

Montague Lambert
The first commander of 'C' Telegraph Troop, 1870.

Charles Tisdall
Transferred from 'B' Troop. Afterwards went to India and died of disease.

Bindon Blood
Joined in 1871. Thought Aldershot was 'rather in the rough'.

Some years later Major C F C Beresford offered this little insight into what many in Aldershot might have thought during those formative years:

> In our Army for some years the Field Telegraph Troop at Aldershot was regarded by many as harmless amusement provided for engineer officers at the expense of the taxpayer, but it was much admired as it marched past in the Long Valley [Aldershot]. For all that it was quietly doing good work in training officers and men, and the result of that training has been fully recognised by Generals who commanded in late expeditions. (3)

Reorganisation in 1877

A reorganisation of 'C' Troop took place in 1877. This was essentially to change from thirty-six miles of cable as previously described, to a combination of cable and 'airline' as it was called. Again this followed Prussian practice, for the Prussians had never had the benefits of Hooper's rubber insulated cable, instead having disastrous results with gutta percha insulated cable. It was unsuitable for the reasons stated earlier, and so they had adopted poled telegraph lines using uninsulated wire. 'C' Troop's change to a mix of cable and airline gave some flexibility to match the circumstances, each method with its pro's and con's. Whilst the line itself was lighter, the poles added considerable weight, and the quest for the perfect field telegraph pole was never satisfactorily resolved.

The reorganisation saw a reduction in the number of cable wagons, the exclusion of the original pontoon wagon (an oddity in a Telegraph Troop, but copying the Prussian telegraph unit, the rationale having been that it would make the Troop self-sufficient when it needed to lay cable across rivers), and the introduction of airline wagons, leaving the total at twenty-four wagons. The overall length of line was thus increased from the original thirty-six miles of cable on twelve wagons to thirty miles of cable carried on ten wagons and thirty miles of airline carried on four wagons, each drawn by a team of six horses.

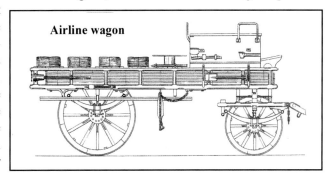

Equipment changed little from then until early in the 20th century and so that, essentially, was what they communicated with for the next twenty years or so. One instrument, invented by Captain Philip Cardew in 1880 when he was Instructor in Telegraphy at Chatham (and known as 'the Edison of the Engineers'), was the vibrating sounder or 'buzzer'. Extremely sensitive, it could overcome many of the deficiencies of field telegraph lines and was frequently used. Many innovations in civilian telegraphy were, however, unsuited to the more robust military environment and, although available, were not introduced to army service.

History changed course – the Scramble for Africa

When 'C' Troop was formed in 1870, at the time of the Franco-Prussian war, the threat to Britain can most tactfully be described as from across the English Channel, and had been for many years. The Troop's original *raison d'être*, its equipment and its organisation, were all geared to the requirements of 19th century European warfare. Its intended role was to connect to some convenient nearby point in the European civil telegraph network that had been developed since the 1840s, and bring the telegraph to the Army HQ, wherever that might be – just as the Prussian and Austrian telegraph units did.

But no sooner had the Troop been formed than history changed course. Along with other European countries, Britain entered an era of imperial expansion, often described by historians as the 'Scramble for Africa', so that for the next thirty years, until the end of the Boer War in 1902, all 'C' Troop operations (and those of its successor, the enlarged Telegraph Battalion, formed in 1884) were conducted in Africa, where the communications scene was quite different.

The European and American experiences, in relatively civilised countries with existing telegraph lines and railways, was in contrast to what turned out to be the British military requirement in the undeveloped countries of the empire where the army field telegraph was destined to be used until the end of the 19th century. The combination of weight of the equipment, dependence on animal transport, and great distances became their main problems, rather than the science of electric telegraphy - as the succeeding chapters will show.

THE FORMATION OF ARMY TELEGRAPH UNITS

The Postal Telegraph Companies

The other thread in the formation of the army's telegraph units was caused somewhat fortuitously by the 'nationalisation' of the private telegraph companies and their absorption into the Post Office.

Early users of the electric telegraph were railways and press agencies (Reuters were registered in 1851), but as the system expanded and equipment improved the telegraph network extended into all sorts of commercial purposes. In 1851 a submarine cable was successfully laid between Dover and Calais, thus connecting Britain with France and other continental systems and accelerating growth. After previous expensive failures, the transatlantic cable opened for traffic in 1866. In America and Britain this rapid growth spawned a profusion of privately owned telegraph companies, whereas in continental Europe the national telegraph networks were, from their inception in the 1840s, mostly developed and owned by the national governments – with the notable exception of systems run by the German company Siemens and Halske which, as explained in chapter 4, became involved in the Indo-European line through Russia.

By 1868 the telegraph network in Britain consisted of some 3,000 telegraph offices, nearly 100,000 miles of wire, and a multiplicity of operating companies, but it was much criticised. The equipment they used was a hotch-potch and often incompatible; there was no control of line routes and London had become festooned with wire going over rooftops in all directions; there were no common standards; charges were high, much higher than continental systems; and there were frequent errors and delays. This inefficient state of affairs was seen to be the result of inter-company rivalry and the absence of a controlling regulatory body. It contrasted sharply with the more efficient services offered by continental, government-owned networks.

Reform in Britain was sought, within and outside parliament, and by the late 1860s the Postmaster General, Frank Scudamore, was involved. He had campaigned for some time for the telegraphs to become an arm of the postal services. Many agreed with him, and this led to decisions in parliament, incorporated in the Telegraph Acts of 1868-69, for the Post Office to acquire, control, and operate most of the various British telegraph companies – a move that these days would be called 'nationalisation'. It was soon to lead to many of the improvements that had been foreseen, helped from 1870 onwards by the invention of many new telegraph devices. (Having in 1868 copied the practice in continental European countries and nationalised the civil telegraph, this arrangement, it might be observed, lasted until 1984, when British Telecom, by then separated from the Post Office, was 'privatised'. This complete reversal in direction was subsequently copied by other European countries!)

There were, however, some early problems resulting from the change, the biggest one being the shortage of skilled manpower to undertake the huge amount of work immediately involved. The civilian telegraph staff of the Post Office were insufficient for their enlarged task.

It was a classic case of a problem becoming an opportunity. Colonel Gosset, at the time the Commanding Royal Engineer at Woolwich, perceived this as a way of getting a good telegraph training for the officers and men of the Royal Engineers, who had hitherto only treated the subject in a rather theoretical manner. (Lieutenant St John had discovered this during the Abyssinian Expedition, and had reported accordingly – see chapter 6.) Gossett therefore proposed the employment of Royal Engineers to assist the Post Office. In this way the army would get a body of thoroughly trained military telegraphists and linemen, and at the same time they would be able to assist the understaffed and overworked civilian telegraph staff of the Post Office. The proposal for the employment of Royal Engineers was submitted to the Secretary of State for War and the Postmaster-General.

The Marquis of Hartington, who was then Postmaster General, accepted the offer eagerly, and, after some preliminary discussion on matters of detail, a Treasury Minute was issued in April 1870, in which it was declared that the arrangements then existing for the employment of Royal Engineers soldiers working for the Ordnance Survey should be followed in the present case, subject to such special rules as the Post Office might lay down.

The 22nd Field Company RE, based at Chatham, was re-roled as a Postal Telegraph Company and placed under the command of Captain Charles Webber. Some of the men forming this new Company had gained experience in telegraph work during the Abyssinian Campaign of 1868, and others had passed through the telegraph course at the School of Military Engineering, but the great majority of the NCOs and men were volunteers, unacquainted with the work they were about to undertake, although all were skilled artificers or clerks.

After a brief training at Chatham, the Company, consisting of three officers and eighty NCOs and men, was moved to London. where at first they were temporarily quartered in the St. John's Wood barracks. The duties to be performed were twofold: firstly, the clerical work associated with telegraph offices including the operation of the instruments; and secondly, the maintenance of the lines in selected parts of the country. A number of the men who had some operating experience were drafted into the various offices, and the remainder were employed in the renewal of the line from Uxbridge to Oxford, which was in a very bad state. They carried out this novel work with enthusiam and by the time the line was completed their training had made them skilled telegraph mechanics.

They had now established their position, their value was duly recognized, and they were met by the civil officials of the department with the utmost cordiality. In the course of the summer their headquarter office was set up in Fountain Court, Liverpool Street, London, which was hired for the purpose by the Post Office.

In September 1871 the 34th Field Company RE was also converted into a Postal Telegraph Company and attached to the Post Office. All ranks drew the regimental pay of their rank, but instead of army Engineer pay they drew Telegraph pay provided by the Post Office. As this was always higher than Engineer pay, there were always sufficient candidates! The two RE Postal Telegraph Companies became technically skilled and very familiar with Post Office practice.

At first the sight of soldiers working on the telegraph lines in London aroused public curiosity. One of the developments in London was to transfer much of the jumble of overhead wire to underground routes. Major Charles Beresford, in his *Records of the Postal Telegraph Companies, R.E.*, described a typical contemporary scene:

> The work in St. Paul's Churchyard consisted in drawing out old and in new cable, and leaving the work temporarily jointed. It was bright summer weather, and as the morning wore on the crowd began to realize the novel sight of redcoats working in a street of the City. They were never tired of staring and crowding round, and sometimes pushing. Indeed it requires some coolness, good temper, and power of apparent abstraction to sit on the pavement with one's feet in the draw box, the cable lying across your knees, and to work carefully with the soldering iron and the spirit lamp, with an eager crowd of bodies surmounted by curious eyes gazing at you, and perhaps one of your men sitting or standing opposite doing his best to prevent the eager spectators meeting their heads in an arch over you and quite obscuring the light. As to the remarks falling round, the more callous and inattentive one is, the better. The little brown tent pitched on the footway is now familiar to the eyes of Londoners. At that time no such convenience for helping the underground constructor existed.

Future Amalgamation

Whilst their formation had been opportunistic, the two Postal Telegraph Companies were to be of considerable advantage in the coming years. Looking briefly into the future, 'C' Troop and the two Postal Telegraph Companies were amalgamated in 1884 to form the Telegraph Battalion, or 'TB' as it was colloquially known. The Telegraph Battalion was split into two Divisions. The 1st Division was the regular army element, essentially the former 'C' Troop, still based at Aldershot, and the 2nd Division was the combination of the two Postal Telegraph Companies with the primary role of working in support of the Post Office in southern England from its London base, and the secondary role of acting as an army telegraph reserve force when needed - as it was going to be on occasions. It was a good arrangement at the time, with many of the telegraphists and linemen being cross-posted at intervals, providing flexibility and a wider range of experience for the telegraphists, although the sedentary life style in civilian telegraph offices drew occasional comment that they lacked the fitness needed for military operations.

* * * * *

THE FORMATION OF ARMY TELEGRAPH UNITS

The 24th Middlesex (Post Office) Rifle Volunteers

A further unit, the 24th Middlesex (Post Office) Rifle Volunteers, consisting of Post Office reservists, also contributed telegraphists. First formed in about 1877 with initially only a postal element, and known at that time as the Post Office Rifle Volunteers, it became known as the 24th Middlesex RV and expanded in 1884 to include a telegraph element, known as 'The Field Telegraph Corps'. Initially fifty strong, with all its members being drawn from the staff of the Central Telegraph Office in London, it expanded in 1885 with a further fifty men drawn from other telegraph offices around the country.

As reservists on a six-year commitment they received, in addition to their Post Office pay, sixpence per day as members of the First Class Army Reserve, and the 24th Middlesex RV was responsible for their training and discipline. When called up for active service the individuals concerned became part of the Telegraph Battalion and then received military pay according to their rank as well as Post Office pay.

A detachment of twenty-six reservists was first deployed in support of the regular army when, during the Nile campaign of 1884-85 to relieve General Gordon in Khartoum, the regular units found themselves overstretched due to the extent of that operation and a concurrent operation in Bechuanaland. Their contribution to the Nile campaign, will be described in chapter 14. Later, a very large number of Post Office reservists, some 450, both telegraphists and linemen, were deployed in support of the regular army in the Boer War, 1899-1902. [4]

Endnotes

1. *Military Telegraphy and Signalling*, by Captain R H Stotherd RE, 13 May 1870. RUSI Journal, Vol 14, pp 312–333.

2. The 'C' Troop Record is held in the archives of the Royal Signals.

3. *The Field Telegraph: Its Use in War etc*, by Major C F C Beresford RE. 9 April 1886. RUSI Journal, Vol XXX, p574.

4. The development of the 24th Middlesex RV is described in more detail in *St Martin's-Le-Grand* (the Post Office's in-house news magazine, named after the location of their head office in London, near St Paul's cathedral, and also near the head office of the present day British Telecom) dated January 1900, when the further need for reservists for the Boer War was described.

Chapter 8

The Ashanti Expedition, 1873-74

Introduction

Following the formation of 'C' Troop in 1870 and the two Postal Telegraph Companies in 1870 and 1871, the first operational deployment arose in 1873. Surprisingly, it was in West Africa, and also surprisingly 'C' Troop did not participate. Telegraph communications for the 1873 Ashanti expedition were undertaken by the Postal Telegraph Companies.

Historical Background

In the second half of the nineteenth century Britain tried to get rid of its colonies on the west coast of Africa – the Gold Coast, Sierra Leone, and Gambia - acquired variously by Britain and other European countries during the seventeenth and eighteenth centuries. They were unprofitable, but the efforts of the government to divest them were countered both by traders who wanted protection of their interests and by humanitarians trying to stop the slave trade.

In the Gold Coast the European settlements were limited to a strip of country near the coast, about forty miles wide. This coastal strip was inhabited by a people called the Fantis, who were generally unwarlike, useless in any fighting role, and even when employed as unskilled labour were unreceptive to discipline and organisation. Inland, and covering all routes to the interior were a number of native states, the principal one being Ashanti. The King of Ashanti, who in 1873 was Koffee Kalkalli, had his capital at Coomassie (also known as Kumassi) about 145 miles inland from Cape Coast Castle. The Ashantis first came in contact with the British in 1806, and in 1824 defeated a British force, killing the Governor of the West African settlements and eight other British officers. Peace was arranged in 1831 but was broken in 1853 and again in 1863 when the Ashantis invaded the colony of the Gold Coast. In 1873, following the withdrawal of the Danish and the Dutch, the British were left in control of the entire coast of what today is Ghana, as well as the hinterland to a depth of about forty miles. Fearing that the British would interfere with their domestic slavery and slave raiding, which had been acquiesced by the Dutch, the Ashantis attacked the British settlements and trading posts in force. The garrison at that time consisted of a detachment of the West Indian Regiment from Sierra Leone and some local police and volunteers, but they were inadequate for the level of violence that had broken out.

On the news of the rising reaching England, assistance was given by ships of the Royal Navy, and parties of marines and naval ratings were landed. The 2nd Battalion of the West Indian Regiment from Barbados was dispatched to West Africa, arriving at the Gold Coast in July 1873. With this reinforcement the local forces were able to prevent the enemy from reaching the coast but the marines and naval detachments suffered severely from fever.

The possibility of the dispatch of a force from England was discussed and, in August 1873, the ambitious Colonel Garnet Wolseley, then thirty-eight years old, was appointed Governor and High Commissioner and sent out in October to report on the feasibility of military operations. He concluded that in the pestilential climate of West Africa – 'the white man's grave' – military operations could only be undertaken with any safety by European troops in the so-called dry season during December, January, and February. An expedition would have to be a quick dash in, overcome the Ashanti army, and a rapid exit. On his recommendation a force of about 1,500 men was sent from England. This included two battalions of infantry (the 23rd Royal Welch Fusiliers, and the 2nd Battalion Rifle Brigade), a detachment of Royal Artillery with four guns, and part of the 28th Company, Royal Engineers tasked principally with opening the route to the interior. Major Robert Home RE was the CRE.

It was only as an afterthought that a telegraph detachment was added to the force, when Colonel Wolseley applied to the War Office for 200 miles of telegraph line and an officer and twenty-five linemen and operators. The delay in sending them led to problems that were to have a detrimental effect on the contribution they were able to make, and their numbers were too few. It was a mistake to be repeated, with similar effect, in several further operations over the next decade, namely the Zulu War of 1879 and the

Egyptian Campaign of 1882. Staff officers were slow to grasp the need for the newly available communications facilities to be included in operational planning procedures at an early stage.

Wolseley took with him staff officers who were to become members of his so-called 'Wolseley Ring' – a clique of selected officers he trusted, and whose careers he developed. They were an interesting lot, some of them to be well-known names in the coming years.

> ### The Wolseley Ring
>
> In 1866 it had become necessary to launch an expedition in Canada, known as the Red River Expedition, and Colonel Garnet Wolseley was placed in charge. He planned the expedition in his usual thorough way, assembling a staff of hand-picked officers – the embryo of the 'Ring'. The expedition was a success, and Wolseley, in his usual immodest style, attributed this to "the fact that it was planned and organised far away from all War Office influence and meddling" – in other words, by him!
>
> On his return to England, Wolseley, who had for years railed about the inefficiency and anachronisms in the early Victorian army, worked with a small military staff under the hand of Lord Cardwell to introduce reforms (the abolition of the purchase of commissions was the best known). These staff offficers were also added to the 'Ring'.
>
> Wolseley's next operation was in Ashanti in 1873 - the one described here - and the 'Ring' again assembled. They included names which were to become well-known over the next thirty years of Victorian empire-building campaigns. Among them were Captain Redvers Buller, then studying at Staff College but plucked away to go to Ashanti; his powers of organisation and leadership had impressed Wolseley in the Red River expedition in Canada. In Ashanti Buller was given the task of intelligence officer, and performed well. (Queen Victoria, who took a close interest in her army, read some of his intelligence reports.) Others in Ashanti who were to become prominent over the next two decades were Captain Henry Brackenbury, Lieutenant Colonel Evelyn Wood and Captain William Butler. Also, arriving later at Wolseley's request to take charge of his lines of communication, was Lieutenant Colonel George Colley, then professor of military administration at the Staff College. To any reader of late-Victorian military history they are well-known names. Major Robert Home, the CRE on the Ashanti expedition, became a member of the Ring and rose to be Colonel before dying prematurely of typhoid.
>
> This collection of 'favourites', who under Wolseley enjoyed special treatment, caused resentment to many in the army generally, who saw their own careers adversely affected. A rival contemporary organisation, the 'Roberts Ring', developed amongst officers who had mostly seen service in India under their leader, Field Marshal Lord Roberts, C-in-C India, who also had his favourites.
>
> This internal army strife was made worse by personal animosity between Wolseley and the Duke of Cambridge, cousin of Queen Victoria, whose little fiefdom of the Horse Guards had been reduced in power by Cardwell's reforms, in which Wolseley had had a hand. Numerous books about the Victorian army elaborate on the subject.

The detail of the operation does not concern this narrative. The main force advanced inland through dense jungle from the coast to Coomassie, the Ashanti capital, which they reached and destroyed on 4 February 1874, although many of the enemy escaped by retreating further into the jungle. The evidence of human sacrifice was everywhere visible, thousands of skulls being piled in a sacrificial grave. Unfortunately the Ashanti army, having scattered, was not destroyed; the king and his symbolic golden stool were not captured; and there was no formal surrender of the Ashanti forces. According to plan, the force then withdrew quickly, their mission only partly accomplished, and many of them racked by fever. There was to be more trouble later, in 1896. Wolseley returned to London to be promoted to Major General, awarded the GCMG and KCB, and to receive the thanks of Parliament, a sword of honour from the City of London, honorary degrees from Oxford and Cambridge, and £25,000 from a grateful nation. [1]

So much for the campaign outline. It is now time to describe the telegraph operation.

The Telegraph Operation

Communications for the expedition were provided by a detachment made up from the 22nd and 34th Postal Telegraph Companies RE, formed in 1870 and 1871 respectively. 'C' Telegraph Troop, the regular, mounted unit formed in 1870, was not deployed because the country was completely unsuitable for their horses and wagons, although it is surprising that they did not contribute any of their experienced personnel. Placed in command of the detachment was the twenty-seven year old Lieutenant Herbert Jekyll RE of the 22nd Postal Telegraph Company, normally based in London and employed on work for the Post Office. His name appears on the list of original members of the Society of Telegraph Engineers, formed in 1871. The detachment consisted of twenty-five NCOs and men but of these only six were clerks (as telegraphists were then described), which was to prove far too few for the number of telegraph offices that were needed and the work involved, the rest being linemen responsible for constructing the line but not operating it (although some could, and did). [2]

They arrived at Cape Coast Castle on HMS *Himalaya* on 9 December 1873, but carrying with them only part of their stores, hastily assembled due to the belated request for telegraph. Their operational task was to provide telegraph communications along the route taken by the troops advancing from Cape Coast Castle inland to Coomassie. There was not yet any submarine cable to West Africa, so there was no strategic rear link to consider.

The telegraph detachment had many problems to contend with, due partly to the belated request for them and partly to the local situation they encountered. It had been decided to land as few of the troops as possible, until the road - a euphemism for what was mostly a track hacked through the jungle, fit only for human porterage - being built by the engineers to Coomassie was well on the way to its destination. As part of this plan, which was adopted for the preservation of the men's health, only part of the telegraph detachment had been sent ashore. Moreover, the telegraph stores ran short, as the remainder of them were aboard HMS *Dromedary* which did not arrive at Cape Coast until 31 December. These stores were landed as soon as that ship arrived, but no sooner were sufficient men and stores available than the transport failed, and they could not be conveyed up country. The cause of this was that on 1 January several infantry battalions had arrived on troopships and landed at the same time as the *Dromedary*, and were given movement priority to the front, absorbing the available transport for some while. Jekyll tried to recruit his own native porters, and in the course of the eight days which followed the arrival of the *Dromedary* succeeded in collecting and despatching three gangs of twenty-five men, each under intelligent headmen. "There was much difficulty in getting these men", he said, "as I had to compete with the Control Department, by whom, apart from their special facilities, large rewards for men were being offered". These, together with the problems of construction, meant that the line could not keep up with the troops advancing to Coomassie, and it eventually terminated about fifty miles short, at Acrofoomu. As with the Abyssinian expedition six years earlier, the problems of these early days, which were to prove recurrent, were transport, stores, and manpower, and above all a lack of integrating the communications into the operational planning from the start.

On arrival Jekyll's first task was to procure telegraph poles. These were available at Beulah, nine miles away from Cape Coast Castle, and there he went immediately on landing. 1,500 bamboo poles, green and freshly cut, each about 16-18 feet long and three inches in diameter, were bought at twopence each. The next job was to transport them to the telegraph hut at Cape Coast Castle and a sapper was put in charge of fifty local women to accomplish this task – something he might not have foreseen as one of his future duties when he took the Queen's shilling. There, various methods of fitting the insulators to the bamboo was tried, and Jekyll describes the solution that was adopted:

> The top of the bamboo was sawn off square, 6 inches above a joint. Four turns of No 11 wire were twisted tightly round the top, and a plug of soft wood was driven into the cavity to fill it completely. A hole was bored down the middle of the plug, into which the insulator was screwed.

Next a gang of fifty unskilled local labourers under a headman were engaged, some preparing the poles as just described, some helping to transport the poles from Beulah, and some helping with constructing and fitting out the hut for telegraph office. Jekyll described the limitations of his local work force:

> Light work, such as preparing bamboo poles and fixing insulators, they soon got accustomed to, and with supervision did very well; but when it came to digging holes, stretching wire, and cutting trees, a great deal of persuasion had to be used, and the constant presence of white men was necessary to keep them at work. In digging holes they soon gave up spades, and preferred to use pick irons, with which they loosened the earth, removing the loose soil with their hands. While working, the man sat on the ground, and worked between his legs till the hole was as deep as the length of his arm. ….. They were slow to learn any kind of work to which they were unaccustomed, even the simplest operation, such as uncoiling wire along the ground,

they were unable to perform for a long time. In carrying loads, they greatly objected to weights in excess of what one man could conveniently take on his head, such as the coils of wire, which weighed upwards of 100 lbs a-piece. The coils were always slung on bamboos, each coil being borne by two men, but they disliked this method of carriage, and would sometimes take it in turns to carry the whole load, whilst at other times the coils would be laid down and abandoned by the road side. ..... None of the skilled or delicate work could be entrusted to them, and their aversion to ladders was such that they could seldom be employed off the ground. ..... They were, moreover, extremely timid. Many deserted after seeing some old bullet marks in a tree, others after being knocked down by lightning shocks, when handling the wire. This incident was productive of some good, inasmuch as it inspired great respect for the wire, which was henceforth regarded as fetish, and never molested by the natives.

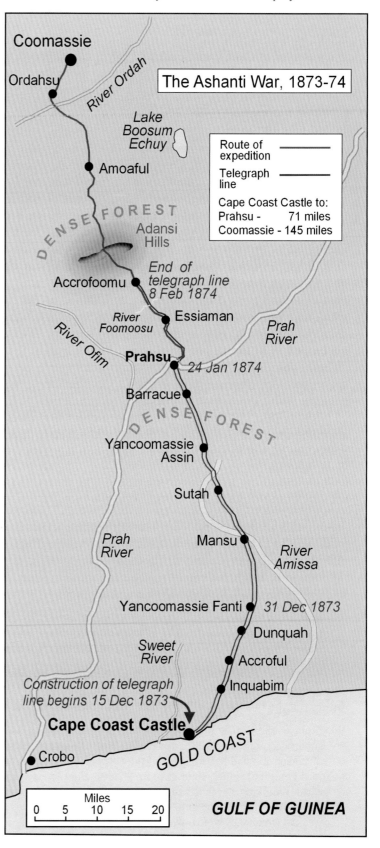

Meanwhile, Jekyll confirmed the operational plan for the telegraph with the Chief of Staff (Colonel G R Greaves), and construction of the telegraph line began, Jekyll noting almost immediately that "a few large orange trees in the street interfered with the course of the wire, and had to be cut down".

Building the line out of Cape Coast Castle was not easy, due to the crooked road, steep hills, and dense undergrowth up to twenty-five feet high which had to be completely cut down. The country for a while became a bit more open and a faster rate of construction was attained temporarily, then became more difficult again as they entered the jungle. Some care was necessary in selecting trees to act as telegraph poles. Those that were swayed by the wind were unsuitable, as this broke the wire. Either they had to be so large that they stood firm, or so small that their tops could be cut off and the trunks left as poles. Some trees were so covered with creepers, parasites, and ant's nests, that clearance became very difficult; others were so soft and spongy that they afforded no hold for the spikes attaching the insulators.

The line itself was No 11 BWG (Birmingham Wire Gauge) galvanised iron wire, which proved very satisfactory and was used for many other purposes as well as the line itself. Jekyll wished he had brought some insulated cable so that the difficult stretches of bush and jungle could be bypassed by cable laid on the ground. From late

THE ASHANTI EXPEDITION, 1873-74

Summary of Telegraph Work

1873
12 December	15 men of Telegraph Detachment landed. Poles bought at Beulah. Stores landed.
13 December	Poles brought in. Stores arranged.
14 December	Office prepared at Government House, Cape Coast Castle.
15 December	Commenced line construction.
22-31 December	Offices opened progressively at Inquabim, Accroful, Dunquah (temporary), Yancoomassie ,Fanti.
31 December	HMS Dromedary with further telegraph stores arrived at Cape Coast Castle.

1874
1 January	Telegraph Detachment landed. Stores from Dromedary landed.
7 January	Stores sent to Mansu now that transport available.
1-24 January	Offices opened progressively at Mansu, Sutah, Yancoomassie Assin, Barraco, Prahsu.
2 February	Office opened at Essiaman.
7 February	Working party fired on near Accrofoomu.
8 February	Office opened at Accrofoomu, Essiaman closed. Five offices now working.
23 February	All offices closed. Instruments and batteries brought back to Coast, line left standing.

January and through February seasonal thunderstorms were frequent and violent, constantly making the line unuseable, the lightning splitting the insulators. Two patterns of insulator had been taken, and one of them proved very defective in design, causing problems. The storms caused many trees to fall, again damaging the line. They also caused electric shocks to those handling the wire, the induced current staying in the wire for some time afterwards. The natives learnt that tampering with the wire brought violent retribution from the heavens above and that, together with their docile nature, prevented any unwanted interference with it. In an attempt to speed things up the labour force was increased to eighty, but only with difficulty as many applicants were unsuitable. They seldom achieved a construction rate of more than two miles per day, although Jekyll had planned for five miles per day.

As construction of the line progressed, telegraph offices were opened along the route, some only temporarily due to shortage of equipment and clerks to operate them, reaching Prahsue on 24 January and Acrofoomue on 8 February. The line was operated using the single current method of working, the simplest and most reliable method. The offices were equipped with Morse recorders, with which the telegraphists were perfectly familiar, the particular instrument being known as Military Direct Writer, and it proved to be very satisfactory. Another type of office instrument was also taken, but was never used, as it was known to be susceptible to the thunderstorms that were prevalent – the Magnetic Alphabet Instrument. They also brought flags and signalling lamps, but in the local conditions these were never used.

There was great difficulty in manning the telegraph offices properly because of shortage of clerks compounded by health problems. Six clerks, as already noted, were quite insufficient, and why so few were taken is not clear - probably a sequel of the belated planning. They had to man their telegraph office entirely on their own for twenty-four hours a day, and Jekyll later reported that he really needed at least sixteen clerks to man the offices properly, as well as orderlies for delivering received messages and other supporting duties. Several offices had to be closed because of lack of staff. Details of the traffic passed were incomplete as several office diaries were lost, but from what was saved it seems that several thousand messages were passed.

Like the rest of the force, the telegraph detachment suffered greatly from disease and fever. A month's exposure to the heat and humidity in an area where tropical disease was prevalent, together with the hard physical work, led to many cases of fever, usually requiring a few day's rest. Several of the clerks found it difficult to undertake their work, unable to get up, or to read or write coherently and legibly. Quinine, which they took

Constructing the line through the jungle.

to combat malaria, caused deafness. (Fortunately Morse sounder instruments were not being used in the telegraph offices.) Sometimes the linemen had sufficient knowledge of the equipment and operating procedures to stand in for the indisposed and under-staffed telegraphists. Jekyll, in his post-expedition report, remarked that "great as were the advantages of having clerks employed in Post Office telegraph offices for the experience they gained, such sedentary work was detrimental to the robustness of their health and the level of general fitness needed on military operations." He suggested they should be rotated for three or four months a year into the regular military environment.

All the delays to constructing and operating the telegraph line prevented it advancing as rapidly as the road that was being built by the engineers and the advancing troops. Ultimately it did not reach further than Accrofoomue, eighty miles from Cape Coast Castle, but fifty miles short of the intended termination at Coomassie. This distance was worked with five offices. Linemen, with some native labourers and tools, stores, and test equipment, were posted at intervals along the line ready for any repair and maintenance tasks. The frequent thunderstorms from late January onwards caused damage to the line, so their services were in constant demand. (3)

Towards the end of the campaign, on 26 January, Jekyll himself, having suffered repeated attacks of fever and contracted malaria, was invalided back to Britain. His duties in command of the telegraph detachment were taken over by Lieutenant Cotter RE, attached from the 28th Company RE which was also involved in the operation.

By the end of January the construction work was over. On 23 February, when the troops withdrew, the telegraphists closed up the telegraph offices, taking instruments, batteries, and unexpended stores with them, but leaving the telegraph line standing. Remnants of the line and coils of wire were found, and used, by a subsequent expedition over twenty years later – a testimony to its construction and the quality of No 11 BWG galvanised wire.

Lieutenant Cotter and the telegraph detachment, the last of the engineer forces to withdraw, embarked on the S.S. *Manitoban* on 4 March 1874 and returned to England, reaching Portsmouth on 20 March 1874.

Conclusion

It was the first operation undertaken by the newly formed telegraph units, and they returned to England having gained much in experience, although many of the problems found in West Africa were unique to that area. In terms simply of the telegraph operation, it was in no way remarkable. That it failed to keep up with the advancing troops was understandable in the circumstances, and Jekyll explained that he had preferred to build as reliably as possible rather than rush the construction, which might possibly lead to failures difficult to repair at vital moments of the operation. This was supported by the CRE on the expedition, Major Robert Home, who said: "Too much praise cannot be given for the way in which this line was made. Lieutenant Jekyll fully guarded against all accidents over which he could possibly exercise any control." Jekyll praised the work of his men and made several comments on their performance. Sergeants Longstaffe and Dowie, the two senior NCOs in the telegraph detachment had performed exceptionally well. The linemen had been well trained in the Post Office and were extremely competent at testing, fault-finding, and repair, and were even able to use and operate the instruments, although not at the speed of the clerks.

From another viewpoint, perhaps the greatest unsung benefit of the operation was that the newly formed army telegraph organisation operated alongside the 'Wolseley Ring'. The influential 'Ring' began to acquaint themselves with, and to use, the telegraph in the field, to break out from the extreme limitations of visual signalling, and to exploit the new method of communication now available to them. Wolseley himself, always striving for efficiency and not constrained by the Victorian army officer's usual aversion to the adoption of technical innovation, quickly developed a very good understanding of the uses and capability of the electric telegraph in field operations - and exploited it to its best advantage in the forthcoming campaigns.

THE ASHANTI EXPEDITION, 1873-74

Endnotes:

1. A contemporary description of the operations, *The Ashanti War (1874)*, in two volumes, was written by Capt Henry Brackenbury RA, Wolseley's Assistant Military Secretary (and one of the Wolseley Ring).

2. Back in England, while recuperating in his house at 2 Morpeth Terrace, London, SW1, Herbert Jekyll wrote a detailed report of the work of the telegraph detachment. It will be found in the *Professional Papers of the RE*, Vol XXIII, New Series, RE Library, Chatham, from which the various quotations have been extracted.

3. Both the sketches of the line construction were drawn by the artist Orlando Norrie and originally published in the *Illustrated London News*.

Colonel Sir Herbert Jekyll KCMG

Lieutenant Herbert Jekyll, as he was when in command of the telegraph detachment in Ashanti, was born in 1846, the son of Captain E J H Jekyll, Grenadier Guards. He came from a family with a long record of sending their sons into the law, the church, or the army. He passed out of the RMA Woolwich in December 1866, winning the sword of honour, and was commissioned into the Royal Engineers. In 1870 he was posted to the 22nd Company RE, then based at Chatham. That same year, the 22nd Company became a Postal Telegraph Company and in its new role moved to London. While serving in that unit Jekyll was appointed to the Ashanti expedition, and his reports of the expedition show that in the previous three years he had acquired a good technical grasp of electric telegraphy.

After the expedition he returned to the Post Office and was appointed to special employment under the Postmaster-General until the end of 1876. That ended his connection with military telegraphy. His next appointment was with Lord Carnarvon, Secretary of State for the Colonies, who asked for his services as private secretary. Jekyll held this appointment until Carnarvon resigned in January 1878. Carnarvon next presided over a Royal Commission to report on the defence of British interests abroad, and Jekyll, by now a Captain, went with him as the secretary.

After staff appointments in London he went to Dublin in 1892 as a Lieutenant Colonel to be secretary to the Lord Lieutenant, Lord Houghton of the Liberal government. In 1895 this tour ended with the change of government. Jekyll remained in Ireland as CRE Cork District until 1897, being promoted to Colonel in 1896.

He next assumed the appointment of secretary to the Royal Commission on the Paris Exhibition, living in Paris, and taking over as Acting Secretary of the British Embassy there. At the beginning of 1901 he was awarded the KCMG, retired from the army, and entered the civil service as Assistant Secretary to the Railway Department of the Board of Trade. He served on important committees dealing with light railways, electrical standardisation and canals. From 1907 he was the first chief of the London Traffic branch.

He retired in 1911 but, with his wife, was active during the 1914-18 War in work for the Order of St. John of Jerusalem, of which he was Secretary-General for ten years and Chancellor from 1911 to 1918. He died in 1933, aged 86. Interestingly, his sister was Gertrude Jekyll, the well-known garden designer who worked in association with the architect Sir Edwin Lutyens.

The Ashanti Medal 1873-74

Ashantee 1874

A medal was struck and awarded to all who had participated in the expedition between 9 June 1873 and 4 February 1874, with the clasp 'Coomassie' awarded to those who were present at Amoaful and the actions between there and Coomassie, as well as those north of the Prah 'engaged in maintaining and protecting the main communications of the main army'.

Chapter 9

The Heliograph

Although properly established army telegraph units now existed, the electric telegraph was unsuitable for tactical operations, where mobility was the prime requirement. Tactical signalling by visual methods still depended on the very limited capability of flags and lamps, brought into service in the late 1860s (chapter 5). But a potent new club in the signaller's bag was about to enter military service - the heliograph (literally meaning *sun writing*). It had greater range, its signalling speed was good, and it was easily portable, but of course it did need sunshine, which in Britain has always been a fickle commodity so enthusiasm was limited.

Coinciding with the expansion of the British Empire, and consequently the greater deployment of the army to sunny climes, mostly Africa and India, the heliograph was to contribute significantly to British army signalling operations for many years to come. It played a prominent part in the forthcoming Afghan and Zulu wars, to be described in the next two chapters; it reached its heyday in the British Army in the 2nd Anglo-Boer War (1899-1902); it was still being used in some theatres in World War I (1914-18); it was still described in the army's manual for Signal Training (All Arms) 1938; and it was even occasionally used in North Africa in World War II (1939-45). Consigned to history since then, and its features probably unfamiliar to most, this chapter describes the instrument and its evolution.

Evolution of the Heliostat

Reflecting the sun to send messages in primitive form had been used by numerous warring factions since ancient times, but it was haphazard and did not permit any coherent form of communication. The first instrument to reflect the sun that was of any practical significance was the heliostat. Although it had been invented some time before, it is usually associated with Carl Gauss, the German mathematician and astronomer, who used it in his work on geodetic survey in 1821. The heliostat was simply a plane glass mirror that reflected a steady beam of sunlight towards a distant point. (It was sometimes known in those early days as the heliotrope although, strictly speaking, that was a similar instrument used for determining the size of the earth.)

After Gauss's work the first practical application was in map survey work, when the heliostat would be located on a prominent feature and act as a beacon for triangulation purposes. The clever part of this otherwise simple instrument was the optical arrangement incorporating a telescope that directed the narrow reflected beam (about half a degree wide) accurately to the distant point, which had to be visible and be identified, and enable the operator to keep it aligned while the sun continuously changed its position in the sky due to the rotation of the earth. One method of doing so is shown in the diagram below. The heliostat was used in India for map survey work as early as 1833 by the Surveyor General of India, Colonel (later Sir) George Everest, after whom the mountain is named.

The instrument comprised a reflecting mirror and a piece of plane glass which were set at exactly 90 degrees to each other. The instrument was aligned so that the sun's rays struck the reflecting assembly as shown in the diagram. The two elements of the ray were divided, so that one was reflected by the mirror to the distant point (angle A), and the other was reflected, at lower intensity, by the plane glass through the telescope to the observer's eyepiece (angle B). Since the two were set at 90 degrees to each other, angle A + angle B was 180 degrees. Because the second glass was not mirrored it was possible to see through it and aim the telescope at the distant point. Thus, the reflection of the sun from the mirror was also directed to the distant point.

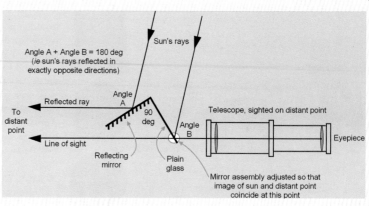

THE HELIOGRAPH

After the heliostat came into use for survey purposes, the Morse code was introduced in 1837, but the instrument's new potential for signalling, simply by interposing some form of shutter in the reflected beam and using it to 'flash' the dots and dashes of Morse code, was not developed in Britain until the early 1850s - and then by a civilian explorer. The reason is probably a combination of the Victorian army officer's lack of interest and imagination in such matters as signalling, and the unreliability of sunshine in Britain and elsewhere in northern Europe where British military interests at the time were mostly concentrated.

It was not until more fertile minds applied themselves to the idea that the signalling solution was reached. One early user was Francis Galton (later Sir Francis Galton), a cousin of Charles Darwin, and a man of prolific intellect. He was not a military man at all but, amongst many other things, he was an explorer, and between 1850-52 he mounted a long and difficult expedition into Damaraland, south-west Africa (now Namibia), the country then little-known. He took with him a number of hand-held heliostats which he used for exchanging simple messages between roving parties of his expedition. He described many aspects of the journey, including signalling, in his book *The Art of Travel,* first published in 1855. It devotes a section to signalling, both by heliostat and flags, and Galton acknowledges that he based his signalling codes on the flashing methods developed by Bolton. His descriptions are most explicit. Whilst Galton's signalling efforts may have been a little amateurish all the basic elements were there, but it still seemed to attract little interest in the subject in army circles. [1]

There is also evidence that the Russians used the heliostat for some form of signalling around Sevastopol during the Crimean War, 1854-56 (see chapter 2). *The Times* of 11 July 1855, in a report from its correspondent in the Crimea, states that: "a long train of provisions came into Sebastopol today, and the mirror telegraph which works by flashes from a mound over the Belbeck was exceedingly busy all forenoon".

The first recorded British military operational use of the heliostat was by Captain E W Begbie of the Indian army in the Dafla expedition in Assam in 1873 - some twenty years after Galton first used it for signalling during his expedition.

Development of the Heliograph

Whilst the heliostat could signal effectively, improvements were possible. In 1869 a small team at the remote telegraph station at Jask, at the mouth of the Persian Gulf, one of the telegraph stations on the recently constructed Indo-European telegraph route (see chapter 4), applied themselves to the problem. It might well be imagined that, with few distractions at that lonely and unattractive outpost, some mental stimulation was therapeutic. Headed up by Mr (later Sir) Henry Mance, a civilian who at that time worked in the Indian Telegraph Department and had been posted to Jask, they concluded that two design changes would make a considerable difference to the heliostat's performance. Firstly, the large and noisy Venetian blind screen, clattering in front of the heliostat's mirror, was an encumbrance, heavy to carry and slow to operate. Secondly, the sighting arrangements could be improved, so that frequent adjustment of the mirror's alignment to take account of the apparent movement of the sun was made easier and quicker for the operator.

They modified the heliostat's design to overcome these disadvantages. To replace the obscuration screen the mirror was no longer fixed, but was pivoted horizontally so as to be vertically moveable, able to be deflected by the operator tapping a small lever so that the narrow reflected beam, just over half a degree in width, was shifted a few degrees, enough to make it visible or not to the distant station. In the 'rest' position the reflection was not visible at the distant station, but when the operator tapped out his message the mechanical movement of the mirror deflected the beam upwards so that it then reflected the sun and the dots and dashes of the Morse code became visible at the distant station. [2] To improve the sighting arrangement a small disc of the reflective silvering was removed from the centre of the mirror, so that the operator could look through the aperture towards the distant station, and a sighting rod was set up about ten yards in front of the mirror (see diagram overleaf). These two key changes to the design of the heliostat increased signalling speed, simplified mirror alignment, and improved portability.

The new instrument became known as the heliograph. The semantics were a little confusing to many at the time who did not understand the difference between a heliostat and a heliograph and used the terms indiscriminately. As just described, the heliostat used a fixed (static) mirror, and the Morse signalling code was transmitted by interrupting the beam in front of the mirror with some form of shutter; the heliograph signalled using a deflecting mirror.

THE HELIOGRAPH

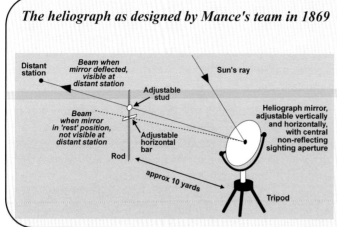

The heliograph as designed by Mance's team in 1869

The sighting arrangement used a rod with two adjustable sighting marks on it, set vertically about 10 yards in front of the heliograph. Looking through the non-reflecting aperture in the centre of the mirror, the stud and the distant station were aligned. The adjustable horizontal bar and the deflection adjustment of the mirror assembly were set, so that in the 'rest' position the reflected beam fell on the bar and when the mirror was deflected the beam fell on the stud, and thus also the distant station. This early method was rather inconvenient, and the instrument was soon developed with a sighting rod fitted as an integral part of the mirror assembly.

The intended application of the heliograph was as a back-up to the permanent electric telegraph system in India. It was put into production at the Indian Telegraph Department's workshop in Karachi in 1869, and although it was the result of a joint effort it was Mance, who worked for the Department and made the production arrangements, to whom it was attributed.

A description of the heliograph's development is in a letter published in the Royal Engineers Journal in 1872 by Major J M Morgan RE, explaining that a number of people, including Mance and himself, were working at Jask on "improving the practical design of something that had been known for quite some time" – namely the heliostat. Morgan claims that it was he who had the idea of making the mirror moveable rather than obscuring the reflected beam by a screen – the basic difference between the heliostat and the heliograph – and also that it was he who had the idea of removing the silvering from a central spot in the mirror so that it could more easily be aligned. He concluded his letter by saying: "I think you will agree with me, after this explanation, that the instrument devised by Mr Mance can hardly be called an 'invention', but an adaptation of a well known means to his end ….". Apart from the British efforts, there is evidence that a Frenchman produced something on a similar principle a few years earlier. [3]

There were proponents of both instruments. The problem with the heliograph was that unless the tripod on which the instrument was mounted was firmly based, the constant tapping by the operator to deflect the mirror as he signalled the Morse code tended to misalign it, and with its narrow beam accurate and firm alignment was essential. Even if firmly mounted, heavy-handed operators could also misalign it as they tapped over-enthusiastically. On the other hand, the problem with the heliostat was that a second tripod was needed for the separate screen in front of the mirror, to the disadvantage of portability, and the movement of the screen tended to make a loud clattering noise. Well-trained operators could overcome the disadvantages of the heliograph and, with the pivoted mirror moving delicately, could signal their messages more quickly then those using the heliostat with its rather cumbersome screen. The heliograph became the preferred option.

The first use of an experimental heliograph in England was carried out by some Royal Engineer officers who, having heard of Major Morgan's experimental design in Jask, attempted not very successfully to replicate it in 1870. [4] The newly-developed heliograph's potential for military signalling was obvious but in Britain there was no rush to bring it into service; despite the need for something better than flags and lamps, it was not even trialled by the army until 1872. The artificiality of military exercises around Aldershot, together with short distances and unreliable sunshine, saw little demand for anything more than the tried and trusted galloper delivering messages by horseback over a distance of a few miles. In India, where part of the British army was stationed, the sunny climate, the topography, and the greater distances contributed to greater effectiveness, and the operational demands injected more realism. For these reasons the element of the British army stationed in India became much more competent in its operation. The heliograph was approved by the Indian government for military use in 1875, but even there it had taken six years!

Certain aspects of Mance's original design of the heliograph were soon found to have a number of shortcomings for military use, and a more robust version of the instrument was developed by the Royal Bengal Engineers at their base at Roorkee under their Instructor in Signalling, Captain George Savage.

The tripod on which it was mounted was strengthened and the head assembly was redesigned. This enabled the second, or duplex mirror (needed to provide a double reflection when the sun was behind the signaller) to be mounted on the same tripod as the primary mirror, thus doing away with a second tripod and improving portability. One very worthwhile further improvement was to replace the sighting rod used by Mance with a sighting arm forming an integral part of the head assembly, described on the next page.

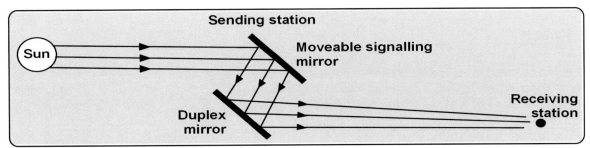

The principle of the duplex mirror, used when the sun was behind the sending signaller

Signalling with the Heliograph

The range that could be achieved by small mirrors was remarkable. The practical rule of thumb was that a mirror could signal over a distance of ten miles for every inch of diameter of the mirror, although of course this depended on atmospheric conditions which in sunny countries were generally best in the morning before dust and heat distortion impaired visibility. Various sizes of mirror were used, between three and ten inches in diameter. The optimum size, balancing range against size and weight, was considered to be a 5-inch mirror, able to signal to fifty miles in good conditions, and military heliographs generally used that size. The size of the mirror makes no difference to the width of the beam. A bigger mirror simply produces a brighter reflection, visible over a greater range, but not a wider beam.

The mirror was made of high quality plane glass, essential to provide a sharp reflected beam of low dispersion. The ray from the heliograph's mirror, simply reflecting the sun (diameter 865,000 miles, distance 93 million miles), makes an angle of 0.53 degrees, which in trigonometrical theory translates into a beam 82 yards wide at a range of five miles. This can be visualised by looking at the disc of the setting sun as it disappears below the horizon; the heliograph's beam reflected by a plane mirror was simply the reverse. In practice, as distance increases, the reflected ray diffuses into a slightly wider beam - explained, as radio communications engineers today will know, by Fresnel's theories on coherence and electromagnetic wave propagation, and light is the visible spectrum of an electromagnetic wave. The practical rule of thumb for the heliographer was that the reflected beam was 16 yards wide at a distance of one mile. This narrow beamwidth made it essential for the instrument to be accurately aligned and set up on a steady base, so for that reason it could not be used aboard ships by the navy, or in due course from balloons when these were used for observation and reporting purposes, firstly in the 2nd Anglo-Boer War in 1900. One advantage of the narrow beam, of course, was that the enemy could rarely intercept it and was unaware that communications were being conducted.

A heliograph station was normally operated by three signallers: No 1 wrote the message on a message form as it was received; No 2, using a telescope if necessary, looked at the transmitting station and called out every letter as it was received; No 3, using signalling codes, flashed back to the sending station any operating messages such as 'received' or 'repeat'. Signalling speed with the heliograph depended of course on the skill of the operators. 10 words per minute was the speed demanded from qualified signallers, but well-trained operators managed 15 words per minute.

A major problem in practice was maintaining skills in regimental signal sections, especially regiments based in Britain. The signallers were often the commanding officer's more intelligent soldiers, potentially his future non-commissioned officers, and generally needed by him for other duties in his regiment. The time needed to maintain signalling skills when measured against the vagaries of sunshine in Britain and the lack of any immediate operational needs, conflicted with what were perceived to be more pressing requirements. For many years, regimental signalling in the British army in the latter part of the 19th century could have been a lot better than it was. [6]

Eventually, evolving from Mance's original design and the Roorkee improvements, the instrument was made by a number of companies in Britain, perhaps the best known being Elliott Bros, with their head office in Charing Cross, London.

The Heliograph, 5-inch, Mark V, shown in the Signal Training (All Arms) 1938 manual appears little different to the instrument described over fifty years earlier in the Manual of Instruction in Army Signalling 1887 shown in the diagram opposite.

Endnotes:

1. The *Art of Travel* by Francis Galton was first published in 1855 and ran to many editions. Galton also read a paper before the British Association in 1858 entitled *Sun Signals for the Use of Travellers (Hand Heliostat)*. Another description of his heliostat, made by Messrs Troughton and Simms, was published in *The Engineer* on 15 October 1858. The Galton Collection is held at the University College, London.

2. Numerous descriptions, probably repeating each other, refer to the heliograph as having an 'oscillating' mirror. This is quite incorrect terminology. Oscillation, as mathematicians and radio engineers will understand, is a form of simple harmonic motion, and is a regular, repetitive process, usually represented in technical diagrams by the trigonometrical sine wave. The heliograph's mirror did not oscillate; it was pivoted horizontally, and deflected in response to the operator's commands as he signalled the Morse code.

3. Further detail on Mance's heliograph will be found in *The Journal of the Royal United Service Institution*, Vol XIX, pp 533 to 548, which records a presentation made on 14 June 1875 by Mr Samuel Goode entitled *Mance's Heliograph, or Sun Telegraph*.

4. Major Morgan's letter was published in the *Royal Engineers Journal*, 1 March 1882, along with a copy of an earlier letter containing all the detail that he had originally written on 11 November 1872.

5. The detail will be found in a letter published in the *Royal Engineers Journal*, 1 February 1882, pp 43-44. The letter was written by Captain John Tisdall, brother of the late Lieutenant George Tisdall, who it will be recalled (chapter 7) was a founder member of 'C' Troop, and subsequently went to India, where he died. The two brothers, and others from 'C' Troop experimented. An interesting addition to John Tisdall's letter was a copy of a letter by Elliott Bros, who made heliographs for the British army, in which they state that: "in France an optician in Paris, whose name we now forget, is credited with having introduced a similar heliograph as far back as 1854-56." So who really designed the heliograph is rather obscure, and the reality is that no one individual can be accredited.

6. The problems of getting good signallers in infantry and cavalry regiments lasted for many years. It was described by Colonel F G Keyser CB, Inspector of Signalling at the Army Signal School, in a presentation he made to the RUSI on 10 February 1893, recorded in the *Journal of the Royal United Service Institution*, Vol XXXVII, pp 245 to 267.

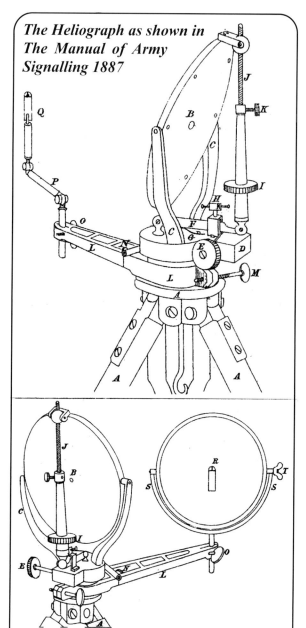

The Heliograph as shown in The Manual of Army Signalling 1887

A Tripod
B Signalling mirror
C 'U' frame for B
D Tangent box
E Tangent screw
F Lever arm
G Spring
H Capstan-headed screw for regulating beat
I Key
J Vertical rod in socket
K Screw for clamping J
L Jointed arm
M Clamping screw for L
N Clamping screw for hinge
O Clamping screw at end of hinge
P Double-jointed sighting rod
Q Sighting vane with sighting spot and cross lines
R Duplex mirror with sighting spot
S U-frame for R
T Clamping screw

Chapter 10

The 2nd Afghan War, 1878-80

The scene now shifts back to India and the army signalling operations that took place there during the 2nd Afghan War, 1878-80. These were still early days in the development of field communications, but tactical signalling in the war attained a new level of competence and made a significant contribution to the hard-fought operations that took place in what was then a very hostile environment. The key to this new level of signalling competence was of course the recently introduced heliograph. It was ideally suited to the conditions around the North West Frontier of India (now Pakistan) and Afghanistan.

Army Communications in 1878

But first it may be instructive to take stock of the wider perspective of army communications that had been reached by late 1878 when the Afghan War started. There were two distinct elements, electric telegraph and visual signalling, and there were contrasts between Britain and India in each of them.

In India there had been attempts by the army to emulate Britain and form telegraph units in the three Presidencies (Bengal, Bombay, and Madras), but it had not happened. The resources needed, it was said by office wallahs in high places, would be better applied to the country's civil telegraph, which would support the army in war when needed and, they continued, the Indian Army had no role outside the country. This theory was the subject of much heated argument with the army, and early in the war it was found to be flawed, not least because the army was operating outside India.

In the absence of army telegraph units the Director General of the Indian Telegraphs, Colonel Daniel Robinson RE, introduced an alternative. He arranged for soldiers from any British unit, not just Royal Engineers, to be attached to telegraph offices throughout India for training in telegraphy. In this way a large number of 'soldier signallers', as they were called, were trained and employed as telegraphists for a tour of duty in civilian telegraph offices. Attracting additional pay when qualified, it was a popular duty away from humdrum barracks routine, and there was no shortage of volunteers. When the Afghan War started, these trained soldier signallers were essential in meeting the demand for telegraphists, and every available soldier signaller, even if from regiments on active service in the war, had to be used. Fundamentally, though, the organisation was inadequate for the political decisions that were taken - not for the first or last time!

It was not until 1876, six years after the British Telegraph Troop was formed, that the Royal Bengal Engineers managed to establish a Field Telegraph Train under Lieutenant George Savage RE, the Instructor in Signalling at their HQ at Roorkee. Not as impressive as its title may sound, it consisted of one officer and two NCOs from the RE, two British 'soldier signallers' as described above, one Indian corporal (naik), ten Indian soldiers, and thirteen mule drivers – a total of twenty-five men and thirteen mules. They carried six miles of cable and the equipment for two telegraph offices and were intended quickly to lay a field cable that kept up with the movement of the army HQ, while the Indian Telegraph Department followed up more slowly with a poled line, the cable then being picked up and the procedure repeated. The Afghan War imminent, it was realised that this was inadequate and in September 1878 two companies of Bengal Sappers were hurriedly equipped as telegraph line construction units - more about them shortly.

The Field Telegraph Train of the Royal Bengal Engineers at Roorkee.

Visual signalling was the province of regimental signallers, but in Britain at that time it had its problems. These were mainly due, as was explained in the previous chapter, to the limitations of British topography and unreliable sunshine, and a consequent lack of real commitment to regimental signalling. With a few enthusiasts leading the way, the British infantry and cavalry regiments serving in the Indian Army had quickly discovered the capability of the recently introduced heliograph, and signal tactics evolved rapidly. (At that time, due to post-Mutiny sensitivity, signal sections were confined to British regiments; Indian and Gurkha units established them later.) In India they were well ahead of their British-based counterparts, helped also by the fact that officers there generally took a greater interest and commanders included it in their operational planning. The contrasting signalling capability between the British and Indian armies was brought home by the concurrent Zulu war of 1879, when to begin with the British Army's signalling performance was abysmal (to be described in the next chapter).

Mance's original design of the heliograph had been modified by the Bengal Engineers at Roorkee, as described in the previous chapter, mainly to produce a more robust instrument better suited to military use, and by 1878 this and Mance's were both in use. They were used extensively and with good effect in the area south of Peshawar during the Jowaki Afridi expedition of November 1877 to January 1878. The unruly Afridi tribe was widespread, but the lawlessness of the clan around the area of Jowaki had incurred the government's displeasure. Apart from its primary punitive purpose, the expedition happened to be a timely signalling exercise for the forthcoming Afghan war.

Captain A S Wynne operating a heliograph during the Jowaki expedition. The artist cannot have understood his subject - the mirror, angled downwards, is not going to reflect the sun.

Cause of the 2nd Afghan War

What caused the war? Afghanistan, apart from its own incessant internal feuding, was in 1878 in a difficult position. It sat between an expanding Russian empire pushing aggressively southwards, and the tribally complicated North West Frontier of Britain's Indian empire (now north west Pakistan), those forces compounded further by religion. The British had maintained an uneasy but not unfriendly relationship with Afghanistan. The Russians, expanding their empire southwards, and with eyes on the country, gained the sympathy of the Afghan leader, Amir Sher Ali. They sent a mission to Kabul, and installed a Resident. The British, to maintain the *status quo* in the face of this perceived diplomatic rebuff, wished to do likewise. A British mission, headed up by General Sir Neville Chamberlain (perhaps better known to posterity as the inventor of the game of snooker), was sent to negotiate with Sher Ali but, on his orders on 21 September 1878, it was turned back at the Khyber Pass, the entry to Afghanistan from India. Interestingly, a member of this mission was Major Oliver St John who, it will be recalled, had as a Lieutenant been involved in the construction of the Indo-European telegraph line in Persia and in 1867 had been the Director of Telegraphs on the Abyssinian Expedition. Still an army officer, he had changed career direction and was now in the political service - a little more about him shortly.

After consultation with London the Viceroy of India, Lord Lytton, then sent an ultimatum to Sher Ali, demanding a favourable reply by 20 November 1878. Lytton had become Viceroy in 1876. A novice to India, he was seen by many as the wrong man for the job and a compliant stooge of Disraeli's government pursuing its much-criticised 'forward policy', and is generally considered to have mishandled the relationship with Afghanistan. The ultimatum was unnecessarily high-handed - an imperious decision reached before a full understanding of the situation, patience, and diplomacy had run their course. Sher Ali was in a difficult position; although not unfriendly to Britain, he was now being rushed into a war. Also, it upset many of the volatile North West Frontier tribal sensitivities which for some years had been carefully and diplomatically nurtured by local British administrators who refused to be drawn into Afghanistan's internal affairs – 'masterly inactivity', as they called it. But no reply came from Sher Ali by the deadline on 20 November, so war it was. In a misguided attempt at 'regime change', an invasion of Afghanistan was ordered.

THE 2ND AFGHAN WAR, 1878-80

The War – Phase 1

There were to be two distinct phases of the war, for reasons that will unfold. Initially, three columns were to advance into Afghanistan, one from Peshawar through the Khyber Pass, another from Kohat through the Kurram Valley, both of these heading for Kabul (see map below); the third column, much further to the south, and with a long approach route, was to advance through the Bolan Pass and Quetta to Kandahar. The Khyber route was the shortest and the only one to be open at all seasons; the entrance to it, near Jamrud, lay within a short distance of the supply base at Peshawar. The Kurram route involved crossing the Shutargardan Pass, 11,500 feet high, closed by snow in winter. The long Bolan route to Kandahar was a winter route only, because marching and resupply in summer would cross the burning Sind desert, impossible for men and animals. Most of the action took place in the north so the narrative will confine itself largely to the Khyber and Kurram columns.

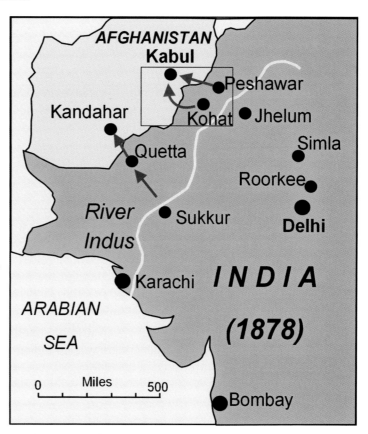

One might ask why there were three routes? Operational capability during the entire war was heavily constrained by the supply and transport arrangements. Railheads were then a long way off, and beyond them everything was moved by animal transport along a small number of long and increasingly rugged routes of limited capacity, many not suitable for wheels. It was for these reasons that separate routes had been chosen. But the transport was not well organised and animals died in their thousands due to both hot and cold extremes of climate and poor husbandry; the source of essential supplies was precarious; and the commissariat arrangements were poor. If the rate of movement seems slow, those were the reasons.

The Peshawar Force, as the Khyber column was called, some 16,000 strong, was under the command of Lieutenant General Sir Samuel Browne VC KCSI CB – best remembered today for his eponymous sword belt. They left Jamrud, then on the border, on 21 November. Without going into any detail of the fighting operations, they advanced through the Khyber Pass and

The War - Phase 1. Outline Plan for Three Columns.

reached the squalid little town of Jalalabad on 20 December 1878 (see map on next page). The country was well suited to the heliograph, signalling stations being deployed as operations demanded, although a turning movement to capture the fort at Ali Masjid suffered from lack of communication between the two elements and became a mess. Generally, heliograph communications were maintained within the force and back to Jamrud and Peshawar during the advance. They consolidated in Jalalabad while the sluggish resupply caught up.

One of the recently formed telegraph companies of the Bengal Engineers, under Lieutenant W F H Stafford RE, was assigned to the Khyber Column. (The other company, under Lieutenant P Haslett RE, went with the Kandahar Force.) The telegraph line was initially only extended from Peshawar to Jamrud, that section having to be constructed for the civil department by the army engineers. The Peshawar Force now established at Jalalabad, telegraph was needed there, but the civil department again found difficulty in delivering, so military telegraph lines were constructed under the orders of the CRE, Colonel Maunsell. The army engineers went to Jalalabad and constructed back towards Jamrud, where after forty miles they eventually met the civilian department's line at Bosawul, near Dakka at the western end of the Khyber Pass and fifty-three miles from Peshawar. The army operated their section of line for some six weeks under this

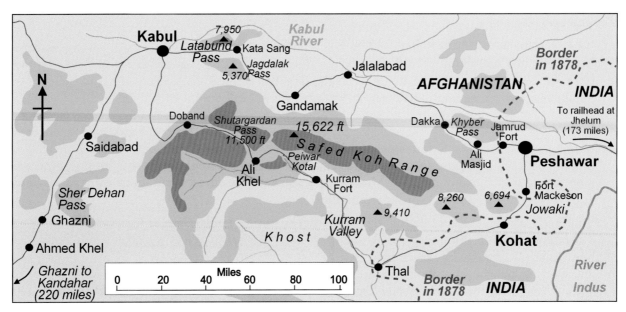

The routes to Kabul through the Khyber Pass and the Kurram Valley.

disjointed arrangement, until eventually the civilian department, under the charge of Mr S P V Luke, a Superintendent of the Indian Telegraph Department, got its line to Jalalabad and took over its operation. The army recovered their line ready for further use. [1]

The telegraph line along this route was, to begin with, not much damaged by the local tribes who were paid handsomely for their neutrality, but the different clans were variously fractious – perhaps by nature, perhaps according to how the money from the political officer's slush fund had been disbursed by the *maliks* (headmen) - and no arrangement could ever be relied on, for duplicity was endemic. As the campaign developed, the line of communication through the Khyber Pass required constant defence and punitive measures against offenders, absorbing an entire Division to protect it. The telegraph line was frequently attacked and long sections of wire removed.

The force stayed at Jalalabad until 12 April 1879. Intelligence, obtained by officers of the political service through paid agents and informers (there was no Indian army intelligence organisation), reached them that Sher Ali had fled northwards, that his ineffective son Yakub Khan had been released from prison and after meteoric promotion was now on the throne, and that the country between Jalalabad and Kabul was, in the usual local style, in a state of anarchy.

The advance to Kabul continued, and thirty-five miles later they reached Gandamak. The telegraph line was extended there by the army, who again managed to keep up with the advancing troops, and they operated the line for about two weeks before the civil line managed to catch up and open their office. The military line was then recovered again, ready for use in the anticipated advance. Altogether the army signallers passed some 5,000 messages over a nine week period. Unfortunately civil and military telegraph procedures were different, and messages had to be manually delivered and retransmitted from separate offices at the changeover point. Together with the duplication of effort in constructing a second line alongside the other and then recovering the first, it proved a frustrating and inefficient method of working. When possible, the two lines were left to work simultaneously, thus clearing a backlog of traffic, but in principle it was a silly arrangement. Nevertheless, soldiers and civilians from the two separate organisations co-operated well on the ground, and after a fashion it worked. A heliograph chain also backed up the vulnerable telegraph line through the Khyber Pass back to Peshawar and thence to the Indian civil telegraph system.

Afridi tribesmen in the Khyber Pass.

Turning to the Kurram column, they had also started their advance on 21 November, from Thal. A smaller force than the Peshawar Field Force, they were commanded by Major General Sir Frederick Roberts VC CB. Their Superintendent of Signalling was Captain A S Wynne of the 51st (East Yorkshire) Regiment (later the King's Own Yorkshire Light Infantry). The experienced Arthur Wynne had been Superintendent of Signalling in the Jowaki Afridi expedition some nine months earlier. He was, as it happened, the brother of Captain Warren Wynne RE who, just a few months later, was going to improvise signalling from the besieged garrison at Eshowe during the concurrent Zulu War in South Africa, to be described in the next chapter.

The Kurram Valley Force's first engagement was on 2 December at the Peiwar Kotal (a *kotal* is the saddle over a range of hills, used as a crossing point), when the heliograph was used with an intermediate station to relay messages between the main column and a turning column as, after a night march that got delayed, they performed a successful right flanking movement to surprise and defeat the well-entrenched Afghan defenders. The signalling enabled the two columns to synchronise their attack, which might otherwise have failed. Having gained possession of the *kotal*, they were able to maintain heliograph signalling back down the valley to Kurram and their line of communication. Further advance was delayed by transport problems.

In addition there were problems with the hostile tribes to the south-west of the Kurram valley, and much time was spent trying to bring them under control. Supporting these operations, and also survey work that was being carried out, proved a busy time for the signallers, during which heliograph communications were provided over an extensive area in the Khost region south and west of Thal. The telegraph line back to Kohat was frequently cut. An advance depot at Ali Khel (a *khel* is a clan area with an associated name) was established, in preparation for the spring advance. Amongst the stores taken there was 120 miles of telegraph wire. On 5 April the Force HQ moved there. At this time peace moves started at Gandamak (see below) so the planned advance was suspended, but Roberts and his staff conducted reconnaissances through the Shutargardan Pass – something that was to prove most useful later.

A notable use of the heliograph occurred in this period. Captain Straton, a signalling officer with the Kurram force, ascended the Safed Koh range of hills which separated the two columns (see map above). He reached the top of the Agam Pass but found that Jalalabad was obscured in a dust storm. He proceeded to the Karaini Peak, over 15,000 ft high, searched with his telescope, found the camp at Gandamak some thirty miles distant, and aligned his heliograph – a Mance model with a 3-inch mirror. His flashing attracted their attention, and within fifteen minutes Generals Roberts and Browne, commanding the two columns, were in communication through Straton's relay station.

Laying the telegraph line in the Kurram Valley. It eventually went as far as Ali Khel. The civilian on the left is Mr Josephs, a Superintendent of the Indian Telegraph Department. The kilted soldier is probably from the 92nd Highlanders. At the collapsible table is the telegraphist with his instruments. On the right is the cable drum being carried by two Indians. (Photo: Nat Army Museum.)

The story somewhat apocryphal, the novel flashing phenomenen now proliferating around the North West Frontier and in eastern Afghanistan had convinced the local tribesmen that the heliograph was a mystic instrument engaged in paying homage to the Sun God. As it happened that year, the snow which usually fell in November held off until February, and unusually fine winter weather prevailed. In-depth tribal analysis concluded that this was the response from on high to the urgent supplications being made, to help the *feringhis* move their supplies. They were unfortunately not duped by the telegraph, and continued to cut the line and remove lengths.

In late April 1879, back at Gandamak, the Khyber Force was approached by Yakub Khan, Sher Ali's successor, to discuss peace terms. The outcome, after much telegraphy between Gandamak and the Indian government's summer HQ at Simla, was the Treaty of Gandamak, signed on 26 May 1879. Amongst other things it included agreement to: perpetual peace and friendship; the establishment of a British Resident in Kabul; the ceding of the Khyber Pass and the Kurram Valley to India; and a permanent telegraph line to be built by the British at their expense to Kabul through the Kurram valley.

All apparently settled, the troops of the Peshawar Force returned to India, marching under terrible conditions in the summer heat, cholera rampant. It was, without exaggeration, known as the 'Death March'. As part of the withdrawal the laboriously constructed telegraph line was dismantled back to Landi Kotal. During this period the line had been cut ninety-eight times and about sixty miles of wire was stolen and never recovered. In the Kurram valley General Roberts completed his reconnaissances and in late July returned to Simla. The Kurram Force was thinned down but remained in the area because, under the Treaty, this part of Afghanistan had been ceded to India and needed policing.

The War Resumes – Phase 2

Under the terms of the Gandamak Treaty, the British Resident moved into Kabul on 26 July 1879. Six weeks later, on 3 September, he and his staff and escort were all murdered by Afghans in the Residency. 'Perpetual peace' had been short-lived. The news was brought by an informant to the political officer at Ali Khel, where the forward HQ of the Kurram Force was located and served by the newly constructed telegraph line through the Kurram valley; it was immediately telegraphed back to Simla. The war was on again, and this time with a vengeance. Major General Roberts, then at Simla, sent a short cipher telegram to his deputy left in charge of the depleted Kurram column with preliminary orders, and hastened back to his command. The dispersed troops re-assembled, reinforced by others. Speed now of the essence, the plan was for the Kurram column to advance over the Shutargardan Pass and occupy Kabul. This route, however, could not be used for supplies after November, so the Khyber route had to be reopened.

New Faces

New faces appeared at this juncture. Arthur Wynne, Roberts' signalling officer, had returned to England after the first phase of the war for health reasons, having been Mentioned in Despatches and awarded the Brevet rank of Major. In England, early the following year, he described his experiences to a meeting of The Royal United Service Institution, providing a good first-hand description of signalling in the war. [2] A new Superintendent of Signalling was appointed, Captain Edward Straton of the 22nd (Cheshire) Regiment, and he, too, was to make a great success of the job. He had come to India with his regiment in 1873, became interested in heliography, and qualified as an instructor in army signalling in early 1878. He volunteered for service in the war and joined the Kurram Field Force in April 1879. His contribution to the signalling operations was outstanding. [3]

At the same time a war correspondent, Howard Hensman, joined the column, and his subsequent book provides a number of descriptions of signalling in the resumed operation. [4] War correspondents had been forbidden by the Indian government after irresponsible reporting in the first phase of the war, but apparently neither General Roberts nor Hensman knew of this, and Hensman was permitted by Roberts to accompany his column. He was for all practical purposes 'embedded', to use the modern connotation, although nobody at that time would have recognised the term. Descriptions by Wynne, Hensman, and Roberts himself will be used to give the flavour of the confident and capable way that signalling was being used to control operations.

Captain Edward Straton.

Visual Signalling Organisation and Tactics

At this point a short description of visual signalling tactics might be interesting. There being no formation (*ie* Brigade and Division) signalling units, the regimental signalling resources were 'pooled' under a Superintendent of Signalling (as he was called in the Indian Army), who was on the staff of the appropriate HQ, although as far as possible regimental signallers were deployed to serve their own units. The formation of such an *ad hoc* body was an arrangement that caused them many mundane administrative difficulties. Nevertheless, the signallers were keen on their work, understood its importance, and saw the results that could be obtained with the heliograph. They often acted on their own initiative when, from their vantage points, they saw things happening that they thought their commander should know about, and alerted him. Deploying the signalling stations to keep up with operations demanded mobility, and it is apparent from contemporary descriptions that the infantry signallers were often mounted. Signallers from cavalry regiments accompanied all reconnaissances; they mostly used the Mance 3-inch heliograph, a light and portable instrument.

Everybody, not just the signallers, was taught to be observant and look for the flash of the heliograph, and draw attention to it – Indian sentries often spotted it. There were to be many cases where troops on the move were known to be in an area, but their exact position and situation were not known. If visual survey by telescope had revealed nothing, the drill for anybody wishing to get in contact was to 'sweep' the generally known area, a sector at a time, and flash the narrow beam of the heliograph with a series of 'dots' (Morse code 'E' – 'where are you?'). Hopefully there would come a response from somewhere, and the stations would align and identify themselves before exchanging messages. The range was usually anything up to about twenty miles, but there were many occasions when greater distances were worked. Signalling was usually best in the clear morning atmosphere, before dust and haze developed. The heliograph beam could penetrate a certain amount of haze, but if the signaller could not see the distant station it was difficult to align the mirror on it.

Captain Straton (centre, in khaki) and signallers of the 72nd Regiment with their heliographs and telescopes. Also shown are two signalling lamps, but their use is not mentioned in contemporary descriptions.

The Capture of Kabul

The advance to Kabul by Roberts' troops, now known as the Kabul Field Force, began on 11 September 1879. The Shutargardan Pass was occupied, and heliograph communications established to Ali Khel. Skirmishing continued along the route, the heliograph frequently used. A major action was fought at Charasiab. As they launched the attack on Kabul on 6 October, the war correspondent Hensman described the scene:

> Not the least important arrangement of the day was that of signalling. Captain Straton had parties of men with General Baker and Major White, and a third batch of signallers was sent to watch the Chardeh

Valley, and the movements of large bodies of tribesman who lined the crests of the range overlooking the camp from the west. Heliograms were exchanged between these points and the headquarters camp, and General Roberts was kept fully informed of all that was happening in these directions. This focussing of all information upon a common centre enabled the General to make his dispositions with accuracy and effect; without the signallers, dangerous delays might have occurred. The heliographing was so thoroughly well done that Sir F Roberts complimented Captain Straton on his arrangements. The only drawback was a succession of small sand storms, which swept across the camp and blotted out everything for the time being.

Readers today might see in this description the embryo of what was nearly a century later to become known as C3 – command, control, and communications on the battlefield.

Kabul was occupied, and the Afghan army melted away, some to fight again another day. In Kabul, proclamations were made, summary trials held, and justice meted out swiftly on the gallows erected in front of the Residency - uninhibited by any of the legal impedimenta that nowadays prevails.

The heliograph chain from Kabul to Peshawar had been established by 19 November. Then the telegraph line, rebuilt from Landi Kotal along its previous route through Jalalabad and Gandamak, was brought up to Kabul, progress being delayed by the usual problem of procuring and transporting poles. There was no source of poles in the barren country between Peshawar and Dakka, but from Dakka onwards good deodar or pine saplings were procurable in villages under the Safed Koh and at places along the Kabul river. (5) Mr Josephs, the Superintendent of the Indian Telegraph Department who had been assigned to the Kurram column and built the telegraph line to Ali Khel, had accompanied General Roberts to Kabul and from there he started to construct the line back towards the working party approaching from Landi Kotal. They met at Jagdalak on 19 November and telegraph communication between Kabul and India was established. General Roberts mentions that:

> Towards the end of November, Mr Luke, the officer in charge of the telegraph department, who had done admirable work during the campaign, reported that communication was established with India. As, however, cutting the telegraph wires was a favourite amusement of the tribesmen, a heliograph was arranged ... *etc*.

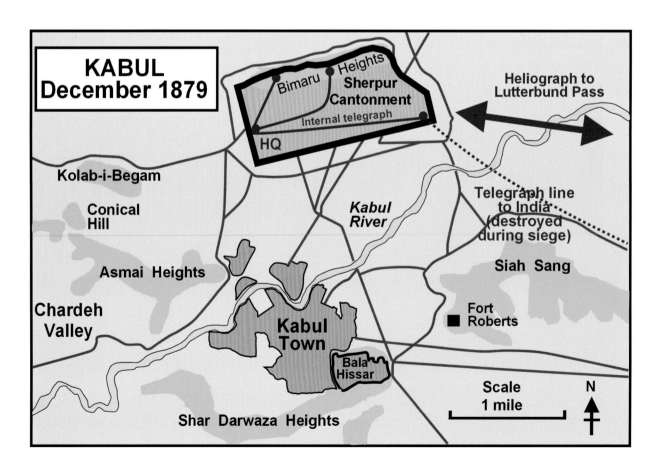

Signalling in Sherpur

Well aware of their vulnerable position, should the Afghan army regroup, the recently promoted Lieutenant General Roberts decided in early November to fortify Sherpur, a large cantonment originally built for the Afghan army but big enough for his entire force. Situated about two miles north of the town of Kabul, with a perimeter some 2,800 x 1,000 yards to defend, its fortifications were strengthened. Within the cantonment an internal communication system of six local telegraph offices and visual signalling was set up. (The telephone, recently invented, would have been ideal, but its use is not mentioned.) Informants duly warned in mid-December of an impending attack, as tribes from a wide area assembled, and the troops, 7,000 in all, concentrated in their well-prepared sanctuary, snow lying on the ground.

One of the reasons for selecting Sherpur was that from there it was possible to communicate by heliograph with the signal station above the Lutterbund Pass – at over 6,000 feet high, quite a chilly place in December, the Sun God not repeating his beneficence of the previous year. From there, the next signalling station was on the heights above Jalalabad, seventy-five miles from Kabul, and thence through two other intermediate heliograph stations, or telegraph depending on the state of the line, down the Khyber Pass route and back to India. The 180 miles from Sherpur to the heliograph station on the church tower at Peshawar could be achieved with four intermediate heliograph stations, all of them having to be heavily protected. Thus, while the siege lasted, even though the telegraph line between Kabul and Pezwan, a distance of fifty miles, had been entirely destroyed on 15 December, Roberts was in communication with India (although cloud sometimes restricted signalling at the high Lutterbund relay station).

The expected and fairly suicidal attack by the Afghans was launched on 23 December 1879; it was easily repulsed from the well-organised defences, with few British and Indian casualties but over 1,000 Afghans killed. The Afghan remnants evaporated into the surrounding countryside, disguising themselves as peaceful citizens, although many were identified through informants. The troops moved out of Sherpur soon after the attack and resumed occupation of Kabul. Visual signalling stations were established in the surrounding hills (see sketch below) at Sherpur, Bimaru Ridge, Bala Hissar (meaning Upper Fort), Siah Sang, and Sher Darwaza, giving excellent coverage for miles around.

After the siege was over and the enemy had dispersed the reconstruction of the line began, and telegraph communication with India was re-established on 8 January 1880. It remained a vulnerable target, and during the time of the force in Kabul it was cut about fifty-seven times and over fifty miles of line was stolen. The heliograph remained an essential back up.

After the siege, while the surrounding area was being brought under control during January 1880, another interesting signalling incident took place. Captain Straton was visiting the Jalalabad area and was at the

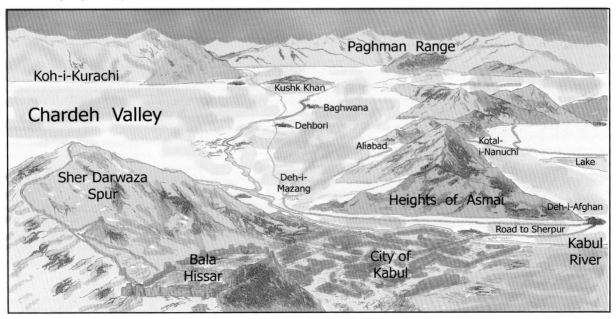

A panorama of Kabul 1879, adapted from an original pencil sketch by Lieutenant C Manners Smith, staff officer to Major General Roberts.

signal station at Ali Boghan, some five miles to the south-east of Jalalabad, when he saw that an enemy force of Mohmands, another hostile tribe, had crossed the Kabul River. He immediately signalled this intelligence to Jalalabad, and a brigade was despatched to intercept. Communications were maintained between Straton at his vantage point, the HQ at Jalalabad, and the intercepting force. The next day Straton saw the enemy again, some 1,500 strong, moving to a position which would cut off the intercepting troops. This new information was quickly signalled, and guns despatched from Jalalabad which shelled the enemy and dispersed them.

The Political Mess

The political situation, meanwhile, had degenerated into a mess; the future direction was vacuous. The Afghan population was hostile, the new Afghan leadership question was unresolved, the army was not big enough to control the country, its resupply was problematic, there were no suitable replacements for the battle-weary troops, and the cost of it all was beyond the means of the Indian government to sustain - all of which might sound tediously familiar. What had been achieved, apart from the withdrawal of the Russian Resident, was highly questionable. Lytton's policy had failed. How these problems were addressed need not be described here. Suffice to say that putting the best possible face on an early withdrawal was the order of the day.

The Kandahar Column

Before that, however, a brief description should be given of the Kandahar column in the south. Their experience had been less exacting. During the first phase of the war they had advanced from Quetta. The Superintendent of Telegraphs was Lieutenant George Savage RE, assisted by Lieutenant Haslett's recently formed telegraph company, and the Superintendent of Signalling was Lieutenant John Dickie RE, all Royal Bengal Engineers. A telegraph line was constructed to Kandahar, but not beyond. The local tribes were not as hostile as in the northern sector and this line, although occasionally cut, gave a reasonably reliable service back to India. Any movement beyond Kandahar depended on the heliograph for communications.

After the Treaty of Gandamak the troops remained in Kandahar because in midsummer it was too hot to move them back across the Sind desert to India. They were reinforced when the war resumed, but no serious action took place. As part of the convoluted withdrawal process subsequently decided, a column under Lt Gen Sir Donald Stewart left Kandahar and headed for Kabul through Ghazni (see map), eventually reaching Kabul in late April 1880 after some quite heavy but successful fighting at Ahmed Khel. They came into contact with forward elements of Roberts' Kabul force on 22 April by means of the heliograph at over thirty miles range from the top of the Sher Dehan Pass just north-east of Ghazni to Saidabad, and sent news of their successful battle. On 5 May Stewart took over command of the Kabul garrison from Roberts.

On the same day they heard in Kabul that a new government under Mr Gladstone was in power in London, the British electorate being disillusioned with Disraeli's questionable ventures in both Zululand and Afghanistan, and that Lord Lytton, the Viceroy of India, had 'resigned'.

In late July 1880 a disaster occurred at Maiwand, some twenty miles west of Kandahar, and the army suffered heavy casualties at the hands of an Afghan force, the detail again not relevant here. The garrison remaining in Kandahar became besieged. A relief column of selected troops, stripped of all non-essentials, was despatched from Kabul under the command of General Roberts in the celebrated 330 mile march from Kabul to Kandahar, completed in the remarkable time of twenty-three days, capturing the public imagination at the time.

Captain Edward Straton, the Superintendent of Signalling, of course went with them. Heliograph communications were used throughout the march to protect and keep control of the column. (Failure to do this had been a prime cause of Stewart being surprised and coming under attack at Ahmed Khel.) As they advanced, intelligence was being gained by political officers scouting ahead, gaining information, and reporting back. Lieutenant Colonel Chapman, the column's chief of staff, described the signalling in support of the intelligence gathering:

> It may be here noticed that the great perfection to which the practice of army signalling had been brought made the heliograph a very important aid in the communicating of intelligence. ..... I refer to this subject under the heading of 'intelligence' as the value of this agency can only be fully appreciated when it is applied in the field, and particularly during operations in mountainous country, to convey rapidly information of importance. [6]

As the relief force approached Kandahar, the garrison there on the lookout but with no precise information, the signallers with the advanced cavalry units made contact from Robat on 27 August at a range of about twenty miles. Hensman, who was with them, described it:

> Kandahar was said to lie, in our line of vision, directly beneath this hill. Captain Straton had brought with him some of his mounted signallers, and at half-past eleven a light was directed towards Kandahar. We could not see the city, even with our telescopes, as a thick haze hung over the country about it, and for quarter of an hour no answer was given. The first signal station was on a low hillock to the left of the road, but Captain Straton took another instrument to the slope of a rocky ridge on the right, whence he could also communicate with the main body of our troops halted for the day at Khel-i-Akhund. He had scarcely left the road than Sergeant Anderson [72nd Highlanders], with the first heliograph, saw a faint flash at Kandahar. It was so weak a glimmer that nothing could be made out, but in a few minutes we read a message: 'Who are you?' The answer given was: 'General Gough and two regiments of cavalry.' Then Captain Straton's light was seen by the signallers in Kandahar who, puzzled by two flashes, asked: 'Where are you?' After this our first station was closed, and the signallers with Captain Straton began sending messages ... .'

Communications with Kandahar were established and details of the situation exchanged. Lieutenant Colonel Oliver St John (mentioned earlier), the political officer, managed to escape from Kandahar and came with a small party to meet them and report the situation. The plan for relieving Kandahar was prepared, and on 1 September the attack was launched. Straton was in on the planning, and arranged for the deployment of his signalling stations.

Captain Straton Killed

Near Kandahar, on 1 September 1880, just as the last battle of the war was being won, Edward Straton was killed. Hensman again described it:

> It was, of course, all-important to stop this shelling of the [Baba Wali] Kotal, now virtually in our hands, and the easiest way was to send a party of signallers up the hillside to the right of the Kotal, whence the news of our rapid success could be flashed down below. Captain Straton with two mounted signallers was with the brigade, and he was ordered to establish a station on the ridge above. ..... He had not gone more than 50 or 60 yards from Generals Ross and Macpherson when a shot was heard, and Captain Straton fell from his horse. A dark figure was then seen to rise from a dip in the ground, fix a bayonet on his rifle, and rush forward. The two signallers, men of the 72nd [later Seaforth] Highlanders, had dismounted by this time, and they fired at 40 yards distance, bringing the Afghan down. The man was bayonetted as he tried to rise. It was discovered that he had already been severely wounded, and could not have hoped to escape. The bullet from his rifle had passed through Straton's heart. The decease of Captain Straton is a great loss to the force; the perfect way in which he controlled the signalling was universally recognised. He never spared himself when hard work had to be done, and the soldiers under him shared his enthusiasm. General Roberts always relied implicitly on him, both on the march and in action, for he knew that if it were possible for heliographing to be done, Captain Straton would have his men in position and his instruments at work.

Panorama of Kandahar, showing the Baba Wali Kotal.

Conclusion

And so, sadly, ends the story of signalling in the 2nd Afghan War. At the start of the chapter it was said that tactical signalling in the war attained a new level of competence, due to the excellent use made of the recently invented heliograph. Let General Roberts, later Commander-in-Chief, India, and ultimately Field Marshal Earl Roberts of Kandahar VC KG KP GCB GCSI GCIE, have the last word, quoted from one of his despatches:

> In Captain Straton Her Majesty's Service has lost a most accomplished intelligent officer, under whose management army signalling, as applied to field service, reached a pitch of perfection probably never before attained.

The Afghan War Medal

The Kandahar Star.

The Star was awarded to all those who took part in the march from Kabul to Kandahar.

It was made from guns captured at the battle of Kandahar, 1 September 1880.

Endnotes:

1. The situation is explained in an article *Memoranda on the Military Telegraph Train in India*, by Maj Gen F R Maunsell published in the Royal Engineers Journal, 1 March 1881. As Col Maunsell he was CRE of the Khyber Column during the war, and his exasperation at the arrangement comes over clearly.
2. *Heliography and Army Signalling Generally* by Maj A S Wynne, 51st LI. RUSI Journal, Vol XXIV, 15 March 1880. After a distinguished career he later became Maj Gen Sir A S Wynne GCB.
3. A more detailed biography of Capt Edward Straton will be found in *The Afghan Campaign of 1878-1880* by S H Shadbolt.
4. Hensman, Howard. *The Afghan War of 1879-80*. Pub London, W H Allen & Co, 1882. Hensman was Special Correspondent of the *Pioneer* (Allahabad), and *The Daily News* (London). He joined Gen Roberts' force when the war resumed in September 1879, and covers the period from the capture of Kabul to the end of the war, giving many first-hand descriptions. Now a rare book.
5. In his address to the Society of Telegraph Engineers *On the Construction and Working of a Military Field Telegraph (based upon experience gained during the campaign in Afghanistan in 1878-80)* Mr Luke, the civilian of the Indian Telegraph Department responsible for the task, enlarged on construction detail of the line, sources and types of pole, insulators, wire, terminal equipment, batteries, logistics, *etc.* Journal of the Society of Telegraph Engineers, 26 May 1881, pp 232-270.
6. RUSI Journal, Vol XXV, 9 Mar 1881. A description of the march from Kabul to Kandahar by Lt Col E F Chapman, Chief of Staff of the Kabul-Kandahar Field Force.

Chapter 11

The Anglo-Zulu War, 1879

Introduction

Overglamourised today, the six-month Zulu war of 1879 was an inglorious conflict, both politically and militarily. Its causes were controversial, and much of its execution was inept. In the early stages of the war the signalling was lamentable. Despite the advances made during the previous decade, and in contrast to the progress in India, the war started with no effective army signalling in place. When, after early defeats, the Telegraph Troop was called for, and sent along with other reinforcements from Britain, there were still problems. Things did not go as they should have done, and the key word that will unfold in this story is *improvisation*.

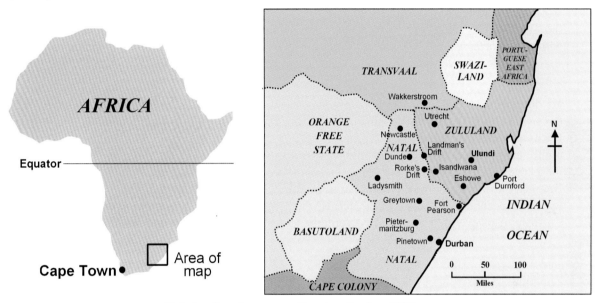

Zululand and surrounding territories, 1879.

Causes of the War

After a period of black tribal warfare in the early part of the 19th century the Zulus, under their redoubtable leader, Shaka, had established their tribal territory. Its boundaries, for the most part, became well-defined. To the south, the Tugela and Buffalo rivers were the border with Natal which had been a British colony since 1843, having been wrested from the Boers who had settled there five years previously. To the north, the Limpopo river was the border with the Swazis. To the west was the boundary with the Boer-led Transvaal; a border had been defined but it had became a disputed area due, characteristically, to aggressive Boer attempts at 'land-grabbing'.

In 1872, after a relatively settled period under the rule of King Mpande, King Cetshwayo had come to the Zulu throne, and tensions with neighbouring territories increased. There were a number of reasons. The ongoing boundary dispute with the Transvaal Boers, and years of prevarication in resolving it, led Cetshwayo to strengthen his army and threaten military action to protect what he rightly saw as his territory. However, fighting between Zulus and Swazis – something of a tribal pastime to display manhood - and other internal affairs, caused concern about Cetshwayo's military intentions. Generally, there was turmoil. It was perceived that lawlessness in Zululand was again getting out of hand, some of it spilling over into Natal. The Governor of Natal asked for an explanation of several incidents, and Cetshwayo responded to the effect that this was interference in the matters of an independent state and the Governor was to mind his own business.

Political decisions now affected events. Following the discovery of diamonds in Kimberley in 1868, the British government's policy towards South Africa - a vacillating dither blowing hot and cold as governments and personalities came and went - changed again (and not for the last time). What had been an unwelcome drain on imperial resources now, with diamonds on the scene, offered a decent financial return. The best way of administering it, the Colonial Office in London decided, was to make South Africa a confederation, uniting all the various black and white states - under British control of course! It was a grandiose dream that ignored the reality of a complex situation, and the plan was to lead directly to both the Zulu War in 1879 and the First Anglo-Boer War the following year.

As the first step in the confederation process a local Natal administrator, Theophilus Shepstone, was called to London, briefed by the Colonial Secretary (Lord Carnarvon), knighted, and sent back with instructions to annex the poorly governed Boer state of the Transvaal. This he did in April 1877. At the same time Sir Bartle Frere, who had proved himself to be an able administrator in India, but like most British politicians understood little of the causes of the years of turbulence in South Africa, was appointed High Commissioner for South Africa and Governor of the Cape. He arrived in Cape Town in March 1877 charged with the task of implementing confederation.

In 1878 a long-delayed Boundary Commission sat, to decide on boundary disputes between the Zulus and the Transvaal Boers. The Commission found in favour of the Zulus - not what Frere wanted to happen, and an embarrassment to his plans, so the findings were not published in the hope that the Zulus would soon commit some further misdemeanour which could be used as the excuse to invade and take over their territory.

He did not have to wait long. Two Zulu wives had been unfaithful to their husbands; their indiscretions discovered, they fled to Natal. Zulus crossed into Natal, took them back to Zululand, and murdered them. It was all small beer but caused disproportionate excitement in Natal. Frere seized on the incident and sent highly exaggerated descriptions to London. On 11 December 1878 the Zulus were called to a meeting at the Lower Tugela Drift, on the border between Natal and Zululand, ostensibly to hear the report of the Boundary Commission. That done, an ultimatum about future arrangements to be imposed on Zululand was unexpectedly delivered. The terms, such as disbanding his army, were quite unacceptable to Cetshwayo, as everybody knew they would be. The Zulus were given one month to comply, or it would be war.

The ultimatum expired on 11 January 1879 so war it was, and on 12 January the British forces invaded Zululand. It was a shabby and disreputable affair, and retribution was soon to follow.

The Natal Colonial Telegraph System

At the start of the war the Natal colonial telegraph system was the only telegraph that the army could use. It was very basic, lagging the development of the Cape Colony telegraph system by a long way, due to the later establishment of Natal as a British colony, lack of money, and generally inferior administration and infrastructure.

Conversely, development of the Cape Colony's telegraph system had progressed well, covering a large area, and by 1877 it stretched eastwards to reach Kokstad, about 900 miles from Cape Town. In contrast, all that Natal had at this time was the original seventy mile route, opened in June 1864 between the port of Durban and the Natal capital, Pietermaritzburg.

Indecisive discussions had taken place for some years to link up with the Cape Colony telegraph system for commercial, political, and military reasons. The Cape Government offered terms to build the route from Kokstad on to Pietermaritzburg, 100 miles, but there was continued vacillation by the Natal government. Late in 1877, the High Commissioner, Sir Bartle Frere, based in Cape Town, and no doubt with an eye on impending strategic requirements, decided it was time to intervene. He and the Lieutenant-Governor of Natal agreed that the Cape's offer to build the route from Kokstad to Pietermaritzburg would be accepted.

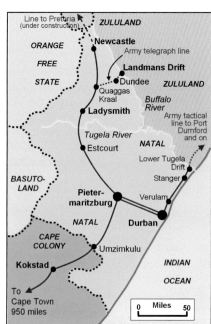

The Natal telegraph system, 1879.

THE ANGLO-ZULU WAR, 1879

That was completed in April 1878, and so the telegraph service between Cape Town and Pietermaritzburg was at last established. Also in April 1878 a second telegraph line was built between Durban and Pietermaritzburg, this one following a slightly different route, beside the new railway line that was under construction. In late 1878, as a result of further railway development, another new telegraph line was built from Durban to Verulam, about twenty miles north of Durban, the work being done by the Natal Public Works Department. The Natal Telegraph Department, although responsible for the colony's telegraph from the time the Natal government took it over from a private company in 1873, still did not have its own telegraph construction party, nor for that matter any great expertise in running a telegraph system, and was assisted by the Cape Telegraph Department.

Meanwhile, for a quite separate reason, a new telegraph line to the Transvaal was required, as Frere had anticipated. The Transvaal had been annexed by Shepstone on 12 April 1877 as the first phase of Lord Carnarvon's ill-conceived plan for the confederation of South African states, but politically things there were not going well and better communications were needed between the High Commissioner in Cape Town and the Transvaal capital, Pretoria. To achieve this it was decided to extend the telegraph from Pietermaritzburg. In August 1878 a new telegraph line was authorised to be built from Natal to the Transvaal, running from Pietermaritzburg *via* Estcourt, Ladysmith, and Newcastle, to its eventual destination, Pretoria, a distance of nearly 400 miles. This was all by cross-country route as the railway had not then even reached Pietermaritzburg. Work was due to begin in early January 1879, but at that point the war was about to start and the construction party was diverted to another task, the extension of the telegraph line from Verulam towards Zululand, before returning to continue the Pretoria line.

A Close Run Thing

With the Zulu war looming, military planning began to influence matters. Major John North Crealock, Military Secretary to the Army commander, Lord Chelmsford, was based in Pietermaritzburg, the Natal capital, and letters between him and the government authorities, beginning in early November 1878, show how Chelmsford was anxious to extend the newly-built Verulam telegraph line on to Stanger (now called KwaDukuza in modern KwaZulu Natal), nearer to the Zululand border at the Tugela river, and that it was agreed that this would be done, the cost to be paid 'from the military chest'. Then a further request was made by Crealock on 28 November, stating that 'the Lieut.-General is very anxious that an extension of the proposed line of telegraph on the coast should be made from Stanger to the Lower Tugela Drift, and he trusts that H.E. the Lieut-Governor will sanction this arrangement'.

While time was rapidly running out and nobody questioned the need, much space in the correspondence between Crealock, the Colonial Secretary, the Colonial Engineer, and others was taken up by the question of cost and who would pay. The extension was eventually sanctioned on 10 December as an 'Imperial charge, pending adjustment when affairs in the Colony become settled.' The cost for the section from Stanger to the Lower Tugela Drift was £1,260, calculated at the rate of 14 miles at £90 per mile. What a time-wasting commotion for fourteen miles of telegraph line! [1]

The construction party, who were preparing to start work on the long telegraph line to Pretoria, were now diverted to this new requirement. With deadlines imminent (the ultimatum was delivered on 11 December 1878, and the Zulu war started on 11 January 1879), the extension to the Lower Tugela Drift, which was to become the base for one of the columns advancing into Zululand, became so rushed that, to speed matters up, the construction party only erected every third telegraph pole. Once the line was up and working, they went back and erected the missing poles. The other lines shown on the telegraph map, to Pretoria and army lines into Zululand, were not built in January 1879 when the war started, and will be described later.

It was all a close run thing, although such a Wellingtonian turn of phrase would be replaced these days by something like 'just-in-time', implying business efficiency. It was only eight months before the war, following fifteen years of procrastination, that the telegraph line between Natal and Cape Town came into service, and it was only immediately before the war, by dint of special effort and the fortuitous availability of the Cape construction party, that the telegraph line was completed to the Lower Tugela Drift. Moreover, there was still no strategic international telegraph communication from South Africa - more about that shortly.

The War Begins

The day after the ultimatum expired on 11 January 1879 the British forces confidently invaded Zululand. They comprised British troops already in South Africa (where they had mostly been involved in a succession of Frontier wars against the Xhosa tribe in the Eastern Cape), augmented by some more sent from England (but not at this stage the Telegraph Troop), some locally enlisted white soldiers, and local native troops known as the Natal Native Contingent, the latter generally a badly-led amateur rabble. Under the command of Lieutenant-General Lord Chelmsford, they advanced in three main columns towards Ulundi (see map).

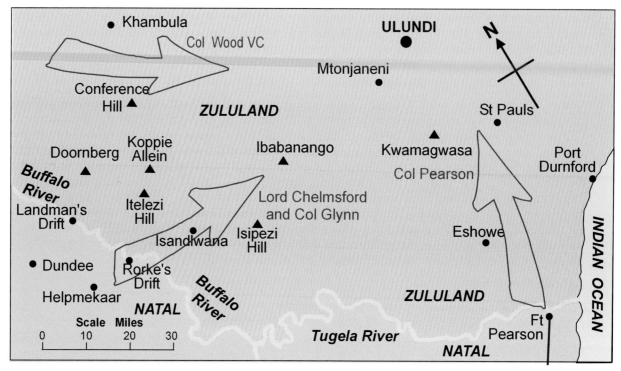

Plan for the first invasion of Zululand, January 1879.

Defeat at Isandlwana

The campaign started badly. The central column, commanded by Colonel Glyn but accompanied by Lord Chelmsford and his staff, suffered the first and best-known defeat, at Isandlwana, when on 22 January 1879 a British regiment and supporting troops, many of them Natal natives, some 1,360 in all, were surprised and killed by the Zulus when scarcely a few miles over the Buffalo river into Zululand. Only a handful escaped.

The massacre occurred when Lord Chelmsford had split the column. The leading party, under Chelmsford's command, advanced further into Zululand while the remainder at Isandlwana waited in a temporary, over-extended position to move forward as soon as a new position was found. In this vulnerable situation, and demonstrating their mobility, the Zulus pounced with complete surprise and devastating effect. The Zulu method of communication was by runner. They were fit, fast and well-trained, and their system was well organised, messages being disseminated down a chain of command rather in the manner of a chain letter. They could be concentrated, moved, or dispersed with remarkable speed, all a part of Shaka's inheritance.

One of the reasons for the defeat at Isandlwana can be attributed to poor British communications - gallopers belatedly carrying garbled messages. Nowhere in the history of the battle does one read anything about signalling, yet there are photographs taken before the invasion of Zululand which show soldiers of the 1/24th Regiment at Helpmakaar with heliographs, so they were not entirely without either the equipment or the signallers.

The day of the battle was dull and the sky was overcast, so the heliograph would have been unreliable. There was also an eclipse of the sun of two-thirds totality at about midday – causing it to become known as the Day of the Dead Moon – although by then events were out of control anyway. But neither of these circumstances actually made any difference, because the movement plan was only finalised late on the preceding day and no signalling arrangements seem to have been made to keep the two elements in communication. Signalling in the British Army did not yet form an essential part of a commander's tactical plan.

It is fruitless to speculate how matters might have been different if, after Lord Chelmsford split the column, the troops advancing into Zululand had been in communication with the those remaining in the unprepared position at Isandlwana. A heliograph message spelling out the danger, or even a flag signal, might have made all the difference. Instead, the wrong deduction was made because, viewed from a distance through a telescope, the tents in the camp at Isandlwana had not been struck as they should have been when the camp came under attack. The heroic defence of Rorke's Drift took place later the same day, but signalling was irrelevant to that epic.

Strategic Communications

Apart from the dearth of tactical communications in this early phase of the war, there was difficulty with strategic communications. At that time there was no submarine cable to South Africa. The news of the dreadful defeat at Isandlwana did not reach London until twenty days after the battle. The message was transmitted from Pietermaritzburg by the recently connected Natal and Cape colony telegraph system to Cape Town, put aboard a ship, and retransmitted by telegraph from St Vincent via Lisbon, thence by another submarine cable to Porthcurno, near Land's End, Cornwall, and finally overland to London, where it arrived early on 12 February. Britain was shocked at the news.

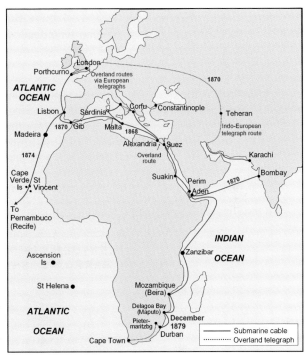

The curious may ask why there was a telegraph from St Vincent, and why was Lisbon involved? The British led the world in submarine cable manufacture and laying at that time, and had secured the contract to connect Portugal with its colony, Brazil. The route was completed in 1874, the cable from Lisbon coming ashore at Madeira and St Vincent in the Cape Verde Islands *en route* to Pernambuco (Recife). Between Cape Town and St Vincent, over half-way to England, messages still had to be carried by ship – a passage that took something like twelve days, and the mailboats usually sailed only once a week.

Principal strategic communications, 1879.

The *Official History of the Zulu War* explains how the news was sent to England:

> Lord Chelmsford, on his arrival at Pietermaritzburg [after the battle], took measures for informing the Secretary of State for War, both by telegraph and despatch, of the Isandhlwana disaster, and at the same time asking for reinforcements. South Africa, not being in telegraphic communication with Europe, the General's message had to be conveyed from Cape Town to St Vincent, the vessel being despatched on the 27th instead of the 28th, in order that the news might be delivered as speedily as possible. [2]

South Africa was not internationally connected by submarine cable until late in 1879 when the Eastern and South African Telegraph Company was formed to link the 3,900 nautical mile route from Aden to Durban. Aden was a landing for the cable route from Britain to India, laid in 1870. South Africa, in London's view a relative backwater, was a lower priority. Eventually the spur from Aden down the East African coast to Durban was laid, making landings at Zanzibar, Mozambique, and Delagoa Bay (now Maputo) on the way, the costs being shared between the governments of Great Britain, Portugal, Natal, and the Cape. Durban was already connected to Cape Town by landline. It was a long and tenuous route, but at last there was a telegraph link between South Africa and London. [3]

Better communications with London in the period leading up to the Zulu war might well have curbed the British High Commissioner, Sir Bartle Frere, and prevented the war anyway. Carnarvon's successor, the recently appointed Colonial Secretary in London, Sir Michael Hicks Beach, had in the previous November complained about Frere to the Prime Minister, Disraeli: "... I cannot really control him without a telegraph …". The absence of timely communications had been a major factor in allowing the pre-war situation to get out of control.

The Right Flank Column

Returning to the Zulu war, on 12 January 1879, the day after the ultimatum expired and the same day that the central column had entered Zululand, the Right Flank or eastern column, under the command of Colonel Charles Pearson, the 3rd Regiment ('The Buffs'), crossed the Tugela river from its base at Fort Pearson. The fort was situated on an eminence on the south bank, adjacent to the lower Tugela Drift, where the recently built colonial telegraph line now kept it in communication with Durban and Pietermaritzburg.

The ensuing advance was slow, due to bad weather, but on 22 January, the same day that, eighty-five miles away and unbeknown to them at the time, the central column was being overwhelmed at the batttle at Isandlwana, the Right Flank column was attacked by a force of about 6,000 Zulus at Inyezane. It was a fierce fight; the Zulus experienced effective British firepower and were repulsed with about four hundred dead, the British suffering only fourteen dead. By 23 January Colonel Pearson and some of his troops got as far as Eshowe, about twenty-five miles into Zululand, intended to be an advanced depot on the route to Cetshwayo's base at Ulundi.

Soon afterwards, on 27 January, a telegram was carried by runner from the telegraph head at Fort Pearson to Colonel Pearson at Eshowe; it was from Sir Bartle Frere, telling of the disaster at Isandlwana but giving no details. The next day, 28 January, another messenger brought a telegram from Lord Chelmsford confirming what Frere had said, and providing some detail. The garrison at Eshowe, although given the option of withdrawing to Natal, decided to stay, and fortified the deserted mission station that they had occupied, known as KwaMondi, just to the east of the present town. The last runner to Eshowe got through on 11 February, carrying the message from Lord Chelmsford that no reinforcements were available. Marauding Zulus prevented further communication. The siege of Eshowe had begun.

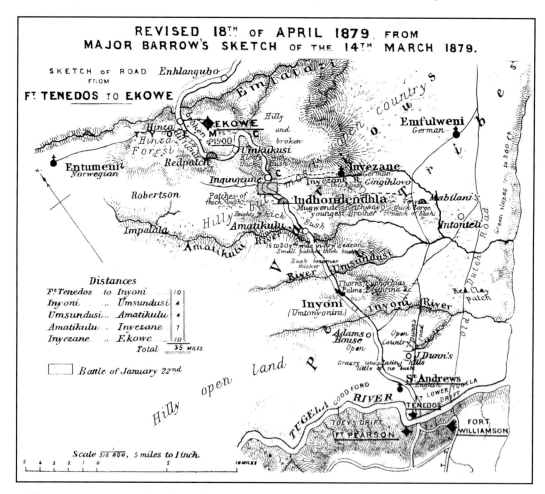

Contemporary map showing the ground between Fort Pearson and Eshowe (Ekowe). Produced by the Intelligence Branch of the Quartermaster General's Department, 1879

Within three weeks of it starting, the campaign had degenerated into an utter mess. Chelmsford sought reinforcements from England, amongst them something to bolster his communications, while in London steps were taken to replace Lord Chelmsford. These matters were going to take several months to accomplish, the sea voyage between England and Durban alone taking nearly four weeks.

Meanwhile, an immediate problem was to try and communicate between Fort Pearson and the besieged garrison in Eshowe. Although a sideline to the main campaign, the way in which this was achieved, by signalling using mirrors, makes an interesting little tale on the theme of improvisation.

Signalling to Eshowe

In anticipation of the war the 2nd Field Company Royal Engineers had been sent from England to reinforce the local troops. On arrival in Durban it was discovered that their stores had been loaded incorrectly and equipment was missing. A small detachment under one of their officers, Lieutenant Charles Haynes, stayed in Durban to attempt to resolve the problem while the main body, who were urgently needed for field engineering tasks, had moved forward to Fort Pearson under the company commander, Captain Warren Wynne. By the time Haynes reached Fort Pearson his Company commander, Wynne, had gone ahead and was now besieged in Eshowe. The resourceful young Haynes decided on his own initiative to do something about communicating with them. Lord Chelmsford and his staff, by then at Fort Pearson, were sceptical.

The besieged mission station at Eshowe, twenty-two miles away to the north and 2,000 feet higher, was in fact visible from the area near Fort Pearson. There were no trained signallers and no heliograph instruments available, either at Fort Pearson or Eshowe, so Haynes improvised with a mirror which he obtained from the nearby Smith's Hotel, a rather grandiose description of the modest refuge for travellers overlooking the lower Tugela drift which, before the war, had been their last point of succour before heading across the river into Zululand. The contemporary map shows the ground between Fort Pearson and Eshowe.

Haynes established his signalling station near the St Andrews Mission Station, shown on the map four miles north of the Tugela from Fort Pearson, and he kept an intriguing little log book recording the messages sent and received. A small, hard-back, ruled cash book measuring about four inches by seven inches, it was presumably the only thing available to him at the time. The St Andrews Mission Station signalling log book is now held in the museum of the Royal Engineers, and it remains perfectly legible. It begins with a diagram sketched by Haynes on the inside cover illustrating the set-up of the signalling station, reproduced below.

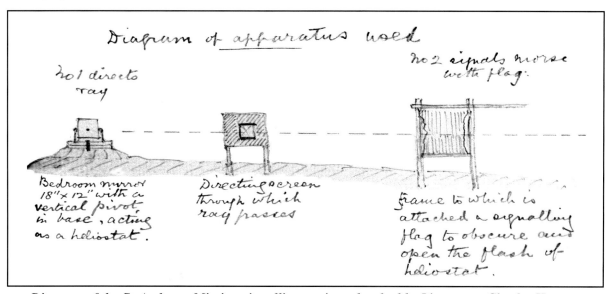

Diagram of the St Andrews Mission signalling station, sketched by Lieutenant Charles Haynes.

The centre spot of the mirror is clearly marked in the original diagram, and Haynes will have removed the silvering from the back of the mirror so that he could look through the centre of it, exactly as if using a heliograph instrument. The mirror, the aperture in the directing screen, and the distant Mission Station at Eshowe were all aligned by looking through the centre spot of the mirror, as though aiming a rifle through a rear sight and a fore sight to the target. The reflection of the sun from the mirror was directed on to the screen, and centred on the aperture. The mirror was now aligned to reflect to the distant station, and assuming reasonable visibility the ray passing through the aperture would be quite easily seen at a distance of twenty-two miles. The Morse code was flashed by interrupting the beam with the signalling flag – not quite the method of use for which the flag was intended, but effective nevertheless. As the angle of the sun continually changed due to the rotation of the earth, the mirror would have to be adjusted every few minutes so that the reflection continued to be directed on to the aperture in the directing screen. As none of them were skilled signallers, the Morse code signalling would have been a bit amateurish.

The log book lists the messages - sent messages on the left-hand pages and received messages on the right-hand pages, all with dates. The messages are simply text – no formal message procedure was used, even though message forms then existed. Answers were often commendably brief – "Yes" or "No". One of the messages signalled to Eshowe was to inform Colonel Pearson of the safe birth of a daughter.

Charles Haynes (right) was Mentioned in a Despatch by Lord Chelmsford dated 10 April 1879, published in the London Gazette of 7 May 1879. "For some little time we had been in communication with Colonel Pearson by means of flashing signals; for this very great assistance to our operations, I am indebted to Lieutenant Haines [sic], R.E., who, despite some failure and discouragement at first, persevered until complete success was attained." Haynes later served in the Bechuanaland Expedition 1884-85, and retired as Colonel C E Haynes CB on the 8th July 1912, although he was was re-employed for war service between 1914-18. He died in 1935, aged 80.

Eshowe Signals Back

It was Sunday 2 March when the flashing signals from the St Andrews Mission station were seen by those at Eshowe, but how were they going to signal back? The following day Captain Warren Wynne described their first attempt in his diary:

> … I determined to try and effect communication by means of a large screen raised above the ground, revolving on horizontal pivots, which, being brought alternately to a horizontal and vertical position in front of the places to be signalled to, should produce dashes and dots through the spaces of time of its appearance.

The screen was made of black canvas stretched across a frame, but he had trouble constructing and erecting it due to stormy weather. It was not successful, and he had to build another. At length, on 10 March, his diary records:

> I put up a new signalling screen in the afternoon, much stronger than the last, which works well. We signalled two or three messages to the Tugela signalling station. Signalling from thence was kept up for two or three hours. Weather fine and sunny.

Capt Warren Wynne

Wynne's method therefore was not a heliograph of any sort; he was not flashing the sun, but simply moving the screen on its horizontal pivots so that it was seen or not according to its position. It was a hybrid form of shutter telegraph, using but one large 'shutter' and signalling the letters using the Morse code. The normal form of shutter telegraph used an array of smaller shutters and a semaphore signalling code to represent the letters, but this method would have been unreadable over the distance involved. The rate of transmission using a single screen and Morse code to signal the letters was exceedingly slow. It was a brave attempt, but not very effective. Sadly Warren Wynne soon afterwards contracted enteric fever, became incapacitated, and died a few weeks later. (4)

Responsibility for signalling was taken over by Captain Henry MacGregor, 29th (later the Worcestershire) Regiment, the DAQMG at Eshowe. (Signalling in those days was the responsibility of the Quartermaster General's department.) A better method of signalling was needed and the breakthrough came with the discovery of a bedroom mirror, found by a servant of one of the staff officers while rummaging in a baggage wagon at Eshowe. He took it to Captain MacGregor, who recognised its potential but was faced with a number of practical problems before it could be turned usefully to signalling back to Charles Haynes.

History Led Astray by Artistic Licence

History has been somewhat led astray by a sketch purporting to show the signalling from the lower Tugela to Eshowe that was published in the *Illustrated London News* on 24 May 1879. It was drawn by their special war correspondent, Melton Prior, who arrived in Durban in late March, and is reproduced below.

This sketch is in complete contradiction to all the contemporary evidence from primary sources - there were no heliographs at Fort Pearson, and Lieutenant Haynes was using a mirror purloined from the nearby Smith's Hotel. The diagram in his signalling log book shows the arrangement. How did this misrepresentation occur?

The answer is in Melton Prior's autobiography, *The Campaigns of a War Correspondent*, relating his curious story of a bad dream in which he imagined he would be killed near the Tugela river. He even claimed to have had a letter from his mother telling him that she, too, had had a similar dream that he had gone to Eshowe with the relieving column and been killed, and that she had seen his funeral. How his mother in England knew about the plan for the relief of Eshowe and managed to communicate by letter in time is beyond belief. In the book he admits to only a brief overnight stay at Fort Pearson to see the troops set off to relieve Eshowe, before he returned in a funk to Durban early the next morning while other correspondents accompanied the advancing force.

" ... He [a Mr Walter Peace, in Durban] ... informed me he was going to the Tugela , whence the troops were to start for the relief of Colonel Pearson at Etchowe. He had already engaged a carriage and offered me a seat, so I had the opportunity of seeing the troops and making a sketch of them starting on the expedition, and I slept the night in the tent of the commissary, General Strickland. He was very kind to us, and gave us breakfast, and then we started back, and I arrived once more in Durban, feeling rather ashamed of my want of pluck.

During the time that the column was operating for the relief of Colonel Pearson I was engaged sketching the arrival of fresh troops from England, and the many interesting scenes to be found in Durban. ..." [5]

During his brief overnight sojourn at Fort Pearson to see the troops depart to relieve Eshowe, and taking into account the time available and his state of mind, it can be concluded that he never went to the signalling station near St Andrews, some four miles away across the Tugela river. He may have been told about 'the heliograph', jumped to the wrong conclusion and, back in the comfort of his hotel in Durban, used his imagination to sketch the heliograph station. It would, after all, have made better 'copy' for the ILN than the real thing!

On 25 March 1879 Lord Chelmsford visited the St Andrews Mission signalling station, accompanied by his Chief of Staff, Lieutenant Colonel John North Crealock. Crealock was an amateur artist, and sketched the occasion.

His sketch, entitled *Signalling to Ekowe by sun flashing*, and dated 25 March 1879, is one of a set now held in the Sherwood Foresters' museum, and is reproduced below. It shows the signalling arrangement in the background, with the mirror, the screen with aperture, and the flag used as a shutter, all in accordance with the sketch shown earlier in Haynes' own signalling log book. The officer looking through the telescope to read the signals from Eshowe is Captain Forbes Lugard Story of the 99th The Duke of Edinburgh's (Lanarkshire) Regiment, one of the regiments forming the Right Flank column. Charles Haynes is one of those named at the bottom of the picture. It can safely be assumed that Crealock's sketch, if less artistic, is the accurate version.

The first was the position of the sun. As the map above shows, the line joining the area of Fort Pearson and Eshowe is almost exactly north-south, and Eshowe is at a latitude of about 29 degrees south. The sun, of course, is always to the north in such southerly latitudes. Thus, using the sun to flash from the lower Tugela northwards to Eshowe, as Haynes did, was simply a direct reflection from the mirror as the sun traversed the sky behind Eshowe. The problem was in the other direction, from Eshowe to the Tugela, where the sun was behind the signalling mirror.

With a proper heliograph instrument this not uncommon situation is accommodated by the use of a second mirror, the duplex mirror, which with careful adjustment provides the means for a double reflection. But there was no heliograph instrument at Eshowe, so how was MacGregor with simply a mirror going to direct the sun's rays south to the Tugela river, and how was he going to flash signals using the Morse code?

As it happened, the equinox, on 21 March, fell in the middle of the period of the siege, and thus one can easily work out the angle of the mirror at Eshowe at midday on that day. By definition, at the equinox, the sun is over the equator, and the angle of the sun to Eshowe would be its latitude, which is 29 degrees south. By simple geometry, the angle between the sun, the mirror, and the distant signalling station at noon was 119 degrees (these angles are shown in the diagram). This obtuse angle increased as the sun was lower in the sky, either side of midday. Had the siege been a month or so later the sun behind them, moving towards its northern solstice, would at that latitude have become too low in the sky to enable this single-mirror method to work. So, to his good fortune, MacGregor could manage it with simply a mirror for what must have been only a few hours either side of midday.

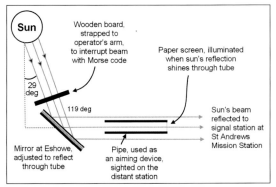

The improvised heliostat at Eshowe. MacGregor's method, using a pipe.

With the help of a piece of piping found in the grounds of the mission station, MacGregor managed to construct a rudimentary heliostat. The arrangement is shown in the diagram. He visually aligned the pipe on the distant signalling station, simply by looking through it, and held it in that position on some sort of stand. He then aligned the mirror so that the reflected ray shone through the pipe. The ray, passing through the pipe, was therefore also aligned on the distant St Andrews signalling station. A wooden board was strapped to the arm of the operator who stood between the mirror and the sun and, by the movement of his arm, interrupted the direct ray to form the dots and dashes of the Morse code.

This method of directing the mirror, ingenious though it was, turned out to be rather impractical, as it took a long time to align the mirror, and then dazzled the unfortunate person required to look back down the pipe to confirm correct alignment. Instead, a piece of paper was fixed to the end of the pipe, which was illuminated when correctly aligned. After a short period of signalling the sun had moved, the mirror had to be adjusted to a new angle, and the process had to be repeated. Eventually another method, using two sighting rods was suggested by Captain G K E Beddoes of the Natal Native Pioneers. This proved more effective and was brought into use. Actually, Haynes' simpler and better method with a directing screen and aperture could have been used, but the people at Eshowe didn't seem to know about it! (6)

The heliostat at Eshowe started signalling on 14 March, and communication was maintained daily with Fort Pearson, weather permitting, until Eshowe was relieved on 3 April by a force sent from Fort Pearson. It had been wonderful improvisation, the primitive communications being established using initiative and ingenuity, basic devices, and a rudimentary knowledge of signalling and the Morse code. (7)

Pause for Thought

After a ten week siege Eshowe was eventually relieved on 3 April, the day after the battle of Gingindlovu when the Zulus were defeated by the relieving force. By 7 April the troops were all back at the Tugela river where they had started on 12 January.

While Eshowe was under siege, the Left Flank Column had been engaged in a number of battles in western Zululand - Intombe Drift, Hlobane, and Khambula, the first two being defeats and the last a convincing victory. Signalling played no part in any of them.

An old photograph taken from the north bank of the Tugela river. Smith's hotel, where Haynes got the mirror from, is on the left; the Ultimatum Tree is the large tree in the centre; and the camp at Fort Pearson is on the hill on the right. The road can be seen going down to the drift; it is is now a modern dual carriageway, crossing the river on a substantial bridge.

Despite giving the British a bloody nose, the Zulus had by this stage of the war suffered about 7,000 casualties – the exact figure is not known – and had come to realise that their strength, speed, mobility and courage were no match against defended positions and modern weapons. Demoralised, they never again carried out massed attacks on the British. But the British, after their mauling, were in no position immediately to advance into Zululand and finish it off. After the relief of Eshowe, Lord Chelmsford returned to Natal on 9 April, to await the reinforcements he had asked for from England and plan the second invasion. There was a pause.

Up to now the signalling had been rudimentary, as between Fort Pearson and Eshowe, or non-existent. What improvements were to be expected in the next phase of the war? Things did get better, but again with much improvisation.

* * * * * * * * * *

Reinforcements are sent, and the war continues

The Telegraph Troop sent to Natal

The reinforcements from England, assembled with haste, and some of them of dubious quality, began to arrive in Natal from late March onwards. The plan now was to regroup and make a second invasion of Zululand.

Amongst the reinforcements, and sent expressly at the request of Lord Chelmsford because of the failure of communications in the first part of the war, was 'C' Telegraph Troop. They were about to participate in their first operation since being formed in 1870. Their manuscript journal, the 'C' Troop Record, as well as reports written by the Troop commander, Major A C Hamilton, give a detailed insight into what they did. [8]

As they prepared to depart from England Hamilton was no doubt wondering how his unit, established nine years earlier in the panic caused by the Franco-Prussian war, and consequently equipped and trained for a role in Europe, where they would never be far from an existing telegraph line, was going to cope with matters in faraway Zululand where things were very different.

The Troop was placed under orders for South Africa on 21 March 1879, almost two months after the defeat at Isandlwana, and a week later they left their base at Aldershot. They consisted of Major A C Hamilton (Troop commander), Lieutenants J C MacGregor (not to be confused with Captain Henry MacGregor of the 29th Regiment who was in Eshowe), J Hare, H B Rich, and F G Bond, Troop Sergeant Major A Lewis, and 209 NCOs and men, with 110 horses. They were accompanied by a surgeon and a veterinary surgeon. On 4 April they sailed from Portsmouth in the SS *Borussia*, a hired transport ship.

Problems were soon encountered. As they entered the Bay of Biscay a storm blew up, lasting several days. The ship had no proper ventilation on the horse deck and rolled 45 degrees. The horses were so distressed that sixteen died before they reached Madeira, and eleven more succumbed before the end of the voyage from the effect of the gale. Eventually, after six weeks cooped up aboard ship, they landed at Durban on 10 May. Here they managed quickly to replace the lost horses with four teams of mules.

At Durban they split into two Sections to support the new operation. It had been decided that the second invasion of Zululand would be carried out by two columns: one in the east, the 1st Division, starting from Fort Pearson as before, to be supported by No 2 Section of 'C' Troop; and the main column, the 2nd Division together with a separate 'Flying Column', advancing into Zululand across the Buffalo river from the south-west, supported by No 1 Section of 'C' Troop.

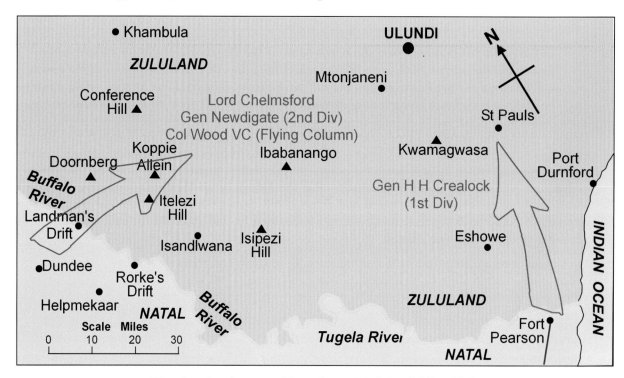

The plan for the second invasion of Zululand, May 1879.

The Western Column

The main operation was carried out by the Western Column, so it is appropriate to begin by describing their story. Lieutenants John MacGregor and Frank Bond and Troop Sergeant Major Lewis headed the Section of 'C' Troop in support of this column, to be joined soon afterwards by Major Hamilton, who had firstly accompanied No 2 Section to Fort Pearson to see them installed there. From Durban an advance party with the new mule teams and fifteen miles of airline went ahead quickly to Ladysmith and thence to Dundee. After about a week, by which time the horses had been fittened up after the long sea voyage, they were followed by the remainder of the Section, which arrived on Sunday 7 June, "having marched 202 miles through a very rough country" says the Troop Record. It was good going after the long voyage that the men and particularly the surviving horses had endured.

At this point the story of the Natal telegraph system should be resumed. When the working party charged with building the line from Pietermaritzburg to Pretoria returned to their original task, after being diverted by the needs of the impending war to extend the telegraph line to the Tugela river, it was found that there was insufficient transport available – it had all been requisitioned for the war instead. In consequence the project proceeded much more slowly than planned. By early May the line had only reached Newcastle, and there, for the time being, it came to a halt. By good chance, and never part of the original plan, the newly built line was turned to good use by the Telegraph Troop - to provide the rearward communications for the western column to Pietermaritzburg.

When the main body of No 1 Section of the Troop arrived at Dundee they found that the lightly equipped advance party had already tapped into the recently built telegraph line at Quagga's Kraal and extended it by airline to Dundee. The next day they laid cable from Dundee to Landman's Drift, on the Buffalo river, the border between Natal and Zululand, where the Force HQ was established initially and where the telegraph head was to remain throughout. Thus, at that point, there was telegraph from the HQs of the two columns, at Landman's Drift and at the Lower Tugela Drift, back to Pietermaritzburg.

The telegraph office at Landman's Drift was kept very busy for the duration of the campaign but it was not possible to extend the telegraph beyond there into Zululand because of shortage of line. In any case it can be assumed that an extended and vulnerable telegraph line running through enemy territory would not have lasted long. Nevertheless, as they advanced into Zululand, the communications of the 2nd Division and Lord Chelmsford's accompanying headquarters were to be hampered by the lack of any electric telegraph beyond Landman's Drift.

Insufficient Line

'C' Troop carried insufficient line to reach the objective of the expedition, King Cetshwayo's base deep in Zululand at Ulundi, eighty-five miles from Landman's Drift. Their normal scale, based on calculation for their intended European role, and after the minor reorganisation in 1878, was thirty miles of airline and thirty miles of cable; it was hopelessly inadequate for the present task.

Hard decisions had been made before they departed. It was foreseen in England that there was insufficient line for the task so eighty miles was taken, but the trade-off in weight and space for the same number of horses and transport capacity meant that they left behind the telegraph office wagons (it was possible to dismount the telegraph equipment and operate an office from tents or other temporary accommodation), and the poles also had to be left behind on the assumption that new ones could be procured locally. Depending on which type of wire and cable they took, which is not recorded, a typical weight would be over 100 lbs per mile of line, so eighty miles would be about four tons – a heavy load to be hauled by animals from Durban to the Zululand border. The poles would increase the weight further.

Two other factors compounded the line problem. As explained, the Troop was divided into two Sections to support the operational plan for two Columns, so their limited resources were shared and neither could reach Ulundi from their connection points with the Natal permanent telegraph system. Also, it had been intended to follow up 'C' Troop with a Postal Telegraph Company, one of the Army's two reserve telegraph units, bringing with it 100 miles of line. This plan was later countermanded and the reserve unit never came, but by the time the additional 100 miles of line arrived the war was over. When it did come, each Section was allocated fifty miles. The poles left behind in England were replaced from local sources, in the case of Number 1 Section, by bamboo poles cut in the forests around Pinetown, on the way to Dundee. An unexplained mystery is why the stores to be used for the telegraph line from Pietermaritzburg to Pretoria, in the process of being built when the war started, and presumably already procured and held in Natal somewhere, was not made available for the Telegraph Troop to use. [9]

Visual Signalling – the Heliograph

With no telegraph beyond Landman's Drift, communications into Zululand were going to have to be provided by visual signalling, principally the heliograph, or else by galloper. Flag signalling, with its limited range, was only going to be of any use for local communications. The topography of Zululand was admirably suited to the heliograph, with prominent hills, far-reaching lines of sight, clear air, and usually plenty of sunshine, although there were long periods of unusually bad weather during critical periods of the operations, just before and after the final battle at Ulundi. (Latter day large commercial forestry plantations have altered the topography since then.)

But there were problems. Firstly, there had been an organisational change to signalling arrangements in 1875, and telegraph and visual signalling, previously both under Royal Engineer control and based at Chatham, had been separated. Visual signalling in infantry and cavalry regiments came under the charge of an Inspector in Army Signalling, and had moved to Aldershot. The Royal Engineers remained responsible for providing telegraph to the HQ, but below that level visual signalling was provided by regimental signallers, whose resources were pooled and came under the control of a Divisional Signalling Officer, appointed when the occasion arose - a rather *ad hoc* arrangement, its obvious defects frequently contributing to inefficient signalling for years to come. At the time of the Zulu war this new arrangement was not effective, and signallers were not well trained. Secondly, the heliograph was a relatively new method of visual signalling, there were not enough of them, and the troops from Britain were not well practised, as explained in Chapter 9.

Under these arrangements, the commander of the 2nd Division, Major General Newdigate, had appointed Lieutenant J H Scott Douglas, 21st Regiment (later Royal Scots Fusiliers), an officer who possessed a 1st Class Instructor's certificate from the School of Army Signalling at Aldershot, as his Divisional Signalling Officer. Well qualified though he was, the means at his disposal, both heliograph instruments and competent signallers, were scant.

It becomes clear from Major Hamilton's reports, written during and shortly after the war that despite the policy of separating visual signalling and electric telegraphy laid down by senior officers in high places in London four years previously, the state of regimental signalling in 1879 was inadequate, and he had to assume the appointment of both Director of Military Telegraphs and Signalling for the entire Force. He describes how, almost up to the time of the final battle at Ulundi on 4 July, the only heliographs available in Zululand were a pair of 3-inch, a pair of 6-inch, and a pair of 10-inch instruments. The only signallers acquainted with the use of the instrument were some men of the 17th Lancers, actually volunteers attached from the 16th Lancers in India who had learnt to use the heliograph there. These few competent signallers had to be augmented by using telegraphists from 'C' Troop, who of course knew the Morse code (used with electric telegraph, heliograph, and flag signalling at that time), although Hamilton explains that "when the work was first commenced under my direction, neither the officers nor signallers of my troop had ever *seen* a heliograph, nor until lately had we any printed instructions on the subject. The instruments were therefore not used to the best advantage, and there was considerable want of skill in reading the signals. This has now [after a few days training] been overcome by practice, and the work is going on with the utmost precision and dispatch. I may mention that messages have lately been transmitted at the rate of 14 words per minute, while the capability of the instrument, as claimed by the inventor and patentee, is only 10 words per minute."

With their initial six heliographs, a chain of stations was established from the forward positions of the army in Zululand, where the parties of 17th Lancers worked the two 3-inch instruments, back to to the telegraph head at Landman's Drift. The 6-inch and 10-inch instruments were operated by telegraphists of 'C' Troop. Their deployment during the advance into Zululand will be described shortly.

Despite the problems with visual signalling into Zululand, the telegraph office at Landman's Drift, handling the traffic for the rear link to Natal while the Column assembled and organised itself, was extremely busy and worked well.

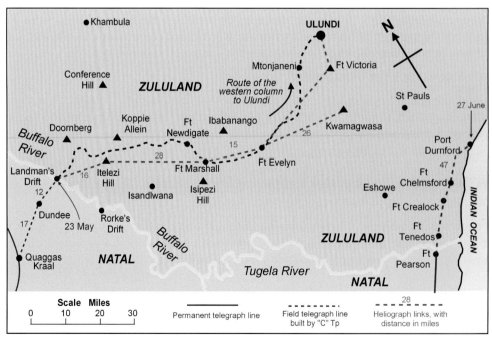

Communications into Zululand, 22 May to 4 July 1879.

Into Zululand

When all was ready, the advance into Zululand started. The 'C' Troop Record describes the activity. "On 12 June Lieutenant MacGregor with a party of signallers and heliographs advanced with General Wood's great convoys which occupied 12 miles of road. Major Hamilton and Lieutenant Bond with another party established a heliograph station on the Doornberg [Thorn Mountain]. As the column advanced, Major Hamilton and the Doornberg party moved to Itelezi mountain and on 24 June opened communication with Lieutenant MacGregor on Ibabanango and with Landman's Drift. By means of these stations and intermediate stations between them and the advanced posts, communication was kept up during the campaign, the highest rate of sending being 15 words per minute. Signal stations were opened at intervals at Fort Newdigate, Fort Marshall, Conference Hill, Fort Evelyn, Kwamagwasa, and Fort Victoria." It is worth noting that the link between Fort Evelyn and Kwamagwasa, 26 miles, was being worked by Mance heliograph instruments with only a 3-inch mirror.

Near the site of the heliograph station at Kwamagwasa Hill today will be found the war graves of two signallers, those of Lieutenant Scott Douglas, the Divisional Signalling Officer, and Corporal Cotter of the 17th Lancers, both killed by Zulus. James Scott Douglas was the eldest son of Sir George Scott Douglas of Kelso who at the time was MP for Roxburghshire, and the headstone records that "he met a soldier's death near this spot while in the execution of a dangerous and important duty as signal officer of Lord Chelmsford's army, ... 1st July 1879." The heliographs unable to operate because of bad weather on 30 June, and escorted by Corporal Cotter, he was carrying important despatches at the time. These despatches were to General Lord Wolseley, who had been sent from England to supersede Chelmsford, and was now in South Africa. (10)

Lt Scott Douglas   *The graves of Lieutenant Scott Douglas and Corporal Cotter, near Kwamagasa.*

For the final battle of Ulundi, Lieutenant MacGregor, equipped with a heliograph, was positioned with the forward troops on a vantage point near the fighting square as it advanced on the Zulus. The battle was soon won, but the weather was deteriorating. MacGregor wondered whether to take the initiative and signal the victory himself, but decided that he should really wait for the official despatch. By the time this arrived the sun had gone and he could not send it for transmission down the heliograph chain to Landman's Drift. In the event it was a war reporter, Archibald Forbes, war correspondent for *The Daily News*, who galloped away, and in quite an epic dash was the first to break the news at the telegraph station at Landman's Drift.

The Eastern Column

The work of breaking the Zulu army having been done by the western Column, what had been happening in the meantime to the eastern Column and No 2 Section of 'C' Troop?

After landing at Durban the Section, under Lieutenants Hare and Rich, marched through the undulating country beside the east coast, covered in sugar plantations, to Fort Pearson on the Tugela river. Their destination was the HQ of the 1st Division commanded by Major General Henry Crealock (not to be confused with his younger brother, Lieutenant Colonel John North Crealock, who was Lord Chelmsford's Military Secretary), where they arrived on 21 May after five day's march.

THE ANGLO-ZULU WAR, 1879

The 1st Division was to advance into Zululand from the east as shown on the map above, but in the campaign as a whole it became a subsidiary effort, lacking clear orders from Lord Chelmsford, and getting nothing by way of motivation from General Crealock, its slow progress earning it the unflattering title of 'Crealock's Crawlers'. The Column was, however, disadvantaged by disease, much of it caused by the rotting carcases of dead transport animals, and the unhealthy climate of the low-lying coastal area where malaria was prevalent. They also suffered transport problems and a liberal share of storm, fire and flood. The rationale for a two-Column strategy can, amongst many other things, be questioned, particularly as it served only to split scarce resources such as transport and communications.

As the Troop commander, Major Hamilton, was based at Landman's Drift, the telegraph Section with the eastern column was attached to 30th Company RE, the 1st Division's field engineering company responsible for building the roads and bridges, under the command of Captain Bindon Blood RE. It will be recalled (chapter 7) that in the early 1870s, as a Lieutenant, Blood had himself served a tour of duty with 'C' Troop as one of its founder members, so was familiar with their work. The Section's task was to extend the telegraph line from Fort Pearson, where the Natal permanent telegraph system ended, and to follow the advance into Zululand, opening telegraph offices where needed on the way. Equipped with half of 'C' Troop's limited line resources, they reached Fort Chelmsford by the 31 May and Port Durnford by 27 June, the line being partly a cable route (ie laid on the ground) due to shortage of overhead line. On 2 July, and again on 4 July, the insulated cable was damaged by grass fires.

The field telegraph kept pace with the movement of the forward troops – not too difficult in this case - and as a result there were much better communications to this Division even though it was operationally less important. The line does not appear to have been tampered with by Zulus, who by now were very wary and keeping their distance.

On 5 July they received a telegram reporting the victory at Ulundi. Although the war was over, there was still work in providing communications while troops were deployed across Zululand.

Change of Command

After the early disasters it had been decided in London that Sir Garnet Wolseley would replace Lord Chelmsford as the Commander. The ambitious Wolseley – by then growing in bombast and arrogance - was delighted. At the end of May he sailed for Natal, where he had previously been Lieutenant-Governor for six months in 1875. Accompanying him on his staff was Major Charles Webber RE who for some years before had been heavily involved in the formation of the Postal Telegraph Companies of the Royal Engineers in London (chapter 7), but was now a DAQMG with logistic responsibilities. Although never one of Wolseley's 'Ring', he clearly caught Wolseley's eye, and in due course served as his Director of Telegraphs in two expeditions, in 1882 to Egypt and in 1884-85 on the unsuccessful Nile expedition to relieve General Gordon in Khartoum.

Wolseley reached Cape Town on 23 June, desperate to be in charge of the action and no doubt to bask in some of the glory of the now inevitable victory, and there he sent the first of a series of telegrams. His ship next called at Port Elizabeth where he fired off more brusque and peremptory telegrams to Chelmsford and others (by means of the Cape and Natal telegraph systems already described) asking for a report on the situation, the general tenor being that he was now in charge and the present incompetents were to do nothing but await his arrival when all would be resolved at a stroke, a cause of some amusement to Chelmsford's staff and presumably also to the telegraphists as this highly combustible material was transmitted down the line.

Wolseley's first telegram reached Landman's Drift and was forwarded, to Chelmsford's headquarters, then near Mtonjaneni. Before the reply could be telegraphed back from Landman's Drift Wolseley reached Durban on 28 June and soon fired off a further telegram to Chelmsford *via* Landman's Drift which reached him on the evening of 2 July, as he was preparing for the battle of Ulundi two days later. It is likely that the death of Lieutenant Scott Douglas, mentioned earlier, occurred as a result of his endeavours to get Chelmsford's reply to Wolseley's first telegram from Mtonjaneni back down the chain to the telegraph station at Landman's Drift.

An extract from a further telegram to Lord Chelmsford from Wolseley sent from Durban on Tuesday 1 July 1879 reads:

THE ANGLO-ZULU WAR, 1879

> Your letter and enclosures of 28th June received. ….. Wish you to unite your force with the First Division, as I strongly object to the present plan of operations with two forces acting independently …
>
> Am now starting, 4 P.M., and join First Division at Port Durnford by sea tomorrow. As soon as I get things in order there I intend to force my way to St. Paul's Mission Station. Communicate news to me daily through Marshall. Send messages in the cypher which you use with Crealock by native messenger across country to First Division. I shall endeavour to communicate with you the same way. Acknowledge receipt of this message immediately by flashing to General Marshall. If you have no cypher with Crealock, send message in French. [11]

The landing at Port Durnford was aborted due to the sea state, so in the event this ploy failed. Wolseley had made desperate attempts to get to the scene of the action in time, but to his chagrin arrived at Fort Pearson by road from Durban only after the battle had been fought and won. There he was handed a telegram received on 5 July announcing Chelmsford's victory at Ulundi the previous day. Wolseley assumed command anyway, and Lord Chelmsford departed on 8 July.

Continuing Communications Work

The war may have ended, but the communications requirement, both electric telegraph and visual signalling, got bigger. The Zulu king, Cetshwayo, had escaped from Ulundi and was being hunted (he was captured a few weeks later), and troops deployed all over Zululand in an attempt both to capture him and to bring order to the country. Communications were needed over a wide area. Zululand was now a safer place, for the Zulus recognised they were beaten and simply wanted to get back to a normal life; there was no subsequent guerrilla warfare.

Just three days before the final battle at Ulundi, more much-needed heliographs had arrived, the Troop Record on 1 July reporting that: "Several 10-inch heliographs received, found to work very well." These were instruments manufactured in the workshops of the Bengal Engineers at Roorkee.

With these additional heliographs an extensive system of signalling was set up across Zululand, and signallers from the infantry and cavalry were employed to work them, "but in very few cases did these attain beyond a mediocrity of skill, owing to a want of systematic training and practice in peace time", observed Major Hamilton in one of his reports. Unfortunately the detail of this deployment is not available, but in the discussion period after a meeting at the Royal United Service Institution on 15 March 1880, Hamilton described it as follows: "We employed the heliograph to a very great extent, having at times 15 or 16 different signalling stations. The greatest distance between the stations was 35 miles, the general distance being something like 20 to 25 miles and sometimes less."

Back at Landman's Drift there were problems with the telegraph. On 7 July a terrific thunderstorm wrecked the lines and fused the instruments. That part of Natal, between the mighty Drakensberg range to the west and the hot and humid Indian Ocean to the east, is often subjected to fierce electrical storms, with vivid lightning and torrential rain. Then a few days later, in Natal, supposedly friendly territory, the local natives stole half a mile of wire near Dundee and later wrecked eight miles of line between Dundee and Quagga's Kraal. The 'C' Troop Record tells how "a party under Lieutenant Bond remained out all night, and the following day repaired the line and drove in 600 head of cattle." It does not elaborate on whether this was done as a punishment. If so, the army must have fed well for a while! Or perhaps it was simply because the cattle kept knocking the airline down while they scratched on the poles, a recurrent hazard for the line.

Meanwhile, the additional 100 miles of line had belatedly arrived, and was allocated fifty miles to each of the two Sections. On 10 July a 12-mile airline built with copper wire on bamboo poles and no insulators was completed to Koppie Allein (Lone Hill) and a telegraph office was opened there. On 12 July a 13-mile line was constructed to Conference Hill. "Owing to defective insulation of the line and want of battery power it was found necessary to put telephones on this line with ordinary [*ie* Morse] sounders; this worked well", said the 'C' Troop Record – something that will be explained in a moment.

On 14 July a line was commenced towards Itelezi and was worked with telephones. On 1 August six miles of cable were paid out from Landman's Drift to Doornberg and "an office and heliograph station established signalling to the Oscarberg over Rorke's Drift by which means Sir Garnet Wolseley, who had arrived there, sent all his messages down and up the country."

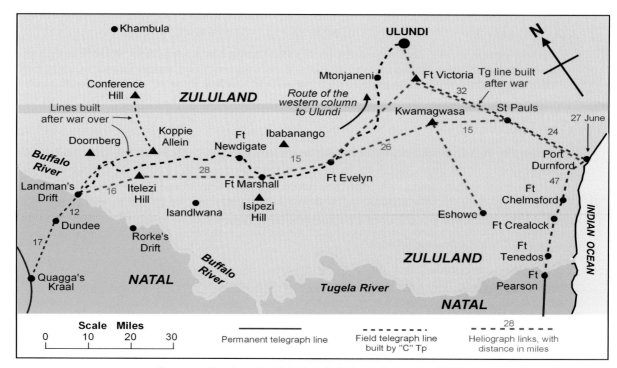

Communications in Zululand, July and August 1879.

On 26 July the eastern Column's supply of additional line was landed at Port Durnford enabling the Section to extend the telegraph communications further into Zululand. By 4 August the line was completed to St Paul's "through dense bush, swamps, and reeds" and the telegraph office there was opened. On 7 August storms wrecked the line and the rivers were impassable, although "2nd Corporal Head and SS [Shoeing Smith] Dawkins showed most distinguished pluck by swimming their horses across a swollen river and examining the line to Port Durnford." (Dawkins later became the Farrier-Sergeant in the Troop.) By 11 August a further twenty miles of line was connected on towards Mtonjaneni, and a telegraph office opened at Fort Victoria (not marked on most contemporary maps). Their supply of line was almost exhausted, but a final mile was laid to a heliograph signalling station which was opened nearby and communications then extended to Mtonjaneni and Ulundi by heliograph. Eventually, by 15 August, they had built 102 miles of line. This was an extremely long line to operate and maintain in tactical conditions - forty miles was considered the practical limit for a military telegraph line.

One disadvantage of such a long line not mentioned in the contemporary reports or descriptions, but which becomes evident from reading post-war requests for better equipment, was that a message could not be telegraphed directly between more than three stations before it had to be re-transmitted manually by the intermediate telegraph office. This was due to the distances involved (much greater than the Europen scene), the poor quality of the line, the insensitivity of the Morse sounder (the telegraph audio receiving device), and the lack of relays which would re-transmit automatically. Such enforced manual re-transmission along the line was the cause of delays to message traffic. [12]

The 'C' Troop record noted that "the health of the men was bad, owing to the malarious swamps, and one man died, and fourteen men were invalided to Durban."

By this stage of the campaign, with the Roorkee heliographs having become available at the beginning of July, a greater number of the telegraphists of 'C' Troop had become adept in their use, and they were used as part of the communications system in eastern Zululand. Deployed between St Paul's, Port Durnford, Eshowe, Kwamagwasa, and Ulundi, heliographs extended the communications when they ran out of line near Mtonjaneni, and also acted as a back up means to the telegraph line, which turned out to be particularly useful when the line was severely damaged by the ferocious storms already mentioned.

Probably because the action had been elsewhere, with the western Column, the good work of this Section in laying and maintaining such a length of line, extending it where needed by heliograph, and operating the telegraph offices, all in adverse conditions, has not received the attention it merits. Captain Bindon Blood, Officer Commanding 30th Company RE with 1st Division, included the work of the telegraph Section in a report he made at the end of the war. He noted that: "The field telegraph was managed throughout the operations by Lieut. J. Hare, R.E., and owing to his energy and good arrangement it was very successful." The line was mostly airline on various poles, some of the service pattern, some 3-inch scantlings from Durban, and some cut in the bush. The main problem with the line was due to interruptions caused by stray cattle and by the carelessness of wagon drivers.

Blood observed that the instruments and batteries sent out with the Telegraph Troop were old and considerably worn, and the equipment generally too heavy, the loaded wagons being unfit for use off roads. He suggested that all telegraph equipment, both airline and cable, should be carried on two-wheeled carts which were much better for rough work than wagons and were used in India with the greatest advantage. He also commented on the paucity of regimental signallers in the units of the 1st Division. [13]

In late August, after Cetshwayo was caught, there was a general withdrawal of troops from Zululand and the telegraph lines were reeled up as the forts were evacuated.

Telephony

The Zulu War saw the first use of the telephone by the British army, the Bell telephone having been introduced in 1876, three years before the war. Telephones at this time would only work over short distances, say a few miles, so they were only used for local point-to-point communications, typically outposts to the local HQ. The thermionic valve, which could amplify the complex waveform of the human voice, and thus enable greater range, was not invented until early in the twentieth century.

The Illustrated London News reporter with the Eastern Column described his experience of it:

> A neighbouring hill, a mile and a quarter distant, on which a vidette [observation post] is stationed, is connected with the fort [Fort Crealock] by telephone. It is the first time that instrument has been used in warfare. It is of the greatest service, as voices are easily recognised by the sound.

The low-key description in the 'C' Troop Record of working Morse sounders using a telephone receiver is the only occasion on which it mentions the use of the telephone in the Zulu War, yet it turned out to be a novel development of military telegraphy. Using the extreme sensitivity of a telephone receiver, or earpiece, and coupling it with the much less sensitive Morse sounder, it became the first operational use of what might be called Morse telephony – an apparent contradiction in terms.

After the war, back in England, when the future development of telephones was being discussed, Major Hamilton wrote a letter describing its use in the Zulu War in more detail:

> In South Africa I made several experiments with the telephone supplied to me by Major (now Lieut. Col.) Webber, R.E., but the only one which had a practical and useful result was on the line between Koppie Allein and Conference Hill.
>
> A line of telegraph between these places was urgently required for a few days, but we had not sufficient material without using some cable of which the insulation was in part burnt off by grass fires and was practically bare wire.
>
> The weather being very dry we were enabled to get a sufficient electric current through this wire to work the troop instruments during the day, but in the early morning, when the dew was on the ground, the current went to earth, but by putting a telephone on, the working of the Morse instruments could be distinctly heard and messages read off in that way.
>
> The amount of faulty cable on the line was 5½ miles, and besides this for the first, second or third day there was three miles of bare wire on the ground for which no poles were, for a time, available.
>
> The rest of the line was bare wire on poles, the whole quantity in circuit being about 23 miles. [14]

It is likely that the telephones supplied by Major Webber were an experimental type of Bell telephone, a modification of what was then known as the rifle range telephone, made by the Telephone Company. After the war, back in England, the newly-discovered technique was developed by a Royal Engineers officer, Captain Philip Cardew, an instructor in telegraphy at Chatham, to produce a new instrument called the vibrating sounder. The extremely sensitive vibrator, or buzzer, as it was also called, came to have considerable application in military telegraphy for more than twenty years, its big disadvantage being that it induced interference in adjacent lines (so had no application in civil telegraph systems). In its primitive form, though, it had its operational roots in the Zulu War, between Conference Hill and Koppie Allein on 12 July 1879, when the telephone receiver overcame the defective insulation of the cable and the damp conditions.

Final Report on Signalling

Major Hamilton's final report, written on 1 October 1879 whilst at Wakkerstroom, addressed a number of signalling issues. (15)

Heliograph

Heliographs came into their own in this war – when, eventually, there were enough of them and signallers learnt how to use them. Of the assortment in use by the end of the war, the preferred instrument was the 5-inch Indian pattern made at Roorkee. They were lighter, more robust, and worked on a better principle than the Mance heliograph. (They mounted the duplex mirror on the same tripod and head assembly as the signalling mirror, whereas Mance used two tripods) Hamilton recommended "that only this one size and pattern should be adopted in the service, at all events for field operations, for they are powerful enough to be easily read with the naked eye at 40 miles distance, and at the same time small enough and light enough for a mounted man to carry without the least inconvenience. The 3-inch Mance heliographs, used by the cavalry, were completely worn out by the end of the campaign."

Flag Signalling

As explained in chapter 5, flag signalling in those days was carried out by a single large flag used to signal letters in the Morse code. The large flags gave greater range but they were too cumbersome and Major Hamilton reported: "in sending long messages the signallers had to relieve each other after every 50 or 60 words, being perfectly exhausted by the exertion required." Field glasses were not issued to signallers, who whenever possible had to borrow them from officers, so the range capability of flag signalling was often not fully exploited. Flag signalling in the war was mainly used to control movement of columns and other local work. It did not have the range needed for the distances in Zululand.

Lamp Signalling

Night signalling with lime light lamps was used in at least two cases. The first was during the march of the 1st Division from Napoleon Hill to Fort Durnford, when the Division was broken up and halted overnight in two camps, between which communication was maintained by this means (the lesson of Isandlwana having apparently been learnt). The second was between Fort Victoria (the terminus of the Fort Pearson and Port Durnford field telegraph line) and Ulundi when, Hamilton reported: "this work was carried on for several hours every night for about a fortnight in a very satisfactory manner. ... The General Commanding [then Wolseley] was encamped at a distance of 10 miles from the terminus of the telegraph line, and but for this all messages arriving after sunset would have had to be forwarded by mounted orderlies in the dark, or have been detained until the following morning."

Other Matters

A range of other matters were criticised: the inadequate scales of signalling equipment and supplementary equipment such as field glasses and telescopes; the different procedures that had developed for visual and telegraph message handling as a consequence of the organisational split in responsibility, disruptive when a message had to be passed between the two methods; the design and weight of signalling stationery such as message forms, and different design of forms between visual and telegraph for essentially the same function.

Finally, after stating that 'the importance of signalling has been strongly exemplified during the late campaign', a number of recommendations were made: to give greater encouragement to the practice of signalling, as is done already to that of musketry; that an officer is detailed in every regiment to superintend the instruction in signalling; a badge of 'crossed flags' for qualified signallers ; signalling to be tested during the regiment's annual General inspection, and a number of other matters aimed at giving the subject more prominence and attention.

Return to England

After the war was over No 1 Section went on to further work in the Transvaal connected with the operation against Sekukuni and the Pedi tribe; the detail need not be described. No 2 Section closed its telegraph offices, recovered all its cable in mid-September, and made its way southwards from Fort Pearson.

The two Sections of 'C' Troop eventually reunited at Pinetown, a garrison town near Durban, on 26 October. To spare them another long voyage, the long-suffering horses were sold by auction at Durban, and on 22 November the Troop embarked on the hired transport SS *Galatea* for England, arriving at Portsmouth on Monday 26 January 1880, "having suffered much from typhoid fever on the voyage." They reached Aldershot at the end of their first overseas operation on 28 January 1880, just about ten months after their departure.

Epilogue

South Africa remained in turmoil after the Zulu war. After the Sekukuni rebellion, the understrength British forces had to deal with other dissident tribes, in Pondoland and Basutoland. Then the Boers in the Transvaal rebelled and the first Anglo-Boer war started in December 1880, as a result of which the Transvaal regained its independence.

The Telegraph Troop returns to Natal in 1881

The Telegraph Troop was to send a detachment back to Natal far sooner than any of them could have imagined. After the disastrous first Anglo-Boer War, a combined unit commanded by Lieutenant Arthur Bagnold comprising elements from 'C' Troop and the two Postal Telegraph Companies returned in April 1881, to repair the telegraph line between Newcastle in Natal and Pretoria. In the Transvaal the line had been destroyed by the Boers and was needed urgently for post-war negotiations with London. The new submarine cable from Durban, not available in the run up to the Zulu War, was by then in operation and the telegraph was used extensively to communicate between London and Pretoria. The work was completed by 2 June. (16)

Zulu War Medals Presented

On 4 April 1881 the 'C' Troop Record tells us that "the Zulu War medals are presented to the Officers, NCOs and men of the Troop by the CRE, Colonel Fitzroy Somerset, on the church parade."

The Zulu War Medal

The medal was awarded to all participants in the series of wars in South Africa from 1877 to 1881 (ie from the last Frontier War in the Eastern Cape, through the Zulu War, to the first Anglo-Boer War). It was unusual in that it did not name battles but merely records the year of service in which the recipient was engaged; thus, for example, there is no bar for 'Rorke's Drift', which is a memorable battle but not individually recorded on a medal or a clasp. The members of 'C' Troop, like those who fought at Rorke's Drift, were awarded the bar '1879'.

The medal ribbon is yellow with black stripes, rather similar to numerous GS medals awarded for operations in Africa. The names and regiments of the recipients are engraved on the edges of the medal in upper case letters. The obverse shows Queen Victoria; the reverse shows a lion with underneath a Zulu shield and four crossed assegais.

There is one such medal in the Royal Signals museum, awarded to Driver Clatworthy. He was also awarded the medal for Tel-el-Kebir in Egypt in 1882, so it is the only medal display in the museum to embrace the two operations in which 'C' Troop participated before it was expanded into the bigger Telegraph Battalion in 1884.

As well as medals, the following letter, addressed to Major Hamilton, was received from the Horse Guards. Perhaps it should be mentioned that most units received such a missive - a pleasant little ritual in the style of a bygone age. 'HRH The Field Marshal' was of course the Duke of Cambridge, cousin of Queen Victoria.

> Sir
>
> I am directed to inform you that His Royal Highness The Field Marshal Commanding in Chief has been much pleased at receiving reports from Major-General Lord Chelmsford GCB and the Hon H. H. Clifford, V.C., K.C.M.G., C.B., respecting the efficient manner in which 'C' Troop Royal Engineers under your orders performed the duties devolving upon them during the late Zulu campaign, and I am at the same time to convey to you the expression of His Royal Highness's appreciation.
>
> I have the honour to be
> Sir
> Your obedient servant
> Sgd: J Grant
> D.A.General. R.E.

'C' Troop Biographies

Major Alexander Charles Hamilton, the Troop Commander, was promoted to Lieutenant Colonel on 21 August 1883. He retired in August 1888 in the honorary rank of Colonel. The direct line of descendants having been broken, he claimed the title to the Lordship of Belhaven and Stenton in 1893, becoming the 10th Lord, and became a Representative Peer for Scotland in the House of Lords. He died in 1920.

Lieutenant John MacGregor was posted out of the Telegraph Troop prior to its return to England, to take over command of a RE Field Company operating in the Transvaal against the tribal leader Sekukuni. He was subsequently appointed Secretary to Major General Sir George Colley, Governor of Natal in 1880. A fine officer, he was killed during the first Anglo-Boer war at the Battle of Ingogo, 8 February 1881, shot in the head while carrying Colley's orders at the front of the battle. Colley himself was killed at the battle of Majuba Hill a few weeks later, on 27 February 1881.

Lieutenant Frank Bond had a distinguished career, retiring in 1919 as Major General Sir Francis Bond, KBE, CB, CMG. After the Zulu War, still in 'C' Troop, he took part in the Egyptian Expedition of 1882 and then served for a long period in India, mainly with the Bengal Sappers and Miners. He served again in South Africa during latter part of the 2nd Anglo-Boer War (appointed CB) before returning once more to India in command of the Madras Sappers and Miners. He was recalled to England during the First World War and became Director of Quartering. After his retirement from the active list he undertook charitable work. He died in 1930.

Lieutenant Henry Bayard Rich took part as a field engineer in the Egyptian Expedition in 1882. He was later promoted Captain and returned to India, where unfortunately he died of a fractured skull as a result of an accident while playing polo at Rawalpindi on 17 November 1884.

Lieutenant John Hare was promoted Captain on 1 January 1880 and retired from the army on 8 September 1886.

Troop Sergeant Major Lewis, obviously a tower of strength in the the Troop, has also disappeared into obscurity, presumably because WO and NCO's records were not as well kept as officer's records. He was specially mentioned by Major Hamilton in his final report: 'Troop Sergeant Major Lewis ... also rendered especially good service in instructing the regimental signallers at various posts in the use of the heliograph, and afterwards marched with Colonel Baker Russell's flying column, in all signalling operations, all of which duties he performed with great energy and intelligence.'

THE ANGLO-ZULU WAR, 1879

Endnotes

1. The correspondence is in the Natal archives, Pietermaritzburg, under the following references: CSO 668, 4307/1878; CSO 669, 4439/1878; CSO 670, 4523/1878 and 4523/1878.

2. Some sources state the message was sent from Madeira, normally the next port of call on the shipping route, used as a coaling station. The nearest telegraph station to Cape Town at that time was St Vincent. The Official History is the source quoted here. *Narrative of the Field Operations, The Zulu War of 1879*. Prepared by the Intelligence Branch of the Quartermaster-General's Department, published 1881.

3. Some sources state 1880. The annual report of the General Manager of Telegraphs for the year 1879, Ministerial Department of Crown Lands and Public Works, in the South African Cape Archives, states: "before the end of the year the submarine cable was completed right through to Aden, and was opened for traffic" - ie in December 1879.

4. Captain Warren Wynne was a much respected officer who had assumed command of the 2nd Field Company RE in England and had been despatched on the campaign at very short notice. He contracted enteric fever on 12 March, and died on 9 April, his 36th birthday, shortly after the relief of Eshowe. He was buried along with other casualties in Euphorbia cemetery near the Tugela river. He left a widow (his second wife, the first having died soon after their marriage) and two young children in England. He was promoted to Brevet Major on 2 April, although he was unaware of this as it was not gazetted until after his death. This improved his widow's pension, and an additional award was made in consideration of his distinguished service, bringing the total to £100 per year. His widow, Lucy, never remarried and drew her pension for sixty-seven years until her death on 2 June 1946, aged 94 (*Supplement to RE Journal*, July 1946, p 82).

    Wynne's personal diary is reproduced in the *History of the Royal Engineers*, Vol 2, p35. His obituary was published in the *RE Journal*, 1 July 1879.

    His letters and diaries were published privately by Lucy Wynne in 1880. They have subsequently been reproduced in a book, *A Widow-Making War*, edited by Howard Whitehouse, published in 1995.

    Interestingly, while Warren Wynne was in Zululand his brother, Major Arthur Wynne (later Maj Gen Sir A S Wynne, GCB) was in Afghanistan serving as Superintendent of Signalling to General Sir Frederick Roberts during the concurrent 2nd Afghan War (see chapter 10). Arthur Wynne subsequently served in Natal in 1881, when he was responsible for a chain of heliograph stations connecting Natal with Pretoria while the telegraph line was being rebuilt, and later in Natal in 1900 during the 2nd Anglo-Boer war under General Buller as Chief of Staff and Divisional commander.

5. Melton Prior. *The Campaigns of a War Correspondent*, pp 90-91, published 1912.

6. Several other sources give additional descriptions. A presentation to the RUSI in London entitled *Heliography and Army Signalling Generally* was given by Maj A S Wynne, 51st LI (Warren Wynne's brother, as described in note 4 above), and recorded in *The Journal of the Royal United Service Institution*, Vol XXIV, 15 March 1880, p 256. Also at that meeting was Major Henry MacGregor, and his description of signalling from Eshowe is included in the record of the meeting.

    Another RE officer in Eshowe was Lieutenant Thomas Ryder Main. He was in charge of a section of the Natal Native Pioneers. He later wrote an autobiography, *The Recollections of Colonel T R Main*, which will be found in the RE Library, Chatham, and it includes a description of signalling in Eshowe.

7. The author visited the area in November 2002. Nowadays, the hill where Fort Pearson once stood is clearly visible right beside the N2 motorway running north from Durban where it crosses the Tugela river on a modern bridge, and it has become a small game reserve. On the other side of the road is the 'Ultimatum Tree' where the ultimatum that led to the war was read to the assembled Zulus on 11 December 1878. It is not the original tree, which in its dying years was finally destroyed by floods in the 1980s, to be replaced by another which looks passably like the original. As for Eshowe, it had become a bustling place served by all the trappings of the present day: there are microwave radio towers, banks with hole-in-the-wall machines from which, with a piece of plastic, one may draw cash from one's account anywhere in the world, there are people using mobile phones, shops selling personal computers and offering computer training, and other shops advertising the installation and repair of satellite television aerials. It was fascinating to muse on what the participants in the Zulu war, struggling with mirrors some 120 years earlier, would have thought about the scene.

8. The *'C' Troop Record* is a contemporary manuscript journal describing many of the Troop's activities, and is held in the archives of the Royal Signals, Blandford. Its record of the Zulu War is a little out of sequence and appears to have been compiled from separate reports by the participants after returning to England, and then written by the scribe in partly the wrong order. However, this minor aberration can easily be unravelled.

Major Hamilton, the Troop Commander, wrote a series of reports during and immediately after the operation. They are reproduced in *Extracts from the Proceedings of the RE Committee 1880*, Appendix 1, RE Library, Chatham.

9. Described to the Royal United Service Institution on 15 February 1884 in a presentation entitled *Field Telegraphs in Recent Campaigns* by the then Lt Col A C Hamilton, formerly the 'C' Troop Commander, and recorded in *The Journal of the Royal United Service Institution*, Vol XXVII, p 332. The short description of the Zulu War in *The Royal Corps of Signals. A History of its Antecedents and Developments* by Maj Gen R F H Nalder (pp 24-25) is inaccurate in its description of the cable taken and its deployment, implying that there had been a miscalculation in the stores requirement. It is also inaccurate about Prior's description of signalling to Eshowe. The above description, from a reliable contemporary source, should have corrected those errors.

10. A biography of James Scott Douglas will be found in *The South African Campaign, 1879* by MacKinnon and Shadbolt, published 1881. An obituary was published in the *Illustrated London News*, 23 August 1879.

11. *Narrative of the Field Operations, The Zulu War of 1879.* p 114.

12. Extracts from the *Proceedings of the RE Committee 1880*, p 129. Major Hamilton, in a post-war letter dated 15 May 1880, explained the problem and asked to be equipped with suitable relay or translator equipment.

13. Extracts from the *Proceedings of the RE Committee 1980*, Appendix II. Bindon Blood's report also includes an interesting account of road building and bridging work.

14. Extract from the *Proceedings of the RE Committee 1880*, p 142. Letter by Major A C Hamilton dated 25 November 1880. Major Webber was at the time commanding a Postal Telegraph Company, and presumably had a source of supply of the new telephones. Webber later came to S Africa with Wolseley, serving on his staff as AA & QMG, with responsibility for the western lines of communication.

15. Extract from the *Proceedings of the RE Committee 1980*, Appendix 1. Report dated 1 October 1879.

16. The work was described in the *Journal of the Society of Telegraph Engineers, 1882,* Vol XI, pp 312-341. *The Organisation and Operations of the Field Telegraph Corps in the Transvaal, 1881* by Lieutenant Arthur H Bagnold RE.

Chapter 12

The Campaign in Egypt, 1882

Historical Background

Egypt had been conquered by the Ottoman Turks in 1517 and had effectively become a province of Turkey. Napoleon invaded in 1798 and the French occupied the country from then until 1802, when it returned to Ottoman control. In 1869 the Suez Canal was opened, built by the French but with the Egyptians owning a large number of shares in the valuable company. In 1875 the Khedive (Viceroy) of Egypt, a hereditary position then occupied by Ismail Pasha, under the nominal suzerainty of the Sultan of Turkey, ran into financial problems due to a combination of mismanagement, prodigality, corruption, and intrigue. Whilst some money had been spent usefully on infrastructure such as railways, telegraph, and canals, huge amounts had been squandered on personal projects such as palaces.

The British Prime Minister, Disraeli, paid off the Khedive's debts and acquired shares in the Suez Canal company, at that time worth some £4 million. It was a minority stake, rather less than half the total capital, but it gave London a large share in controlling its strategic interest in the route to the principal asset in its empire, India. Until then Britain had no interest in becoming involved with Egypt.

Things went from bad to worse for Khedive Ismail, and in 1876 he became bankrupt. To protect their financial and strategic interests, Britain and France found themselves drawn into the affairs of Egypt and took a leading role in its administration and financial control – an unenviable task in a regime then characterised by inefficiency and corruption. The Sultan of Turkey deposed Khedive Ismail in 1879 (by means of a cryptic telegram addressed to the 'ex-Khedive'), and he was exiled to Naples. His son, Tewfik, was installed in his place. Captain Evelyn Baring (later Lord Cromer) and a Frenchman, M de Blignières, were appointed as Controllers-General. The many economies the new ruler, Tewfik, was forced to make caused discontent, none more so than amongst the army where officers were dismissed without pay. This led to a nationalist movement headed by one of them, Colonel Arabi, better known as Arabi Pasha, who stirred up resentment against the foreign interference; it was principally directed at the Turkish clique that was believed to control the new Khedive, but was more widely construed as anti-European.

Between 1881 and 1882 turmoil spread and Colonel Arabi led a revolt. Britain and France became involved because of their financial and strategic interests, and supported the Khedive's rule and Turkish suzerainty. The Turks refused to take action, so France and Britain 'demonstrated' their fleets off Alexandria but that served only to inflame matters further. In Alexandria some fifty Europeans were murdered, so gunboat diplomacy was controversially invoked, and the British navy bombarded the harbour forts. The French, at the time concerned primarily about German ambitions towards their own country, abstained from warlike action in Egypt. Britain alone was left to restore order, although many in England disliked the idea.

Typically, attitudes polarised. In Egypt the British were seen as bullying aggressors about to invade and take over their country, and in Britain the Egyptians were described as terrorists who had reneged on commitments and (conveniently forgetting the bombardment of Alexandria) murdered Europeans, making it necessary for military action to restore law and order. Thus the inept diplomacy led from riots to ultimatums, and then to a British military campaign in 1882 to defeat Colonel Arabi and his supporters. 'C' Telegraph Troop took part in the campaign; it was their second operation. The situation to face them there was entirely different from that in Zululand.

The Campaign

The expeditionary force sent to Egypt to crush Arabi's rebellion was an Army Corps of two Divisions and a Cavalry Division, under the command of Lieutenant-General Sir Garnet Wolseley GCB. The troops were drawn from England and various overseas stations including India, the force totalling nearly 40,000 men. The numbers were overwhelming and the campaign that followed was short and decisive. Wolseley's orders were to support the authority of the Khedive by suppressing the revolt under Colonel Arabi, and he was given a free hand as to how he did it – just the sort of situation Sir Garnet revelled in.

THE EGYPTIAN CAMPAIGN, 1882

Arabi had arranged his defences on the assumption that the force would land at Alexandria. In a masterly deception plan, Wolseley feinted at nearby Aboukir Bay but caught him by surprise and instead seized and defended the Suez Canal, landing most of his force at Ismailia, halfway along the Canal, and establishing the Force Headquarters there. That he was able to do this depended on good communications – but not military ones, as will be explained shortly. Meanwhile, the telegraph line from Egypt to Syria and on to Turkey was cut by the British troops.

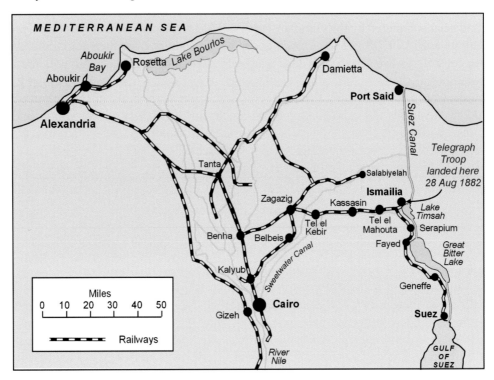

The Telegraph Troops

Lieutenant Colonel C E Webber was appointed the expedition's AQMG for Telegraphy (reflecting the staff organisation of those days, when communications were a responsibility of the QMG's department). Charles Webber, it will be recalled, had commanded the 22nd Postal Telegraph Company when it was formed in 1870, had accompanied Wolseley to South Africa in 1879 and, in a role dealing mainly with administration and logistics, had served on his staff in Natal and the Transvaal immediately after the Zulu War, so was well-known to him. In 1882 Webber was also the President of the Society of Telegraph Engineers, so in August and September, while most of the Society's members were enjoying their summer holiday and probably did not notice his absence, he was recalled by Wolseley to serve in Egypt. Back in London in November 1882 he did, however, tell the members all about it at a meeting of the Society in November.

'C' Telegraph Troop received its warning order on 8 July. The Troop was reinforced by two officers and sixty-two NCOs and men of the 22nd and 34th Postal Telegraph Companies. Altogether six officers, 184 NCOs and men, and sixty-five horses and wagons, with eighty miles of field cable and 300 miles of airline set off for Egypt, the lessons of the Zulu War and insufficient line having apparently been digested. The officers were: Major Sir A Mackworth ('C' Troop Commander), Lieutenants H J Foster, F G Bond, and R W Anstruther, all from 'C' Troop, and Captain M D Whitmore and Lieutenant R L Hippisley from the Postal Telegraph Companies. As well as Lieutenant Frank Bond, Troop Sergeant Major A J Lewis and many others in the Troop had served in the Zulu War in 1879. There was also one officer from the Army Medical Department, Surgeon C P Turner. In addition, a contingent was sent from India.

Captain Mortimer Whitmore was despatched to Woolwich arsenal to supervise the supply and loading of additional stores. Also on board the same ship were other engineer units with large quantities of stores, and the stevedores at the dock sandwiched the equipment wherever they could fit it to trim the ship properly. The ship was trimmed but its freight was in a muddle, which was to become a great problem when it was time to unload. The troops embarked for Egypt on the hired transport SS *Oxenholme* at the West India

Docks, London on 9 August 1882. The ship made an exceedingly slow passage and anchored at Ismailia, which had been seized on 20 August, at 4.30 pm on 28 August, well after the majority of the force had landed and deployed. As soon as they arrived they found that their services were urgently needed.

The Indian contingent brought the Mule Telegraph Train of the Madras Sappers and Miners under Lieutenant J E Dickie as Superintendent of Army Signalling. They brought with them heliostats for visual signalling and ten miles of cable. The heliostats turned out to be unsuccessful due to the flat country and the effects of the mirage caused by the heat. They were not called upon to undertake any technical work, although during the subsequent advance they laid a back up cable but it never needed to be used. This was the first time that telegraph troops of the Indian Army had operated in support of the British Army in what may be called a European expedition. There had of course been a joint Anglo-Indian force in the Abyssinian Expedition (described in chapter 6), but that expedition preceded any formal telegraph organisation in both armies, and on that occasion the contribution from India had been beset by problems.

The Egyptian Postal and Telegraph Administration

Egypt had developed a network of telegraph and railway infrastructure in the Nile Delta – an area generally known as Lower Egypt. The Egyptian State Telegraph Department had apparently never troubled itself to produce a full circuit arrangement, so the exact detail of such things as number of wires on a route, and their circuit allocation, was not available to the British beforehand; only the general coverage was known from a map provided by the Intelligence Department. The map is not reproduced here, but relevant parts will be explained when necessary.

The State Telegraph department was staffed mostly by Egyptians, but with some Europeans of different nationalities – British, French, Italians, and Germans. The telegraphists were mainly Egyptian, and about three-quarters of them could only operate the telegraph in their own language; those who could operate in other languages could do so only slowly – some seven to eight words per minute, which was quite inadequate for a working circuit. The linemen were nearly all Egyptian, with little technical knowledge, unable even to test a line or prepare the batteries.

With the outbreak of anti-European rioting that had taken place on 11 June, most of the European staff had left and, with its supervisory staff gone, the State Telegraph department was in an even greater degree of disorganisation than usual. Most of the Egyptian telegraphists sympathised with the rebellion and were of course averse to the British invasion. The aim of the expedition was to crush the rebellion and not to destroy or take over the Egyptian infrastructure, which it was recognised was their property. Thus although at the start it was the Egyptians who damaged their own telegraph lines and equipment, to prevent the British using them, they later became quite co-operative when it was realised what the British policy was, that everything was going to be handed back to them to continue as before, that their jobs were safe, and the hysteria surrounding Arabi's revolt had been quelled.

Strategic Communications

The British-owned Eastern Telegraph Company maintained and operated its own lines in Egypt, connecting Suez, Cairo, and Alexandria. This was the overland section of the submarine cable route that now went from Alexandria westwards through the Mediterranean to Malta, Gibraltar, Lisbon, and London, and eastwards from Suez down the Red Sea to Aden, Bombay (installed in 1870, as described in chapter 4), and the east coast of Africa (1879). A map of the relevant telegraph lines in Egypt is shown on the next page.

The Eastern Telegraph Company's overland line between Alexandria and Suez was constructed of two wires on iron poles, running alongside the railway on the east of the Nile through Tinta, Benha, and Kalyub to Cairo. It returned from the Cairo telegraph office to Benha, and then ran, again beside the railway, through Zagazig to Nefisha junction (just west of Ismailia), where it turned south to Suez. Rather than rely on the Egyptian State Telegraph to operate and maintain its strategically and commercially important line, these responsibilities were undertaken by the Eastern Telegraph Company itself.

The line was well constructed but tended to suffer from loss of insulation caused by early morning dew on the metal poles. This was the dichotomy: metal poles were stronger and, despite the initial expense, lasted longer, but they were susceptible to conducting current to earth when damp, and this of course adversely affected the transmission quality of the line. Wooden poles were cheaper in the first instance, and in damp conditions tended not to leak current to earth so much, but went rotten or were attacked by white ants, so did not last as long and needed much more maintenance – something that was often neglected.

THE EGYPTIAN CAMPAIGN, 1882

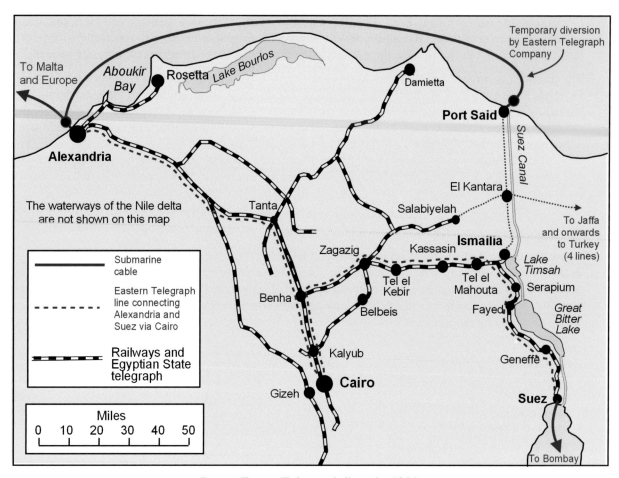

Lower Egypt. Telegraph lines in 1882.

On 11 July 1882 the rebels cut the Eastern Telegraph landline through Egypt interconnecting the submarine cables landing at Suez and Alexandria. Since the original Indo-European route had been opened in 1870 the number of cables to other destinations had grown, and the landline was a vital telegraph artery handling all the traffic to numerous countries in the Far East, Australia, New Zealand, and from Aden down the east coast of Africa to South Africa. The result of the rebel action was that all this traffic was diverted from Aden to Bombay and thence through the Indo-European route through Persia and Russia, increasing the traffic load on that route enormously. Fortunately there was sufficient capacity, and these strategic links coped with the additional traffic. The only threat was that a concurrent disturbance at Kashan, in Persia, meant that the lines there were threatened by rioting citizens, but fortunately no damage was done.

Even before then, with unrest in Egypt increasing as Arabi fermented discontent, and knowing its land lines through Cairo to be vulnerable, the Eastern Telegraph Company took it upon itself to make its communications more robust. This initiative, which was to be crucial to the successful operation of the rear link for the expeditionary force, but is rarely mentioned in descriptions of the campaign and its communications, is worth recording. It was described by Mr W T Ansell of the Eastern Telegraph Company at the meeting of the Society of Telegraph Engineers in November 1882, when their President, Lieutenant Colonel Charles Webber, was reporting on the army telegraph work undertaken during the campaign. Apparently somewhat peeved by Webber's failure to mention anything about the Eastern Telegraph Company's role, Ansell felt it appropriate at the end of the lecture to describe their work. This is what he said:

> The communication was kept up in the first place entirely by submarine cable, and but for the forethought of my directors, which enabled them at a moment's notice, I may say, to lay a cable from Alexandria to Port Said, Sir Garnet Wolseley would have found himself a long way off from his headquarter base [at Ismailia]. It may be of interest if I explain how this came about. When the events of 11th June [riots in Alexandria and the murder of British subjects] startled everybody, it was clearly seen we were on the eve of some very great

events. The Eastern Telegraph Company had one of their ships in the Indian Ocean; she had come into Aden for stores and coals. My directors immediately ordered her up to Suez, and to make her way through the Canal if she could get through. It was felt a very touch-and-go matter then as it was doubtful whether the Egyptians would stop the passage of the Canal. However, they did not do so, and the *Chiltern* [a cable laying ship] got through, and made the best of her way to Alexandria. There she was converted into a floating station, and until long after the bombardment the whole of the staff of the Eastern Company were located on board the vessel, did all the work, and kept open the line of communication.

The cable which was laid from Alexandria to Port Said came about in this way. It was only a day or two before the general commanding-in-chief [Wolseley] left England that he intimated the desirability of laying cable communications between Port Said and Alexandria. The Eastern Telegraph Company happened to have a cable ship, the *John Pender*, engaged upon some works on the coast of Portugal, and she made the best of her way, according to her orders, to Alexandria, and she was fortunate enough to be able to lay that cable from Alexandria to Port Said almost within a day or two from the time when the use of it was required by the general commanding-in-chief. [1]

Thus it was important that the line running along the length of the Suez canal from Port Said to Suez was secured, not just for communication with the Army HQ at Ismailia during the campaign, but for the ongoing telegraph link between Britain and India and other eastern destinations. The Eastern Telegraph Company had commercial as well as military consequences to consider.

The securing, operation, and maintenance of the telegraph line alongside the Suez Canal between Port Said and Suez became of great strategic importance, although to begin with there were no army telegraph troops there to undertake the task. The first to arrive in Egypt were forces of the Indian contingent, and the staff of the Eastern Telegraph Company rendered great assistance to them, repairing the line after it had been damaged by Egyptians. The navy, assisted by marines, secured and then guarded the canal itself, and small parties were deployed at posts along its length, from which they patrolled regularly. The Suez Canal Company were, to begin with, wary of British military intentions, but as the campaign progressed, and it was made clear to them that their property and livelihood were not under threat, they became friendly and quite co-operative.

Later, when 'C' Troop arrived, some of them, mostly the soldiers of the Postal Telegraph Companies, were assigned to the work of repairing and operating the line between Port Said and Suez. On this task they worked under Lieutenant Colonel Charles Webber, while the remainder worked under Major Sir Arthur Mackworth in close support of the fighting force as it advanced from Ismailia to Cairo under Sir Garnet Wolseley.

The episode demonstrates the haphazard way in which the strategic communications to support the army were handled in those days, a concomitant of the staff organisation of the Victorian army. Despite improvements to capability brought about by the technical innovation of the Victorian era, there was simply no higher level function in the badly organised staff to deal with these things. There was no signals staff forming part of any headquarters. This was to continue until the end of the century. The organisation of army signalling did not develop on a broad enough front. In the case of the campaign in Egypt in 1882, had it not been for the initiative taken by the Eastern Telegraph Company, the rear link communications for the force would have been extremely vulnerable, and their landline through Lower Egypt would almost certainly have been destroyed and could not have been reached to repair it.

The Field Telegraph

Turning now to the main campaign, on arrival at Ismailia after their slow passage Major Mackworth learnt that the arrival of the field telegraph troops had been anxiously awaited. The advanced brigade, under Major General Graham, had pushed forward to Kassassin, about twenty-two miles from Ismailia, and was without communications to the base at Ismailia. The telegraph line alongside the railway had been cut by the retreating rebels and until the Telegraph Troop arrived there was nobody with the necessary skills or equipment to repair it. There was an urgent need to open up communications with Kassasin as soon as possible but the badly loaded stores delayed matters considerably, and it took from 6.00 pm on the day of landing, 28 August, until the following evening, working hard in the stifling heat, to unload and sort out but a part of the equipment. Many other ships had been badly loaded, so this was a lesson for the future, although it is is doubtful if it was much registered by the dockers in England.

On the morning of 30 August 1882 Sir Arthur Mackworth went by train to Kassassin to conduct a reconnaissance. Although he makes no mention of it, he must have suffered in the tremendous heat in the Egyptian desert in August. Near Tel el Mahouta the train derailed and blocked the line, but he continued by foot and discovered that on the north side of the line there were two wires on iron poles belonging to the Eastern Telegraph Company (not presently being used by them as the circuits had been diverted along the Suez Canal, as described above), and on the south side three wires on wooden poles belonging to the Egyptian Government. All five had been damaged by the enemy, but the Eastern Telegraph line less so than the other, which had several long gaps in it where both wires and poles had been removed. At the stations there were neither batteries nor telegraph instruments.

As a result of his inspection Sir Arthur decided to use the Eastern Telegraph line, at least initially, and then to use the best portions of either to get one wire through without delay. Using his field cable was at this point not an option; it was still deep in the hold of the ship, and it was to be another two weeks before all the equipment was unloaded, "although Captain Whitmore and Lieut. Anstruther with plenty of men worked with all their might at them", he reported afterwards.

He also saw that the pressure on the single railway track was such that it was hopeless to depend upon it for the transport of telegraph stores, and so he decided to get what stores had been unloaded up to Tel el Mahouta that night and, working outwards from Tel el Mahouta, make two separate circuits, one to Ismailia and the other to Kassassin. When this was done the circuits would be joined together.

Sir Arthur returned from his reconnaissance later that day, 30 August, and the plan was made. Lieutenants Mortimer Whitmore and Ralph Anstruther were to continue unloading the ship, and then Anstruther was to work under Lieutenant Colonel Webber with responsibility for the repair and operation of the line between Port Said, Ismailia and Suez, leaving Mackworth free to attend to the field telegraph in support of Sir Garnet Wolseley's operations and the rear link from Wolseley's forward HQ to Ismailia. Lieutenants Frank Bond, Richard Hippisley, and Hubert Foster were to work with Sir Arthur, while Lieutenant John Dickie and Lieutenant P B Baldwin of the Indian contingent were to lay a separate back-up cable south of the Sweetwater Canal. Hippisley was sent ahead immediately with a light detachment to make a start on repairing the line before darkness, while Sir Arthur and the mounted portions of No 1 and No 2 Sections followed up more slowly. The dismounted portion, under Lieutenant Bond was ordered to follow as soon as possible in the morning by train. No 3 and No 4 Sections and the detachments of the Postal Telegraph Companies were left behind and were assigned the task of working on the line Port Said-Ismailia-Suez.

Sir Arthur described their march to Tel el Mahouta:

> ..... Our march over the desert was more picturesque than agreeable. A glorious moon but the air sultry and the dust stifling; also the wretched horses were still feeling the effects of the voyage, and the draught through sand fetlock deep was very heavy, but everybody put a good face on it, and by daybreak on the 31st we were in camp at Tel el Mahouta ... . (2)

Lieutenant Bond and his men arrived at 11.00 am and got to work. Lieutenant Hippisley and a working party went on to Kassassin by the same train that had brought Bond, to work back to Tel el Mahouta. Sir Arthur worked with another party to meet him. Bond went back to Ismailia to work up from there, and Lieutenant Foster worked to meet him. By 4.30 pm that day, 31 August, they got their first message through from Kassassin to Tel el Mahouta, and by midday the next day, 1 September, they were through from Kassassin to Ismailia.

There was a lot of work to catch up on. Communication was difficult and uncertain. At times there were heavy blocks in the telegraph lines and much impatience on the part of senders, some of whom seemed to expect the sort of service they got in England. Indeed, one senior officer complained bitterly that he had dispatched three simultaneous messages from Ismailia to Kassassin, one by boat on the canal, the second by the railway, and the third by telegraph, and that they had arrived at their destination in precisely that order! But 'C' Troop pressed on, and in time order was produced out of chaos. It would, of course, have been better if they had been sent with the first troops to deploy, rather than the last, but as mentioned before, staff planning had not yet got used to the idea of getting communications established at an early stage.

During the next five days, to 6 September, they established offices at Mahsamah railway station, near Kassassin, at Mahsamah Camp where the Cavalry Division was based, and at General Graham's headquarters in Kassassin Lock. Also, under instructions from Lieutenant Colonel Webber, they repaired the three wires on the Egyptian poles using their 3-strand galvanised wire and field poles to replace the damaged portion. It was arranged that the top wire of the three should be the express wire, to be worked with the Morse recorder ('inker'); the second wire as a circuit for intermediate offices, worked with sounders, and the third wire was for railway traffic, with the single needle telegraph instrument. This arrangement was carried out and completed by 6 September.

During this period also, Lieutenant Anstruther and his men had opened offices at Serapeum, Fayed, and Geneffe on the Suez-Ismailia line; and the mounted portion of Nos. 1 and 2 Sections were employed in bringing up more stores from Ismailia to Tel el Mahouta and Kassassin.

Lieutenant Frank Bond was attached to the Cavalry Division with a small party of mounted telegraphists, to accompany it as it advanced, to repair or destroy telegraph lines, or to seize and work offices. He and Foster also superintended the laying of short branches of line that were required to connect up headquarters of brigades.

On 7 September the stores of Nos 1 and 2 Sections were moved from Tel el Mahouta to Kassassin, to form a depot there. Sir Arthur was then warned to be ready to accompany General Graham and provide communications for a reconnaissance he intended to make next day into the cultivated valley on the south side of the canal. He was advised not to take a cable wagon as the ground was cut up by irrigation ditches, so a drum of cable was rigged up on a hand barrow, with another to follow with reserves of cable.

Before daybreak on the 8th they set off on the reconnaissance, the party consisting of one cart under Lieutenant Foster paying out the cable, a second cart with the cable reserve following under Lieutenant Bond, and an escort of six mounted men of 'C' Troop under Troop Sergeant Major Lewis protecting them. Sir Arthur recorded that they reached their first objective at 6.00 am and

> ... were 'through' to Kassassin the moment we made earth, working with the Theiler sounder and the telephone. General Graham sent a message to General Willis and shortly after, the enemy appearing in considerable force, sent another message, and at the same time told me to reel up and retire. This we did as well as we could, but I sorely repented having used the tip-cart, because, as it was not provided with automatic winding gear as the wagon would have been, the cable had to be reeled up by hand, which was necessarily a slow operation.

The enemy advanced rapidly and opened fire on the detachment from about 600 yards. At this point Sir Arthur was ordered by General Graham to cut the cable and withdraw with the rest of the reconnaisssance party.

In the afternoon Sir Arthur obtained permission to go out and recover the cable:

> ... and accordingly at 3 p.m. (the hottest part of the day, when the enemy were generally taking their siesta) we started reeling up the cable from the home end. Lieut. Foster as before was in charge of the job, and this time he had the cable wagon with automatic winding gear, which enabled him to work at a trot. Sergeant-Major Lewis and I divided our little escort of 9 men between us, and scouted to both flanks. Our excellent *hakim* [medic] Turner accompanied us in case he was wanted. We succeeded in reeling up to within a short distance of the point where we had cut the cable in the morning, when finding that the enemy had removed the remainder of it, and we were about 400 yards from a picket of theirs, and thinking that a few yards further might tempt him to open fire on us, I gave the order to retire with what we had saved. The portion we lost was less than half a drum or about 400 yards out of the 3½ miles we had payed out in the morning.

The next morning, the 9th, there were reports that the enemy were advancing in force, so as a preparatory measure the horses were harnessed before sitting down to breakfast. During breakfast a shell passed just over the tent they were in, and burst a short distance away without causing any damage. Sir Arthur mounted his horse when another shell arrived "passing close over my horse's loins and bursting at his side. Lieut. Foster, who was coming out of the tent at the moment, judiciously threw himself on his face and escaped. with a good smothering of sand. Surgeon Turner also nearly came to grief as he persisted in finishing off neatly a puttie which he was putting on at the time". More shells arrived, the horses were unpicketed, and all took cover safely under the canal bank. Sir Arthur commented that "it is difficult to understand how so hot a fire in so confined a space could have been so harmless, but the deep sand prevented the fragments of the shells spreading much".

THE EGYPTIAN CAMPAIGN, 1882

While this barrage was in progress, telegraphists under Corporal Chapple were steadily working in a tent with shells bursting around them, the tent hit more than once by fragments while they cooly sent their messages. Quite an exciting time for them! It was the first time 'C' Troop had been in the forefront of an action. (A later similar claim by a Section of what by then was the Telegraph Battalion at the attack on MacNeill's zariba in Suakin in 1885 must be rejected.)

On 12 September, Lieutenant Anstruther and No 3 Section joined Sir Arthur at Kassassin for the march on Tel el Kebir. They were to follow the left centre of the advance on the north side of the canal with a field cable, and the Indian mule telegraph train (under Lieutenants Dickie and Baldwin), which had been placed under Sir Arthur's command, was to do the same for Sir Henry Macpherson's advance on the south side. (Major General Sir Henry Macpherson VC KB, commanded the Indian contingent.)

Just before dusk a detachment of 'C' Troop under Corporal Elsmore worked under the direction of the Brigade Major (Fraser) to lay out a line of telegraph poles which were to guide the night advance of the Highland Brigade – a novel use for telegraph poles! This line of poles extended over two miles in front of the British outposts, and as they were on foot, and likely be surprised by the enemy's cavalry, it was a potentially dangerous task.

At 7.00 pm on 12 September they began to advance towards Tel el Kebir, reeling out line after having connected it to the permanent line along the railway to keep them in communication with Ismailia. Foster was in charge of this. Bond was with the Cavalry Division, Anstruther was to follow up the Indian cable line with an overhead line, if required, and Hippisley was to repair the wires along the railway towards Tel el Kebir as far as circumstances would permit.

Sir Arthur describes how they moved across the desert, steering by the stars as they laid out the cable (luminous compasses were not developed until some three years later). At this stage they were leading the advance as the infantry did not start until later, and thought it prudent to wait. At midnight the commander, Sir Garnet Wolseley, came up and sent a message through the line to Sir H Macpherson. After a pause they moved forward again and in the breaking dawn could discern troops just to their right. To Sir Arthur's pleasure they turned out to be the Headquarters staff, "so that we had hit off our position to a T".

In daylight the enemy now started shelling them, with one or two "unpleasantly near us". Sir Garnet ordered them to retire. During the period some messages from England for Sir Garnet were received over the cable, and Sir Arthur delivered these to him personally. He then received orders to go with all speed to Tel el Kebir station, about three miles away and so "I left the main body to follow at the ordinary pace and told Lieut. Foster to bring on the cable wagon and a cart with reserve cable at a good round trot". They completed the three miles cable-laying in half an hour, finding it quicker to let the cable take its chance, and mend it on the few occasions that it broke due to the excessive speed.

Things continued apace. They bustled through the lines of Tel el Kebir hardly stopping to look at the results of the battle there, and reached the railway station about 8.00 am. There they established a field telegraph office in the saloon carriage which Arabi had travelled in the day before, and on making earth were immediately through to Kassassin. Sir Arthur took some pride in this:

> We had ten miles of cable out – which means 12 drums, and consequently 11 joints, in addition to the five or six joints due to the cable breaking in the last three miles, and I think it scores to the credit of the detachment who worked the cable that in spite of these numerous joints, and the hurry and excitement under which they were made, (and some of them in darkness) every joint was honestly finished off that there was no fault along the line from the time it was laid until it was taken up the same evening.

> As soon as we were ready Sir Garnet's Military Secretary handed me the message for transmission to the Queen, the Secretary of State for War, and others, announcing the victory. These were sent off at 8.30 a.m. and at 9.15 we received Her Majesty's reply.

> I believe this to have been the first occasion on which a British General has been able to telegraph the news of his victory from the actual field of battle, …

The telegraphist who sent and received these messages was 2nd Corporal W F Seggie who, later, in recognition of his service, went to Windsor Castle to receive his campaign medal from Queen Victoria personally.

It was certainly a vast improvement on the length of time it had taken to get the news of the defeat at Isandlwana to London during the Zulu War three-and-a-half years before, but it was also due to the foresight of the Eastern Telegraph Company described earlier.

THE EGYPTIAN CAMPAIGN, 1882

The field telegraph to Tel el Kebir was extremely busy all day, using the Theiler sounder and telephone. This was similar to the arrangement used in the closing stages of 'C' Troop's operations in the Zulu War, and had become the method of transmission when the line was long or its insulation quality was poor. The conventional Morse sounder used on its own was not sensitive enough to detect the low level of current flowing.

By the evening the permanent line had been repaired by the team working under Lieutenant Hippisley and the circuit was transferred to that. The cable was then reeled in ready for any further move.

The next morning, the rebels in wild retreat, Sir Arthur received orders to take four parties of telegraphists to man forward offices at Zagazig, Benha, Belbeis and Kalyub. Sir Garnet Wolseley came to the telegraph office and 'talked' to several senior officers (using the telegraphist at each end to conduct a one-to-one conversation). After a delay due to problems with the railway, Wolseley, his headquarters staff and others, and the elements of 'C' Troop moved forward by train. Lieutenants Hippisley and Anstruther accompanied Sir Arthur on the train, and Lieutenant Foster stayed at Tel el Kebir. Lieutenant Bond had gone on with the Cavalry Division to Belbeis the previous night. They reached Zagazig at 9.30 pm, and had a heavy night's work in the telegraph office there.

The next morning, continuing the now rapid advance to Cairo, the Highland Brigade started for Benha. Sir Arthur left one of his detachments at Zagazig, arranged for another to go on the first train to Belbeis, and then himself took the first train to Benha, leaving Hippisley and Anstruther to follow in the next train with the remainder. Sir Arthur describes how, in a dense fog:

> … the ADC and I rode on the engine in charge of the driver who was a native, and as it was not certain when we might find the rails taken up and could hardly see ten yards in front of the engine we had a little pleasant excitement but nothing came of it, and we entered Benha peaceably enough, closely followed by the train conveying the HQ staff.
>
> I was able to report the telegraph offices ready for use and open to Cairo and Zagazig very shortly after we arrived, and almost immediately received a telegram for Sir Garnet informing him that the Cavalry had entered Cairo without opposition that morning. A train was immediately prepared to convey the Guards and the Headquarters staff on to Cairo, and (leaving Lieut. Anstruther and a party of telegraphers at Benha) I took Lieut. Hippisley and the fourth party on with me by this train. We entered Cairo at 9.15 a.m. and found that Lieut. Bond had already seized the railway station telegraph office, which I believe he did with only three men of 'C' Troop, disarming an Egyptian guard of about twenty men, placing one of his men over the telegraph office … and sending the third to report to the General of Cavalry what he had done.
>
> We soon got the wire to work here, Lieut. Hippisley taking charge, while I went to the central office in the town and replaced the native clerks by our own men.

Meanwhile, back in the base area near Kassassin, the interface for message traffic between the military and the strategic communications, the telegraph system was being overwhelmed. Lieutenant Colonel Webber described the conditions in his lecture to the Society of Telegraph Engineers in November:

> The Kassassin office was a very curious scene. In the first place, the heat in the tent was intense; the flies covered almost everything, the faces and hands of the clerks [telegraphists], and every part of the instruments and batteries. The batteries were every now and then giving out on account of the intense heat. Many people were coming to the office paying for messages, with the appeal 'Do take my message – pay anything for it', and from the front mounted messengers were coming in with affectionate messages from various members of the army to their friends at home, telling them they were well, and that they had escaped, and so forth. … As the news [of the victory] flashed all over the country, every affectionate relation and loving heart wanted to know how their dear one was, and sent a telegram; and these poured in from the other end till at last messages came from Port Said and Ismailia to say that they did not know what to do, that the line was blocked, and that there were some 20 hours delay in traffic, which you can readily imagine was the fact.

The short campaign was virtually over. Sir Arthur describes how Bond and Hippisley were both exhausted after their exertions. Hippisley had to be left in Malta on the way home, seriously ill, but was to make a full recovery. (He later became Director of Army Telegraphs in the Boer War.) Sir Arthur concluded by expressing his thanks to the Troop:

> … This may not be the proper occasion for it, but I find it difficult to refrain from expressing my gratitude to each one and all of our fellows who were with me for their hearty and loyal co-operation, and I can feel proud of having commanded the Field Telegraph Corps in the first campaign in which they actually had to work under fire.

THE EGYPTIAN CAMPAIGN, 1882

Assessment of the Communications

So how good were the communications? Visual signalling had been unsuccessful. One staff officer, who had served in the Zulu War three years previously, wrote that: "The Corps of Signallers [*ie* regimental signallers formed into a pool under a signalling officer] were no use at all. This was due to no fault of theirs, but to the mirage. South Africa is the country of all others for signalling. The frequent hills and mountains render only a few posts necessary, and the mirage is rare. But in Egypt, besides this nuisance, the signallers must be stationed at short intervals, owing to the flatness of the country, and consequently are of little use". The signallers of the Indian contingent under Lieutenant Dickie, who had brought heliostats with them, found the same problem. Visual signalling thus contributed little to the campaign, low-level communications resorting to messengers and gallopers.

The telegraph was far from perfect. It was not best suited to the fast moving campaign, when visual methods would in theory have been better, but it had provided a very useful service in difficult conditions with numerous problems to be overcome.

Back in England

'C' Troop returned to Portsmouth on 20 October, and the men from the two Postal Telegraph Companies dispersed and returned to their units. On 18 November the 'C' Troop Record states that: "The Troop without horses and waggons proceeds by rail to London to take part in Her Majesty's review of the troops lately returned from Egypt." And on 21 November, "Lieut. Col. Sir A. W. Mackworth, Sergeant Buck (22nd [Postal Telegraph] Company), and 2nd Corporal W. Seggie are decorated with the Egyptian war medal at Windsor by Her Majesty The Queen; the remainder of the Troop [are decorated] at Aldershot by the C.C.R.E." It is surprising how quickly things such as campaign medals were struck and presented in those days.

Other honours were distributed. Sir Garnet Wolseley was promoted General, received the thanks of parliament and a grant of £30,000, and was created a baron.

Lieutenant Colonel C E Webber was created a CB and was awarded the Order of the Medjidie 3rd Class. Webber was to serve once more with Wolseley, on the Nile Expedition of 1884-85. For his service Major Sir Arthur Mackworth was gazetted Brevet Lieutenant Colonel for meritorious service in the field, and also awarded the Order of the Medjidie 3rd Class and the Khedive's Star.

It was the second but also the last operation to be undertaken by 'C' Troop and the two Postal Telegraph Companies before the three units, with the aim of creating greater organisational flexibility, were amalgamated into the Telegraph Battalion in 1884.

Endnotes

1. *Journal of the Society of Telegraph Engineers*, 23 November 1882, p 547.
2. *The Field Telegraph Corps in Egypt.* The Royal Engineers Journal, vol 112, 1882, pp 269-272.

THE EGYPTIAN CAMPAIGN, 1882

Colonel Sir Arthur Mackworth CB (Baronet)

Sir Arthur Mackworth was born in 1842 and inherited the title from his father a few weeks before his tenth birthday in 1852, becoming the sixth baronet.

A little delving into an old copy of Burke's Peerage reveals a bit about the Mackworths. The family was of considerable antiquity in Shropshire and Derbyshire (where the village of Mackworth will be found a few miles north-west of Derby), and one of its members fought at Poitiers in 1352. Sir Francis Mackworth, a distinguished Royalist, fought on the side of Charles I. Colonel Humphrey Mackworth, a man of considerable note, was MP for Salop, Governor for Shrewsbury, and one of Cromwell's Council. The Baronetcy was created in 1776, as Mackworth of Gnoll Castle, Glamorganshire, and the 1st baronet sat as MP for Cardiff.

He married Alice Cubitt in Paris in 1865, and between 1866 and 1887 this union brought forth seven sons and five daughters – a progeny that might well have been curtailed in 1867 when Sir Arthur nearly died in a sailing accident in the River Medway when a sudden strong wind capsized their boat and one of the crew of three drowned.

Sir Arthur retired from the army in the rank of Colonel to become prominent in Monmouthshire as the Honorary Colonel of the 1st Battalion the Monmouthshire Regiment, a JP, as a Deputy Lieutenant, and as Chairman of its Territorial Forces. He was awarded the CB in 1897. He died in March 1914.

His fourth son, Harry Mackworth, followed in Sir Arthur's footsteps by joining the Telegraph Battalion, and finished his service in the Royal Signals in 1923 as Colonel Sir Harry Mackworth.

The Campaign Medal.

The Sphinx is shown on an ornamental pedestal with the word Egypt above and 1882 below. The head of the Queen on the obverse is similar to that on the Ashanti Medal. Two bars were originally issued with this medal - 'Alexandria 11th July' (a naval engagement), and 'Tel-el-Kebir'.

The Khedive's Star.

All troops who took part in this and subsequent Egyptian campaigns received the Khedive's Star, a dull bronze medal of five points. On the reverse, in the recessed centre, is the Khedive's monogram T.M. - Tewfik Mohammed - surmounted by a crown and crescent.

The Order of the Medjidie.

The Order was instituted by the Sultan of Turkey, and from the Crimean War onwards it was awarded to many thousands of officers of the British Army and Navy. Five classes of the Order were issued, according to the rank of the recipient, 1st Class being the highest. The medal consists of a silver star of seven points, between which are seven small crescents and stars of five points.

Chapter 13

The Bechuanaland Expedition, 1884-85

Introduction

In March 1884 an organisational change took place, as described briefly at the end of chapter 7. 'C' Telegraph Troop and the two Postal Telegraph Companies, 22nd and 34th, were amalgamated. Initially called the RE Telegraph Corps, the title soon changed into the Telegraph Battalion, or the 'TB' as it was often informally known. It was structured into two Divisions. 'C' Troop, the regular army unit, still at Aldershot in direct support of the Army Corps, formed the 1st Division, while the two Postal Telegraph Companies, still routinely working in support of the Post Office but providing an operational reserve when needed, formed the 2nd Division. The prime reason for the reorganisation was to provide a more integrated unit with greater flexibility for cross-posting and training.

It was a timely change, for in the next two years events demanded the deployment of the full resources of the Telegraph Battalion, and more. The first of these was an expedition to Bechuanaland later in 1884.

Historical Background

The turbulent history of South Africa was still being played out. At that time Bechuanaland (now Botswana) - a large arid area to the north-west of Cape Colony, and the home of several impoverished, squabbling, Seswana-speaking tribes - was ruled by its own chiefs, although under strong British influence. However, for several decades it had suffered intrusions and much aggravation from Transvaal Boers, and the tribal chiefs had asked for British protection.

The First Anglo-Boer War in 1881, with its ignominious result for Britain, had resulted in the Boers regaining control of the Transvaal. Following that came a renewal of Boer activity in Bechuanaland. Boer freebooters were trekking into Bechuanaland and had set up self-declared 'republics', Goshen and Stellaland, in tracts of land there. [1] Having suffered too many problems in South Africa in recent years, the British government was loath to get involved.

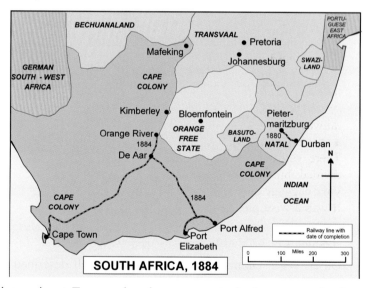

In the Cape, Cecil Rhodes, the arch-imperialist, with an eye on future opportunities, wanted to secure Bechuanaland as an extension of the Cape Colony. His objective was to bypass the intransigent Transvaal and ensure control of a communication route for trade with other territories further north - Matabeleland, Mashonaland, and beyond. Within a few years these tribal territories became Rhodesia, controlled by the British South Africa Company, headed up of course by Rhodes!

Two events stirred the British government into action over Bechuanaland. Political agreements (the London Convention) about its border and the two so-called 'republics' made in London in early 1884 with President Kruger of the Transvaal, the arch-Boer, were subsequently not observed by Kruger, so the government's view hardened. To add to that, the 'Scramble for Africa' took on a new dimension when, in August 1884, Bismarck grabbed territory to the west of Bechuanaland - what was to become German South-West Africa (now Namibia). Germany was sympathetic to the Boers, so the strategic position of Bechuanaland, lying between German South-West Africa and the Transvaal, was radically changed. The British government now supported Rhodes' arguments, plucked up the courage to challenge Kruger, and ordered an expedition to reclaim the two 'republics'. [2]

THE BECHUANALAND EXPEDITION

The Expedition

When the British government decided in November 1884 to send the expedition into Bechuanaland, Lieutenant Colonel Sir Charles Warren was chosen to command it. Then aged forty-four, he had been commissioned into the Royal Engineers in 1857 but had followed an unconventional career path. In 1876 he had been sent to South Africa by the Colonial Office as a boundary commissioner to delimit the boundary between the Orange Free State and Griqualand West, an area west of Kimberley. As a result of that work the nearby town of Warrenton was named after him. He had remained in South Africa in various other roles, including an investigation of 'native questions' in Bechuanaland and as administrator of Griqualand West. He returned to military duties in England in 1880. Thus experienced in the situation in South Africa, if not in tactical operations, he landed in Cape Town in December 1884. Fortunately for him and his expeditionary force the stand off with Kruger was going to fizzle out without any fighting. [3]

The expeditionary force consisted of some 4,000 men, partly imperial troops and partly locally recruited volunteers. Their route from the Cape Colony into Bechuanaland was going to take them well beyond the limits of the existing Cape telegraph system, so a detachment from the Telegraph Battalion was provided to extend the colonial system and keep them in communication. Other Royal Engineer elements including a field company and a balloon section were also deployed.

The detachment of the Telegraph Battalion comprised a regular section from the 1st Division reinforced by men of the 2nd Division, the former Postal Telegraph Companies. Captain Richard Jelf, at that time in command of the 1st Division, was appointed Director of Military Telegraphs, while the detachment of fifty-five men was commanded by Lieutenant Ian Anstruther, who had taken part in the 1882 expedition to Egypt. Jelf was later to comment how well the soldiers from the two Divisions worked together. They took with them only a limited amount of stores. Essential additional line and reserve stores were, for some reason, to follow in another ship. As usual there was a delay in this, with the result that during the course of the expedition there were delays to the construction of the line. Also, as they discovered on arrival, the number of men sent initially was insufficient for the planned task and a second detachment of forty-five men from England was ordered. The reader will by now recognise this as a familiar set of defects.

They landed at Cape Town on 19 December 1884, and spent two days unloading, drawing additional stores, and getting everything into order. During that time Captain Jelf met up with Mr James Sivewright, the General Manager of the Cape Telegraph Department, to make arrangements. They left Cape Town by special train at 6.30 pm on Sunday 21 December for the Orange River station, which at that time was the railway terminus. (The line was still under construction; it reached Kimberley the following year.) Being a special train it stopped for periods as required, making the forty-hour journey a pleasant undertaking. There were some steep sections of the railway as it climbed away from the coastal plain, when the train was divided, but Jelf followed the good old army principle of never allowing yourself to get separated from your baggage and refused to have his troops separated from the wagons containing their telegraph stores. It was a wise precaution, for Sir Charles Warren himself, also coming up shortly afterwards by another train with his baggage and certain luxuries for a party on Christmas Day, found on arrival that his baggage was missing, and Christmas dinner in the veld was not the celebration that he had intended!

On arrival at Orange River station, which Jelf described as "a very disagreeable place" (most places he arrived at during the expedition received a similar description), they unloaded their stores.

> The heavy telegraph wagons had all to be skidded down off the backs of the trucks, with any material that could be found at hand. Anstruther managed it admirably, and the men worked with a will. Then, in the cool of the afternoon we started, I with a small advanced party in two Cape wagons with 10 mules each, and he an hour afterwards with the men. Crossing the Orange River about one-and-a-half miles from the railway terminus was an appalling manoeuvre. The river being very low the wagons had to come down the almost perpendicular side of the drift [ford] on to the pont, which in its turn could only get about half across the river before it was so shallow, and we had to take a second desperate plunge off the pont into the rocky bottom of the river. We got over in due course and found the camp near the river, but with nothing else to recommend it. We had to pitch our camp in a blinding dust storm. ... Anstruther and the main body came swinging in at about eight, singing their songs as usual.
>
> ..... The saving point of this camp is the Orange River, in which the whole army (1,300 I think there are, already here) bathe every day. We drink the same water (from a few hundred yards higher up), and though said to be perfectly wholesome, it is always the colour of grey powder, with a sand in suspension, which never drops to the bottom however long kept. Anyhow, the men and all of us are in splendid health. [4]

They spent the next few days gathering in more equipment - mules, horses, harness, wagons, carts, and native drivers. Then, as the first of the troops to move forward, they left their temporary camp to move on to Barkly West. This was the place where the colonial permanent telegraph line ended, and from where they had to construct and operate the field telegraph line northwards along the route of the expedition - a line that eventually was going to be 350 miles long.

It was a hot march over rugged roads and rough ground, with some "terrific drifts", notably at the Modder River. They were very impressed by the way their wagons stood up to the harsh treatment, particularly the much-maligned airline wagon: "I believe the long distance between the fore and hind wheels is the very essence of getting over the ground in this country", said Jelf, in those equestrian-oriented days. They arrived at Barkly West and crossed the deep Vaal River by pont. The town, they found, was "quite a cheerful little place, with real green banks and hotels, the first we had seen since leaving Cape Town three weeks before." It had become a sort of health and weekend recreation resort for the nearby diamond mining town of Kimberley. (5)

The telegraph detachment during the expedition.

The troops, however, were denied the bright lights of Barkly West; they had to go a further six miles and camp outside the town, near the Vaal River. (Vaal is literally translated as 'tawny' in Afrikaans.) While encamped there they ran a temporary telegraph line to the permanent telegraph office in Barkly West, where the Cape telegraph system ended. Here, again, they drank the river water, which "was extraordinarily thick, but never seemed the least unwholesome, and the men were as healthy there as everywhere else", reported Jelf. Water supply was to be a problem for the expedition in the barren land they traversed, but they were extremely fortunate to avoid the problems of enteric fever which killed so many who drank from rivers in the Boer War fifteen years later.

Then the expedition advanced, and they began to construct the telegraph line northwards. Their route was through Taungs and Vryburg, the capital of the so-called 'republic' of Stellaland. Ultimately, although they did not know it at the time, the expedition was to go on to Mafeking. The terrain was monotonous, generally flat and featureless scrubland with a few hillocks.

The telegraph section worked well up to the front of the advancing expeditionary force. The procedure they adopted was for regimental signallers working with heliographs to

'The telegraph section of the Royal Engineers on the march.' A contemporary sketch by Mr Julius Price, a member of Methuen's Horse.

accompany the leading troops, maintaining communication with the following troops through relay stations sited on the few hillocks. It appears that the signallers did a good job, also maintaining certain links to the flanks. The telegraph line followed as quickly as possible behind and was constructed almost as rapidly as the column advanced - when the stores were available! They had not been able to bring sufficient reserve stores from England on the ship with them, and there was subsequent delay in getting the line and insulators they needed for the distance involved - reminiscent of 'C' Troop's similar problem in the Zulu war five years previously. Eventually Lieutenant Anstruther and his team started to construct the line from Barkly on 14 January 1885, and the first twenty miles were completed in four days.

By 23 January all the stores needed had arrived and, in Jelf's words:

> ..... from that moment things have gone, I rejoice to say, like clockwork. We were given a troop of Methuen's Horse as escort, 12 bullock wagons (besides our 8 section wagons and carts), a flock of sheep, a butcher, a baker ..... and off we trekked. Behind us the whole army was waiting to come on, as soon as we had found the road practicable and laid our line. We were even ahead of the Intelligence Department officers, so we had to prospect the road, find water (which was pretty scarce), and lay the telegraph line in an almost unsurveyed country, which was rather a big job. However, we had great luck, all went excellently well, and we ran into Taungs, at the rate of about 8 miles a day for 5 days, on January 31.
>
> Taungs is a rather disgusting place, the capital of Mankaroane's country [the local tribal chief]. It consists of an immense collection of Kaffir kraals, interspersed with a few low drinking stores ...

Despite its lack of attraction, Taungs (meaning 'the place of the lion') became the advanced base, and troops and stores were pushed forward to it. Jelf described how he managed to get the local inhabitants to "make me 500 stay pegs (corresponding to 'pins, tent, wood, large', which Woolwich would not allow me) at a halfpenny apiece, out of stinkwood. They answer admirably, and save cutting up valuable poles."

After three days at Taungs they set off for the next objective, Vryburg, the capital of Stellaland, forty-five miles away. (6) As they advanced they were having to drop off numbers of men to maintain and operate the line, thus progressively diminishing their numbers available for construction. By 9 February, further line stores having at last been received, they were able to be in telegraph communication with Cape Town, a distance of 725 miles. There was unfortunately a problem with a submarine cable which prevented communication with London until 24 February. When that was restored, they were in communication with London and "able to receive Reuter's dispatches regularly."

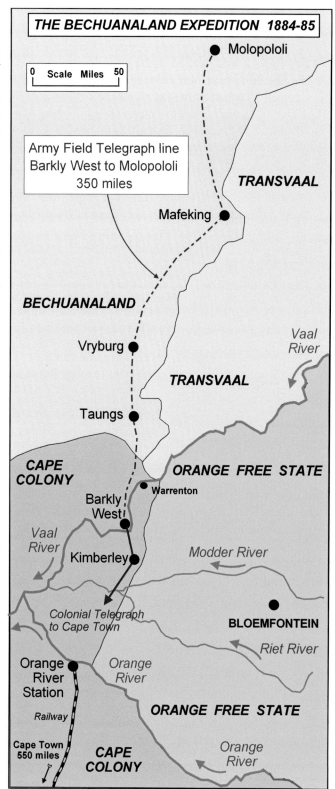

They were also now reinforced by the second section sent from England under Lieutenant Heath, who had found some difficulty in reaching them at Vryburg. "Heath has just turned up with a strong detachment for us, not before they are wanted, and all delighted to come out. He got stuck at the Modder river by floods, and only got past by floating his wagons and stores over - a very smart performance", said Jelf.

By 2 April the force had reached Mafeking, originally a kraal called *Mafikeng* occupied by the Baralong tribe, the name meaning 'place of boulders'. It had grown in a manner similar to Vryburg, with Boer mercenaries being rewarded with a large area of land, which in 1882 they unilaterally declared to be the 'republic' of Goshen. The expedition headquarters was established there, now connected by a 220-mile field telegraph line back to Barkly West and the Cape Colony telegraph system, and thence to Cape Town and by submarine cable to England. Moreover, it was, in Jelf's words, "working without a hitch".

'Laying a telegraph wire over the side of a rocky gorge'. Another contemporary sketch by Julius Price.

Now, with all the telegraph stores they needed, a further demand arose, explained by Jelf:

> ..... on June 1 I was called upon with a scratch section made up of volunteer pioneers, mounted riflemen,, a civilian or two pressed into service and enlisted, some natives, and what I could scrape together from my already existing fourteen offices, to carry an extension of the line to Molopolole, 110 miles further north, and through a most waterless country. I gave this job to Heath, who had had no construction so far, and accompanied him myself, leaving Anstruther here. Heath did the job capitally, though the country was very difficult - hilly, covered with rocks in many places, and with thick bush through which he had to cut his way, and the 110 miles with five offices were completed and open for traffic in three weeks.

Thus the field telegraph line they had constructed from Barkly West was now 350 miles long, and providing an excellent service. But Jelf questioned the way ahead. The levels of traffic were low, and the telegraphists were generally under-employed. Moreover, as they well knew from their contact with England, the Nile Expedition - to relieve General Gordon in Khartoum, described in the next chapter - was underway and telegraph resources for this expedition were in very short supply. They felt at this point that they should be redeployed there. Moreover, Jelf commented, "they have the certainty of medals and things to look forward to, which the supine character of the Boers in this country has done *us* out of."

The detailed withdrawal of the expeditionary forces, and in particular the telegraph detachment, is not of particular interest. The field telegraph line was left in good working order and handed over to the Cape Colony telegraph department. It had been a well-handled telegraph operation, but it was conducted neither under any hostile action to destroy the lines nor pressure of telegraph traffic. The problems had been logistics, as was recurrent in those early days - getting line stores beyond their normal scale for distances that were beyond the somewhat blinkered vision of European-oriented planning, and then getting them to the distant scene of action quickly due to their weight and the limitations of transport, both by sea and land. Telegraphy had advanced considerably, but transport was going to have to await mechanisation in the early part of the 20th century.

To complete the political story, in 1885 the northern part of Bechuanaland was made a British Protectorate, and the southern part was made a Crown Colony. It eventually became the independent republic of Botswana in 1966.

THE BECHUANALAND EXPEDITION

Endnotes

1. The term 'freebooter' is derived from the Dutch word *vrijbuiter* used by the Boers. They were also often described as 'filibusters', the general meaning being one who engages in unauthorised warfare against a foreign state.

2. The political situation is covered succinctly but in more detail in *The Scramble for Africa* by Thomas Pakenham, pp 376 to 380.

3. Warren's unconventional career later included dabbling in politics and as Chief Commissioner of the Metropolitan Police. After controversy there he reverted to the military. He retired in October 1899 but soon afterwards was recalled and appointed as Commander of a Division sent to South Africa at the start of the 2nd Anglo-Boer War. Without either higher command or tactical experience, it was an inappropriate appointment. He was out of his depth, and his ineptitude was in large measure responsible for the tragic circumstances of the disastrous battle of Spion Kop. He returned to England soon afterwards. His biography will be found in the *Dictionary of National Biography*.

4. Jelf described the expedition in a number of reports reproduced in *The Royal Engineers Journal*, this extract published on 2 February 1885, with several others over succeeding months.

5. Barkly West had a colourful history, caused by diamond mining. It was originally known as Klipdrift, in Afrikaans meaning literally 'stone ford', a crossing point on the Vaal river. In 1870, when the diggers arrived to search for diamonds there were various claims by individuals to ownership of this now potentially attractive patch of land. Their claims were overtaken by the Orange Free State which claimed the land on the south side of the river as as part of their territory, and likewise the Transvaal claimed the north side. The chaos that ensued between rival claimants need not be elaborated here. After arbitration the land was ceded to the Griquas, who then turned to the British in Cape Colony for protection, and the land was ceded to the British. The Governor of the Cape Colony, Sir Henry Barkly, visited at the end of 1870 to establish various formalities in this new and controversial British acquisition. One of them, to reinforce its now British credentials, was to rename the town as Barkly West.

6. Vryburg, or literally 'the town of liberty', has an interesting little history. In 1882, during an inter-tribal war, one of the tribal leaders had employed Boer mercenaries to help him. Having won the war with their help he awarded a substantial tract of land to them, which they divided it into 416 farms. The mercenaries then decided to form their own state. While they were discussing this one evening over a camp fire in the veld, a comet was seen in the night sky, and the decision was made - that their new 'republic' would be called Stellaland!

Chapter 14

The Nile Expedition, 1884-85

Historical Background

Following the disasters inflicted on Egyptian troops by the Mahdi's army in the Sudan in 1883 the British government decided in December that year to evacuate the Sudan, and on Sir Evelyn Baring's advice from Egypt a senior British officer was sent to Khartoum to organise it. [1] The selection fell on Major General Charles Gordon, a former Governor-General of the Sudan with a reputation as a successful administrator of primitive people and a good commander of irregular troops, although his wayward character and sometimes eccentric behaviour were not unknown. His task was to report on the situation and to organise any subsequent evacuation of remaining Egyptian forces, but by the end of May 1884, exhibiting his wayward streak, he had refused to withdraw and had let himself become besieged in Khartoum.

The expedition sent to rescue him was controversial from start to finish. In the first place, many had thought Gordon the wrong man for the job, and Gladstone's government was unwilling to undertake a costly expedition simply to extricate him. They had no interest in defeating the Mahdi and occupying the Sudan, and anyway Gordon had exceeded his orders, which were to rescue the Egyptian garrisons threatened by the Mahdi's hordes. His mission at the outset had been well-nigh impossible but he could easily have escaped earlier, although he refused to do so out of misguided interest to protect those he had been sent to save. Put perhaps somewhat simplistically, he had 'gone native'. Nevertheless, many at the time saw his action as heroic. The government came under pressure of public opinion, and in the end relented and belatedly authorised an expedition on 5 August 1884, but their time-wasting vacillation had fatal impact on its outcome.

Outline Plan

Having decided to mount an expedition there was further controversy about the route. There were two options: either up the Nile from Cairo to Khartoum, a distance of 1,600 miles through the many cataracts that obstructed navigation of the river south of Wadi Halfa, or a shorter route across the barren desert from the port of Suakin on the Red Sea to Berber on the Nile, a distance of 280 miles, and then along the Nile valley to Khartoum, a further 200 miles. Although shorter in distance, there were numerous problems associated with that route: it was through hostile territory controlled by the Mahdi's sympathiser, Osman Digna; there were problems with the supply of water for a large number of troops; and it would be necessary to construct a railway to keep the force supplied. Nevertheless, in the measured opinion of many it was the better option. General Lord Wolseley, who had been selected to command the expedition (another controversial decision), had drawn up contingency plans some months earlier while the government was dithering. He favoured the Nile route and so, despite time now being of the essence, that was the way they went. Many problems were to arise from an underestimation of the difficulties of navigating the Nile in the Sudan, causing more time to be lost.

What would have happened if the Suakin-Berber route was chosen is now a matter of conjecture. From the point of view of telegraph communications, the Nile route was entirely through Egypt and those parts of the Sudan not controlled by the Mahdi's forces, and there was a telegraph line of sorts. Although badly maintained, it was quickly brought up to standard, as will be explained. On the Suakin-Berber route the telegraph line had been destroyed, and passed through desert controlled by the Hadendoa tribe who were the Mahdi's adherents under Osman Digna. It is extremely unlikely that a satisfactory telegraph service could have been constructed and maintained over such a length of line passing through hostile territory. But, not unusually at the time, it appears that the telegraph was not a factor considered by anybody and it seems to have had no influence on the choice of route, yet in the event it was to prove vital. The official *History of the Sudan Campaign*, prepared by Colonel H E Colville, includes the various reports and opinions that were considered, and nowhere is telegraph communication mentioned. [2]

The political decision to go having been made on 5 August, the force assembled in Cairo during September 1884. It is not intended to describe the campaign in any detail, but the outline plan was simple. Having advanced southwards through Egypt to the Sudan border, a strong column of infantry in small boats was to work up the river from Wadi Halfa, accompanied by mounted troops moving overland on the banks. If the column was delayed, a Camel Corps was to leave the Nile at Korti and strike across the Bayuda Desert to Metemma. From there a small detachment was to be sent to Khartoum in steamers sent down by Gordon and these troops would sustain the defence of Khartoum until the arrival of the main body in Spring 1885. It was not to be.

It is well-known that they failed to reach Khartoum in time, and that Gordon was killed in Khartoum on 26 January 1885, a few days before the forward elements of the expeditionary force were in a position to rescue him. More controversy followed. Were they to carry on and 'teach the Mahdi a lesson', or were they to retreat ignominiously, the mission having failed? At the end of a long and precarious line of communication, equipment in short supply (even boots had worn out and there were no replacements), they were in no position to continue. Also, the British government was at the same time alarmed by the unconnected development of threats by Russia on the Indian border (which they conveniently exaggerated, to distract attention from the Sudan). After some political face-saving chicanery, the message eventually arrived from London in mid-April 1885 that they were to withdraw.

The expedition had failed in its purpose. Its planning and execution were not perfect, and more vital time was lost as a result. Inevitably there were unforeseen problems. There were many transport and logistic difficulties caused by the long and complicated line of communication, and towards the end, as they got near Khartoum, there were battles. These matters will simply be mentioned in the narrative where relevant. [3]

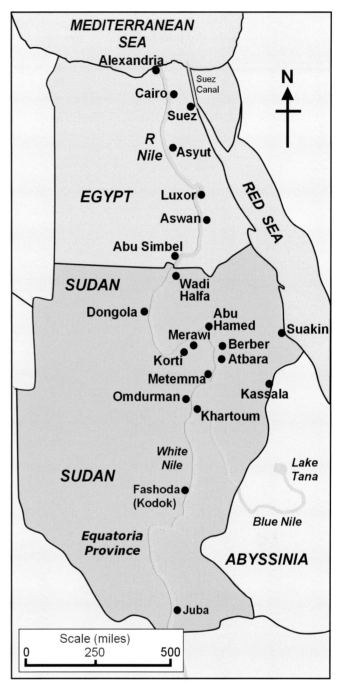

Egypt and the Sudan, 1884.

The Telegraph Battalion

Now it is time to turn to the description of the telegraph operation. The very idea of trying to maintain communications for the expedition with its various elements stretched along the length of the Nile for over 1,200 miles, and with London, made it a unique undertaking for the army field telegraph at the time, and the success they achieved made it a remarkable operation in the history of military communications.

Since the 1882 campaign in Egypt, organisational changes had been made to the army's telegraph troops and in 1884 the Telegraph Battalion had been formed - an amalgamation of 'C' Telegraph Troop and the two Postal Telegraph Companies, as described at the beginning of chapter 13. This amalgamation for war purposes was the result of the recommendations of a committee presided over by Lieutenant Colonel Richard Harrison CB RE in 1881, intended to produce a more flexible and efficient organisation, with greater interchange of personnel and skills. It was this new organisation that was going to support the Nile

Expedition. Also available for the first time, to supplement them, were the telegraphists of the 24th Middlesex Rifle Volunteers, a unit comprising volunteer civilian Post Office reservists who did annual military training, similar to the modern Territorial Army (see description below.).

Drawing on their limited resources remaining in Aldershot (part of them were with the expedition in Bechuanaland at the time, as described in the previous chapter), the 1st Division of the Telegraph Battalion provided No 4 Section consisting of three officers, Captain F W Bennet (who as a Lieutenant had served with the 8th Railway Company RE in Egypt in the 1882 campaign), and Lieutenants G A Tower and A M Stuart), together with sixty-eight NCOs and men. They left Gravesend on 3 September 1884 and arrived in Alexandria on 15 September, taking with them their established equipment – twenty miles of cable, twenty miles of airline, and equipment for six telegraph offices. Horses and wagons, inappropriate for this expedition, were not brought. This scale of equipment was of course quite insufficient for the special nature of this task and was supplemented by additional line and telegraph equipment brought out by subsequent detachments. These additional stores were readily available although as usual in those days, transport was going to be a problem, particularly in the upper reaches of the Nile.

The 24th Middlesex (Post Office) Rifle Volunteers

The 24th Middlesex (Post Office) Rifle Volunteers evolved from an organisation called the Post Office Rifles, originally formed because of Fenian troubles in 1867. With a security role in England, the volunteer members were called 'Special Constables'. Prominent in leading them was Major J L du Plat Taylor, who was also Private Secretary to the Postmaster General.

In about 1877, as an extension of their role, and after a number of changes of name and organisation, the then Lieutenant Colonel du Plat Taylor proposed to the government that an Army Postal Corps should be formed to undertake postal duties for any force sent abroad on active service. His idea was not taken up at the time, due it seems to lack of funds, but in July 1882, when the bombardment of Alexandria took place, leading to the subsequent short campaign in Egypt, he renewed his offer, and this time it was accepted. Authority was given on 18 July 1882, and two officers and 100 men were enlisted from the Post Office into the Army Reserve. Of these, two officers and fifty men went to Egypt, sailing on 8 August 1882, returning after the short campaign on 23 October. Their duties had been entirely in connection with postal services.

Following that, in 1884, Colonel du Plat Taylor suggested that a company of telegraphists from the Post Office should also form part of the 24th Middlesex RV. As a result, authority was given for the enlistment in the Reserve of the Royal Engineers of fifty Post Office telegraphists, and volunteers were immediately forthcoming. This number was increased in 1885 to 100. They eventually became 'L' Company (Royal Engineer Telegraph Reserve) of the 24th Middlesex(Post Office) Rifle Volunteers, sometimes referred to as the Field Telegraph Corps.

It was not long before their services were required. Because a large proportion of the First Division of the Telegraph Battalion were deployed in Bechuanaland at the time, they were called upon for the Nile Expedition and in September 1884 twelve men went. Afterwards Sergeant Parish, Corporal Hopgood, and Sappers Isles and Tee were specially commended by the Commander-in-Chief. (Parish and Tee had actually served in the 1882 expedition in the postal service.) In March 1885, a further detachment of twelve was sent to Suakin, returning to England six months later.

The conditions of service of these men were that they enlisted into the army and served one day 'with the colours', then passed into the 'First Class Reserve', receiving pay in peacetime as members of that body. When called up for active service they then served with the Telegraph Battalion, receiving military pay according to rank, as well as their Post Office pay. In April 1885 they expanded to 100 men, half drawn from the Central Telegraph Office in London, and half from the various provincial telegraph offices. They were re-named as the Royal Engineer Telegraph Reserve in April 1889.

Their zenith came in the Boer War, 1899-1902, when they were desperately needed. They volunteered and served in large numbers.

The most eminent Post Office reservist seems to have been Sapper Tee, mentioned above. He first served in a postal capacity in Egypt in 1882. Having apparently afterwards qualified as a telegraphist, he next served in the rank of Corporal in the Nile Expedition of 1884, and was Mentioned in Despatches. He then served, for the third time, as CSM W G Tee in the Boer War.

Captain A H Bagnold from the 2nd Division led the next detachment, arriving in Alexandria a week after No 4 Section, with 100 miles more line and the equipment for eight additional telegraph offices. Arthur Bagnold, it will be recalled, had commanded a detachment of 'C' Troop sent to Natal and the Transvaal in 1881 in the aftermath of the first Anglo-Boer war to rebuild the telegraph line from Natal through the Transvaal to Pretoria. Having unloaded his stores, Bagnold went by rail to Cairo and on to Asyut, and then by steamer to Aswan, arriving there in early November. [4]

Lieutenant C Hill and a detachment of thirty-seven NCOs and men left England on 17 September and arrived in Alexandria on 29 September, bringing with them 160 miles of line. Captain C K Wood, with Sergeant Strevens and twelve NCOs and men of the 24th Middlesex RV, arrived in Alexandria on 8 October, bringing with them 176 miles of line but no poles, and various other equipment. A few stragglers joined soon afterwards.

Altogether there were six officers and 127 NCOs and men, of whom sixty-three were office telegraphists, thirty-eight were linemen, twenty-two were drivers, and four were administrative staff. Sixty were from the 1st Division (*ie* regular), fifty-five were from the 2nd Division (the Postal Telegraph companies), and the remainder were reservists from the Post Office. They had with them 456 miles of line and a large set of equipment such as telegraph instruments, batteries, and stationery - the paperwork not to be underestimated, for a huge number of messages were to be sent.

Colonel C E Webber CB was appointed as the Director of Army Telegraphs. [5] As narrated in earlier chapters, he had served for many years with the Postal Telegraph companies and had then, as a Major, accompanied General Wolseley to South Africa as AA&QMG for the lines of communication at the end of the Zulu War in the period 1879-80. He had also, in the rank of Lieutenant Colonel, been Director of Army Telegraphs with Wolseley in the Egyptian Campaign of 1882, while he was President of the Society of Telegraph Engineers, of which he had been one of the founder members. Wolseley, who arrived in Cairo on 9 September 1884, noted in his personal campaign journal on Wednesday 17 September 1884 that "Webber arrived". [6]

Communications Planning

A telegraph route between Cairo and Khartoum already existed, albeit very precariously over much of its length for one reason or another, and followed the Nile nearly all the way. Its construction had started in 1863, in the time of Khedive Ismael. South of Wadi Halfa, on the border between the two countries, it came under the administration of the Sudan Telegraphs, and Khartoum was connected by 1868. In 1884, when the expedition arrived, the entire telegraph system in Egypt and the Sudan was extremely poorly managed, operated, and maintained; inordinate delays in traffic handling occurred in consequence. Many are the stories that described the inefficiency in all facets of the work. In the Sudan parts of it had been destroyed and not rebuilt, including the line to Suakin. The line between Berber and Khartoum had been cut just before the expedition arrived in Egypt, and Gordon had become incommunicado.

The first task of Colonel Webber was to make arrangements with the Egyptian Telegraph administration. This proved to be a frustrating encounter with unhelpful bureaucracy, although the impression gained from reading much of his correspondence is that Webber himself seems to have been a rather pedantic character, and that cannot have been helpful. Financial matters assumed great importance and the Egyptian administration seemed primarily concerned with extorting the maximum profit from the situation. Eventually terms were agreed.

Four lines left the Cairo telegraph office heading south. One local line branched off westwards soon afterwards to Fayoum. Another was an omnibus wire for railway traffic to Asyut where both it and the railway terminated. The third served numerous intermediate stations between Cairo and Aswan, where it too terminated. The fourth wire was the through line, and beyond Aswan the telegraph depended on this single line to Wadi Halfa and on southwards through the Sudan. After the difficult negotiations with the Egyptian Telegraph department, arrangements were made for the army to rent this line for their exclusive use. This one wire was to be the lifeline for the expeditionary force as it moved southwards towards Khartoum, extending eventually to a distance of 1,273 miles, but much improvement was going to be needed to the line and the operation of the telegraph.

In Cairo all the lines had originally been routed through Khedive Ismael's palace, ensuring that he had control of the entire telegraph system in his despotic country, and all the traffic that passed over it was under the control of his own operators. The Khedive in 1884 (Tewfik) permitted the army to terminate the Sudan wire in its own office, which was set up in accommodation provided by the Egyptian Railway and Telegraph Administration under the chairmanship of Mr Le Mesurier. Only two of the Egyptian telegraphists were found to be up to the standard for operating a busy line with many stations, so the Cairo office was staffed mostly by telegraphists of the 2nd Division under Sergeant Parish of the 24th Middlesex RV. They worked very long hours.

The strategic communication back to England was in the hands of the Eastern Telegraph Company. They provided and maintained their own overland route in Egypt between Suez, Cairo, and Alexandria. Suez was connected by submarine cable down the Red Sea to a number of landing points including the Sudanese port of Suakin, and then on to Aden, Bombay and other eastern destinations. From Alexandria the prime route to England was a further submarine cable via Malta and Gibraltar to Lisbon, and from Lisbon to Porthcurno near Falmouth and thence to London. Other routes from Malta existed across Europe. The service provided by the Eastern Telegraph Company and its telegraphists proved to be excellent.

One of Webber's early decisions was to set up a 'clearing house' in Cairo, and it was to be the first time that one on such a scale was set up by the army. The term is no longer used but in those days the military telegraph was often the only communication facility available so it became a multi-user circuit, and non-military users had to pay; this included the press and all private telegrams, British or Arabic. The clearing house combined the functions of a signal centre, where messages were handed in and collected, with the collection and accounting of money. It was also the interface with the strategic communications provided by the Eastern Telegraph Company, and further financial transaction took place.

A Mr Oatway was provided by the British Post Office to run the financial side of the clearing house and produce the accounts. The financial situation in Egypt at the time, as might be imagined, was extremely confusing. Oatway's task was complicated firstly by the number of different agencies he had to deal with – the separate Egyptian and Sudan Telegraph Administrations, both subdivided into civil and military accounts, the Khedive's account, the Eastern Telegraph Company, the British government, and numerous others. There was a miscellaneous assortment of coinage in everyday use - Egyptian, Turkish, English, French, Italian, and Maria Theresa dollars. Each telegraph office along the line, mostly manned by a few telegraphists, had to collect money for non-military traffic. Collecting the money of many different coinages, and accounting for it, was no easy task. [7]

The Line to Wadi Halfa

Throughout September British troops advanced up the Nile from Cairo, passing through Asyut and Aswan to Wadi Halfa, the movement this far being by the existing railway between Cairo and Aswan, and beyond Aswan by Thomas Cook's Nile steamers which were able to reach Wadi Halfa. General Lord Wolseley himself started from Cairo on 27 September and reached Wadi Halfa on 5 October.

Supporting this initial deployment two officers and fifty-six NCOs and men of the Telegraph Battalion left Cairo for the south on 18 September, taking on the stores that had gone straight from the port of Alexandria to Asyut. As they moved southwards during late September and early October much work to the existing line had to be undertaken by the line parties, to improve its construction and to enable telegraph offices to be located to army requirements. This main telegraph line ran along the west bank of the Nile.

In support of the army's advance, army telegraph offices were opened at Cairo, Asyut, Aswan, Shellal, and Korosko on 1 October 1884 (see map below). Aswan to Shellal was a branch line running beside a short railway on the east bank of the river that was used to convey freight up the first cataract. These were followed on 6 October by offices at Wadi Halfa, Gemai, and Sarras.

At Asyut a loop of over half a mile was constructed from the main telegraph line on the west bank to the town on the east bank. The permanent telegraph office in the town was attacked by swarms of mosquitos, apparently much more lethal than Dervishes, disabling the telegraphists in a few days. The office there was then moved to a diabeah (a Nile river boat) which inexplicably seemed to improve matters.

During the winter frequent interruptions to service on the section between Cairo and Asyut were to occur at night due to moisture in the atmosphere in the lower Nile and the poor insulation quality of the line.

THE NILE EXPEDITION, 1884-85

Over this section, which was still euphemistically 'maintained' by the Egyptian Telegraph administration, it was found necessary to increase the battery power over the circuit at night, from seventeen to twenty-eight batteries, and even that was often not enough to operate the circuit as so much current leaked away. Further south the dry climate was to prove extremely beneficial to the operation of the telegraph over long distances.

At Aswan, Captain Arthur Bagnold was kept busy. Having assembled and sorted his stores, he hired a house on the east bank in which the military telegraph office was established, the main line being on the west. This gave rise to an interesting situation. The Egyptian Telegraph administration had constructed an overhead loop from one bank to the other about two miles upstream of the town, selecting two unusually high bluffs close to the river, about 250 feet high, and on these they placed masts about eighty feet high, an overall height of 330 feet above the river. The masts were almost a mile apart. The strain on the wire was immense. Poles were placed on an intermediate island to try and absorb some of the strain. This new-found wonder of the world had to be replaced by an armoured submarine cable, and the task fell to Bagnold. He borrowed a theodolite from the Egyptian administration to measure the distance but found that the instrument was deficient of its cross-hairs! Bagnold, a well-trained engineer, improvised; a piece of human hair was substituted, and it sufficed. As a result of his work the loop was then constructed from flat-bottomed boats lashed alongside a naval steam pinnace, using two single armoured submarine mining cables, and it worked well thereafter.

Between Aswan and Wadi Halfa, 210 miles, the existing line was in an extremely decrepit state due to years of neglect. That this section of line had worked at all was due largely to the dry climate in upper Egypt and its beneficial effect on the insulation of the line. Many of the wooden telegraph poles were

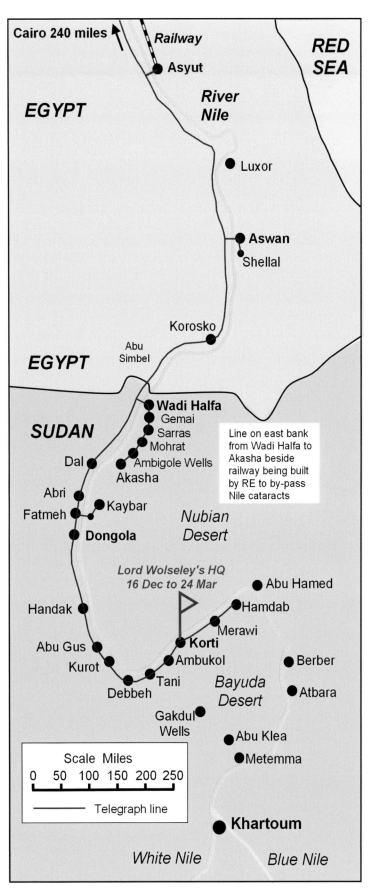

The telegraph line through Upper Egypt and into the Sudan.

honeycombed by the ravages of white ants, which had feasted heavily on them, and were in imminent danger of collapsing. Some poles had to be replaced and others realigned, there were many badly made joints in the wire, insulators had to be replaced, and a great deal of other work had to be done. Some of the poles had been stumped (ie cut off at ground level and the shortened pole re-set in the ground) several times before, so the line was lowered to such a degree that camels could not pass underneath! The work to bring the line up to standard was undertaken by Lieutenant Hill with a working party comprising Sergeant Jones and twelve men of the Telegraph Battalion and local Egyptian staff of the post office administration, the latter working well and showing none of the bureaucratic indifference that afflicted their superiors in Cairo.

The army working party lived on a diabeah as they followed the line south along the course of the river, and twelve camels were used for the work on land. Some labour was taken compulsorily at the villages along the way, which was the usual practice in Egypt for repairing the telegraph line. By the time they had reached Wadi Halfa they had put in 78 new poles, stumped and reset 587 poles, replaced 166 defective insulators, and so on. It was to be time and effort well-spent for the line subsequently proved reliable, and was later to carry a huge volume of traffic with little further maintenance needed. The hard work of the linemen at this initial stage was to be replaced over the coming months by the hard work of the telegraphists.

Captain Bennet arrived at Wadi Halfa ('Bloody Halfway', as it became known to the soldiers with their sights set on Khartoum) on 6 October. As at Aswan the telegraph office was moved from the west bank, where the main telegraph line ran, into the town on the east bank, and a temporary cable loop was laid across the river. The Nile was 1,000 yards wide at that point. Submarine cable not available, field telegraph cable had to be used, a purpose for which it was never intended, and it could not for long withstand the pressure of thirty feet of water and the force of the current. It had to be replaced five times, and when the last available length of field cable was in use, the single-armoured submarine cable, as used at Aswan, arrived and was laid on 25 November.

Meanwhile, in mid-November, destined for Wadi Halfa, Arthur Bagnold continued up the river from Aswan in two 'whalers' (boats specially designed for river work, and brought out by the expedition) with his baggage and a considerable amount of stores. With him were Sergeants Strevens and Seggie, and sixteen men. (Seggie, it will be recalled, had served in the 1882 expedition, and was the telegraphist who sent the message from Tel el Kebir to Queen Victoria.) On 30 November, just thirty miles from Wadi Halfa, they overtook the party repairing the line on the west bank under Lieutenant Hill.

The chaotic scene at the advanced base at Wadi Halfa was described by Colonel William Butler, one of Wolseley's senior staff officers:

> The shore was covered with the vast preparation of the coming campaign. Under a sun which blazed fiercely overhead, soldiers, sailors, black men and yellow men, horses, camels, steam-engines, heads of departments, piles of food and forage, newspaper correspondents, sick men, Arabs, and generals seemed to be all thrown together. [8]

By the time Arthur Bagnold arrived at Wadi Halfa, Captain Bennet was in hospital with typhoid fever and Bagnold found he had plenty to do:

> I fixed up translating sets and did a good deal of squaring up in general. Colonel Webber left for the South on the 7th Dec, with Corporals Page and and Brown, all on camels. The telegraphic requirements of the railway had to be met, connections to the boat repairing stations on the river made, and again stores concentrated as they came up in driblets from the North. It was not until the 27th December that I embarked at Gemai [to Dongola] with 4 whalers, my small native sailing boat, and a small 'nuggar'.

Into the Sudan

The telegraph line from Wadi Halfa on along the course of the Nile to Berber had been constructed in 1868 and Berber had become a hub for the Sudan telegraph system. At that time lines had been constructed on to Khartoum, and across the desert to Kassala and Suakin. [9] A rebellion took place in 1883 and the Kassala and Suakin lines were destroyed by the Hadendowa tribesmen of the eastern Sudan. These primitive men did not understand how the telegraph worked, but they knew it carried messages which were used by their oppressors, so they set about destroying it, taking revenge also on the unfortunate Sudanese linemen who maintained it. These linemen had been soldiers in a regiment which had been decimated in a previous battle and subsequently disbanded, but they had been re-employed and

THE NILE EXPEDITION, 1884-85

successfully trained as linemen. The Hadendowas showed them no mercy when caught. They thought that by cutting the line and putting the two ends into the lineman's ears, he would speak like a telegraph instrument and it would tell them what the messages were saying. When this proved not to be the case, some of the poor linemen were tied to telegraph poles with the wire and burned alive. [10]

The stretch of the Nile between Wadi Halfa and Abri (see map) was difficult to navigate due to a series of cataracts, causing considerable river transport problems. On the east bank of the Nile running south from Wadi Halfa was the remains of an old railway which had been started in 1873 with the idea then of circumventing this difficult stretch of river, and with the intention of going to Debbeh. By 1877 the railway line had only reached Sarras, although the route as far as Ambigole had been mostly prepared, but it had been ineptly managed, was over-spent, and the project was abandoned. The railway that was built had fallen into a delapidated and useless state. To help overcome their river transport problems, it had been decided that the expeditionary force would refurbish and extend it. A railway company of the Royal Engineers undertook the work, firstly renovating the original railway, then extending it by constructing a new railway to Akasha, which they reached by August 1885. With time of the essence, this project took far too long. By the time it was completed the expedition was withdrawing, and the railway line was used for that unintended purpose.

To provide telegraph facilities along this railway route on the east bank, Sergeant Tubb with a detachment ran a field telegraph line from Wadi Halfa to Sarras, Mohrat and Ambigole, just over twenty-eight miles, and from Sarras to Semneh, nine miles, but this took an inordinate time due to lack of transport and the rocky nature of the ground, progress being only three miles a day. Lieutenant Hill continued the field telegraph line from Mohrat to Akasha, thirty miles. These two lines, running alongside the railway line from Wadi Halfa to Akasha, although constructed only as field telegraph lines, gave excellent service without any problems.

While work on the railway was being undertaken the advance south from Wadi Halfa into the Sudan continued by river, in the face of many navigational difficulties. The small boats that were used, and the problems their crews faced in passing through the cataracts and other obstacles on the river, together with the logistic and transport difficulties created by the bottleneck, is interesting but not part of this story. Suffice to say that progress was very slow. The transport difficulties from now on were to affect the movement forward of the heavy telegraph equipment such as the spare line and poles. Very little of it got beyond Wadi Halfa, and then only after long delays.

The advance of the army telegraph continued south from Wadi Halfa, and on 4 October Lieutenant Stuart with eight telegraphists and some stores left there by river boats, tasked with going forward

Nile Expedition 1884-85. Telegraph Offices

Office	Date Opened	Date Closed	From Cairo (miles)
Cairo	1 Oct 1884	1 Aug 1885	0
Asyut	1 Oct 1884	1 Aug 1885	240
Aswan	1 Oct 1884	1 Aug 1885	620
Shellal	1 Oct 1884	1 Aug 1885	626
Korosko	1 Oct 1884	1 Aug 1885	730
Wadi Halfa	6 Oct 1884	1 Aug 1885	830
Gemai	6 Oct 1884	1 Aug 1885	846
Sarras	6 Oct 1884	1 Aug 1885	863
Semneh	8 Dec 1884	1 Aug 1885	872
Mohrat	4 Dec 1884	1 Aug 1885	883
Ambigole Wells	31 Mar 1885	1 Aug 1885	895
Tanjore Road	3 Jun 1885	1 Aug 1885	903
Akasha	31 Mar 1885	1 Aug 1885	913
Ambigole	28 Dec 1884	25 Mar 1885	907
Dal	21 Oct 1884	27 Jul 1885	930
Abri	22 Apr 1885	23 Jul 1885	955
Kaybar	9 Jun 1885	17 Jul 1885	
Little Fatmeh	9 Jun 1885	17 Jul 1885	
Fatmeh	23 Nov 1884	17 Jul 1885	1,039
Dongola	5 Nov 1884	5 Jul 1885	1,076
Handak	5 Apr 1885	3 Jun 1885	1,126
Abu Gus	29 Mar 1885	31 May 1885	1,159
Kurot	26 Mar 1885	1 Jun 1885	1,172
Debbeh	1 Dec 1884	30 May 1885	1,176
Tani	26 Mar 1885	27 May 1885	1,216
Ambukol	10 Nov 1884	16 Dec 1884	1,210
Korti	17 Dec 1884	26 Mar 1885	1,223
Merawi	12 Nov 1884	25 May 1885	1,253
Hamdab	21 Jan 1885	30 Jan 1885	1,273

THE NILE EXPEDITION, 1884-85

to Dongola, the capital of the province of the same name. After an eventful voyage upstream through cataracts, taking eighteen days to cover sixty-five miles, they opened an office on the west bank at the foot of Dal cataract on 21 October; the office was later moved three miles upstream to be near the camp of Major General Sir Evelyn Wood VC KCB who was in charge of the Line of Communications. [11]

Lord Wolseley used the telegraph offices to keep in contact with Cairo and London as he progressed up the Nile. At Dal, on Tuesday 28 October he noted in his journal: "At Dal we have opened a telegraph office, so I had telegrams today and sent off replies".

The office at Fatmeh was described by Webber, who left Wadi Halfa in early December and travelled overland with a small party on camels, as being the most palatial in the country. "The walls were made of mud, 10 feet high; the floor was mud; the table was actually of wrought wood, and the waiting room had a divan with cushions on it."

Wolseley himself reached Dongola on 3 November, where he was met with due ceremony by the mudir (provincial governor). Lieutenant Stuart, at the vanguard of the telegraph troops, reached there on 5 November and took over the Dongola telegraph office which was up to then run by the Sudan administration. Once into the Sudan the entire staff of the Sudan telegraph administration were handed over to the control of the army. Fifty-eight in all, they were an assortment of telegraphists, linemen, line guards and camel men.

Arthur Bagnold followed in late December, and also described his journey upstream through the turbulent cataracts. It took thirty-two days to cover the 205 miles upstream from Gemai to Fatmeh, with numerous stops to repair the boats, before they eventually arrived at Dongola, where Bagnold was to remain until the campaign ended. During this journey one boat laden with wire and other telegraph equipment overturned while trying to navigate a cataract. The stores were all lost, and one soldier of the Telegraph Battalion, Sapper Dowding, was drowned after he fell overboard. [12]

Between Wadi Halfa and Debbeh the main telegraph line on the west bank had been constructed using Siemens iron poles with No 8 wire, and this section was in relatively good order except in a few places where sand drifts almost covered the poles. There was a good reason for this uncharacteristic extravagance on iron telegraph poles. The local tribesmen, unacquainted with electricity and the leakage of current to earth, had developed the idea that Allah had placed the original wooden poles in the desert for their convenience. They proved excellent material for constructing huts and, when chopped into small pieces, were admirable during cold nights for camp fires and brewing coffee. After all, why leave them to be eaten by the white ants? It was not long before the telegraph wire lay on the ground, but it usually continued to work in this state so long as there was no rain. When iron poles were introduced, they had no such alternative practical utility and tended to be unmolested by the tribesmen.

Beyond Debbeh the iron poles had been destroyed or taken away and the line was broken. This section from Debbeh to Korti had been temporarily repaired as far as Merawi by Major H H Kitchener and men supplied by the mudir of Dongola. [13] Kitchener was at the time employed as an intelligence officer and was scouting ahead of the advancing troops with a bodyguard of local Arabs, reporting back by various means. He had become a fluent Arabic speaker and often, in the course of his intelligence-gathering duties, dressed up and apparently passed off as an Arab – something that is difficult to imagine, with his tall stature and piercing blue eyes! Having been a subaltern in 'C' Troop in 1873, he understood both the importance of the telegraph and its repair and maintenance. As far as possible his advance party recovered the broken iron poles or used the branches of trees to raise the wires to a suitable height, recovering as much wire as possible. When he reached the scene later, Captain Bennet remarked that "the line was in a very shaky state till repaired and strengthened as far as possible by our own men".

The method of carrying coils of telegraph wire overland was novel. The transport consisted of donkeys, and the way of loading the wire on to them was to stand the hind feet in the coil and then lift it up over the tail on to the donkey's back. (Fourteen years later, building telegraph lines in the Sudan after the Omdurman campaign, the recently-arrived Lieutenant J S Liddell was faced with the same problem and was racking his brains about how to transport the telegraph wire on donkeys without proper saddlery when the then Major General Lord Kitchener rode up and showed him how to do it. Liddell was much impressed by the practical know-how of so eminent a senior officer.)

On 16 November Sergeant Holland and a party of eight men started repairing the line from Debbeh to Korti, then continuing on to Merawi. As soon as more telegraphists became available, they were sent on ahead. The Debbeh telegraph office was opened on 1 December. Lieutenant Stuart arrived at Korti on 14 December with the leading troops, and moved the existing office from the village to the army camp about a mile away. Korti was to become Lord Wolseley's headquarters for over three months, and the relocated Korti telegraph office opened on 17 December.

No architectural masterpiece, it was described by Colonel Webber, who as DAT accompanied Lord Wolseley.

> It was constructed with straw walls, a material favoured in that country for cool residence, but also very favourable for the admission of dust and wind; and anyone who spent a day or night (and our telegraph clerks spent many days and many nights) in that office would have felt it extremely uncomfortable. The wind was always blowing, the dust was always driving through the walls, the candles at night were always flickering, paper blowing about, and everything that was antagonistic to good telegraphy was in force.

Yet it was from here, 1,223 miles over that single wire to Cairo, that many important messages were to be exchanged with Cairo and London at crucial points in the operation still to come. Wolseley, as commander of the expedition, had been forbidden by Lord Hartington (Secretary of State for War, in London) from personally advancing further, as he would no longer be in telegraph communication – a point missed by some of his critics who castigated him for staying in Korti when the action was to be further south. Wolseley remained at Korti until 24 March 1885 when the expedition started to withdraw.

Lord Wolseley's camp at Korti.

The troops spent Christmas 1894 at Korti, and the scene was described by the author Thomas Archer:

> The camp at Korti was a refreshing and even a picturesque place, laid out in broad avenues, kept well watered by coolies from the adjacent villages, who were paid for the work. On the high bank above the river there were evening assemblies of officers and men to listen to the band of the Sussex Regiment. Korti was, in fact, a pleasant and welcome resting-place after the severe labour and hardship of the river and the desert. The men's spirits rose as soon as they had begun to recruit their strength there. All day singing and laughter might

be heard in camp; and at night, when the slow water of the Nile shimmered in the bright moonlight, the echo of their choruses might be heard from the opposite bank, and the dark figures of the men be seen, through a veil of smoke, moving about the camp-fires, perhaps with the silhouettes of two or three camels in the background.

On Christmas night, the day having been observed with no small amount of jollification, but soberly and decent withal, the whole force, from the General in command downward, assembled in a vast circle round an open-air stage erected by the Royal Engineers. There were no lack of entertainers, and excellent music, burlesque orations, comic, sentimental, and patriotic songs, part-songs and choruses, nigger dances, and serenades, topical songs, and other amusing performances were kept going with much spirit in accordance with a regular programme, to which both officers and men contributed, several volunteers filling up the intervals during which the performers had to be brought forward, among them a blue jacket from the Nassif-el-Kheirr, whose appearance on the platform elicited unbounded applause. It was a striking scene, a creditable and highly enjoyable performance; and probably no finer body of troops was ever assembled, though they were hardened by toil, sunburnt, scarred, and, truth to tell, rather shabbily clad in worn, and many in ragged, uniforms.

Lord Wolseley was there all the time with cheery looks and words and sprightly confident bearing; and the entertainment was to have been repeated on New Year's night, but no more time could be spared for rest or amusement. Tidings of Gordon still holding on, but of the Mahdi gathering his forces, were received. Rumours, many, and often conflicting, at all events tended to show that there must be a rapid move towards Khartum, and it was determined to push forward a column across the Bayuda territory to Metammeh, and at the same time to pursue the journey by the Nile with a force under General Brackenbury. (14)

From Korti, in January, the advancing forces were split, under separate commanders, into the River Column, which continued to follow the course of the Nile but only reached the area of Abu Hamed, and the Desert Column, which made a dash across the Bayuda Desert towards Metemma culminating in the battles of Abu Klea and nearby Abu Kru. Communications with these columns were mainly by messenger, and took several days.

Lieutenant Stuart was assigned to extend the line for the River Column - something of a relief apparently, for he explained in a letter to his mother written on 9 January 1885 that: "when I left Korti my boss Webber was there straightening things out generally at the office and finding that things were not as they ought to be entirely, so I was not too sorry to evaporate for a little." Cobbling things up from almost nothing, he extended the line from Merawi to Hamdab, recovering what he could of the damaged wire of the old telegraph line to Berber, broken and lying around, and using native labour to cut poles from the mimosa trees in the area and erect the line. The wire was tied to the poles by pieces of old rag. In the very dry atmosphere, it worked! Stuart opened the Hamdab office on 21 January 1885 – five days before Gordon was killed. "My telegraph office is inside the fort which is on the bank of the river", he wrote in a letter to his father. He also wrote a letter on the back of an army telegraph form (Army Book 295B in those days) to his brother Ramsay, saying: "Hamdab is a rather wild place - the mountains squeeze you into the river and the crocodiles drive you out again".

The telegraph was described by Colonel (later Major General) H Brackenbury (who took over as the River Column commander after Major General Earle was killed a few weeks later at Kirbekan):

> Only a few days before we left Hamdab, the telegraph was extended to an office in our fort there. This was a great boon as it saved much time and labour hitherto expended in sending our messages to and receiving them from Abu Dom [aka Merawi], nineteen miles away. The country was quite impracticable for heliographic signalling owing to the absence of marked high hills, and the presence of a succession of low ridges. (15)

The Desert Column could perhaps have been kept in communication, at least some of the way, by wire laid on the dry desert and used with the recently-developed vibrating sounder – but due to the transport difficulties there was no wire. Heliograph was used by regimental signallers intermittently but without much success. Stuart mentions in one of his letters that he was sent to assist in trying to establish a chain of heliograph stations to Gakdul Wells, but it came to nought.

Hamdab was as far as the expedition's military telegraph ultimately reached, although the remnants of the line had continued to Khartoum. Lieutenant Stuart, having been in the vanguard during the advance, remained at the southern end of the line until the withdrawal began. (16)

THE NILE EXPEDITION, 1884-85

Working the Line

This completed the establishment of the main telegraph line from Cairo, numerous intermediate offices having been set up to provide communications for the all-important logistic operation along this very extended line of communication. The opening of telegraph offices had kept up with the advance of the army and Lord Wolseley's HQ. The telegraph line from Cairo to Hamdab was 1,273 miles long. A table showing the offices, their dates of opening and closing, and their distance from Cairo, was shown above. For clarity, not all these offices are shown on the map, but their positions can be interpolated from the table. Twenty-nine offices are shown in the table but they opened and closed according to operational requirements. Never more than twenty-one were open at any one time.

The circuit arrangement for the army telegraphs in January and February 1885 is shown in the diagram. (17) The arrangement of telegraph offices into main line and sub-offices is shown, there being nine main offices between Cairo and Korti, with translators (telegraph repeaters) at Asyut, Aswan, and Wadi Halfa.

The layout having been established, Captain Wood took charge of the section from Cairo to Wadi Halfa, Captain Bennet took charge of the section south from Wadi Halfa, Captain Bagnold was in charge of the Dongola section, and Captain George Tower (promoted since the expedition began) took charge of the Korti section. Lieutenant Hill was in charge of the section accompanying the extension of the railway to Akasha. Captain Wood thoroughly inspected the section of the main line between Cairo and Aswan, which was not well maintained by the Egyptian administration, and managed to effect slight improvements. He was also responsible for the clearing house in Cairo, mentioned earlier. Captain Bennet was struck down with enteric fever from 6 November to 20 December, but recovered and resumed his duties. Between 21 January and 28 March he made his way up the river and inspected the entire line and visited all the telegraph offices from Wadi Halfa to Merawi.

Although they had the equipment for double current duplex working, thus improving the effective traffic capacity of the line, its poor technical quality restricted them to single current simplex and the existing heavy traffic on the line gave them no opportunity to experiment. The poor state of insulation of the line between Asyut and Cairo often prevented work at night over that section. (18) When that occurred the intermediate stations with traffic for each other, by watching and listening to the traffic, could come at once into work and begin transmitting. Sometimes, when very important messages had to be passed, the vibrating sounder was used to get through to Cairo. The operation of the main line with nine stations on it was extremely efficiently handled by the telegraphists, and their skill and stamina over the period of the expedition was of a very high quality.

OFFICES			LINES	
Main Line	Sub-Offices	Miles	Main	Branch
Cairo		0	●	
Asyut		240	●T	
Aswan		620	●T	■
	Shellal			■
Korosko		730	●	
Halfa		830	●T	●
	Gemai			●
	Sarras			●
	Mohrat			●
	Ambigole			●
	Akasha			■
Dal		930	●	■
	Fatmeh			■
Dongola		1,076	●	■
Debbeh		1,176	●	
Korti		1,223	●	■
	Merawi	1,253		
	Hamdab	1,273		■

● Single current set with relay and local sounder.

■ Morse printer (inker, which records Morse code on a paper tape).

T Translator. A sensitive relay which, with a local battery, regenerates and retransmits the signal.

Lord Wolseley used the telegraph extensively, communicating with his own staff stretched out along the Nile, as well as to Cairo and London. On Wednesday 26 November he records that "I had a talk with Buller over the wire". On Friday 28 November: "Out riding at 6.30 a.m. : had a good gallop to the Abu Fatmeh telegraph station which is about 4 miles off. We had no wire enough to run it here, so we use the heliograph for the gap. I hope to have the necessary amount of wire in a few days when the R.E. arrive in their boats". There is no doubt that Wolseley knew all he needed to about the working of his telegraph system.

When there were important matters, a special procedure was followed. In his characteristically tortuous prose, Webber explains:

At vital moments, when it was necessary that our General's message should be in the hands of the Secretary of State for War long before any other that might contain reference to the same subject, it was obligatory to make it impossible that any miscarriage might occur. ... Awaiting the preparation of the message [reporting the fall of Khartoum and the death of Gordon, see below], all work to and from Korti was stopped. Preparatory CQ's [all station calls] were sent calling the officers and clerks-in-charge at each station to stand at 'attention'; they were then told, when the message was ready, to stand clear, and to watch the working of the translators [telegraph repeaters] from a distance. The key at Korti was set to work, and until the whole message had been received in Cairo and repeated, and had been handed over to the Eastern Telegraph Company, no interruption or interference was allowed to take place.

Webber continued by describing the strain he felt:

Only 58 CQ's were sent during the whole time, but it was found necessary to control, direct, and guide the business by conversations on the wire. Most of these were carried on at night with the various persons responsible. ... The personal anxiety and strain of ever being mentally present, and of having to realise daily, if not hourly, what was going on in working and maintaining that 1,150 miles of precarious telegraphs, I will not deny, was very great, and sometimes when unnecessary difficulties harassed one, and particularly when health failed, almost exceeded the limits of endurance. But there was one good thing to stand by, the value of which every telegraph manager will understand, and that was the loyalty of the staff under me.

Others used the telegraph heavily. Major General Sir Evelyn Wood, the General of Communications, sent upwards of 3,000 telegrams in six months, all written by his own hand, and most of them more than fifty words. The total number of messages sent over the telegraph line in the ten month period it was used by the army was 125,936. Thousands of messages were transmitted in Arabic, and hundreds in cipher. Sometimes at Cairo 16,000 to 17,000 words were received in a night, a high proportion being press messages. Press correspondents were a decided nuisance in their desire to transmit their voluminous 'copy', much of it verbose and repetitive, but they paid for it, quite literally, and they had to wait until all military traffic had been cleared.

The Press, Censorship, and Security

Censorship of military telegraph lines during operations was mentioned in a previous chapter, which described the official steps that had been taken to control them and how it was introduced during the campaign in Egypt in 1882. Censorship again took place in the Gordon Relief Expedition. The press traffic has already been mentioned, and censorship of their despatches was applied.

Wolseley had little time for the war correspondents who accompanied the expedition – "those goose quill fellows", he exclaimed disparagingly during one tirade in his journal on 21 December 1884. In mid-October, while the expedition was deploying up the Nile, the correspondents had sent a joint message back to Britain saying that owing to the manner in which the censorship was exercised over their messages and the delay in transmitting them, it was quite useless to continue sending any. Wolseley (so he says, but more likely it was Webber) investigated their complaint and recorded in his journal:

I have asked these high-toned Gentlemen to state particulars & let me know which of their messages have been delayed & in what injurious manner they have been altered by the military censor. The fact is they have little or nothing to send home and they wish to account for the meagerness [sic] of their messages. As long as they were with Evelyn Wood [Maj Gen Sir Evelyn Wood] they received ample information of all that was going on, he being prepared to go to any lengths with them as long as they praised him & made much of him in their telegrams. It was through him that most of our cypher messages became known.

Wolseley had a few days earlier in his journal been critical of "Wood's vanity and self-seeking".

On this campaign there was an added dimension to censorship. As well as the military traffic, the telegraph line was used by Arabs, as of course it mostly had been before the army took it over. Their traffic, which was transmitted in Arabic by the operators of the Egypt and Sudan administrations, was both a source of revenue and a security hazard, for the British operators did not understand and could not read the language. This revenue stream was much enhanced by the mudir of Dongola, a highly religious man who, presumably to keep his minions in place, larded all his messages with lengthy strictures from the Koran. Charged by the word, his telegrams became very profitable.

The security problem was more difficult. Something had to be put in place to prevent Arab telegraphists, should they be so inclined, using the line to transmit messages with security implications. Webber said that it was known that emissaries of the Mahdi did try to send telegrams, but it was also known that they did not reach their destination. The solution to this problem was to use the services of a Mr William MacCollough who, for somebody of apparently Scottish extraction, possessed the highly unusual combination of linguistic and telegraphic skills that enabled him to read Morse code messages while they were being transmitted in Arabic. MacCollough was in fact a superintendent of the Egyptian Telegraph department, one of a number of Europeans to hold such appointments. [19] He was stationed at one of the telegraph offices on the main line, and over the six-month period that the line operated, whatever the time of day or night, was called upon to monitor 'on line' every Arabic message that was being passed. In the period, this amounted to over 5,000 messages.

Apart from measures such as this there was the more fundamental problem of controlling rumour. Curiosity by those toiling up the Nile and cut off from the outer world about the progress of the expedition and plans and events affecting it must have reached a high level. One can easily imagine that the rumour mill was rampant. The officers of the Telegraph Battalion were often quizzed around camp fires about what was going on. Pumped for news, the reticence of the telegraphists in withholding confidential information was admirable. Wolseley, in his journal, describes numerous occasions when he 'talked' to Sir Evelyn Baring in Cairo and to subordinate commanders spread along the line of communication – people such as Major General Sir Redvers Buller who was his Chief of Staff and later in command of the Desert Column, Major General Sir Evelyn Wood and Colonel Sir Charles Wilson. 'Talking' meant that the senior officers went to their respective local telegraph offices and, with the telegraphist acting as their interlocutor, conversed informally on a personal basis about the current situation, problems, intentions, and so on. This, of course, passed *en clair*, and could be read by all the intermediate telegraphists. Much of what was said was inevitably confidential, and would have been welcome grist to those hungry for news. The telegraphists gained an enviable professional reputation for not disclosing any of it to anybody. For example, when the River Column (then near Abu Hamed under the command of Brigadier General H Brackenbury, and following the course of the Nile in what was intended to be the main thrust to Khartoum) was recalled at the beginning of the withdrawal, that fact was known to every telegraphist for two days before it became known to the troops and the war correspondents. Without this confidentiality of the telegraphists, the rules of censorship would have been worthless.

Later on, the Reuters agent in Cairo, a Mr Schnitzler, compiled a daily news bulletin, and this was transmitted to all the army telegraph offices and a copy displayed outside. The soldiers, eager for news, would gather round and usually one of them would be delegated to read it out aloud. They were delighted to get the news from home.

On 29 January 1885 Wolseley, then at Korti, and in the aftermath of the battles against the Mahdi's troops at Abu Klea and Abu Kru (when a selected column of troops – the Desert Column - left the Nile valley and thus also the telegraph line, and struck out across the Bayuda Desert towards Metemma), describes in his journal how he sent and received numerous telegrams, adding "the telegraph is very well worked here by Webber. Yesterday there was very naturally a great deal done by it, as some of the messages were very long that went to the newspapers. The work done amounted to 16833 words sent from here, 310 transmitted through here, & 1485 words received. That sent over a single wire for about 1400 miles is very creditable."

The Fall of Khartoum

On Wednesday 4 February the news reached Wolseley at Korti of the fall of Khartoum and the death of Gordon on 26 January. "I telegraphed the news at length to Lord Hartington [Secretary of State for War]", Wolseley wrote in his journal, "and asked for instructions [about whether to continue the expedition and capture Khartoum, or withdraw]. What a business there will be in London over the news! I told him I was keeping back all press telegrams, & would do so until I heard from him." This very important message would have been one of those entitled to the 'Clear the Line' treatment described above by Webber. The journal entry continues with a further diatribe about Prime Minister Gladstone.

The government's reply, received by Wolseley at Korti by telegram on 6 February, was surprising. He was to protect from the Mahdi the districts 'now undisturbed', he was not to withdraw, they were going to support him in every possible way, and they were going to send troops to Suakin (with a view to opening up a railway and telegraph line of communication from Suakin to Berber). Wolseley replied and, suspicious of the motive, questioned their policy, his view being that it was "merely undertaken for party

THE NILE EXPEDITION, 1884-85

purposes to keep Mr Gladstone in office. In fact the Cabinet have today realised that nothing could save them except a spirited policy, and their telegram to me is the result". Deep in the Sudan, at the end of this long telegraph line that served him so well, he recognised the political chicanery now being unfolded. [20]

He was right to be suspicious, for a futile expedition was soon to be launched to Suakin (in which the Telegraph Battalion also participated, to be described in the next chapter), only to be withdrawn after a few months having achieved nothing. In mid-April, the hue and cry at home having died down, the government's vacillating policy was now changed to the pragmatic decision to withdraw from the Sudan, leaving the country and the Mahdi to their own devices. Meanwhile, for over two months, the expeditionary force hung around the upper reaches of the Nile between Dongola and Merawi getting demoralised by the uncertainty, dreading the onset of a summer to be spent roasting in the desert, and awaiting a renewed campaign in the autumn.

Morale suffered. Several senior officers were sent home, some with health problems, and some had mental breakdowns. Lieutenant Stuart reveals what they thought in a letter to his father written from Korti on 15 February, describing their disgust at the political shenanigans. In the midst of this, Wolseley's journal on 20 February reveals: "That fellow Webber R.E. begged to see me today and had the coolness to tell me that the education of his children demanded his presence at home. I said he ought to have thought about that before he came out here and that he could only leave this by leaving the Army. I am sorry to see an English Gentleman behaving in this way." It was not Webber's finest hour, but it was probably a combination of stress and health problems, and much of the steam let off in Wolseley's journal has to be taken with a pinch of salt. Wolseley himself was under pressure at this time, the expedition having failed.

Whatever ensued, Webber, suffering from dysentery, left Wolseley at Korti and returned to Cairo. In his presentation to the Society of Telegraph Engineers in November 1885, Webber pursued a different line, saying that it had become evident to Lord Wolseley that the conditions for maintaining the line had become so precarious (presumably because, as Director of Army Telegraphs, he had told him so) that better arrangements were required, and it was decided to "move the Egyptian government to transfer that part of the system to military charge; and in March I was despatched from Korti to carry out this transfer." Presumably Webber was referring to the unsatisfactory situation in lower Egypt where the Egyptian telegraph administration was still responsible for 'maintenance' of the unreliable section of the telegraph line between Cairo and Asyut. Waiting for a renewed campaign in the autumn, as at that time was the situation, it made sense to use the time profitably to improve this weakness.

Webber's journey to Cairo took twenty-one days "in spite of shipwreck [when a number of men aboard the boat were drowned] and the consequent loss of many worldly goods, contrary winds, sand banks, and the doctor's prohibitions to travel". Owing to several unusual difficulties, which he does not specify, it took until 1 May to complete the changes with the Egyptian administration, whereby the control of lines, offices, material, staff, and collection of revenue in Egypt was brought under military direction. By then the arrangements were of little value, as withdrawal had been ordered. Webber also mentions that his eyesight was temporarily impaired. He was replaced as DAT by Major H F Turner, who had been sent to Suakin for the expedition being launched from there. Turner left Suakin on 5 April to take up the appointment in Cairo, and was promoted to Lieutenant Colonel. Colonel Webber handed over the appointment of Director of Army Telegraphs on 27 May 1885. No mention of this is made in Captain Bennet's description of the telegraph operations, nor in Captain Bagnold's; both maintained a diplomatic silence. Turner remained DAT until 31 July 1885. Webber returned to England and retired from the army on 22 July 1885, being granted the honorary rank of Major General on retirement, but he received no mention in Wolseley's despatches.

Withdrawal

Returning to Korti and the campaign, where the troops were being kept waiting and wondering, by the middle of March the British government was seeking a face-saving formula for what today would be called an 'exit strategy'. Suitably exaggerated, it used the Russian threat to Afghanistan as a good excuse for withdrawing. Wolseley received a number of 'Secret and Confidential' telegrams from Lord Hartington advising him of the government's cooling ardour for further action in the Sudan, and he caustically observed in his journal: "This dodge of heading telegrams 'Secret and Confidential' and giving them no number [formal telegraph messages were numbered in sequence] and signing them, not 'War Secretary' but 'Hartington', is done to avoid having ever to produce them and to avoid having to send them to the Queen to see".

THE NILE EXPEDITION, 1884-85

On Monday 16 March Wolseley received a telegram from the War Office advising him, in making arrangements for the summer quarters for the troops, to take into consideration the likelihood of a change of policy on the part of the government. He observed in his journal that "I have always regarded this as very possible, almost probable".

Wolseley left Korti on 24 March and reached Dongola on 27 March. Consequently the telegraph office at Korti closed on 26 March and Merawi became the terminal office. This reduced the number of offices on the main line, which now ran between Cairo and Dongola. The line was now divided at Dongola, there being much less traffic to go on south. Between Dongola and Merawi intermediate offices were open at Handak, Abu Gus, and Tani.

At Dongola Wolseley received a telegram from Lord Hartington, Secretary of State for War, saying it would be better if he were at Cairo, and his presence might be needed at Suakin where things were not going satisfactorily. After a further exchange of telegrams, Wolseley agreed and set off for Cairo on 30 March, getting there on 7 April. It is interesting to read his journal and note how well he kept in communication with everything, sending and receiving messages through the various telegraph offices as he moved back to Cairo. The telegraph system clearly worked very well. He was even in telegraph communication with his wife in England, using their private cipher. It is rather difficult to imagine Lady Louisa Wolseley ('Loo' in his journal) sitting at her desk at home with a copy of her husband's *Soldiers Pocket Book* which told her how to do it, enciphering and deciphering her private telegrams.

On 16 April Wolseley received a confidential letter from Lord Hartington informing him of the government's wish to get out of the Sudan, including Suakin - as Wolseley had anticipated ever since he had reported the fall of Khartoum and the death of Gordon, and in accordance with the government's original policy. On Tuesday 21 April this was confirmed by a telegram telling him the government was going to announce the withdrawal in parliament later that day.

On 28 April, and now in Cairo, Wolseley spent two hours in a long 'talk' on the telegraph line with Major General Sir Redvers Buller (who had taken over command of the Desert Column following the death of Sir Herbert Stewart on 16 February) in the Merawi office, giving him the outline plan and instructions for the withdrawal. The telegraphists conducting this conversation on their behalf, and those at the intermediate offices, must have enjoyed another revealing interchange. The next day Wolseley set sail from Suez for Suakin.

Closing the Telegraph Offices

The withdrawal from the Sudan began, its progress able to be followed by the dates of closure of the telegraph offices. The closures began with Merawi on 26 May, and by 3 June all offices south of Dongola were closed. Captain Tower took charge of the Korti section and supervised the closure of offices and the movement of men and stores north to Akasha, where the railway line now reached. Fatmeh became an important point as the withdrawing troops headed for the southern end of the new railway at Akasha, so the Fatmeh office was placed on the main line at this point, bringing the number of main line offices back up to nine.

Communications between Fatmeh and Kaybar also became important during the withdrawal. A heliograph link had been tried, operated by regimental signallers, but due to heavy volume of traffic, dust storms, and the fatigue experienced by the heliographers walking up the mountain where the helio signalling station was situated, it did not work satisfactorily. On 8 and 9 June a bare wire was laid on the ground, there being no poles available, and the 23½ mile line was worked very satisfactorily using vibrating sounders ('buzzers') and only two batteries! Captain Bagnold remarked that: "The inventor of the 'buzzer', Captain Cardew, will, I am sure, be pleased to hear of this success, which has also shown once more how telegraphy has come to the rescue of [visual] signalling when this latter system is in unskilled hands and is the least pressed".

The retreating troops used the wire as a guide by night, picking it up and running it through their hands as they plodded on. When the Dervishes reached Abri in December 1885 the most advanced telegraph office on the east bank was at Akasha, but a fortified post had been established at Ginnis and communication with it was needed. Lieutenant Stuart ran a bare wire for thirty miles through Firket and Mograka to Ginnis. The Dervishes did not know of its existence. A connection was also made with bare wire from the west bank opposite Akasha to the permanent line at Ukma, thus giving Akasha two lines to Halfa. The Dervishes attacked the post at Ambigol Wells. The garrison resisted fiercely, and due to their communications with Wadi Halfa, reinforcements were sent, arriving on 4 Dec 1885.

Captain Bagnold at Dongola also described, once Sir Redvers Buller had gone, how the town had become deserted and he moved the telegraph office from the town to the army camp. "I expect we now have the best office left on the line", he said. "It was built by our men some time ago in anticipation of the move. It consists of eight mud-brick pillars about 12 feet in height, and forming a square of 26 feet sides, and supporting a strong roof of palm logs, old telegraph poles, and covered with several layers of durra stalk matting. The spaces between the pillars are similarly filled, leaving a clear space of one foot for ventilation all around under the eaves. A shady porch gives public access to the counter, and the interior is divided into the office and the battery room and store. In the hottest weather this building is comparatively cool."

He also comments on the working of the line. "Let it not be supposed that our line to Cairo has ever been worked double current, or duplex, in spite of apparatus having been sent out for that purpose. D.C. [double current] working gives little or no advantage in this electrically beautiful climate. ... To work a line with nine offices on it duplex is practically an impossibility; it also complicates the translators [telegraph repeaters]. If some enterprising individual wishes to exercise his ingenuity on the improvement of military telegraph apparatus, let him work out a trustworthy moderate speed automatic which would enable the station to work at a steady rate of, say, 45 words a minute or something just over what a clerk can read. I believe the thing is to be done quite easily without anything like the complications of the Wheatstone." The Wheatstone was an automatic telegraph equipment, used by the Post Office but not considered robust enough for field telegraph.

*The Dongola clock,
in the Royal Signals museum.*

When he moved to the temporary office Arthur Bagnold took with him the clock from the permanent telegraph office in Dongola. His contemporary description of events understandably maintains a discreet silence about liberating the clock, but it found its way to England and, after a period in the officer's mess at the old Signal Training Centre, is now on display in the Royal Signals Museum. Inscribed on plaques below it are the names of the officers and sergeant-majors of the Telegraph Battalion and the campaigns they took part in.

By 27 July all the offices south of Akasha had been closed, and the men and stores recovered to Wadi Halfa. "It seems a waste", said Arthur Bagnold, "to abandon all those lovely Siemens iron poles between Halfa and Debbeh to the Mahdi – but what can one do? The army is short of transport, and the iron poles won't float."

The plan for withdrawal and repatriation of part of the Telegraph Section had now been made. Captain Bagnold and Lieutenant Stuart with sixty NCOs and men were to stay in Egypt for a while, their task being to maintain the communications between Cairo, Asyut, Aswan, Korosko, and Wadi Halfa. They were to form part of a force being left behind under the command of General Stephenson to protect Egypt from any invasion by the Mahdi's troops, and Arthur Bagnold was in charge of the telegraph for the force. He was based in Cairo and Stuart was based with the forward Headquarters at Aswan. [21]

The remainder were to return to England. As they withdrew from Wadi Halfa back to Cairo the appropriate adjustments to manpower were made. In Cairo there was a stocktake of equipment, then those for England left Alexandria on the S.S. *Vancouver*, arriving at Portsmouth on 31 August 1885, having been away for nearly a year. Some did not return: one man had died of drowning during the hazardous ascent of the cataracts, and five had died from that scourge of those times – enteric fever. Twelve others had been invalided home.

But the operation was not over for those who remained. Things in some areas were at this stage of the withdrawal in a little disarray, and came the occasion when General Stephenson in Cairo wished to send a cipher message to a battalion commander at Wadi Halfa but the intended recipient at Wadi Halfa did not have the correct code book. "What shall I do?" said Stephenson in Cairo. Bagnold ascertained that the message was both confidential and important but required only a simple reply – 'Yes' or 'No'. After some rumination he found a solution.

"It so happened", wrote Bagnold later, "that I had at Wadi Halfa a telegraph clerk, Sapper John Conolly, who could send so fast that the other clerks [telegraphists] on the line could not read his sending, and as a Morse reader by sound he could receive and write down clearly as fast as any of his fellows could send clearly. I asked him on the wire, 'How many words a minute can you take down?' and he replied 'Anything you like to send, as long as the sending is good'. I told him that I should want him on line at 9.00 pm. My instruction mechanic,

Captain Arthur Bagnold.

Sapper Davey, and I rigged up a Morse inker [normally a receiving device] as an automatic transmitter, the contact being made between a piece of platinum wire and and the milled roller which normally propelled the Morse paper strip on which messages were received. I then marked out the message accurately in Morse characters on a strip and cut them out carefully with a penknife and a sharp bradawl. Running the strip through the inker, I succeeded in arranging that its speed could be varied between 30 and 50 words a minute. At the appointed time I cleared the Cairo office and the line, leaving only Sergeant W. F. Seggie R.E. in the office [22], and I instructed Halfa to leave only Conolly in that office. We passed the strip through at 47 words a minute and asked Conolly if he had got the message. He replied 'Some of it. Try again.' We passed the strip through again, reducing speed a little, and Conolly said 'Not all'. Again the strip went through at a slightly less speed and Conolly was satisfied. The message was delivered by Conolly himself at Halfa and in subsequent conversation with clerks on the line, I ascertained they knew nothing of its contents. I believe this is the earliest instance of automatic transmission by the Telegraph organisation of any Army."

Bagnold added that: "Conolly was a bit of a wild Irishman, but he was a genius in his way and a good fellow. I regret to record that when in the Army Reserve and employed at Siemens Telegraph Works at Charlton, in June 1889, he touched a pair of high tension terminals and fell dead to the floor". [23]

Bagnold and his men were there until June 1887, when the line was handed back to the Egyptian administration.

The telegraph troops who remained in Egypt under Captain Bagnold and Lieutenant Stuart.

THE NILE EXPEDITION, 1884-85

The Success of the Telegraph Operation

The telegraph operation to support the expedition had been an amazing success. Had it not been, the expedition, even though it failed in its main purpose, could never have achieved what it did. The number of troops spread along a line of communication of some 1,200 miles through Egypt and deep into the Sudan could never have been commanded or sustained without an efficient communication system, and a relatively small band of dedicated and highly capable telegraphists. Between Egypt and Britain they were, of course, well supported by the Eastern Telegraph Company, both its operators and its telegraph system.

Captain Bennet had things to say on the subject. "It is natural to suppose that only those who understand and are acquainted with the details of telegraph working, and were constantly superintending this work, can fully appreciate the enormous pressure that was at times put upon us, and the eagerness and anxiety which all hands evinced to cope with it in a satisfactory manner. That the work was done in a thoroughly satisfactory manner I make no doubt, and I can only say that to have done so over a line 1,200 miles long, with nine offices in circuit, and the most important messages passing from end to end of this long line, to have done this in such a trying climate must be classes among one of the greatest telegraphic feats of the day." He continues by praising the telegraphists, many of whom worked independently in small numbers in isolated positions without complaint. "Over and over again I have heard most satisfactory accounts from independent officers of their zeal, intelligence, and courtesy."

Those of the Telegraph Battalion who took part in the Gordon Relief Expedition in Egypt and the Sudan in 1884-85 deserved to be well pleased with their efforts. Captain Ferdinando Bennet and Captain Arthur Bagnold were both mentioned in despatches, and both were promoted to Brevet Major for their service during the expedition. Colonel Webber, despite his altercation with Wolseley, was promoted to the honorary rank of Major General on retirement.

The medal awarded for the Nile Expedition was the same medal and ribbon that had been awarded for Tel el Kebir in 1882. The clasp 'The Nile, 1884-85' was awarded to all who had served served south of Aswan on or before 7 March 1885, so most of the troops of the Telegraph Battalion will have been awarded this. Other clasps were awarded for Abu Klea and Kirbekan, which were battles beyond the limits of the telegraph line so none of the Telegraph Battalion will have qualified for these. It is anomalous that the medal itself retains the date of the 1882 campaign but the clasp is dated 1884-85.

Two further clasps to the same medal were awarded in 1885 for those engaged in Suakin, 'Suakin 1885' and 'Tofrek'. The Telegraph Battalion was to be deployed there, and some will have been awarded these clasps -see next chapter.

In addition, all troops who took part in the Egyptian campaigns were awarded the Khedive's Star, a dull bronze medal of five points with a plain dark blue ribbon.

THE NILE EXPEDITION, 1884-85

Biographies

Major General Charles Edmund Webber CB

Born in 6 Sep 1838, he was the son of the Reverend T Webber of Leekfield, Ireland. He graduated from the RMA Woolwich and was commissioned into the Royal Engineers in April 1855. He served in India during the Mutiny 1857-58 with the 21st Company RE. In 1858, under Sir Hugh Rose, he was in charge of a ladder party at the storming of Jhansie, leading an assault, and was three times Mentioned in Despatches. In 1866, at the time of the Austro-Prussian war he was attached to the Prussian Army as an observer, and reported on their field telegraph equipment and organisation.

In the period from 1869 to 1879 he was heavily involved in the formation, and subsequently the work of the Postal Telegraph Companies. There are a number of documents in the BT archives in which he is seen to be leading the Army side of negotiations with the Post Office about the formation, organisation, use, and terms of service of the Postal Telegraph Companies. He subsequently spent many years in London, becoming OC 22nd (Postal Telegraph) Company in 1871, and was promoted Major in July 1872.

He served under General Sir Garnet Wolseley on three occasions. Firstly in South Africa in 1879 as AA&QMG, where he dealt with logistic matters in the lines of communication, and stayed there with Wolseley during the subsequent operations in the Transvaal. He was promoted Lieutenant Colonel in January 1880, and was Director of Army Telegraphs with Wolseley in the Egyptian Campaign of 1882 (awarded CB). He was then promoted Colonel in January 1884, and again served with Wolseley on the Nile Expedition in 1884-85.

While still a Captain he was instrumental in forming the Society of Telegraph Engineers in 1871, and became a founder member (along with Captain Philip Colomb RN, Major Richard Stotherd RE, Major Frank Bolton, and others). He was President of the Society in 1882 (being called away during his presidency in August and September to act as Director of Army Telegraphs in the Egyptian campaign).

He retired from the army on 22 July 1885 in the rank of Colonel, having given notice of his retirement to Wolseley during the course of the Nile Expedition, but was given the honorary rank of Major General on retirement.

After retirement from the army, and well-qualified, he became Managing Director of the Anglo-American Brush Electric Light Corporation, in which capacity he became heavily involved in the introduction of electric lighting to London. He became a consultant to several other engineering companies. He died quite suddenly at Margate on 23 September 1904.

Colonel A H Bagnold CB CMG

Arthur Henry Bagnold was a well-known and well-liked character in the days of 'C' Telegraph Troop and the Telegraph Battalion. The son of a Major General, he was educated at Cheltenham and the RMA Woolwich, and was commissioned into the Royal Engineers in December 1872, joining 'C' Troop for a brief tour of duty in 1877. In 1880 he was appointed to the Postal Telegraph Companies, and in 1881 led a joint detachment of 'C' Troop and postal telegraphists to Natal and the Transvaal, where they repaired a long telegraph line between Newcastle and Pretoria that had been destroyed in the First Anglo-Boer War, 1880-81. In September 1884 he went to Egypt, initially to take part in the Gordon Relief Expedition to the Sudan, but stayed there until November 1887, taking over as Director of Telegraphs of the Egyptian Frontier Field Force. In 1887, turning for a while from telegraphy to Egyptology, he used his other engineering skills to carry out a salvage operation on the 100-ton statue of *Rameses II* at Memphis, near Cairo. From 1891-96 he was an instructor at the School of Military Engineering, and during this period he was on the Council of the Institution of Electrical Engineers, which he had first joined in 1877 when it was the Society of Telegraph Engineers. That posting ended his direct involvement with the army's field telegraph, and his later career on the active list was in a wider range of engineering appointments. He was appointed CB in 1907 and retired in 1911.

On the outbreak of war in 1914 he was re-employed, becoming involved in the development of wireless telegraphy, and was placed in charge of the Signal Experimental Establishment that was built at Woolwich (the forerunner of what was later SRDE Christchurch and RSRE Malvern). His valuable work in this appointment was recognised by the award of the CMG. After the war he retired to his home in Shooters Hill, near Woolwich, and amused himself with further inventions in his well-equipped workshop there. He died in December 1943, in his ninetieth year.

Arthur Bagnold had married in 1888, and had a son and a daughter. His son, Ralph Bagnold, later Brigadier R A Bagnold OBE, joined the Royal Engineers in 1915, served in the RE Signal Service, and transferred to the Royal Signals when the Corps was formed in 1920. Ralph Bagnold was a founder member of the Long Range Desert Group, received the Gold Medal for his exploration of the North African Desert, and was a Fellow of the Royal Society.

THE NILE EXPEDITION, 1884-85

Endnotes

1. Sir Evelyn Baring, later Lord Cromer, Agent, Consul-General, and Minister Plenipotentiary in Egypt 1883-1907. Formerly an army officer, he attended staff college in 1867 and was closely associated with Cardwell's army reforms.

2. *History of the Sudan Campaign* (2 vols) by H E Colville, published by the Intelligence Department of the War Office a few years after the campaign.

3. Many books (most of them now out of print but available through specialist libraries or second hand booksellers) have been written, some soon afterwards by the participants giving contemporary descriptions, and some later by authors with access to archive material. See bibliography.

4. In 1935 Arthur Bagnold wrote his personal *Reminiscences of the Nile Expedition* for Lt Col E W C Sandes RE who was at that time compiling the history of *The Royal Engineers in Egypt and the Sudan*. The various quotations attributed to him are taken from this source.

5. Col Webber later wrote of his experiences, published in the *Journal of the Society of Telegraph Engineers*, Vol XIV, 1885, pp 452-474. The various quotations attributed to him are taken from this source.

6. Wolseley's journal contains many pungent criticisms about all and sundry, particularly Prime Minister Gladstone and his Cabinet. These daily effusions were despatched by instalments to his wife in England for safe keeping. The campaign journal covering the period of the Gordon Relief Expedition is reproduced in the book *In Relief of Gordon* by Adrian Preston, pub 1967. A way of 'letting off steam', the journal reveals a bombastic character, who found it difficult to temper his innermost often accurate opinions with a touch of graciousness and diplomacy.

7. Figures were produced by Col Webber and Mr Oatway. With the intervening passage of time and monetary inflation they are perhaps now rather meaningless, but the total revenue of message traffic handled on the military line was £23,000. Some 80,670 messages were handled during the seven month period ending 30 April 1885.

8. *The Campaign of the Cataracts* by Col William Butler, pub Samson Low, London, 1887. Butler was one of the Wolseley 'Ring' since their days on the 1870 Red River Expedition, and was placed in charge of the expedition's whaler boats.

9. These were the lines that were considered, and rejected, when planning rearward communications for the Abyssinian expedition. See Chapter 6.

10. Some accounts erroneously describe this as happening during the course of the expedition. These events occurred, as the narrative states, during unrest in 1883, before the expedition arrived.

11. A detailed description will be found in the files of Maj Gen A M Stuart, held in the archives of King's College, London. Stuart corresponded with Arthur Bagnold in 1935 when they were providing their accounts of the expedition to Lt Col Sandes, author of the *The Royal Engineers in Egypt and the Sudan*.

12. Described in detail by Bagnold in his Reminiscences.

13. Later to be Sirdar of the Egyptian Army, and eventually Field Marshal Earl Kitchener of Khartoum. He will feature prominently in Chapter 17.

14. Extract from *The War in Egypt and Sudan,* Vol 3, pp 263-64, by Thomas Archer, pub 1886.

15. From *The River Column* by Maj Gen H Brackenbury, p 52, pub 1885.

16. Stuart's work is described in general terms in letters from him to his family in England. The letters of Maj Gen A M Stuart are held in the archives of King's College, London. (A copy of Bagnold's Reminiscences will be found in the same source.)

17. Based on Col Webber's presentation to the Society of Telegraph Engineers on 12 Nov 1885. *Journal of the Society of Telegraph Engineers*, 1885, pp 453-475.

18. Details of all 'outages' were reported to the Egyptian administration, measurements of insulation resistance were taken and the reasons proven, but it was of little avail. Webber quoted the total time of circuit interruptions as '40 days of 24 hours in six months' (ie about 20%).

19. MacCollough is mentioned as one of the superintendents in Webber's description of events in the 1882 Campaign, *Journal of the Society of Telegraph Engineers*, 23 Nov 1882.

20. A good description of the political manoeuvres in London will be found in *The Scramble for Africa* by Thomas Pakenham, Chapter 15.

21. Stuart later became Maj Gen Sir Andrew Stuart KCMG CB, although he did not serve again with the Telegraph Battalion.
22. Sergeant Seggie was the telegraphist who, as 2nd Corporal Seggie, at Tel el Kebir in 1882, had sent the message reporting the victory directly from the battlefield to London, and received the reply from Queen Victoria forty-five minutes later. See Chapter 12.
23. Quoted from Arthur Bagnold's *Reminiscences of the Nile Expedition*.

Chapter 15

Suakin, 1885

Historical Background

The story of Suakin begins with Osman Digna. His ancestors were believed to be Kurds who had settled in Suakin in the sixteenth century and had intermarried with the local Hadendowa and Arteiga tribes. Osman Digna himself was born in Suakin in about 1840 and was the offspring of a Levantine father and a Hadendowa mother. His family were ostensibly merchants, actually slave dealers, so Osman Digna travelled the Sudan on this nefarious business, during which time he came under the influence of the Mahdi. When the British interfered with his livelihood he was all too ready to throw in his lot with the Mahdi against the infidels. He kept his fiefdom in the eastern Sudan in a state of ferment for years, and in late 1883, at Trinkitat and Et Teb, trounced Egyptian expeditions sent to deal with him. He then invested Tokar. It was decided that a British force must be sent to defeat him, as the port of Suakin was strategically and politically an important place if Britain was to retain control of the Sudan – something that in 1884 the British government seemed unable to come to any firm decision about. As part of controlling the rest of the Sudan it was intended to build a railway from Suakin to Berber, on the Nile, and thence to Khartoum.

After General Gordon had arrived in Khartoum in January 1884 (see previous chapter) a British force, consisting of two brigades under Major General Sir Gerald Graham VC KCB, was sent from Cairo and arrived at Suakin on 22 February 1884. Landing at Trinkitat, they secured a victory over Osman Digna's forces at Et Teb, forcing him to abandon Tokar. On 13 March there was a further battle at Tamai. However, on 28 March 1884, further demonstrating the British government's lack of clear purpose, the campaign in Suakin was brought to a close and most of the troops withdrew, leaving only a small garrison there. The field telegraph was not deployed with this expedition.

When Gordon became besieged in Khartoum in May 1884 and a rescue was contemplated, political and strategic muddle continued. The controversy about routes to Khartoum took place, as explained in the previous chapter. It was still thought that Suakin would be required for the expedition, so engineers returned there on 1 July 1884. The harbour and other facilities were improved, and a local small-gauge railway was built. Meanwhile Osman Digna's remaining forces, not quashed in the earlier battles, were all around Suakin and remained a threat. The decision to select the Nile route was made in August 1884, but the work at Suakin continued nevertheless.

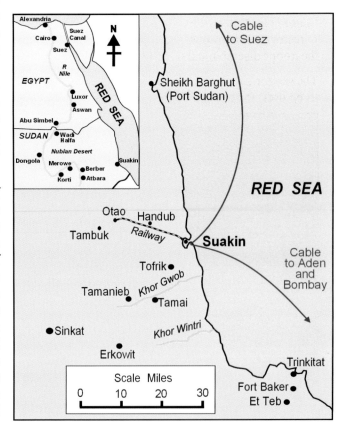

After the fall of Khartoum and the death of General Gordon on 26 January 1885, the government decided, in their continuing state of confusion, to despatch a second expedition to Suakin, again under the command of General Graham. His orders this time were to destroy the forces of Osman Digna, and to construct a railway from Suakin to Berber so as to be able to resupply Wolseley's main force on the Nile. Graham arrived back in Suakin on 12 March 1885. This time he had forces twice as powerful as those he had commanded there the previous year, including contingents from India, and he set about the task which he could have completed then but for the vacillation and indecision of Gladstone's government. Now that Gordon was dead, were they withdrawing, the mission for which they went there having failed, with political repercussions for the already weakened government? Or, as an act of defiance to regain some political capital, were they aiming to re-conquer the Sudan? The government could not seem to decide.

The Telegraph

Amongst Graham's new force in March 1885 was a section of the Telegraph Battalion, later reinforced by a further section, and it is time to turn to their story. They had been placed under orders for General Graham's second venture to Suakin soon after the decision to mount the expedition was taken in February 1885. By that time the Telegraph Battalion had already deployed troops to South Africa earlier in 1884 for the Bechuanaland Expedition, and they were still there (chapter 13). They had then sent more to Egypt, including reservists, in September 1884, under Lord Wolseley for the Nile Expedition, as described in the previous chapter. Now, with both these deployments still in progress, field telegraph resources for the new commitment were getting even more stretched.

No 2 Section of the Telegraph Battalion was prepared for Suakin, under the command of Lieutenant H E M Lindsay. (Henry Lindsay had been with Arthur Bagnold to Natal and the Transvaal in 1882, to restore that telegraph line after the first Anglo-Boer War). Because of the independent nature of the operation the section was reinforced above its normal establishment of one officer and fifty-three NCOs and men. Lieutenant Bowles and additional NCOs and men were posted in, to bring their strength up to two officers and sixty NCOs and men, made up with forty-three men from the 1st Division (ie regulars) and seventeen men from the 2nd Division (the former Postal Telegraph Companies). Twenty-one of these men had less than one year's service, and many of the others were also inexperienced. They sailed from the Albert Docks in London on the SS *Queen* on 16 February 1885, taking with them their normal section complement of five wagons and thirty-seven horses, twenty miles of wire, and three telegraph offices. To use the oft-recurring observation, this was going to be inadequate for the task.

Apart from the inadequate strength and inexperience of the field telegraph troops, one might spare a sympathetic thought for the horses who got shipped around the world on these ventures. Being February, this lot would be in their thick winter coats, and would arrive in three weeks time, after what was often a ghastly journey in a ship not properly equipped for the transport of animals, into the oppressive heat of the Red Sea port of Suakin.

Major H F Turner, at that time in command of the 34th Postal Telegraph Company at Bristol, was appointed Director of Telegraphs, and accompanied the section on the *Queen*. They arrived at Suakin on 7 March, and the troops disembarked on to Quarantine Island, where the army HQ was situated, taking two days to unload.

The section's task was expected to be the construction of a telegraph route from Suakin to Berber, 200 miles, to be used for both military telegraph traffic and the operation of the railway that it was planned to build.

A contemporary sketch of Suakin, 1885. The southern part, seen from the sea.

A large quantity of reserve stores were needed, just as they had been six months earlier with the Nile expedition. 140 miles of 3/16 galvanised iron wire, other stores, and the equipment for a further eight telegraph offices (bringing the total to eleven) were shipped on 23 February on the SS *Romeo*. 80 miles of No 14 hardened copper wire, considered by Major Turner to be better suited for the task envisaged, was shipped on the SS *Argo* on 4 March.

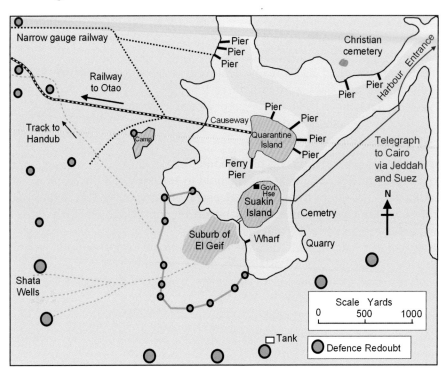

Suakin and its defences, 1885.

The section's first work, on 10 March, was to construct a 3½ mile airline from the Headquarters camp through the town to the office of the Eastern Telegraph Company. This line was equipped with Morse sounders and two telegraphists operated the terminal at the army HQ. They operated under the orders of the military censor who vetted all outgoing telegraph traffic. The Eastern Telegraph office was connected by submarine cable to India (Bombay) and Egypt (and thence to London, as well as to the telegraph offices of the Nile Expeditionary Force), thus providing the strategic rear links as described in the previous chapter.

In the period 11-21 March the section was busy setting up local communications within the Suakin area, connecting outposts into the defensive system and such places as the base hospital on Quarantine Island and the auxiliary hospital at H Redoubt. Some of these were telegraph, but they had also brought Ader telephones and some of the local users were connected by these, which they operated themselves.

The Battle of McNeill's Zariba

On 22 March Major General Sir John McNeill, VC KCB KCMG, commanding the 2nd Brigade (the force in Suakin consisted of two brigades) advanced south with 4,000 men towards Osman Digna's stronghold at Tamai. The CRE, Colonel J B Edwards CB, had suggested that they should be in communication by field telegraph, and accordingly a detachment of No 2 Section of the Telegraph Battalion under Lieutenant Lindsay, consisting of sixteen men, twelve horses, one airline wagon fitted to take ninety-six drums of outpost cable, and a water cart advanced with the force.

McNeill's task was to build two zaribas, one five miles and the other eight miles from Suakin. (A zariba was a defensive position, usually protected simply with locally cut thorn and scrub.) Major Turner superintended the laying of the telegraph airline that accompanied McNeill's force – a frustrating business as clumsy camels continually knocked it down. A vibrating sounder had been joined up at Quarantine Island on the end of the stretch of airline built the previous day, and at daylight, when the march began, this was extended by outpost cable, laid on the ground by the side of the Indian square as it advanced. The cable was reeled off the wagon, and the detachment tested it at intervals as they advanced. For some unexplained reason No 2 Section did not have the proper D5 field cable, so the flimsy outpost cable (so called because it was only intended to connect a sentry at an outpost to the main defensive position by telephone, a distance perhaps of half a mile) and had been borrowed in Suakin from a Field Company RE. It was not strong enough, and was unsuitable for the task – a problem that was adversely to affect telegraph communications in the weeks to come.

Progress was slow, due to the nature of the country and the need to advance defensively because of the threat of attack. McNeill realised he could not accomplish his intended task in time so, after five-and-threequarter miles, at about midday, he used the telegraph link to communicate back to General Graham in Suakin and get permission to build a zariba at Tofrek instead. Whilst this work was in progress, at about 1.00 pm, the telegraph terminal was moved into a part of the zariba, where it temporarily sat on top of some empty ammunition boxes. At 1.40 pm Major Turner indulged himself by sending a little self-congratulatory message to the CRE saying: "We have kept up communication so far with perfect success". True, yet hardly an astounding feat!

With the work of building the zariba only partially completed, they were attacked in mid-afternoon while the troops were tired, the heat was oppressive, and defensive arrangements such as fields of fire had not yet been ordered. The cavalry outposts had given warning, but they had not been deployed far enough forward and they galloped back into camp with the Dervishes swarming after them in a vast mass. The men of the Telegraph section rushed into the zariba, bringing a wagon with them and helped to man the defences. The vibrating sounder had been knocked off the ammunition boxes and trodden into the ground in the general rush. Almost immediately, in scenes of terrible confusion, the enemy were within the partially completed defences, and hand-to-hand fighting took place. It lasted about twenty minutes, a brief period crowded with instances of cool bravery, wild bewilderment, and fanatical desperation. When the smoke cleared, it was a shambles. The battle of Tofrek, or the battle of McNeill's Zariba as it is sometimes called, left 100 British and 1,000 Dervishes dead in under half-an-hour.

'The telegraph tent during an action.'
A sketch depicting the battle of McNeill's zariba, originally published in The Illustrated London News.

During the battle a new vibrating sounder was connected and Lance Corporal Bent, the telegraphist in charge, got in contact with Quarantine Island. General McNeill then sent a message to the Chief of Staff at Suakin. The message was written by McNeill using Lieutenant Lindsay's back as a desk. He was soon able to report that the attack had been repulsed. During the sending of messages Lance Corporal Bent had sometimes to leave his tent and assist with the defence when the enemy appeared to be closing in on the tent. During the course of the battle and immediately afterwards some twenty to thirty messages passed between McNeill's zariba and the HQ in Suakin – two of them from war correspondents with the column. The news of the attack was in some English papers the following day, 23 March – a 'scoop', if indeed that expression had by then been invented. At 7.00 pm the line was cut by the enemy and telegraph communication ceased.

Wolseley's journal reveals exchanges of telegrams about the battle between himself, then near Korti on the Nile, and Graham in Suakin. It is remarkable that the two were able to communicate effectively between Suakin and Korti, *via* the Red Sea submarine cable, Cairo, and the tenuous long line down the Nile for over a thousand miles – a total distance of some 2,000 miles. The direct distance between Suakin and Korti was about 380 miles. (Wolseley, it should be added, recorded in his journal in his inimical blistering style, that he was generally unimpressed with Graham's progress in Suakin so far.)

On 23 March the drivers and long-suffering horses spent the day burying dead Dervishes and camels, and later that day the detachment returned to Suakin, leaving six men in the zariba. On the way back the line was repaired but failed again soon after, the unsuitable, easily-damaged outpost type of cable apparently being to blame. Much credit though is due to Lieutenant Lindsay and the members of the section who performed so well at McNeills's zariba.

On 26 March Lieutenant Bowles and another party accompanied a convoy to the same zariba trying to repair the line. They laid three miles of new cable, but the line was never to be relied on. The cable was repeatedly cut by the enemy and lengths taken away.

Reinforcements Arrive

On 27 March construction of the telegraph line for the railway began. It was built of copper wire on fifteen-feet high fir poles, the standard piece of telegraph equipment. Three wires were put up along the route initially, but it was soon found that the poles could only successfully carry two wires. Three miles were constructed, as far as Western Redoubt.

For the next few weeks lines were built around the Suakin area to meet the requirements of the troops. Telegraph offices were open twenty-four hours a day. Altogether there were nine telegraph offices around the defended area, and 7,342 messages were sent.

In the meantime it had been decided that further reinforcements for the field telegraph in Suakin were needed, and men and material had been assembled in England. The SS *Somerset* sailed from England on 25 March carrying another 200 miles of copper airline, equipment for nine more telegraph offices, fifty miles of D5 field cable (urgently needed to replace the inadequate outpost cable), 2,325 fifteen-feet fir poles, and other stores. The *Somerset*, however, broke down in the Bay of Biscay, and crawled into Gibraltar for repair. The stores had to be trans-shipped and did not reach Suakin until May.

On 1 April more stores were sent - 100 miles of copper wire, batteries, and other telegraph equipment. In addition an order was placed with Messrs Siemens and Co for 6,000 light iron tubular poles; these were ready by May and reached Suakin but were not landed. Altogether material for 560 miles of airline, fifty miles of cable, and twenty-three telegraph offices were sent. It was also intended to send bamboo poles just received from China, but these were diverted to meet the urgent requirement of the Bechuanaland expedition who were also having difficulty obtaining all the stores they needed.

Another section of the Telegraph Battalion, the last one, was also sent. No 3 Section sailed from the Albert Dock in the SS *Goorka*, under the command of Captain C F C Beresford, with its normal scale of equipment but no wagons and only two riding horses. Because of the current pressure on resources sixteen men from the 24th Middlesex RV (Post Office reservists) and a number of completely inexperienced men were sent, in all forty-five NCOs and men. On arrival they would bring the total strength in Suakin up to the Director of Army Telegraphs (DAT), three other officers, and 105 NCOs and men.

Charley Beresford was an inspirational character. Commissioned in 1865, he had served in Ireland, Canada, Bermuda, and in 'C' Troop. Of Irish stock, strongly built, and imbued with an Irish sense of humour, he was a sailor, an oarsman, and an outstanding horseman. In Suakin, where the living and working conditions were dire, he proved to be an outstanding leader of men. This comes through in his records of the expedition, quoted extensively in this chapter. Although appointed DAT, he was very 'hands on'. He knew all his men, he cared for them, and he led them from the front in trying circumstances. He was later to command the Telegraph Battalion.

On 5 April, Major Henry Turner had left Suakin to assume the appointment of DAT in Cairo, to replace Colonel Webber as DAT in Egypt (the events described in the previous chapter), and Lieutenant Lindsay temporarily took over the appointment of DAT. When No 3 Section arrived in Suakin on 21 April the appointment of DAT transferred to Captain Beresford, then the senior telegraph officer. Over the next few days the newly-arrived No 3 Section took over the Suakin local communications, leaving No 2 Section free to accompany the troops advancing as the railway was built.

The Railway Telegraph Line

On 25 April, after action to clear the immediate area of Osman Digna's forces in the area of Tamai, work began to extend the railway. The railway, intended to go to Berber, was initially routed through Handoub to Otao, just over eighteen miles from the coast. As things were to turn out, that was as far as it got. On 29 April No 2 Section left Handoub and went on to Ohao, the ultimate limit of the field telegraph. The telegraph line kept up with the troops at railhead, although communications proved difficult due to the constant damage being caused by the carelessness of the troops and the motley collection of railway construction workers, navvies, and camp followers of many nationalities, as well as the transport animals and their often useless drivers. (Eleven mules and four Maltese drivers were attached to the two Telegraph sections.) The telegraph offices were located in tents beside the railway but were closed and brought in at night to defended areas.

Osman Digna's men attacked the telegraph line almost every night, cutting the poles down and carrying some away – apart from disrupting communications, it was thought that they were attracted by the ornamental brass-headed ebonite insulators, and the copper wire had many uses. A number of patrols comprising infantry and telegraph troops were organised to try and catch them, but none ever were. After one of these nocturnal excursions Captain Beresford wrote that: "It was a beautiful night with full moon, delightful temperature, and still as death except for the caterwauling of jackals and the chirps of the crickets. It was a fine opportunity for star-gazing as we lay on the sand, but I confined my attention to the 'north star' and the 'southern cross', unfailing friends to keep you straight".

Protecting just eighteen miles of telegraph line was hard enough. It is difficult to imagine how anybody thought that a 200-mile telegraph line across the desert to Berber was ever going to survive and provide reliable communications without first destroying all of Osman Digna's forces and controlling the ground completely. (The same could be said for the intended railway line, as that was attacked as well.)

In late April Captain Beresford noted: "Weather getting warm, 95 degrees in the shade today. Several of our men on the sick list but nothing serious. They are chiefly among the PO Volunteers; the change in mode of living has slightly disagreed with their internal mechanism".

On 3 May an expedition was planned to attack an enemy force at Tamanib. The much-awaited fifty miles of D5 cable, originally sent in the *Somerset*, had by now reached Suakin on the *Argo*. It was immediately unloaded and detachments from No 2 Section with wagons and horses were ordered back from Otao to Suakin ready to accompany the Tamanib foray. The expedition to Tamanib was then cancelled.

Interior of the first zariba on the route to Handoub. Note the heliograph station on the tower. A sketch from a contemporary book, The History of the War in the Soudan, by J Grant.

Sergeant Attwood died in the hospital ship *Ganges*, anchored in the harbour, on 4 May, and was buried in the new cemetery at Suakin, his being the first grave there. His coffin was carried on one of the wire wagons. Lieutenant Lindsay was struck down by fever on 12 May. He was sent to the *Ganges* and was evacuated home on 22 May. He was replaced by Lieutenant Buckland of the 24th Field Company RE who took charge of the detachment at Otao. Sickness due to the climate and unhygienic conditions was increasingly taking a toll of the men.

Withdrawal from Suakin

Earlier, on 21 April, while in Korti, Lord Wolseley had received a secret telegram from Lord Hartington, Secretary of State for War, telling of the government's intention to suspend any further action against the Mahdi, to withdraw the Nile expeditionary force from the Sudan, and to suspend the construction of the Suakin – Berber railway. Wolseley's reaction was predictable, and more vitriol was poured on the government in his journal: "How I wish that England had been ruled for the last five years by some real

Terminus of the railway at Quarantine Island.

A grainy old photograph from the Royal Signals archives of No 2 and 3 Sections at Suakin, 1885.

statesman instead of the scarecrow vestryman Mr Gladstone". He was instructed to return from the Nile Expedition to Cairo, and from there he visited Suakin from 2 to 19 May, after which he returned to Cairo and eventually England.

The evacuation from Suakin began shortly afterwards, rumours about aborting the expedition having been rife for some time, with adverse effect on morale. In an attempt to keep the troops motivated, a gymkhana was arranged by the Indian Contingent. Captain Beresford and Lieutenant Lindsay both won prizes in the mounted events. Beresford himself was always a keen horseman. He describes an incident in Suakin on his mare 'Kathleen' (presumably one of the two riding horses that accompanied No 3 Section to Suakin): "The railway leaves Quarantine Island by a causeway about 200 yards long, and a roadway for other traffic runs alongside the rails. When riding Kathleen across it, there is always an uncertainty as to whether we get to the far end, as she loathes camels, and objects strongly to railway trucks and engines. We were caught between strings of both yesterday, and our fate balanced between camels, trains, and the sea for some moments".

Then on 16 May, while Wolseley was still there, the not unexpected order came. The telegraph line from Otao to Handub was recovered, the copper wire, according to Beresford, "is of course no further use for telegraph work. It might be melted down into Victoria Crosses, so we will bring it home". Fever was taking a greater hold, and on 23 May Corporals Humphries and Fearn, and Sapper Newlands were invalided home.

The 'Fuzzy-Wigs', as the Hadendowa tribesmen were called on account of their hair style, made another raid on the line, cutting down six poles, but "Corporal Dowling and party got it through very smartly early next morning". By 24 May all the lines had been recovered back through Handoub to the defended garrison area.

Various re-arrangements of the lines in the garrison were then made. All the reserve stores which had been assembled for the telegraph line to Berber were handed into the ordnance store and were eventually returned to Woolwich. The sections re-equipped and reorganised themselves to their normal scale. Six telegraphists and one lineman were detailed to remain behind with the small force left to garrison Suakin.

Captain Beresford gave his opinion of Quarantine Island, where they now awaited a ship. "Quarantine Island I put down as the worst smelling place in creation. It is at present inhabited by horses, mules, navvies, railway contractors, Arabs, coolies, naval transport, camels, telegraph battalion, hospitals, headquarters staff, &c, &c. Every available spot not occupied by a human being is a rubbish heap or a pile of tibbin [camel's food]".

The remainder of the two sections and Captain Beresford embarked on the SS *Persian Monarch* on 31 May, having spent from 4.30 am to 3.00 pm loading the ship with their stores, "the men working like slaves under the tropical sun", and sailed for Suez. "Never was a spot left with such few regrets ..... and never was there a spot less deserving of parting tears". Their thoughts were with Corporal Todd and Sappers Blessley, Adams, Brooks, Hall, Stevens, and Mock who had to stay behind to maintain three telegraph offices still required by the remaining garrison. They returned sometime later, several invalided home, and Corporal Todd died of fever on board ship. "I hear on all sides praises of the manner in which our men have done their duty during the campaign, under exceptionally trying circumstances", said Beresford.

Like the other troops they had done their job, but the expedition had been futile from the start. Its cost was £3 million, with nothing to show for it. The aborted railway line was another fiasco, its cost put at more than £865,000. The 1885 Suakin episode is a largely forgotten sideshow in the bigger drama of the British government's inept and indecisive handling of the affairs of the Sudan at the time. But to restrict the narrative to tales of the telegraph, a number of conclusions can be made.

Telegraph Performance

The field telegraph troops performed well in extremely trying conditions, although many of them were inexperienced. Captain Beresford remarked that "they suffered severely from sickness and death due to disease, attributable to the climate and in their case also to their adverse environment, and their unremitting work in offices and on the lines in the heat of the day".

The deployment of the telegraph was well implemented, if unexceptional, and raised no particular problems except the unsuitability of the outpost cable they were forced to borrow and use. It was the cause of too much unreliability in the operation of circuits. Why the telegraph sections were neither equipped nor had the foresight initially to take the more suitable D5 cable remains unexplained. When fifty miles of D5 eventually arrived, it was too late to be of any use.

The vibrating sounder, Captain Philip Cardew's recent invention, proved of great value as a field instrument. Although the interference it caused with adjacent circuits was sometimes a problem (which is why it had no place in permanent civil telegraph systems), it was on other occasions the only instrument that could be used successfully on unsatisfactory lines due to its extreme sensitivity.

The copper wire ordered by Major Turner was found useful for long lines that were not liable to be broken but proved difficult to manipulate. "If bound in tightly it is very apt to break; if bound loosely it runs through, and is consequently hard to regulate; it is useless for stays or other purposes, and is expensive. Three-eighteen strand-iron wire is, in my opinion, far more serviceable", said Captain Beresford. Iron poles were the best answer for long lines of semi-permanent communication, but they were expensive and not easily transportable in tactical situations. The fifteen-feet fir pole was found not to be able to carry more than two lines. The search for the perfect field telegraph pole was to continue for many years – indeed it was never found. Captain Beresford also observed that "all linemen should be able to ride".

The level of telegraph traffic was very high, as it was always to be during operations, reinforcing the need for well-trained telegraphists, and plenty of them. The Suakin office was open for eighty-two days and dealt with 7,162 messages, the Headquarter office was open for the same length of time and handled 6,785 messsages, the Quarantine Island office was open for eighty-one days and handled 13,952 messages, and the Handoub office was open for thirty-three days dealing with 3,732 messages. There were a number of other less important offices, opened from time to time, where records are not available, so the overall total was higher. Messages were typically about sixteen words in length. On 15 May at Quarantine Island 810 messages were sent and received on two instruments, considerably exceeding the traffic level on the busiest hand-keyed telegraph circuits in England. "All office telegraphists should know how to connect their instruments and batteries, and be trained in construction work and removal of faults", said Beresford. One imagines that this observation arose from the high number of inexperienced men sent with the sections, including Post Office reservists who probably never had to do such things in permanent telegraph offices.

A number of telephones had been taken and were set up at places such as the HQ, the hospital ship *Ganges*, and the ordnance stores, and other administrative places connected by short lines. They were operated by the users, thus saving further load on the telegraph system, and the arrangements worked well. The telephones were of the Ader type.

The benefits of combining the electric telegraph with visual signalling so that they were complementary were again demonstrated, but the lesson – one learnt in the Zulu War in 1879, soon after visual and electric telegraph were separately organised, was quickly lost again.

Their voyage home suffered various delays, and it took nearly two months from leaving the cess pit of Suakin on 31 May to reach Portsmouth on 23 July 1885. Captain Beresford described the often purgatorial journey with humour, and his words are reproduced in the following pages. Those readers who feel they have at some time endured inconvenient, extended, and uncomfortable journeys in the course of serving the Crown might like to read it – and be thankful!

SUAKIN, 1885

The Passage Home from Suakin

This description of the return journey to England of the two sections of the Telegraph Battalion from Suakin, which took over two months and much loading and unloading of their equipment, was written by Captain Beresford and published in the Military Telegraph Bulletin.

27th May, 1885.

To our intense joy the telegraph section got their orders to embark for Suez two days ago, and we embarked yesterday on board the *Persian Monarch*. Never was a spot left with such few regrets as Quarantine Island, Suakin, and never was there a spot less deserving of parting tears than that 'tight little island' where, for the past few weeks, we have been living over the remains of departed Arabs.

We had no small amount of work embarking all our equipment, the men working like slaves from 4.30am to 3.00pm under a tropical sun. There was no help for it, the job had to be carried out; and, as the ship had her nose towards home, there was all the greater alacrity shown. I hear on all sides praises of the manner in which our men have done their duty during the campaign, under exceptionally trying circumstances. Curiously enough, we find ourselves again on board with our old friends of the *Ghoorka*, i.e. the 17th Company Commissariat and Transport Corps, under Capt. Collard. I am sorry to add, however, that on comparing notes we find that many who came out with us are not returning. Some have been invalided home before this, and a few have been buried at Suakin. Among these latter was Private Mitchell, whom we remember as singing at our concerts during our former voyage the well-encored ditty of 'Brown Upside Down'.

The *Persian Monarch* is a much smaller vessel than the *Goorkha*, and we are now carrying 420 men and 330 horses. The upper decks are choked up with our telegraph equipment, waggons, poles, drums of wire, &c., so we are pretty well jammed. We miss a good deal, and I am sure the horses miss the indigestible morsels provided for them by the fair sex on our former voyage.

There was no sleeping below last night, the heat was too great. It was, however, delightful on deck, with a strong head wind all night blowing away the miasma of Suakin. No one who had not tried it can imagine the luxury of sitting in this breeze, and getting the spray over you, after sweltering in Quarantine Island for a bit.

SUEZ, 7th June.

We got into the docks at Suez early on the morning of the 4th, but were not disembarked till the afternoon of the following day. The interval was spent by the *Persian Monarch* in taking in coal, and we had great hopes that we should have gone on to England in her. These hopes were dispelled by orders to disembark on the afternoon of the 5th, and to go under canvas.

It was a busy afternoon getting out all our stuff, and we did not get into camp till 9 00pm, too late to pitch tents. The camp is pitched on a sandy plain, and is about one mile west of the town. We are now tolerably settled down, and waiting orders to re-embark for England, and looking forward to another happy afternoon getting our equipment on board again. The climate here is a delightful change from Suakin; though hot, it is dry, and the nights fresh, not to say cold, towards morning. We hope everyone is picking up, but, after all the hard work, we have a good many not altogether up to the mark. The hospital is close by, and looks pleasant with its gardens and shrubs. The principal medical officer tells me that the sick are pouring in from Suakin, and that there are many cases of enteric - the effect, no doubt, of hard work during the heat of the day. European troops forget what a tropical sun is, and are very apt to treat it with the same disrespect as if it was behind a London fog. But it is not, and has a way of letting you know it when you least expect it. Lord Wolseley paid a flying visit to Suez on the 4th and 5th, in order to receive the Duke of Connaught, who arrived on the latter date on his way home from India.

11th June.

We are still at Suez, waiting for a ship, of which there appears no immediate prospect. The weather is pleasant, with the exception of the sand storms, which blow up nearly every afternoon, and of which we get the full benefit, being camped on a sandy desert. The S.S. *Erin* came in three days ago with the 10th and 11th Companies Royal Engineers on board; we hoped to have found room on her, but were doomed to disappointment. However, we are not sorry now, as the Canal is blocked by a sunken dredger near the Bitter Lakes, and the *Erin* has in consequence been stopped halfway up, and is likely to remain there for ten days or so. A fresh channel has to be dug round the sunken vessel. We have started a rifle range, which gives us some amusement.

The 24th Company Royal Engineers, which landed from the *Loch Ard* a few days ago, has gone up to Cairo, but all its horses (39) are now attached to No 3 section, the Telegraph Battalion. Having only two mounted NCOs and one driver in the section, we are obliged to turn all the telegraphists on to stable duty, a new experience for them, but very useful. Suez is not a very lively town, though the hotels are generally full of passengers for India, or those homeward bound. The bazaar is the usual collection of miserable little stores and grog shops, where much poison is sold for the delectation of Tommy Atkins.

The suburbs consist of collections of flat-roofed mud hovels, inhabited by natives, donkeys, dogs, and goats; but along the banks of the Sweet Water Canal the scenery is enlivened by palm trees and numerous little gardens, where the 'fellah' works away like a nigger. Near where the horses water there is a small mud enclosure where the natives say their prayers. This they do standing, kneeling, and on all fours, kissing the ground, and pointing for Mecca.

ALEXANDRIA, June 19th.

Here we are, another stage on our homeward journey, having, with pleasure, said good-bye to the desert of Suez. Our last night there we celebrated by a camp fire, which passed off with much *éclat*, and the echoes of the desert were wakened by songs from our most distinguished vocalists - Shoeing-smith Reid bearing off the palm, to judge from the calls upon him by the audience.

Fuel for a bonfire is not always at hand in an African desert, but, fortunately, the Commissariat and Transport Corps were able to supply us with a number of old horse troughs. Some of these troughs appeared to take the form of 'Tibbin' when in flames; but doubtless this was the effect of mirage. The 17th was a busy day, and towards evening the press of work was at its height, when we had only 15 men and 40 minutes to hoist six waggons on to railway trucks. These trucks had high sides which would not fold down, and there was no platform. The uncceremonious way in which we bundled our air-line waggons aboard, and pitchforked the Arabi cart and water-cart on top of all, was delightful. But the job had to be done, and was done in the time; and the men were also done, for it was a tough bit of work, with a thermometer standing not far off 100 degrees.

We started for Alexandria at 7.30pm, i.e., 2 officers and 79 men of the Telegraph Battalion, with 78 horses, and the 17th Company Commissariat and Transport Corps.

Our route lay through Zagazig, and not round by Cairo. The darkness prevented our seeing much of the 'Land of Goshen', which we passed through, but we had a good look at the Delta, which was crossed in daylight.

The train hardly rivalled the *Flying Dutchman* as it took sixteen hours on the journey, landing us in Alexandria about 11.30 on the 18th.

We got our gear out as soon as possible, and marched off to the 'Red Barracks', which are about half-a-mile from the railway station. Here the men soon shook down into comfortable quarters, and we seemed to have got once more into the regions of civilisation.

We are short-handed with the horses, though our telegraph clerks make excellent cavalry soldiers. Corporal Stewart is as much at home on a bare-backed horse as at his circuit in Edinburgh. I feel sure our secretary at the General Post Office, and the officer commanding 24th Middlesex, will be glad to hear of the creditable way in which their men have taken to their new duties. During the last few days at Suez we had a lot of natives told off to help us; this help chiefly consisted in sitting down in the shade of the horse and gently rubbing his leg with a little bit of rag.

I am sorry to say we have had much sickness. We buried poor driver Newton, who died of dysentery, on the 15th. He died early in the morning, and at 6.00pm both sections paraded to attend his funeral. We carried the coffin, covered by a Union Jack, on a wire waggon to the cemetery, which is a walled enclosure in the desert, and about half-a-mile from our camp.

We have also had to leave several men behind in hospital at Suez, among whom there were two or three serious cases. There is no doubt the climate of Suakin was more formidable than Osman Digna.

Here we are still in the Red Barracks, shared by our men with fleas, which are like the sand by the sea shore for multitude; otherwise we are in good quarters, and an excellent climate. Waiting for embarkation is all very well in its way, but when it lasts from day to day for four weeks it becomes a chronic complaint; our Kharkee [sic] coats have long become white with washing, in order to give ourselves some air of respectability in this centre of civilisation.

I much regret to have to record the death of Shoeing-smith Ventham, whom we left behind in hospital at Suez, down with enteric. He was a fine, powerful young fellow. The first I knew of his ailing was, when dis-embarking at Suez, he asked to be excused working as he felt so ill. He had to sleep out that night with the rest of the men in the desert, as we arrived on the camping ground so late. The next morning he went into hospital, and never came out again.

I am glad to say that the latest accounts report Sapper Richardson as doing well.

Regiments are now coming up from the Nile. The men of the Camel Corps embark this week for England, and the Black Watch has just arrived in Cairo. [Both of these had seen action near Khartoum.]

Alexandria is by no means bad quarters, and for explorers is a mine of wealth. All the rubbish heaps, the remains of old cities and palaces, are full of relics of thousands of years back, and one has only got to dig to find them. But this digging is attended, I believe, with a certain amount of risk, as the old enteric fever of ages gone by is also bottled up here, and catches hold of you if you let him out. I hear many ladies anxious to gather curios for their boudoirs have felt the consequences.

S.S "QUEEN", 10th July, 1885.

We are at last homeward bound on the good ship *Queen*, an old friend of the Telegraph Battalion, as she carried out No. 2 section from England last February. Our fellow-passengers are the Light Camel Corps from the Nile, under the command of Colonel McCalmont, the 17th Company C. and T. Corps, and a few of what are known in military parlance as 'details', or various small detachments of different branches of the service - odds and ends, so to speak.

The 'Camelry' is a composite corps made up of companies, each of which consists of about 40 men from one of the light cavalry regiments serving at present at home. All being picked men, they are a very fine lot, though, no doubt, one now sees the fittest who have survived the climate of the Upper Nile. During the last nine months they have had a varied experience and a hard life, having been almost to the furthest point reached by our troops. As a fighting body, the Camelry are unique, for, besides being cavalry soldiers, they now have been taught to ride and take care of camels, and, in addition, are trained to fight as infantry.

The two sections Telegraph Battalion commenced embarking at Alexandria at 6.00am yesterday morning, slinging the horses on board being the first operation. The horses were brought up one by one, made fast in the slings, and hoisted from the wharf to the upper deck, then lowered down through a hatchway to the main deck, where four or five men were ready to receive and place them in the stalls.

We sailed in the afternoon and had a little excitement just before starting, in watching a chase after two of our stokers, who, at the last moment, thought they would let the *Queen* go without them, but who were captured in the streets of Alexandria by a party of police and soldiers sent in pursuit. Having secured our stokers, we weighed anchor at 6.30, and, steaming out of harbour, so one saw the last of the Palace of Meks, the Pharos Lighthouse, and the minarets of Alexandria.

13th July.

We anchored this morning at 9.00am in the Great Harbour of Valetta, and, as usual, were immediately surrounded by swarms of boats carrying every conceivable kind of merchandise, from monkeys to lace flounces for ladies' petticoats, and from singing canaries to cakes of tobacco. As there is no limit to what Tommy Atkins is ready to buy on his homeward journey, there was a brisk business done, and we have now a menagerie of no mean pretensions on board.

Foremost among the quadrupeds is an ornamentally clipped white Egyptian donkey, belonging to Lord Charles Beresford. It is quite a baby, and I regret to say the 'Vet' has pronounced him to be ill with mumps. He can only be tempted by his native 'tibbin' (chopped straw) or beans, and won't look at oats or hay; indeed, he prefers cocoa-nut matting to such European provender.

We have also on board monkeys, dogs of all ages and descriptions, goats from the Soudan, antelopes from the same parts, piping canaries from Malta, &c. We set sail at 3.00pm, and do not touch again till Portsmouth is reached.

17th July.

To-day we felt it decidedly cooler, as we had a breeze from the Spanish mountains, which we could see snow-capped in the distance. At 6.00pm. we passed Gibraltar, and were no doubt reported home by the signal station. The view of the Rock from the east is a remarkable one, and from its resemblance to a recumbent figure is known by the Spaniards as 'The Corpse'. As it stood out in shadow against the evening sky it had a striking effect.

23rd July.

After an uneventful time in the Bay of Biscay and Channel, we anchored at Spithead at midnight on the 22nd. This morning we had a fine view of the Channel Fleet, as at an early hour it arrived to do honour to the Princess Beatrice on her wedding day. [Princess Beatrice, daughter of Queen Victoria, was to be widowed some ten years later, when her husband died during the 2nd Ashanti expedition, described in chapter 16.] The men-of-war took up their station close round our ship, and soon afterwards made themselves gay with bunting.

At 9.00am we went alongside the wharf at the Dockyard, and immediately began to disembark. ....

At Aldershot, which we reached at 6.30pm, we received the agreeable intelligence that instead of going to the Royal Engineers' quarters in the South Camp, we were to get our baggage out, and march off to Bourley Bottom, some two miles off across the Long Valley, and there encamp, in order that the horses might be placed in quarantine. .....

Colonel C F C Beresford

Charles Frederick Cobbe Beresford was born of Anglo-Irish extraction in 1844, the elder son of the Reverend Charles Beresford and great grandson of the Rt Hon John Beresford MP, brother of the 1st Marquess of Waterford. He joined the Royal Engineers in 1865.

Tall – he was known as ' the long 'un ' - and of strong physique, he was a renowned sportsman, excelling at sailing, rowing, and riding. After leaving Chatham he saw service in Ireland, Canada, and Bermuda. He was posted to the Post Office Telegraphs in 1873 before joining 'C' Telegraph Troop at Aldershot in 1873 for four years. With field telegraph in its infancy, he was an ardent supporter of the alliance between military and civil telegraphs that then existed.

He graduated from the staff college in 1880, and then saw service in the Transvaal in 1881. Afterwards, he served in the Intelligence Branch in the War Office until 1883. He then rejoined the Telegraph Battalion, during which time he took part in the short campaign in Suakin.

Later that year he was promoted to Major and commanded the 2nd Division of the Telegraph Battalion for four years, with his HQ in Leicester Square. He gained a reputation as an inspirational leader - clever, quick-witted and humorous. This led to him being given command of the 1st Division of the Telegraph Battalion in 1889. It was he who suggested that Mercury be their badge, a choice that has continued to the present day.

In 1892 he was posted to Ceylon (Sri Lanka) for three years. In the rank of Colonel, he next served as CRE in Plymouth, and in 1898 he was appointed to the staff of the Chief Engineer, Ireland. He retired in 1902, eventually moving to Camberley to be near his old haunts in Aldershot.

In 1910 his elder son, Captain Charles Beresford RE, was killed while trying to stop a runaway horse. During the early years of the Great War he was Colonel of the Frimley Volunteers. He died in December 1925 at the age of 81.

Chapter 16

The Second Ashanti Expedition, 1895-96

Historical Background

Since the campaign of 1873-74 (chapter 8) the Ashantis had ignored the terms of the treaty that had been agreed with them. They remained a threat to neighbouring tribes on account of slave raiding, and they indulged in domestic slavery and human sacrifice, creating trouble amongst the tribes friendly to the British. The name of their capital, Kumassi (previously anglicised as Coomassie), meant appropriately 'the death place'. Also, they were hindering the development of profitable trade with the region.

By 1890 the 'Scramble for Africa' by European countries was well under way. The French from the Ivory Coast to the west and the Germans from Togoland to the east began to make treaties with the native kings and chiefs. To prevent encroachment to the north of the British settlements, the Ashantis were approached with the suggestion that a British Resident should be stationed at Kumassi, and that the British would control the foreign relations of the country, which would thus become a British Protectorate. The Ashanti king, Prempeh, refused to accept this idea.

Accordingly, in October 1895, it was decided to send another military expedition to Kumassi with the objective of establishing a British Protectorate, thus heading off the French challenge to the territory, and putting a stop to the Ashantis various uncivilised practices.

The Expedition

The expedition was raised with commendable speed. It was commanded by Colonel Sir Francis Scott KCMG CB, an experienced officer, having served in the Crimean War and the Indian Mutiny. Most usefully of all in the present circumstances, he had served in the first Ashanti Expedition of 1873-74, in the latter stage of which he commanded his regiment, the 42nd Highlanders (later the Black Watch - an appropriate name, perhaps), for which he had been awarded the CB and promotion to Brevet Lieutenant Colonel.

The force now being sent, about 2,000 strong, consisted of the 2nd West Yorkshire Regiment and a Special Service Battalion made up of contingents from several regiments, with a wing of the 2nd West India Regiment, a strong body of local constabulary, and a Royal Artillery battery of four 7-pounder guns. There was also a body of local scouts, organized and commanded by Major R S S Baden Powell, 13th Hussars, whose task after they crossed the River Prah and entered Ashanti territory was to cover the advance of the main body. [1] With much field engineering to be done, Major H M Sinclair RE was appointed CRE of the expedition with four officers and a party of thirty NCOs drawn from RE units at Aldershot. There was in addition a detachment of the Telegraph Battalion, of whom more shortly.

Generally, the expedition was well-planned and well-founded, taking good account of the problems met by the earlier 1873-74 expedition. Sir Francis Scott's previous experience on that expedition must have contributed. In particular there was good medical support to deal with the expected problems of fever,

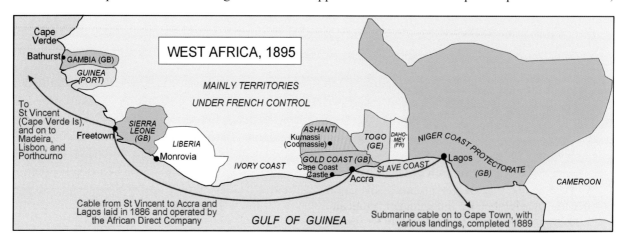

particularly malaria, and strict hygiene and sanitation discipline was observed; despite this, health was again to prove a major problem. There was no wheeled transport as there were still no roads inland, nor pack animals as there was no forage for them, and the tsetse fly and horse sickness - certain early death to beasts of burden - were rampant. It was recognised that carriage of stores would depend on human porterage, and the best possible arrangements for this were made in advance. This reflected the fact that it was the first operation since the army changed its arrangements for administrative organisation and duties in 1888, when the Army Service Corps was formed, and the supplies and transport were much better organised then in the past. [2]

Prince Henry of Battenburg, the uncle of the late Earl Mountbatten of Burma, had been appointed Military Secretary to Sir Francis Scott, and brought two donkeys with him, believing them to be immune from the tsetse fly. Whatever became of the donkeys after this misjudgement is not recorded, but the unfortunate Prince Henry was one of those to die of fever during the campaign.

The route to be followed - the same as that of the 1873-74 expedition, for there was no other sensible choice - was divided into two parts. The first part, just about halfway, was to Prahsu, on the river Prah, and still just in Gold Coast territory. This river consisted of a series of rapids which made navigation impossible, and was usually crossed in canoes. The local Government maintained a reasonable track from Cape Coast Castle to Prahsu. North of the Prah, and into Ashanti territory, the road to Kumassi degenerated to little more than a scarcely passable footpath, in need of considerable improvement. Immediately on arrival, ahead of the main body, Major Hugh Sinclair and his field engineer party, using local labour, started improving the route to Prahsu and bridging the Prah and other streams, as well as providing safe water supplies. At selected camp sites temporary hutted accommodation was built, some of them serving as hospitals, to try and preserve the health of the men.

Telegraph Operation

The Telegraph Battalion at Aldershot received orders on 13 November 1895 for a detachment to proceed with the Expeditionary Force. Commanding the detachment was Captain Reginald Curtis, who was appointed Director of Telegraphs, with Lieutenant Duncan MacInnes as second-in-command. Curtis, then thirty-two, had served some years earlier in the Egyptian army. MacInnes was born at Hamilton, Ontario on 19 July 1870, the younger son of the Honourable Donald MacInnes, of Hamilton, who had moved to Canada from Scotland in his early life and became a Member of the Senate. Duncan MacInnes entered the Royal Military College,

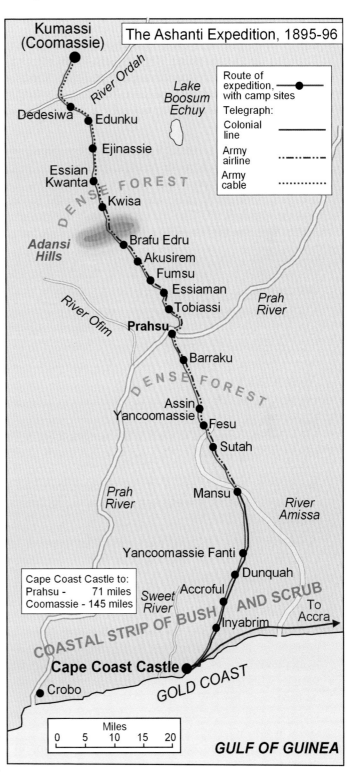

The route to Kumassi.

THE SECOND ASHANTI EXPEDITION, 1895-96

Kingston, Canada, and passed out at the head of his year with Sword of Honour and Gold Medal, obtaining a commission in the Royal Engineers on 16 July 1891. The remainder of the detachment consisted of thirty-two NCOs and men, of whom the 1st Division (the regular element) provided twenty-one and the 2nd Division (the Postal Telegraph Companies) provided eleven. The telegraph operation also benefited from the experience of the 1873-74 expedition, and in the intervening period military telegraphy had advanced as well, so they were altogether better organised, better equipped, and better manned. [3]

As it was known that only native porters would be available to transport their stores, everything as far as possible was packed in 45-pound packages before shipment. Lists describing the contents of each package were sent out to facilitate sorting and unloading - the confusions experienced with shipping telegraph equipment on earlier expeditions were thus minimised. The detachment embarked at Liverpool on the SS *Bathurst* on 23 November 1895 (except seven men who followed in the SS *Loando*). Also on board were Sir Francis Scott, commanding the expedition, and most of his headquarters staff. Cape Coast Castle was reached on 13 December, and unloading began immediately. By midday on the 14th nearly all the telegraph material was on shore and sorted out – a distinct improvement on earlier expeditions, despite the relatively short notice.

The initial task was to establish telegraph communication between Cape Coast Castle and Prahsu, seventy-two miles inland, where the headquarters of the expedition was to be established. From the experience of the 1873-74 expedition it was known that the dense forest would impede the construction of an airline, so a large proportion of cable was taken. Insulated cable, laid on the ground, could bypass difficult stretches and speed things up. The detachment took with them sixty miles of cable (type D7, which weighed about eighty-five pounds per mile) and 100 miles of airline complete with equipment for three offices.

The colonial government had meanwhile anticipated their requirement and had built a good line from Cape Coast Castle to Mansu, half-way to Prahsu, considerably simplifying the task of the army telegraph detachment, and communication that far had been opened on the 13th.

A submarine cable had been laid to West Africa in 1886 from Cape St Vincent in the Cape Verde Islands, giving a route to England via Madeira, Lisbon, and Porthcurno, and at the time of the 1895 expedition it was operated by the Eastern Telegraph Company, so in contrast to the earlier Ashanti expedition in 1873-74 there was now a strategic rear link. The submarine cable landed at Accra, and an overland line connected Accra with Cape Coast Castle. [4]

The local operators on this permanent line were all natives, and neither their rate of work nor their skill were good enough for the pressure of a military campaign. To improve matters two of the army detachment's telegraphists were sent to each of the telegraph offices at Mansu, Cape Coast Castle, and Accra, where they remained until the end of the operation. A small clearing house was established at Cape Coast Castle, but as the Eastern Telegraph Company at Accra undertook the recovery of all overseas charges against the government or press, the financial transactions to be undertaken by the army's clearing house were not extensive – nothing like as complicated as the problems faced by the clearing house in Cairo during the Nile Expedition of 1884-85.

A working party of 150 natives was taken over on 14 December, the day after landing, and on the 15th Captain Curtis, Lieutenant MacInnes and sixteen men marched from Cape Coast Castle with as much cable and equipment as the native working party could carry. This amount was not very great, but no other form of transport was available. Consequently, after the essential loads of fourteen days' rations, blankets, kit bags, reserve ammunition, etc, there were not many porters left for telegraph stores.

Each labourer working for the expeditionary force, and there were several thousand of them, wore an armband to identify which working gang he belonged to, and was issued with a label with a small chain to hang round his neck. These labels had been brought from England and proved invaluable in payment of the various working parties, as they were the only means of recognizing the men. They were paid threepence 'subsistence' per day in part payment of their wage. Based on the high desertion rate of the previous expedition, it was a 'drip feed' to keep them interested, in anticipation of the final pay off which would come at the end! However cunning the strategy, this daily payment was a constant and very tedious task for the officers, and the need to carry large sums of money in small change (mostly threepenny bits!) became irksome. At the end of the operations each labourer received 1s. 3d. per day for the time he had been at work.

Mansu, where the colonial line had been extended to, was reached on 18 December, and on the next day work commenced. One party went on ahead with cable, averaging ten to twelve miles a day, a moving office equipped with a vibrating sounder being at the cable head. Another party constructing airline followed in the rear, replacing the cable as fast as possible, and averaging about three miles a day. The great difficulty with the cable was to have enough of it to continue working, and parties of twenty-five men were constantly sent back to bring up more supplies of cable and equipment from Mansu, where the general transport of the expedition had concentrated nearly all the telegraph stores.

Baden-Powell described the scene as he advanced towards Mansu:

Constructing the line through the jungle. A sketch by Major Baden-Powell.

> At nearly every ten miles we came upon a rest-camp, in a more or less completed condition, for occupation by the British troops when they come marching up. With no little labour, bush has been cleared away for many hundred yards, and huts have been built up of bamboo frames, with trellis sides and palm-thatched roofs. Within them tables, seats, and bed-places have been made of split bamboos - accommodation sufficient for three hundred men, with a complement of officers. Store sheds are being quickly filled with food and ammunition. ... And then, as far as Mansu, half-way to the Prah, the telegraph runs near the path but taking a more direct line through the bush by a track recently cut out with much and heavy labour [the colonial line as far as Mansu]. After Mansu the 'fetish cord', as the natives call it [the army cable], no longer hangs on poles, but lies along the ground close to the path. It is the mere field cable of the Engineers that now takes the place of the more permanent line; and as we press forward, we at length overtake Captain Curtis, R.E., working himself, like his men, half-stripped, and laying out his line at the phenomenal rate of two and a half miles an hour. This in itself is a record that would be hard to beat when all the difficulties of country, climate, and circumstances are taken into consideration. [5]

As had been found by the earlier expedition, the work of constructing airline was very difficult. The path, only eight feet wide and very tortuous, was bounded on either side by dense jungle and enormous trees. In only two or three places on the entire route was it possible to erect more than three poles 'in the straight' consecutively. Curtis described how "the creepers and overhanging foliage necessitated a great amount of clearing, and the clinging plants on the trees were infested with most truculent ants, who attacked the men hammering in spikes with much energy and virulence". Natives were used for the carriage of tools, drums, etc, and to a certain extent in the construction of the line, but the soldiers had to exercise constant supervision. None of this was unexpected; Lieutenant Jekyll had reported similarly in the 1873 Ashanti expedition.

The cable reached Prahsu and communication opened with the Cape Coast on 22 December. The airline reached Prahsu on 1 January 1896, and single current sets

The cable detachment at Prahsu, 23 December 1896. Sergeant Low is standing 3rd from left, Captain Curtis is seated at the front. See endnote (6).

were established. Now well into the pestilential jungle, the work after 27 December was much impeded by a high proportion of the men in the working party being down with fever.

In several places insulator brackets and other relics left by the previous expedition in 1874 were found in the trees, and at Prahsu, owing to a shortage of material at the time, some of the old wire, a few coils having been found on the banks of the Prah, and surprisingly not purloined by the natives in the intervening twenty-two years for any number of other purposes, was used to prolong the new line.

The Expedition's Headquarter staff established themselves at Prahsu, connected by the telegraph to the base at Coast Castle and also with the head of the column as they advanced along what had somewhat derisively come to be called 'the Great North Road'. The communications working efficiently, but not all the telegrams from Cape Coast Castle were welcome at Prahsu; one demanded to know the immediate reason why the men's pay list had been sent to them in manuscript instead of on Army Form O1729!

The system of 'diplexing' the line by adding vibrators and separators to the single current sets was most useful, especially when there was much 'press work' in hand. This enabled two messages to be sent simultaneously in the same direction.

The strain on the men employed on the maintenance of the line was very severe; two linemen were posted at about every twelve miles with four to six natives to carry ladders and other equipment. These men worked outwards from their post, but gradually, as fever became more prevalent, there was often only one man to more than twenty miles, and repairs had to be entrusted to natives, or convoys moving along the road. The faults were caused by trees constantly falling across the wire; by heavy storms or tornadoes, which brought the line down in long stretches; by fast growing creepers making 'earth', and very frequently by the natives who, anxious for ornaments, removed portions of the wire to be made into bracelets. (6)

The telegraphists, who perhaps had the most trying work of anyone on the expedition, were located in unhealthy places with very poor accommodation. Fever frequently attacked one out of two telegraphists at an office, leaving the survivor to work from dawn till well after dark. Cases occurred, however, of men leaving hospital temporarily during the day to assist their comrades, returning to their beds as the fever increased towards evening.

On 30 December orders were received to cross the river Prah, and extend the line beyond. The work of construction became even harder. Communication was opened to Tobiassi by 11am on the 31st, the cable becoming available by the construction of the airline being pushed on as quickly as possible.

The table shows the progress of the line. The cable with moving office was well ahead of the main column, the front being covered by Baden-Powell's levy and the Houssas.

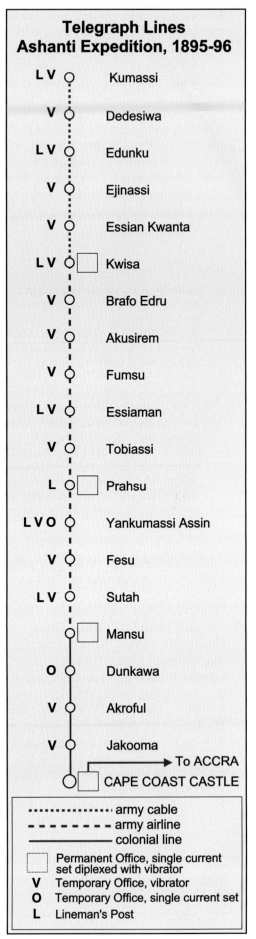

THE SECOND ASHANTI EXPEDITION, 1895-96

At Brafo Edru the advancing headquarters staff caught up with the cable head, and after this point the moving office accompanied Sir Francis Scott, keeping pace with the troops as they marched.

Captain Curtis, who had been with the cable detachment up to Akusirem, was sent to the base from there with fever, leaving that party to continue the advance under the supervision of Sergeant Low. Lieutenant MacInnes was superintending the construction of the airline following behind, and it ultimately reached Kwisa.

The leading troops made their entry into Kumassi on 17 January 1896. Baden-Powell described the scene. The Ashantis, after much prevarication, finally decided that now was not the time to make a futile last stand, native drums rumbling out the message of peace not war. All assembled on what might be called the parade ground, although it was usually the theatre for human sacrifice, the evidence of skulls and bones around. Prempeh, decorated with a gold and brown tiara, beads and golden nuggets, sat on his throne surrounded by his courtiers, a velvet umbrella held over him. The palaver and surrender was about to begin when suddenly Baden-Powell recorded:

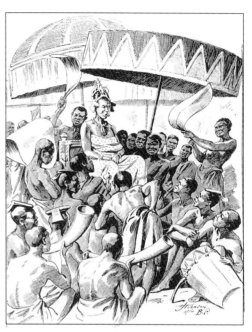

> ... a little party of our force comes hurriedly across the ground, three white soldiers with four natives carrying a reel and winding off the field telegraph; and thus within a few minutes practically of the arrival of the advanced force in Kumassi, the fact would have been known at home had not the previous day's tornado destroyed the line in sundry places. But this feat has not been performed without cost. Captain Curtis is in hospital with fever, as also are many of his men; and it is a fact worthy of record how, in spite of this and of the heavy work connected with the laying and the working of the line, its completion has been carried out with such rapidity and efficiency.

Prempeh watches the troops arrive.

The soldier leading the line party that got to Kumassi at the same time as the leading troops was Sergeant James Low, and owing to very heavy storms on the 17th and 18th, the line to Cape Coast Castle had indeed been damaged. Communication to the coast was not re-established until the 19th.

As the troops returned to the coast the line and offices at Kumassi, Kwisa, and Prahsu were handed over to the colonial Telegraph Department. By means of vibrators and separators communication was maintained without interfering with other ongoing traffic by 'teeing in' when required during the march back. Clerks with movable sets accompanied, respectively, the Headquarters, the Special Service Corps, and the West Yorkshire Regiment; and every evening many messages were sent and received at the temporary halting places.

The last man of the telegraph detachment embarked on the 7 February, and all were back in England on 26 February 1896. In no case did their fever prove fatal.

The total number of messages passing over military wires, reported Curtis, was 13,192, "the value being about £1,200". It seems strange today that military telegraph operations in remote places such as the Gold Coast, where there were many operational problems to contend with, were encumbered with such matters as collecting money to use their facilities – a relic of the days when there were no other means of communicating.

The End of the Campaign

King Prempeh was deposed and exiled to the Seychelles. A British Resident was appointed and a British Protectorate declared over the country. Prempeh was allowed to return to Kumassi in 1923.

The following year, Lieutenant Duncan MacInnes returned to the colony and planned and constructed a fort at Kumassi. Built of stone, and rectangular in form with flanking towers at the four corners, it contained a house for the Resident and barracks for the guard. The defences were soon afterwards tested by a siege of many thousands of Ashantis, but the garrison held out until relieved. The fortifications were described by the commander of the relieving force "as the best defensive positions he saw in West Africa". [8]

THE SECOND ASHANTI EXPEDITION, 1895-96

Endnotes

1. Robert Baden-Powell (who came to be regarded by many of his contemporaries as rather too much of a self-publicist) wrote a book describing the expedition, *The Downfall of Prempeh*. He was also an amateur artist and produced many sketches of the expedition, including the two in this chapter. Later he commanded the garrison at the siege of Mafeking (1899-1900) during the Boer War. He is probably best remembered today as the founder of the Boy Scout movement.

2. The work of the detachment of the Army Service Corps is described in the Journal of the Royal United Service Institution, Vol XL, pp969-998, *To Kumassi and back with the Ashanti Expeditionary Force*, by Lt Col E W D Ward CB.

3. The description of the detachment's work is largely based on the account written after the operation by Curtis, and published in the Royal Engineers Journal, 1 June 1896, pp 122-125.

4. The route from Lisbon to Pernambuco (now Recife) in Brazil, via Madeira and St Vincent, had been laid by the British Telegraph Construction Company for the Portuguese government in 1874. The cable from St Vincent to West Africa had been laid in 1886 by the West African Telegraph Company, with landings at Bathurst in Sierra Leone, Accra, and on. This company was absorbed into the Eastern Telegraph Company in 1889, for operational and economic reasons, when the West African cable route was extended down the coast to Cape Town by the latter Company, thus providing a second route to South Africa. (The original route to South Africa, brought into service in December 1879, was down the east coast of Africa from Aden to Durban.)

5. From *The Downfall of Prempeh* by Robert Baden-Powell.

6. Apart from being quite unsuitable for the West African climate, this dress is rather surprising, as the British Army adopted khaki in 1895. However, it is as contemporarily described by Curtis and as shown in the photograph. All ranks of Royal Engineers on the expedition wore blue puggarees on their helmets, but the Telegraph Battalion NCOs and men wore blue serge frocks (as worn in peace by the 2nd Division, Telegraph Battalion), the other RE wearing scarlet. It must have been the last operation that they were dressed in this manner. They wore khaki in the Boer War, which started in 1899.

7. This reminds the author of the time he was serving in Kenya in 1961, when D10 field cable was introduced and used there for the first time on a training exercise. The novel shiny black plastic insulation immediately attracted the attention of the local female populace, and there were numerous unexplained line faults during the hours of darkness, with stretches of cable 'disappearing'. Next day the cause was apparent. The local ladies were spotted wearing ornate shiny black plastic bracelets and, on the more ambitious, were some quite stylish necklaces.

8. Colonel (later General) Sir James Willcocks, who described it in *From Kabul to Kumassi*.

The Campaign Medal

The campaign medal, made of gun metal, was in the shape of an unusual four-pointed star, together with the cross of St Andrew, and was apparently designed by the widow of Prince Henry of Battenburg, who died during the campaign, and was Queen Victoria's youngest daughter, Princess Beatrice. It was awarded to all who took part in the expedition between 7 December 1895 and 17 January 1896. The medal was issued unnamed, although the Colonel of the West Yorkshire Regiment arranged for the medals issued to his regiment to be named at his own expense.

King Prempeh's Chair

King Prempeh's chair (not the throne) was given to Captain Curtis by Sir Francis Scott, the commander of the expedition, in recognition of the work of the telegraph detachment. After his retirement in 1919, Major General Sir Reginald Curtis, KCMG CB DSO presented the chair to the Royal Corps of Signals in 1921, a year before he died. Brigadier-General Edmund Godfrey-Faussett, then the Commandant of the Signal Training Centre at Maresfield in Sussex, arranged for the chair to be accepted on behalf of the Corps by Captain (Quartermaster) John L Low – the very man who twenty-five years earlier, as Sergeant Low, had laid the cable into Kumassi and established the telegraph office there. The chair is now in the Royal Signals museum.

Biographies

Major General Sir Reginald Salmond Curtis KCMG CB DSO

Reginald Curtis was educated at Cheltenham College and the RMA Woolwich, and was commissioned into the Royal Engineers in 1883.

He served with the Egyptian Army from 1890 to 1893, taking part in the campaign in the Eastern Sudan in 1891, including the capture of Tokar, for which he was awarded the Order of the Medjidie 4th class. For his service in Ashanti 1895-96 he was awarded the brevet of Major and was Mentioned in Despatches. He was sent on special service to the Falkland Islands in 1899, under the Admiralty.

Later in 1899, the Anglo-Boer War in South Africa having started, he became ADC to the Maj Gen RE in Cape Town. In August 1900 he became Assistant Director of Telegraphs for a brief period, seeing operational service in the Orange River Colony and the Transvaal. Late in 1900 he became Chief Staff Officer of the South African Constabulary (which was then headed up by Maj Gen R S S Baden Powell, with whom he had served in Ashanti). He was awarded the DSO in April 1901, and was also awarded the brevet of Lieutenant Colonel. In 1905 he took over as Inspector General of the South African Constabulary, holding that appointment until the end of 1908.

He was appointed Commandant of the Army Signal School in 1912, and the following year was promoted to Colonel, awarded the CMG, and appointed AAG RE at the War Office. He was appointed CB in 1915, promoted to Maj Gen in 1916, and awarded the KCMG in 1917, when he was appointed Maj Gen Aldershot Command. He retired in 1920, and died on 11 January 1922.

Brigadier-General Duncan Sayre MacInnes CMG DSO

Duncan MacInnes was born on 19 July 1870 at Hamilton, Ontario, Canada, and was the younger son of the Honourable Donald MacInnes who emigrated to Canada from Scotland in early life. He entered the Royal Military College, Kingston, Canada, from which he passed out at the head of his year with Sword of Honour and Gold Medal, obtaining a commission as Second Lieutenant in the Corps of Royal Engineers on 16 July 1891.

After the Ashanti expedition Duncan MacInnes was employed in the erection of the fort at Kumassi, built for the defence of the garrison. Its plan and execution reflected great credit on him; in 1900 it was besieged by fifteen thousand Ashantis, and held out until relieved. He also acted for a time as Resident at Kumassi.

During the Boer War he was involved in the defence of Kimberley, for which he was Mentioned in Despatches. Later in that war he was again Mentioned in Despatches and was created a Companion of the Distinguished Service Order.

In October 1902 he was married, at Montreal, Canada, and was then employed with the South African Constabulary for two years. From 1905 to 1908 he held staff appointments in Canada, first as DAQMG at Halifax, Nova Scotia, and afterwards as DAAG with the Canadian Dominion Forces. He returned to England and attended the Staff College, followed by a number of staff appointments.

He served in World War I, taking part in the retreat from Mons, and then in the War Office in staff appointments until 1916, rising to the rank of Colonel. In 1916 and 1917 he was Director of Aeronautical Equipment, with the rank of Brigadier General, and was created a CMG. He went to France in March 1917 as CRE to the 42nd Division. After nine months he was appointed Inspector of Mines. On 23 May 1918 he was killed in an explosion while engaged in this duty, and was buried in the Military Cemetery at Etaples, close to the sea, on 25 May 1918. During the war he had been twice Mentioned in Despatches, and had the Russian Order of St. Stanislaus, with the French Legion of Honour (Croix d'Officier), in 1917.

Captain (QM) J L Low OBE DCM MSM

James Lindsay Low was born at Portsmouth in 1865, and joined the Telegraph Battalion in 1885. By the time of the Ashanti expedition he had been promoted to Sergeant. A few years after that he served in the Boer War, during which he was awarded the Distinguished Conduct Medal. He retired from the army in the rank of RQMS in 1906, to become the landlord of the Station Hotel at Littlestone, Kent.

In 1914, at the outbreak of war, he rejoined and was commissioned as Quartermaster of the Signal Training Centre, first at Bedford then at Maresfield, Sussex. He retired again in 1922 and became a sub-postmaster, finally retiring from this work in 1955 at the age of ninety, and died the following year. His son joined the newly formed Royal Signals in 1921, rising to the rank of Lieutenant Colonel on retirement in 1954.

Chapter 17

The Reconquest of the Sudan, 1896-98

Historical Background

During the years that followed the failure of the Nile Expedition of 1884-85 there were insistent demands to avenge the death of General Gordon in Khartoum and to reconquer the Sudan. Whatever the swings and roundabouts of foreign policy and personal opinions, it was money that influenced events; the British government had resolved that any action to this end should be undertaken and paid for by the Egyptian government, putting a damper on any excessive imperial enthusiam. Nevertheless, the will was there, and the two leading figures in achieving it were going to be the British Agent and Consul General in Egypt, Lord Cromer, and the Sirdar (Commander-in-Chief) of the Egyptian Army, Colonel Herbert Kitchener. Both of them, it will be recalled, had featured in the Nile Expedition (chapter 14).

Colonel Kitchener in his early days as Sirdar. True to form, it appears as though somebody is about to get a red card! From a portrait by von Herkomer.

Kitchener, a lifelong bachelor, whose appointment as Sirdar at the young age of forty-one had been controversial, was known for his arrogance and aloofness and was not popular amongst his peers in Cairo society – no social graces, and no gentleman, they thought. Perhaps less well known were his parsimonious qualities and his capacity as a senior officer to interfere with junior officers' work. Everything was sourced after great investigation into the cheapest possible price and method of procurement; equipment was repaired, and repaired again; and nothing was ever thrown away or discarded until it finally fell to worthless pieces. This was all part of serving in the impecunious Egyptian Army, but Kitchener carried it to extremes. As for inability to delegate staff work properly and an incorrigible urge to stick his finger into detail, this was made worse for his Staff Officer of Telegraphs, who will be introduced shortly, by the fact that, having himself served in 'C' Telegraph Troop RE as a young Lieutenant, Kitchener did know a bit about telegraphy.

Nevertheless, he was a man of great energy, and since becoming Sirdar in 1892 he had recruited and trained a new Egyptian army which bore no resemblance to the earlier rabble. They were led by a cadre of British officers, all interviewed personally by Kitchener before being appointed, married officers not acceptable because wives were a distraction and anyway added to the cost base. Every British officer, whatever his rank in the British Army, was granted at least the rank of Bimbashi (Major) in the Egyptian Army, and paid accordingly, so there was financial inducement for junior officers to serve there.

Under this regime the Egyptian Army had been built up to fourteen infantry battalion (eight Egyptian and six Sudanese, the latter consisting mostly of former slaves only too happy at their new existence), three batteries of light artillery, four squadrons of cavalry, and some supply and transport units. Another element that Kitchener had built up was an Intelligence Department. (As Major Kitchener he had himself been an intelligence officer during the 1884-85 Nile Expedition.) Again on very slender resources, he was kept well informed by agents of the situation in the Sudan.

One deficiency in Kitchener's new Egyptian army, however, was that there were no engineer troops – too costly to train, equip, and maintain! When it came to engineering support, such as building and operating railways and telegraph lines, vital components for supporting operations in a country without roads and communications, Kitchener was going to depend on a small band of relatively junior British officers – Kitchener's 'Band of Boys', as they were known. They were given great powers of initiative and independent command to recruit and implement their plans, but finance was another matter. Their budgets were more than tight, and the cost of labour and equipment was closely scrutinised by Kitchener personally.

By 1896 the Egyptian Army was in a position to undertake limited operations against the Sudan. Meanwhile, as part of what has subsequently been described as the 'Scramble for Africa' - the territorial expansion into Africa by European powers - the French were advancing from the west with eyes on the southern province of the Sudan, the Italians had attempted to invade Abyssinia and had been defeated, and the Sudanese were threatening to attack the Italians there. As a first step to reconquer the Sudan, and to warn off others who might be contemplating a takeover, it was decided that a 'demonstration' up the Nile by the Egyptian government, and the reconquest of the Sudan's northern province of Dongola, might be a suitable diversion.

Thus the Dongola expedition was sanctioned, and £E500,000 was voted to cover the cost of the operation (the cost ultimately was paid by the British government). It was decided that Egyptian troops would undertake the expedition. Kitchener, by now a Brigadier-General, was controversially appointed the commander of the expedition. 'Too young', 'too junior', 'too inexperienced in field command', 'not distinguished enough', opined critics in the upper echelons of military society in London. Their views mattered not, for he had strong support from the Consul General in Cairo, Lord Cromer, previously Sir Evelyn Baring and himself a former army officer (and not universally popular either - 'Over-Baring' had been his nickname). As far as Cromer was concerned, Kitchener was the man to get the job done.

The Staff Officer of Telegraphs

It was at this point that the twenty-five years old Lieutenant Graham Manifold joined the Egyptian Army, seconded from the Telegraph Battalion in Aldershot to take up the appointment of Staff Officer of Telegraphs. He had followed a conventional path into the army. Born in June 1871, the son of a Surgeon-General in the Army Medical Department, he was educated at St Paul's School and the RMA Woolwich, and commissioned into the Royal Engineers in February 1891. After training courses at Chatham he was posted to the Telegraph Battalion in 1893, an appointment that was going to mould his future career. He was a fine horseman – in 1895 he won the first prize for riding and jumping at the Royal Military Tournament (which later became the King's Cup at the Royal Horse Show). Surprisingly, as we shall see, his equestrian skill was something that from time to time was going to be a prime asset in his new appointment. Towards the end of 1895, as a promising young officer, he was selected by Kitchener as one of his 'Band of Boys'. He arrived at Wadi Halfa at the beginning of April 1896, now holding the rank of Bimbashi. There he found he had much to contend with - not just the exigencies of the job but, as the introductory description has presaged, with the heavy-handed and tight-fisted Kitchener himself. Nevertheless, Graham Manifold survived.

Lieutenant Graham Manifold, 1895

The scene set, on now to the expedition itself.

The Dongola Expedition, 1896

The Campaign

The main body of the Dervish garrison of the Dongola province was in the town of Dongola itself, but it was reported that there were 3,000 men further north in the area of Akasha. Kitchener was well aware

THE RECONQUEST OF THE SUDAN, 1896-98

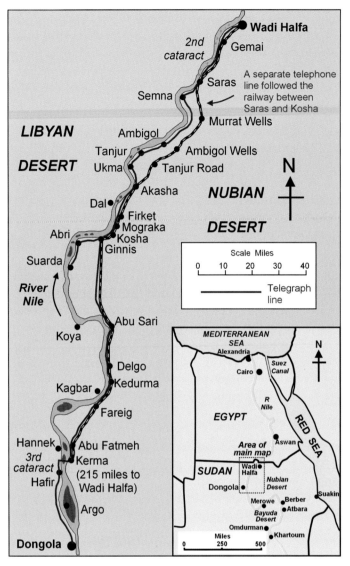

The Nile from Wadi Halfa to Dongola, 1896.

that he was going to face the same problems of transport and supply as the Nile expedition of 1884-85. The plan which evolved was that, while the main body (in total 20,000 men) waited at Wadi Halfa, a small force would advance rapidly and seize Akasha, and supplies for a further advance would be accumulated there. Scouts would be deployed to east and west of the main column. As quickly as possible, the railway and telegraph would be extended south from Akasha to Firket. Upstream (south) of Firket the Nile was navigable in small boats after the level rose, as it did there annually in August.

The Dongola campaign opened on 16 March 1896, and on the 20th the advance column entered Akasha, having met no opposition. This news reached the Khalifa (the Mahdi's successor) at Omdurman and Dervish forces began to concentrate at Firket and other places, while Kitchener consolidated his position in Akasha during May. By the end of May the railway line had been extended to Ambigol Wells, twenty-three miles short of Akasha. Dervish forces probed towards Akasha, but by then it was well fortified. They remained in the area of Firket, and a well planned and executed attack was undertaken by troops who advanced from Akasha on 6 June by river and desert routes. The essence of the attack was surprise, and this was achieved. The Dervishes fought in their usual brave and fanatical way but were overpowered by a superior force. Two hours after the attack started at 5.30 am it was all over, and a decisive victory had been won. In twenty-four hours more than fifty miles of the Nile valley had been cleared. This provided the means for the railway and the telegraph to be extended on towards Dongola. It is now time to turn to the story of the telegraph.

The Telegraph

As staff officer of telegraphs, Graham Manifold's first task was to construct a telegraph line from Sarras to Akasha, fifty-five miles, extending the existing line which ran from Wadi Halfa to Sarras. He had been told by Kitchener that an engineer and construction party, on loan from the Egyptian State Telegraphs, were on their way up the Nile with everything necessary. On 14 April the party arrived, consisting of Mr Paoletti, a Maltese engineer who worked for the Egyptian Telegraph Department, and eighteen native linemen. They brought with them twenty miles of assorted wire, a number of light 16-feet poles, and some small bobbin insulators (designed to be fastened to the pole by screws), but not an instrument of any sort with which to send or receive messages. Manifold got in touch with Kitchener, who told him to use some obsolete telephones which would be found in the railway stores. However, Manifold bravely managed to convince the Sirdar that a few Morse instruments were essential, and also the men to operate them. A request for four trained telegraphists was accordingly sent to the State Telegraph department in Cairo. The Sirdar would not entertain any proposal to provide special airline or field cable at £E25 per mile, and he restricted the order for instruments to the number required for the line to Akasha only.

- 181 -

Part 1

Pending the arrival of the Morse instruments, Manifold contrived, with the assistance of two RE plumbers from the 2nd (Fortress) Company, to make a couple of vibrator ('buzzer') instruments from some old bells, a sheet of brass and a biscuit tin, and also to patch up some of the antiquated magneto-bell telephones, hidden away among the railway stores. With Paoletti and the linemen and stores, he left Wadi Halfa for Sarras by train on 15 April to lay a field telegraph wire along the line of march of the army which closely followed the course of the river, and a railway telephone wire (later increased to two) along the railway line. The field telegraph wire was laid first. The men and stores were divided between an advanced party under Manifold to lay the wire on the ground as rapidly as possible, and a rear party under Paoletti to erect the poles and strain the line (*ie* tension it correctly). Transport was so scarce that only twelve camels could be allotted to the wire party and ten to the poling party, but nevertheless, leaving Sarras on the 17th, the wire party brought the line through fifty-three miles of rough country to Akasha by 20 April, and finished with only thirty yards of wire in hand. Kitchener was presumably pleased at this economy.

Writing of the start from Sarras, Manifold said:

> We paraded at 5.00 am and began to lay the line on the ground. At rail-head, five miles on, I got an escort of twenty men from Fathi Bey [Lt Col Fathi Bey commanded the 7th Egyptian battalion, which formed the rail-head guard. All the other battalions were commanded by British officers] and, at 1.00 pm, halted in the hills and rested under the shade of my umbrella for one and a half hours. The road had been awful in places - boulders, and almost impassable. We laid at one mile an hour. The natives drank gallons of water. I had a sausage for lunch, and it became cooked in the shade of my umbrella. We reached Wadi Atiri about 5.30 pm, after a fearful descent from the hills, and at 6 pm I joined up the 'buzzer' and sent a message reporting the line open. Thirteen miles of such heavy line in this country is good going for one day. I left Paoletti at Sarras to follow on, putting up the poles. [1]

The method of laying was simple. A revolving drum having been unloaded from a camel and placed on the ground, a man took the end of the wire and walked away with it. After he had gone fifty yards, another man picked up the wire near the drum, and then another, and so on until 500 yards had been pulled out by ten men.

Captain Hilliard Atteridge of the London Irish Rifles, and also a special correspondent of the Daily Chronicle, described some of the work:

> ... The telegraph was carried beyond Sarras in the third week of April, being rapidly laid by Lt Manifold RE along the convoy route.

> Laying a telegraph line was in one way easy work. There was no trouble about insulation. The dry sand was almost as good as a sheet of glass for the purpose, so a strong galvanised wire was simply laid along the ground. Later on it was raised on poles to keep it out of the way of accident from passing camels, but even then no insulators were used. Lying on the ground the wire gave far better results than the most carefully insulated English lines. The difficulty was really to get it to 'earth', and the only way of doing this was to run some yards of the wire into the river. All the land was too dry.

> A telephone wire was afterwards laid along the railway. In case of either of these wires being cut there was an organised system of [visual] signalling. Jebel Bringo, a high mountain on the Batn-el-Hugar near Murat Wells, could communicate with Sarras, and Sarras passed on the signals by wire to Halfa. Even if that wire failed it could signal through Gemai. There is always good sunshine for the heliostat [he probably meant heliograph] among the desert hills.

> But the wires were never cut and no attempt was made by the enemy on the convoy routes or the railway line. [2]

Poling of the line to Akasha was not completed until the end of May owing to the scarcity of transport, and for six weeks parts of the line lay unprotected on the rocky ground and exposed to rough usage by camel convoys. Morse signalling could not be undertaken, because neither instruments nor operators were yet available. Manifold's demands for more Egyptian telegraphists were met in Cairo by advertising in the local Arabic Press, and a heterogeneous collection of Levantines was sent up the Nile. Active service in the Sudan, at £E12 per month, had little attraction for these men, and they envied the employees of the Egyptian State Telegraphs, who were paid £E20 per month.

For some time they used soldiers from the North Staffordshire Regiment (based at Wadi Halfa as part of a force guarding the Egyptian border) who had received instruction as telegraphists at the Malta Telegraph School. Although they were deficient in electrical knowledge, these men rendered valuable service and soon became efficient operators, probably finding the work more interesting than the tedious business of guarding the border.

The provision of a railway telephone line from Sarras southwards was taken in hand on 11 May. A line was laid on the ground to Murrat Wells, and onwards to Ambigol Wells, and communication began with instruments made from parts of old telephones discovered in Wadi Halfa. In addition, a wire was added to the existing Wadi Halfa - Sarras line. Captain Percy Girouard, in charge of the railway construction, had secured the Sirdar's approval for a separate railway telephone system, although Manifold did not want to expend eighty-five miles of wire on it and had no liking for telephones. Eventually, some new telephones arrived from Egypt, and communication along the railway was improved.

On 2 June, an intensely hot day, when the thermometer registered 130 degrees in his tent, Manifold opened a special telegraph office for the Sirdar at Ukma, near Akasha. The Sirdar was impatient to know the meaning of a cipher message from Lord Cromer. Major Wingate (intelligence staff officer) had the code book, and no one knew where he was. Manifold received the following message from the Sirdar: "Dear Manifold. This line has broken down again. For God's sake do something to get in communication with Wingate, wherever he is, and get the message from Lord Cromer deciphered. H H K."

"There was a great scurry," said Manifold, "and we telegraphed everywhere to ask about Wingate. That night, the line was broken by camels. About 10 pm, I was turned out of bed to hunt Wingate again, and next morning I was at Ukma at dawn to find the Sirdar still Wingate-hunting. I started to go back towards Sarras, met Wingate at last, and rode into Ambigol about 2.30 pm. The last ten miles were almost impossible - no air, no direct sun, but a haze. My jacket and saddle were soaking with sweat. Twice I stopped under a tree, and drank some warm water and ate a biscuit. Ambigol was deserted except for some Arabs. I sent for several buckets of water and had three poured over my head, and then went out and put the line right. Suddenly, thunder began, then a fearful whirlwind of dust, and then tremendous rain. My line was blown flat as far as I could see. After mending it, I rode into Ambigol Wells, a hot, steamy ride, and I was wet inside and out. Pritchard gave me some soup, and I went to bed and slept, and then had dinner. About 11 pm, the train came in with the news that the telegraph line was all washed down between Wadi Halfa and Sarras." They were suffering freak weather conditions that nobody had planned for; the normal annual rainfall at Wadi Halfa was nil.

Great pressure was thrown on the telegraph staff after the battle of Firket on 7 June, in which Manifold acted as a galloper (mounted messenger), and had his horse shot dead beneath him at a range of about 250 yards. Nearly 12,000 words passed over the wire from Akasha, including the Sirdar's official narrative of the operations and his recommendations thereon. Fortunately they had been loaned some telegraphists by the Telegraph Department a few days earlier.

Manifold was not allowed to lay a line to Firket during the battle, but an office was established there on 9 June. "This is not a bad place," he wrote on 17 June, "but the braying of the donkeys is almost intolerable. Our camp extends for about two and a half miles along the river bank, and every quarter of an hour a feu-de-joie of brays goes down the line." On 1 July, he added, "I have a complete set of telephones from station to station all along the railway between Wadi Halfa and Akasha, eighty-seven miles, eight stations. They work

The Battle of Firket, 7 June 1896, when Manifold, acting as a galloper, had his horse shot dead beneath him. A contemporary picture by R Caton Woodville.

beautifully, and all the telegraphs also are in good working order. The great anxiety now is the cholera. To-day, there is a case at Wadi Halfa. It is pretty warm here - 118 degrees in my tent."

The telegraph line was extended to Kosha, and, when transport became available, to Suarda. At the end of July, four NCOs arrived from the 1st Division Telegraph Battalion at Aldershot. These men - Sergeants Brewster and Kilburn, and Corporals Dennett and Hensler - were invaluable to Manifold, who had been ploughing his lonely furrow with limited resources and a mostly unskilled workforce, for they had energy and initiative and understood their work. Their greater knowledge of the operation of the telegraph eliminated many of the electrical faults that the telegraphists of the North Staffordshire Regiment did not understand. Sergeant Brewster took charge of the Headquarters Telegraph office in Kosha, and Sergeant Kilburn superintended the maintenance of the line between Wadi Halfa and Akasha. Corporal Hensler became the superintending clerk of the mobile Headquarter office as it moved forward during the advance. Mr Paoletti was able to return to Cairo with some of his linemen, whose duties could now be carried out by locally trained men.

The telegraph and telephone lines suffered severely during the bad weather in August. At one time there were sixty miles of telephone line to be re-erected, all of which had been working perfectly on the previous day. The storm on 27 August caused great dislocation of the telegraph system. The poles were removed from several miles of line by a torrent flowing down a *khor* (gorge) from Murrat Wells, and a rush of water at Atiri led to a complete interruption of both vibrator and sounder signals between Sarras and Akasha for two days. Altogether, seven miles of telegraph, and twelve miles of telephone line, were destroyed by these floods. However, the damage was repaired; and on 7 September the line was taken to Delgo in preparation for the general advance on Dongola. On 13 September an office was opened at Kedurma, on the 14th another at Fareig, and on the 19th, during the action at Hafir, the telegraph reached Kerma. No less than 100 miles of bare wire lay on the ground; yet the sounder signals came through perfectly, though the vibrator signals were unsatisfactory. The line stretched for 215 miles to Wadi Halfa, and the circuit was worked direct to that place.

After the successful action at Hafir on 19 September 1896, when a pocket of Dervishes were routed, mostly by gunboats on the river, Kitchener transferred his troops across the Nile and advanced up the left (west) bank. Consequently, it was necessary for the telegraph to follow him by means of a submarine cable across the river and a landline towards Dongola. Two cables had been prepared in Egypt for crossings at points where the width of the Nile, at that season, was estimated to be about three-quarters of a mile, but one was damaged in transit up the river and the other did not arrive at Kerma until 26 September. It seems that Kitchener himself had determined the length of this cable by map measurement, without due allowance for sagging and the effect of current, with the result that it proved to be too short. He may have been influenced,

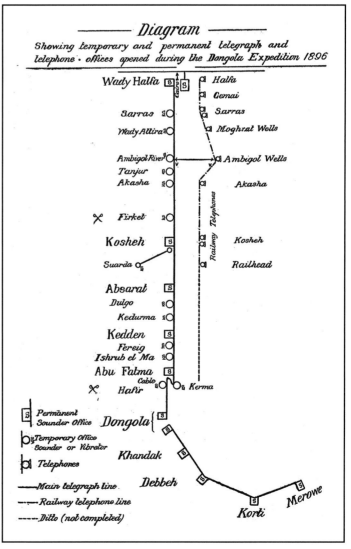

Telegraph circuit diagram drawn by Lieutenant Manifold.

as he so often was, by considerations of economy. "I got orders to lay the cable at Kerma, instead of at Dongola", wrote Manifold, "and as I was of the opinion that the cable was not long enough, I went up to report. I saw the Sirdar, who said that it would be better to cross at Kerma, so at 5.00 am on the 27th, I started making arrangements. A steamer and troops had been placed at my disposal for laying the cable, and we crossed over and made the shore-end fast to a tree, and anchored the tree. I decided to run the cable over the bow of the ship, and gave orders to the native *Rheis,* or Master, to go stern first, straight across the stream, so that he might avoid cutting the cable with the stern wheel. Of course, the idiot went bow first! Also, he did not go straight across, but allowed the current to carry him downstream. The cable flew out like lightning. Two of my R.E. stood in the coil and cleared it as it strained. They had to twist each coil over their heads, and a slip might have been fatal. I stood in the bow, where the line went into the water, and a man poured water continually on the fender, over which the cable ran, to prevent it from catching fire. I told the *Rheis* to make for an island about 100 yards from the mainland and there I landed the remainder of the cable and put it in a boat to complete the crossing, but the end failed to reach the other side by about seventy yards."

Connection was completed by running an overhead line from the shore to the mast of the boat, and afterwards, a line was taken up the west bank for a distance of thirty-five miles to Dongola, the line laid by Corporal Dennett. At Dongola the telegraph office was established in the ruins of the old telegraph office used by the Nile Expedition in 1885. It was then found that through communication was still possible with Wadi Halfa, 250 miles distant, although 130 miles of bare wire lay on the ground.

At the beginning of November, during Kitchener's absence in England, Manifold extended the telegraph to Debbeh, Korti, and Merowe, a distance of 180 miles south of Dongola. In addition, there were 250 miles of main telegraph line from Wadi Halfa to Dongola, seventy miles of branch lines, and 130 miles of railway line – 630 miles of line altogether. The total number of telegraph offices opened between Wadi Halfa and Dongola was nineteen, but many of these were temporary and used only for a few days.

Young telegraphists were recruited in Cairo at £E3 a month, and trained by Corporal Hensler of the Telegraph Battalion in a telegraph school at Wadi Halfa. As these youths gradually replaced some of the more highly paid telegraphists of the Egyptian State Telegraphs, a much greater measure of economy was achieved – no doubt pleasing the Sirdar.

What next? The Sirdar was very pleased – for a start the cost of the operation had only been £E715,000, and then the casualties had been few. He could afford, quite uncharacteristically, to dispense some leisure. One of the Band of Boys described their feelings:

> Although it was confidently expected that an advance to Khartoum would ultimately take place ... it was in the dim future. ... In the meantime why not a little leave instead of stewing in a grass hut at some ruined, steaming, fly-blown place on the Nile, watching the men building mud huts, which certainly would be cooler, but not very enticing to a man who feels run down? Another effect of the reaction is that one notices how beastly the food is, how monotonous and dreary the country, how hard it is to amuse oneself; one longs for the sight of fresh green vegetables, fresh butter, to find that one's things are not always smothered in sand, to get out of sight and smell and sound of a camel, to see a real green field. ... The Sudan campaigners had to stand this test of discipline and endurance a good deal during the three year's campaign. [3]

Manifold had worked hard to supervise the work of his partially trained men, and was granted a period of leave in England. Between arriving in April 1896 and his departure on leave in January 1897, he covered more than 5,000 miles by land and water. His trials were many and varied, but he had the satisfaction of exercising an independent and responsible command for a relatively junior officer, and of knowing that he had succeeded in providing an efficient line of telegraph communication in the reconquered province of Dongola. For his work thus far he was Mentioned in Despatches, and awarded the Order of the Medjidie, 4th class - a Turkish medal awarded by the Khedive of Egypt.

The Order of the Medjidie.

THE RECONQUEST OF THE SUDAN, 1896-98

On to Khartoum, 1897-98

Following the reconquest of the province of Dongola, Lord Cromer and Kitchener had convinced the British government that the reconquest of the Sudan could not be delayed. Egypt would never be secure until the power of the Khalifa was crushed. It was known that a French expedition under Commandant Marchand was already advancing up a tributary of the Upper Congo river towards the southern Sudan, guaranteed to cause excitement in London. The momentum gained by the Dongola Expedition must be continued, it was argued, but the task was going to be beyond the limited resources of the Egyptian Army alone. The British government agreed. Kitchener, promoted and decorated as Major General Sir Herbert Kitchener KCB KCMG as a reward for the success in Dongola, thus set his sights on Khartoum.

Although Kitchener had consolidated his position south of Dongola, he could not depend on the Nile as a line of communication. He needed a railway line, and they would have to build it. Like Wolseley before him at the time of the Gordon Relief Expedition, he rejected the idea of a railway line from Suakin to Berber; he considered this route to be too costly, too difficult to build, and also exposed to frequent attack, as British attempts to build a line from Suakin in 1885 had shown (chapter 14). He decided that a better route was going to be from Wadi Halfa straight across the Nubian Desert to Abu Hamed. There were many railway 'experts' who said it could not be done. On 1 January 1897 the first sleepers of the historic Desert Railway were laid at Wadi Halfa. [4] The campaign was going to be a long haul, but Kitchener's plan was to ensure that he had a reliable line of communication to supply as large a force as he needed, and the rate at which that could be constructed would largely dictate the length of the campaign. In the event it was to take Kitchener twenty-one months, until September 1898, for the Khalifa's forces to be crushed at Omdurman.

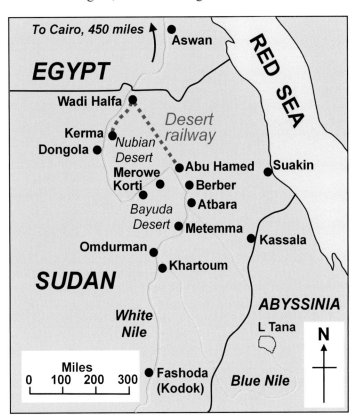

The route to Khartoum, 1897-98

The telegraph had played an important part in the reconquest of Dongola. It was now going to be vital for the operations to Khartoum, and Kitchener was told by Lord Cromer never to leave the end of the wire. But these were huge distances along which to construct new telegraph lines and then operate them with a small and relatively unskilled and inexperienced locally engaged work force. The handful of NCOs on loan from the Telegraph Battalion in Aldershot, each working independently, were going to be given tasks and responsibilities well beyond anything they would have undertaken normally, and they responded magnificently. The twenty-six year old Bimbashi Graham Manifold continued as the Staff Officer of Telegraphs, having returned from leave in June 1897. He was pleased to see the telegraph system he had set up in Dongola still in good condition, and greatly surprised that £2,000 had been allotted for the provision of additional instruments, wire, and poles for further extensions.

By the middle of July 1897 the desert railway had reached a point about 100 miles from Wadi Halfa, just under half-way to Abu Hamed, and Kitchener decided that it was necessary to clear the Dervishes from Abu Hamed so that railway construction could proceed safely.

His plan was to send a mobile column upstream from near Merowe, which could be well-supplied from Dongola now that the railway to Kerma was operating efficiently, and the level of the Nile would rise during July and August enabling it to be used for transport upstream to Abu Hamed. Surprise was essential, to prevent Abu Hamed being reinforced from the stronger Dervish garrison at Berber.

The leader of the mobile column charged with capturing and securing Abu Hamed was Major General Hunter. The date having been kept a closely guarded secret, he set off from Kassinger, near Merowe, on 29 July 1897. He was following much the same route as that taken twelve years earlier by the River Column of the Gordon Relief Expedition in 1885. Graham Manifold's next task was to follow in the wake of Hunter's column, extending the telegraph line along the course of the Nile from its present termination at Merowe.

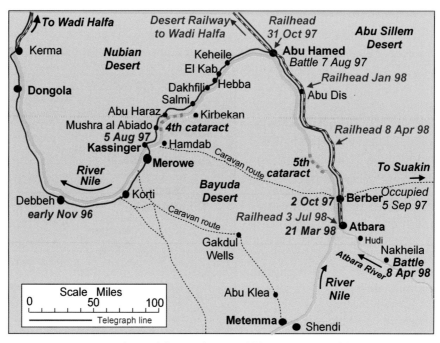

Dongola to Atbara, August 1897 to April 1898.

Marching mostly by night and resting in the heat of the day, Hunter's column covered 118 miles in seven and a half days, at the hottest time of the year – some achievement, and a rate of progress that Manifold, constructing the telegraph line behind them, could not match. The battle of Abu Hamed was fought on 7 August 1897, Hunter emerging victorious. The bodies of many dead Dervishes were thrown in the river, the ground being too hard to dig the number of graves needed. Downstream at Merowe, where his HQ was situated, Kitchener's first intimation that something had happened was the sight of corpses floating downstream. He was much relieved when, on closer inspection, they turned out to be Dervishes.

The first hard news of the capture of Abu Hamed that reached Kitchener's HQ was a message carried by camel rider to the telegraph with rail-head in the middle of the Nubian Desert, and then sent by cipher telegram, *via* Wadi Halfa and back along the Nile through Dongola to Kitchener at Merowe.

For tactical reasons General Hunter had advanced along the right bank (the north bank at this point), but the main telegraph line from Dongola came down the Nile on the left bank so the first task was to take the line across the river. The level was low at that time, and the 650 yard crossing at Kassinger was made on 26 July without any problem. On this occasion, Kitchener, remembering the problems at Kerma caused by his parsimonious interference, had ordered 1,000 yards of submarine cable.

Manifold was next told to be ready to lay the line, but was advised that that there would be no camels for transport as they were all needed for the attack on Abu Hamed by General Hunter's column. "The local sheikhs will like, no doubt, to provide you with donkeys", said Kitchener. A number of decrepit animals were assembled, but without pack saddles. "How do I carry wire on them? ", said Manifold. Kitchener quickly demonstrated. He led a donkey over the 110-pound coil of wire lying on the ground, until its hind legs were inside the coil, he then had the coil hoisted over the animal's rump, so that the wire encircled its body and rested on its back. This was a trick he had learnt in 1885 during the Nile Expedition, when the then Major Kitchener, intelligence officer, was accompanying the leading troops and a detachment of the Telegraph Battalion up the Nile from Debbeh to Hamdab, and were having to rebuild the line as they went, with similar transport problems. However, Manifold reported a few days later: "The donkeys are getting rather tired at playing hoops, and have taken to lying down with a flop as soon as loaded! "

Having acquired some boat transport to carry the poles and some of the wire, Manifold started from Kassinger on 5 August 1897, together with sixty-five donkeys. He laid the line through Mushra el Abiado, Shebabit, and Abu Hariz to Hosh el Geref, opposite Birti. It was a wearisome process as each coil had to be unloaded from the donkeys and unwound by hand. Speed was essential, so the wire was laid straight on the ground. At Mushra el Abiado, after eighteen miles of desert marching, Manifold wrote: "My donkeys are very tired, and I dare say several will die tonight. This place is awful – barren rocks and no shade. The big cataract roars by my bedside."

While the gunboats and other river craft supporting the column ploughed their way through the Fourth Cataract, the telegraph detachment plodded steadily onwards along the river bank, laying their bare wire through Salmi, Dakhfili, El Kab and Keheili to Abu Hamed, which Hunter had captured on 7 August. The route was not poled until later, due to time and transport problems; the bare wire was simply laid on the ground, and in the dry climate it worked. By the time they reached Abu Hamed, a new requirement to continue the line to Berber was ordered.

Soon after the capture of Abu Hamed by Hunter's force an unexpected bonus had come Kitchener's way. He learnt that, surprisingly, the Khalifa had abandoned Berber, and was concentrating his troops at Omdurman, near Khartoum. Dare he take the bait? He could supply Abu Hamed when the desert railway reached there, possibly in two months time, but in the meantime how was he going to supply his forces in Berber if he occupied it, 130 miles further up the Nile? He could only put a small force there, which might find it difficult to defend itself. Yet it was a strategically vital point. It was the centre of many trade routes, including the route to the port of Suakin, and the moral effect of occupying it would be great. After consulting Lord Cromer by telegraph, he occupied Berber on 5 September 1897, less than a month after capturing Abu Hamed.

Then there was another interesting turn of events. Osman Digna, the rebel leader in the eastern Sudan, mentioned in chapter 15 in connection with the Suakin Expedition of 1885 and still there causing trouble in his fiefdom, decided to throw in his lot with the Dervishes around Metemma, and moved his forces there. The effect of this was to leave the eastern Sudan clear of enemy forces and enable Kitchener to take steps to open up the route from Berber to the port of Suakin. Kitchener was greatly assisted by the lack of any coherent strategy by the Dervishes.

Construction of the desert railway continued apace, and it reached Berber on 31 October 1897. This completely changed the situation of the supply line to the front. From Wadi Halfa by the Nile route through Dongola and Merowe the journey to Abu Hamed was eighteen days and, dependent on the state of the river, its capacity extremely limited. By the new desert railway, across the Nubian Desert, it was twenty-four hours, and with much greater capacity. It was with that intention, of course, that the desert railway had been built in the first place. It is also worth noting that in Egypt the railway line that ran from Cairo to Asyut at the time of the Nile Expedition of 1884-85 had by now been extended to Aswan, so that the only section that still depended on the river was between Aswan and Wadi Halfa, and over that section the Nile was navigable by steamer throughout the year. [5]

All of this threw many new challenges to providing telegraph lines for the rapid advance that Kitchener's force had made. Telegraph was needed for the further advance from Abu Hamed down to Berber, then from Wadi Halfa along the railway to Abu Hamed, and also to Suakin and Kassala along the newly-opened route from Berber to Suakin. (A telegraph line had existed between Berber and these two places, but was destroyed in 1883 by the Mahdi's forces, and had never been rebuilt). With their limited resources this all took time.

The new line to Berber was taken in hand by Corporal Dennett of the Telegraph Battalion, who reached there on 2 October 1897, and established contact with Kassinger through 236 miles of bare wire lying on the ground. The line worked in the daytime, but not at night when the insulating property of the dry and sandy desert deteriorated and communications over that distance became impossible. Even so, it was remarkable.

Meanwhile Corporal Lewis had been left in charge of the poling party to Abu Hamed. Their rate of progress was influenced mostly by the supply of poles, which depended on the amount of water transport coming upstream from Merowe. Under this constraint, the poling party was sometimes as much as 150 miles behind the wire-laying party and did not reach Abu Hamed until December, continuing on to Berber in January 1898. It was a long and toilsome business, although no longer in the extreme heat of the summer months.

A line was laid also from rail-head in the Nubian Desert to Abu Hamed. As the construction of the railway progressed it was accompanied by a telephone and telegraph line which was built by a NCO deputed by Manifold. These two lines were on one line of poles running alongside the railway. The telephone service was from station to station, but the telegraph connected only the more important stations along the route, including the watering stations.

The army continued its advance to Atbara, and by the end of January 1898 the general situation was that there were three brigades of the Egyptian Army with supporting cavalry and artillery along the Nile between Atbara and Abu Hamed, concentrated mostly around Berber, and the newly-arrived British brigade concentrated at Abu Dis, thirty miles south of Abu Hamed. At this time the Desert Railway had also reached Abu Dis, and the route from Berber to Suakin had been opened. Then intelligence reported that the Dervishes in Metemma intended to attack. Their position consolidated, they awaited the attack from the Dervishes.

This came eventually on 8 April 1898, at the battle of Atbara (actually at Nakheila, some thirty miles up the Atbara river), the details of which need not be recounted here. It was a resounding defeat for the Dervishes, who suffered a high level of casualties. British and Egyptian casualties were relatively light.

Manifold, the fine horseman, acted as a galloper during the battle of Atbara River on 8 April 1898. Immediately after the victory, he was hailed by the Sirdar, who gave him a telegram to Lord Cromer and ordered him to ride as fast as possible to the telegraph office at Fort Atbara, 32 miles away. "I left Nakheila at 9.45 am," he wrote, "and rode at a canter, with occasional halts to mop my brow. At Abadar, which I reached at 11.20, I had an extraordinary reception. The Jaalin [an Arab tribe who hated the defeated Arabs and their leader Mahmud because he had massacred a large number of them] shouted: 'What news?' I replied, '*Mahmud Khalass*' (Mahmud is finished). I told the Arabs that we had caught Mahmud, and ordered the Sheikh to pursue. The Jaalin ran alongside my horse, leaping and shouting. The din was indescribable. Every man fired his rifle into the air, and my horse became almost unmanageable. At 12.15 pm I rode into Hudi, and arrived at Fort Atbara at 2 pm. The dispatch was in Cromer's hands at 2.38, and I had a receipt at 2.50! Then I went to bed. At 4 pm, the first press messages came in; at 7 pm more press messages; and at 7.30 pm the remainder of the Sirdar's dispatches, which had started by ordinary messenger only one and a half hours after me. I have worked right through the knees of my breeches, and my boots are giving way at the stitching."

The only army that could prevent the advance to Khartoum had been destroyed, and the country up to the sixth cataract had been cleared. Kitchener was now in a most satisfactory position, but was unable to undertake further operations until the Nile rose again, in July or August. The armies rested in the summer heat, and those who had been on active service for a long time took some leave. Construction of the desert railway continued, and it reached Atbara in July.

The transmission of press messages bore heavily on the telegraph operators. Although the Sirdar disliked publicity, he had relaxed to some extent the rigid regulations which had governed the movements of press correspondents, and the result was an ever increasing pile of messages for transmission. In 1896, Kitchener had maintained that, as the Dongola campaign was purely an Egyptian affair, no British correspondents were necessary. He added that the newspapers often sent out unsuitable men, and that he could not afford to supply transport to those who came. At first the unfortunate correspondents were forbidden to proceed beyond rail-head, but later they were prohibited only from going out on reconnaissances, joining or preceding the firing line in general actions, or approaching the Sirdar. They were advised that, in the intervals between campaigns, they might, with advantage, return to Cairo. In some of these restrictions, Kitchener was supported by Lord Cromer, but agitation in Fleet Street brought about a modification of the regulations affecting press correspondents in the Sudan, and in the end the correspondents came to be tolerated, if not welcomed, by the Sirdar, and proved themselves agreeable companions in the field. [6]

THE RECONQUEST OF THE SUDAN, 1896-98

On 12 April, four days after the battle of Atbara, Manifold went to Suakin to arrange for the construction of the telegraph line between there and Berber, his journey taking eleven days. (The Hadendowas had been won over to the Egyptian side, and Osman Digna no longer enjoyed their loyalty. He had in any case moved out of the area to Metemma, so the telegraph line was no longer attacked in the way it had been in 1885). The work was started 1 May 1898 by Sapper May. Manifold then went to Tokar and inspected the line just started from there to Kassala under the supervision of Corporal Lewis. By 23 June, fifty-two miles of the Berber line and seventy-seven miles of the Kassala line had been completed, by 3 August four-fifths of the line to Berber and 167 miles to Kassala had been completed, and both lines were completed soon after the final battle of Omdurman (2 September 1898). It is a noteworthy achievement that the lines were constructed by these two soldiers working independently in charge of native linemen, with little experience of the country and its inhabitants, or knowledge of the language, each overcoming numerous difficulties during their often lonely time in unmapped desert in the height of the summer months. Eventually, the telegraph line from Suakin was to reach Kassala on 24 September 1898, but the line to Berber was not completed until 22 November.

Manifold meanwhile had left the Red Sea litttoral and gone to Cairo, where in mid-June he was informed by the Sirdar that new plans were afoot. The telegraph was to be extended immediately from Atbara to Metemma, and an additional wire was to be provided between Abu Hamed and Atbara. The extension to Metemma, on the other bank, was going to require the Nile to be crossed. Manifold went to Suez and superintended the loading of 1,200 yards of submarine cable, which was sent to Atbara, arriving there as one of the first consignment of stores to reach that place after the railway was extended there on 4 July. By that date, in preparation, Sergeant Dennett (promoted after the battle of Atbara) had laid a bare wire extension along the left bank to Metemma and then poled it, but it could not be connected until the arrival of the submarine cable.

On 7 July the laying of the submarine cable was attempted at Atbara, just downstream from the confluence of the Nile and Atbara rivers. The cable was laid in a big coil on the deck of one of the gunboats, and paid out as the boat steamed slowly across to the left bank. The cable became entangled and a knot formed. The anchor was dropped, but the anchor chain broke, and it became necessary to cut the cable with an axe. The gunboat returned to the right bank with only 600 yards of cable. Bits of cable were recovered, the remains untangled, and repairs made. The next attempt to lay was made on 9 July, slightly upstream of the confluence this time, via a small island in mid-stream, and was successfully accomplished. But it was only a temporary solution, because the cable had been damaged. Kitchener was unhappy, saying that "he couldn't afford to buy new cable *every* day." He realised, however, that another cable was needed; it was ordered on 27 July and, amazingly, arrived at Atbara by train on 9 August.

The next complication was the Atbara river crossing. The source of the Atbara River, like the Blue Nile itself, is in the mountains of Abyssinia (now Ethiopia), and the river is fed by annual rains in those mountains. The river bed is often dry, with only pools of water, as it was at the time of the battle, so the line at that time was simply poled over the dry river bed. The annual flood was due in late July, when the river would become several hundred yards wide, so in preparation, Manifold attempted an overhead line in one span from bank to bank. This was undertaken in mid-July using two composite poles, each forty-eight feet high, but was achieved only with great difficulty. The floods arrived, and the Atbara river became navigable, but the new overhead crossing only lasted a week before being carried away by the mast of one of the boats now navigating the river.

Manifold described what happened next. "Saat-el [His Excellency, *ie* Kitchener] has got overhead wire on the brain, and today I was told that I should soon be ordered to build two high towers on the banks of the Atbara, so I said 'Why not let the 2nd Company [RE] do it?', and Saat-el was delighted with that solution. Major Arkwright [OC 2 Fd Coy RE] is now wrestling with the towers and trying to find some materials for them."

Meanwhile, the overhead line over the Atbara had to be re-established on the 48-feet poles. The next day Manifold had some boats organised and was just shoving off from the bank to do the job when Kitchener came running and demanded to be taken aboard. Manifold, whose heart must have sunk, continued: "We steamed slowly to the *Melik*, paying out the wire and keeping it out of the water. Then the drum was passed on to the *Melik* and the wire came through a snatch-block at the mast-head. The *Tahra* came up on the other side and took the drum aboard, and off we went again and got safely across. Saat-el was very pleased. We strained the wire taut and worked at once through to Metemma."

On 10 August, the new cable having arrived the previous day, preparations to replace the submarine cable across the Nile began. This time a home-made drum had been constructed to assist in the paying out of the cable. "I have found on counting the number of turns", wrote Manifold, "that the company has sent us 60 yards short, and I am doubtful whether 1,140 yards is enough. Even 1,200 yards seems too little, but Saat-el insisted on it." What had happened, it later transpired, was that Kitchener had not agreed with Manifold's estimate of the necessary length of cable, 1,200 yards, and had ordered Lieutenant Pritchard, the Survey Officer, to check the demand. "I made theodolite observations", said Pritchard later, "and then added amounts for height of banks and depth of water, for irregularities of the river-bed, for the impossibility of laying the cable straight from shore to shore, and finally five per cent for contingencies. The Sirdar looked at the calculations and said, 'But they *must* lay the cable straight; and you have included five per cent for contingencies when you have already allowed for every contingency' – and he struck out the five per cent". Kitchener's interference with the work of capable junior officers, and his inability to organise and delegate staff work to subordinates, were incorrigible failings.

On 11 August the attempt to re-lay the cable took place. A staff officer came aboard the boat but said that Kitchener was not coming – much to the relief of all concerned, it must be imagined - so they started to lay the cable. Manifold continues: "It paid off beautifully from the drum, but the ship could not steam properly against the swift current, and we went across in a wide curve. When we were below the point of the island, I realised that there was not sufficient cable to complete the crossing, so I shouted to Gordon to drop the anchor. It was overboard within half a minute but, before it could grip, the last turn of the cable had whipped over the side.. Then Saat-el appeared in the *Tahra*, furious with 'Monkey' [Gordon] for starting without him, but the orderly had told Gordon to 'start at once'. Saat-el is unapproachable on delicate business this morning. It is a pity. More cable should have been ordered." That

The second attempt to lay the cable across the Nile at Atbara, the cable now on a home-made drum. The sketch is from the Royal Signals archives, probably originally published in The Illustrated London News.

cable was now useless. Yet more unnecessary parsimony and high level interference had cost them dear, not just in a trifling amount of additional money, but more importantly, as will be seen shortly, in the very purpose for which the cable was made – the provision of communications!

The construction of the two tripod towers, each eighty feet high, for an improved crossing of the Atbara river was completed on 17 August, and two wires were slung across the river. For a time the telegraph worked to Metemma by way of the newly-raised lines over the Atbara river, the repaired cable across the Nile via the island just upstream from the confluence, and then by the poled line built by Sergeant Dennett on the left bank.

With telegraph operating satisfactorily all the way between Wadi Halfa and Atbara along the Nile route through Dongola and also the main route following the railway, and the lines to Suakin and Kassala under construction, it now remained to complete the remaining 100 miles from Metemma to Khartoum.

The field telegraph kept the Sirdar in communication with Cairo and every post along the Nile, and in close touch with his railways and river transport. It was an important factor in the organisation of the campaign.

Now it was time for the final act, the capture of Khartoum. The concentration of forces ready to attack Omdurman was carefully planned. During the build-up period more British troops were sent forward as reinforcements, and the Egyptian Army element was also reinforced. With the solid line of communication that had been built up, it was possible to keep the increased number well supplied. The railway reached Atbara on 3 July 1898. The telegraph line was extended from Metemma to Wad Hamid, forty miles, by bare wire laid on the ground. During July and early August troops and their supplies, in anticipation of the annual rise in water level, were shipped upstream to Wad Hamid, and the surrounding area, below the sixth cataract. By 23 August, 23,000 men had been assembled just below the sixth cataract, 1,260 miles from Cairo and sixty miles from Khartoum. The telegraph line from Metemma had been extended to Wad Hamid, and during August it was poled by Sergeant Dennett – very hot work in the midsummer heat.

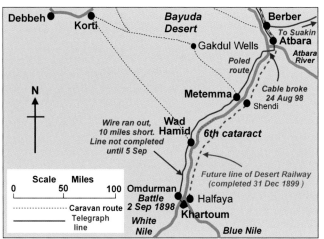

From Atbara to Khartoum, 1898.

The troops advanced from Wad Hamid to Omdurman, where the Dervishes awaited them, moving into position on 31 August. The battle of Omdurman was fought on 2 September 1898, and need not be recounted here, save to say that the Dervishes fought bravely but their strategy, tactics, and weaponry were all primitive. They were overwhelmed and massacred by modern weapons. The Khalifa's army ceased to exist, although the Khalifa himself escaped, to be caught and killed a few months later. The troops entered Khartoum later the same day. [7]

It is interesting to read the accounts of the battle, when 23,000 men – a British Division of two Brigades, and an Egyptian Division of four Brigades, both with supporting cavalry and artillery, as well as support from gunboats on the Nile – took to the field, and to realise that even in 1898, signalling played little or no part in command and control on the battlefield. Visual signalling was under battalion control, and there were neither Brigade nor Division Signal Sections. The heliograph was used, but in a very limited way. There were no telephones connected by field cable. Kitchener had secured the services of a number of his 'Band of Boys' (Manifold was one of them), and he employed them as gallopers – just as Wellington had done at Waterloo. Messages were carried to and fro by officers on horseback; it seemed as though this was still the most trusted and efficient method of tactical communication. Just over a year later, in South Africa, when the Boer war started, the enemy and weaponry were much more capable, the battlefield was on a much larger scale, and tactical communications were found to be extremely deficient.

Also, for a number of reasons, telegraph communications broke down a few days before the battle of Omdurman, and were not successfully restored until 7 September, the lack of news causing consternation in London. What happened?

On 20 August Manifold left Atbara and advanced to Wad Hamid, where on 24 August he was awoken to be given the a message that the cable across the Nile at Atbara had broken. (This was the cable that had been temporarily repaired.) Then a storm arose. This brought down stretches of the poled line to Metemma, and also prevented the boat crossing the Nile at Atbara with messages that could no longer be sent by the broken cable. On the 24th a party was also sent south from Wad Hamid to extend the line towards Khartoum, but by the 26th they were running short of wire and no more could be found – some twenty-five miles were missing. By the 29th, about thirty miles from Khartoum, they only had just over eleven miles of wire left. The 31st August brought more heavy storms. Some more wire was found but by 1 September, the day before the battle, they ended up ten miles too short, with no more wire available. It is inexplicable. Having had several months to prepare for the final assault, where had things gone wrong? If they had enough wire, where was it? There is no ready answer. Meanwhile, the Staff Officer of Telegraphs was required to act as a galloper!

Telegrams sent on 30 August were the last to reach England before the battle. No more were sent until 3 September, the day after the battle, when a message was carried by steamer from Khartoum to the telegraph office at Wad Hamid. The poled line from Wad Hamid to Cairo was unbroken, except for the damaged cable at Atbara. They had planned that bare wire on the ground from Wad Hamed to Khartoum would suffice initially, until it could be raised on poles soon after, but the unexpected and unusually heavy rain had scuppered that plan. As for the shortage of wire to reach Khartoum, it remains unexplained. A working telegraph line did not reach Khartoum until 7 September, five days after the battle.

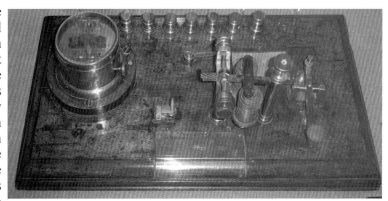

The Khalifa also employed the electric telegraph, a small local system connecting Omdurman with several key points in Khartoum. It worked until the day before the battle of Omdurman, when part of it was destroyed by a preliminary bombardment from gunboats. When Sergeant Dennett entered the northern outskirts of Omdurman after the battle, he noticed the telegraph line that had been used by the Khalifa's troops, and following it, he came to the terminal station in a hut in the Khor Shambat. There he found the terminal set and a captive Egyptian

The telegraph set discovered at Omdurman by Sergeant Dennett. It is now displayed in the Royal Signals museum.

telegraphist, who had been working the line up to the day of the bombardment, when communications had ceased after the line had been broken. The telegraph circuit had included the arsenal and the dockyard in Khartoum, the crossing of the Nile being by cable to Makran Point. The original set of instruments, consisting of a single current set and a graphite wheel recorder, had been supplied by Messsrs Chadburn in 1872; subsequently the recorder had been replaced by a relay provided by Messrs Siemens to an order from Gordon Pasha in 1878. [8] The most useful part of that system was the crossing over the Nile between Omdurman and Khartoum, and when communications were established between Khartoum and the outside world, Sergeant Dennett, remembering the fiasco at Atbara, used it!

And so ends the story of the telegraph operations to support the reconquest of the Sudan. Lieutenant Graham Manifold resigned his appointment with the Egyptian Army and left at the end of 1898. He had done a sterling job, taking on huge responsibility for one of his rank and experience, but it is easy to imagine that it was time to return to civilisation, distance oneself from the frustrations of Egypt and the Sudan, and further one's career in the mainstream. He handed over the Sudan Telegraphs to Lieutenant

Campaign Medals

For those who took part in the many scattered operations for the reconquest of the Sudan there were two campaign medals. The Queen's Sudan medal of 1896-97, with no bars, the medal ribbon being half yellow and half black, with thin red dividing stripe, and the Khedive's Sudan medal 1896-1908 with fifteen bars – Firket, Hafir, Abu Hamed, The Atbara, and Khartoum, being the battles mentioned in this narrative. Very few telegraph troops will have been awarded this medal – Graham Manifold and the small team of NCOs from the Telegraph Battalion who supported him. Manifold was also awarded the Order of the Osmanieh (on the right).

J S Liddell on 29 December 1898. Manifold was awarded the Order of the Osmanieh for his work.

Sergeant Dennett, whose work had been outstanding, stayed in the Sudan. On terminating his military service he joined the Sudan Postal Telegraph Department, and rose to be a Superintending Engineer before his retirement in 1921.

Some years after the conquest, when matters had stabilised and the Sudanese responded to the new regime, the army continued to play a leading role in the development of the telegraph throughout the country.

Maj Gen Sir Michael Graham Egerton Bowman-Manifold KBE CB CMG DSO (1871 - 1940)

Graham Manifold next served in South Africa during the Boer War, when he was in charge of the railway telegraphs, working under another of the former 'Band of Boys', Major Percy Girouard, who had masterminded the desert railway. For his work there Manifold was awarded the DSO. In 1901 he added the prefix 'Bowman' to his name. After that he served, perhaps inevitably, with 'K' Company RE (formed out of a reorganised Telegraph Battalion) on telegraph work with the Post Office in southern England, attended staff college, and held several general staff appointments. He also qualified as an interpreter in Russian.

When the First World War broke out, then a Major, he commanded 'A' Signal Company in Aldershot. The Company was deployed with the BEF to France and he was responsible for communications during the retreat from Mons, and the battles of the Marne, the Aisne, and first Ypres. He was then recalled to England to train the Army's expanding signal units. In 1915, in the rank of Lieutenant Colonel, he was appointed Director of Army Signals of the Mediterranean Expeditionary Force, and was Director of Army Signals throughout the Gallipoli campaign. He was next appointed Director of Signals to the Egyptian Expeditionary Force in the rank of Colonel. In 1917 he served under Lord Allenby during the advance from Egypt through Sinai to Syria. One of Allenby's despatches described how he 'overcame all difficulties in the great and complicated system of communications, and had never failed to solve the difficult problems caused by rapid and long advances – an organiser and administrator of great ability'. During the war he was Mentioned in Despatches eight times, awarded the CB and CMG, reached the rank of Brigadier-General, and in 1919 he was knighted. After the war he was appointed an Instructor at the Staff College, Camberley, attaining the rank of Major-General.

With many of his senior rank seeking appointments in a reduced post-war army, and well-qualified for other things, he retired in 1921 and took up new employment with the Marconi Company, and later Cable and Wireless Ltd, acting as a negotiator in foreign business until 1934 when, aged sixty-three, he became President of his local branch of the British Legion.

He had married Kathleen Brandreth, the daughter of Admiral Sir Thomas Brandreth, in 1899, and they had two sons and a daughter. A few months before his death in March 1940, in his sixty-ninth year, he received the sad news that one of his sons, John, a Lieutenant Commander RN, the navigating officer of HMS *Exeter*, had been killed in action at the Battle of the River Plate.

Endnotes

1. Lt Manifold kept a diary, and in 1934 wrote notes of his experiences to Lt Col E W C Sandes, the author of *The Royal Engineers in Egypt and the Sudan.*.
2. *The Dongola Expedition of 1896*, by Captain A Hilliard Attridge, London Irish Rifles and Special Correspondent of the Daily Chronicle. RUSI Journal, Vol XLI, June 1897.
3. *Sudan Campaign 1896-1899* by 'An Officer' (actually Lieutenant H L Pritchard RE, involved with railway work, and later to become Maj Gen H L Pritchard CB CMG DSO), pub 1899, London.
4. The story of the survey and construction of the Desert Railway is very interesting but beyond the scope of this narrative. Good descriptions are given in *The Royal Engineers in Egypt and the Sudan* by Lt Col E W C Sandes, and in *The Sudan Railways* by Lt Col W E Longfield.
5. The section of the railway between Luxor and Aswan proved to be defective. Girouard was sent to inspect it and reported in scathing terms on the standard of construction and the work needed to bring it up to proper

standard. The upshot of all that was that he met Lord Cromer and was offered the highly-paid appointment of President of the Egyptian State Railways and Alexandria Harbour. With Kitchener's blessing (the main work of the Desert Railway having been completed), Percy Girouard accepted, and was replaced by his deputy.

6. Several wrote books afterwards. Winston Churchill wrote *The River War*, and G W Steevens wrote *With Kitchener to Khartoum* (Steevens later died of enteric fever in 1900 during the siege of Ladysmith, Boer War). The Hon Hubert Howard, correspondent of *The Times* and the *New York Herald*, was killed by 'friendly fire' when a gunboat shelled the Mahdi's tomb in Khartoum after the town was captured.

7. A good description of the battle will be found in *The River War* by Winston Churchill, who was then a war correspondent.

8. This set of instruments was saved by Sergeant Dennett and presented to the Telegraph Battalion RE. It is now displayed in the Royal Signals Museum.

Bibliography

Books providing background history to the signalling narrative are listed below. Few of them are presently in print, but most can be obtained from second-hand booksellers, using the internet to search. Others are held in specialist libraries.

General Background - Historical

Barnett, Correlli. *Britain and Her Army.* Pub London, Cassell & Co, 1970. ISBN 0-304-35710-3.

> The book covers the period 1509 to 1970, but chapters 12 to 14 give a good description of the military, political, and social background of the late-Victorian period, underlying the stagnation that led to the general lack of interest in army communications prevailing at the time. Available in paperback.

Farwell, Byron. *Queen Victoria's Little Wars.* Pub London, Allen Lane, 1973.

> A general historical description of various campaigns and expeditions in Victorian times.

Pakenham, Thomas. *The Scramble for Africa.* Pub London, Weidenfeld & Nicolson, 1991. ISBN 0-349-10449-2

> Provides a good background to the expansion of their empires in Africa by European countries in the period 1876 to 1912. Available in paperback.

Spiers, Edward M. *The late Victorian Army, 1868-1902.* Pub Manchester University Press, 1992. ISBN 0-7190-2659-8.

> A description of the British Army coinciding with the principal period covered in the narrative, from Cardwell's reforms until the end of the Boer War.

General Background - Technical

Although the technology of the time was fairly basic, some terms are used throughout the narrative that a few readers will not fully understand. Developments in telegraphy, and some technical explanations will be found in these books.

Beauchamp, K G. *History of Telegraphy.* Pub Institution of Electrical Engineers, History of Technology series, ISBN 0-85296-792-6.

> This book covers the subject comprehensively. It records the growth of telegraphy over two centuries, depicting the discoveries and ingenuity of the experimenters and engineers involved, the equipment they designed and built, the organizations that used it, and its applications, including army applications.

HMSO. *Instruction in Army Telegraphy and Telephony.* Pub 1914.

> The contemporary army manual of telegraphy. Gives a good description of all the instruments and their method of working that are described in the narrative. Not generally available, but copies are held in places such as the Royal Signals archives.

Herbert, T E. *Telegraphy - a Detailed Exposition of the Telegraph System of the British Post Office.* First edition published 1906, London, Isaac Pitman & Sons, Ltd.

> Various editions published up to 1946. Gives technical detail of telegraph instruments.

Chapter 1. Early Telegraphs.

Wilson, Geoffrey. *The Old Telegraphs.* Phillimore & Co Ltd, London, 1976. ISBN 0-900592-79-6.

> An interesting and informative book about signalling by visual telegraph methods in the period immediately preceding the electric telegraph.

Mallinson, Howard. *Send it by Semaphore - the Old Telegraphs During the Wars with France.* Crowood Press, 2005. ISBN 1-86126-734-7.

> Describes the first telegraph age, when relatively great advances in communication speed and distance were made using semaphore and shutter methods, preceding the electric telegraph.

Bright, Charles. *Submarine Telegraphs - Their History, Construction, and Working.* Pub London, Crosby Lockwood and Son, 1898.

> An authoritative description of the evolution of submarine cable technology and developments - the key to international telegraphy in the second half of the 19th century - by a leading telegraph engineer of the day.

Chapter 2. The Crimean War.

Numerous books have been written. The two below have been published recently.

Warner, Philip. *The Crimean War: A Reappraisal.*

Ponting, Clive. *The Crimean War: The Truth Behind the Myth.*

Chapter 3. Telegraph Developments in India.

Heathcote. T A. *The Indian Army: The Garrison of British Imperial India, 1822-1922.* Pub David & Charles, London, 1974.

> A very comprehensive description of many aspects of military organisation and life in India at the time the British controlled it.

Chapter 4. The Telegraph Route to India, 1862 - 1870

Goldsmid, Col Sir F J. *Telegraph and Travel.* Pub Macmillan and Co, 1874.

> A comprehensive description of the planning and construction, as well as the politics and diplomacy, of the Indo-European telegraph route by the Director General of the project in the period 1865-70. Now a rare book.

Dickson, W K. *The Life of Major-General Sir Robert Murdoch Smith.* Pub Edinburgh, Wm Blackwood, 1901.

> A biography by his son-in-law, mostly concerning Murdoch Smith's work in Persia with the Indo-European telegraph, which lasted for over twenty years, and his interests in Persian antiquities. Now a rare book.

Wright, Denis. *The English Amongst the Persians.* 2nd edition, pub I B Tauris, 2001. ISBN 1-86064-638-7.

> First published in 1977, Sir Denis Wright pursued a diplomatic career. He served in Iran (Persia) in 1953, to re-establish broken diplomatic relations, returning there from 1963-71 as the British Ambassador. Well-placed to understand the fraught relationship that has existed between the two countries over many years, he describes the political background at the time the Indo-European telegraph was built.

Chapter 5. Signalling Developments in the 1860s

McLennan, John F. *Memoir of Thomas Drummond*, Edinburgh. Edmonston and Douglas, 1868.

O'Brien, Barry. *The Life of Thomas Drummond.* London, Kegan Paul Trench and Co, 1889.

Chapter 6. The Abyssinian Expedition, 1867-68

Holland and Hozier. *Record of the Expedition to Abyssinia.* Pub War Office, London, 1870.

> The definitive and very detailed account of all aspects of the expedition by two staff officers who participated; in two volumes, with an entire chapter on signalling and telegraph. Now a rare and expensive book.

Bates, Darrell. *The Abyssinian Difficulty.* Pub Oxford University Press, 1979. ISBN 0-19-211747-5.

> Sir Darrell Bates spent some time in Ethiopia during wartime service in the King's African Rifles. He subsequently pursued a diplomatic career, and in retirement wrote several books. This one is well-researched and well-written, tending to cover the history and the wider picture rather than the military campaign. A recent reprint in paperback by the Naval and Military Press should be easily obtainable.

Hozier, Captain Henry. *The British Expedition to Abyssinia.* MacMillan and Co, London, 1869.

> By the same author as the War Office Record of the Expedition (above). Hozier was Assistant Military Secretary to General Sir Robert Napier, commander of the expedition, and draws on personal experience and official documents. A paperback reprint is available from the Naval & Military Press.

Myatt, Frederick. *The March to Magdala.* Pub London, Leo Cooper, 1970. ISBN 0-85052-026-6.

> A very readable account describing both the wider history of events as well as a good insight into the expedition and personalities. Easily obtainable second-hand.

Chapter 7. The Formation of Army Telegraph Units, 1870-72.

> There are no particular books relevant to this topic. Various documents in the Royal Engineers library, Chatham, and the Royal Signals archives, Blandford, give detailed information.

Chapter 8. The Ashanti Expedition, 1873-74

Brackenbury, Henry. *The Ashanti War - a Narrative.* Pub Edinburgh and London, Wm Blackwood and Sons, 1874.

> In two volumes, prepared from official documents by permission of the expedition commander, Major General Sir Garnet Wolseley. The author, Henry Brackenbury, at the time a Captain in the Royal Artillery and Wolseley's Assistant Military Secretary on the expedition, became one of the 'Wolseley Ring'.

Chapter 9. The Heliograph

> The best descriptions are given in the Army Signalling Manuals of the period. The Army Signalling Manual 1877 is held in the Royal Signals archives.

Chapter 10. The Second Afghan War, 1878-80

> The war, largely forgotten until recent events, was a political foreign policy bungle that resolved nothing. That apart, it saw the first successful use of organised tactical signalling for a conflict of that size and over considerable distances.

Intelligence Branch, Army HQ, Simla. *The Second Afghan War 1878-80, Official Account.* Published London, John Murray, 1908.

> Compiled shortly after the war from original documents, but regarded then as a confidential work and not published until 1907, after revision to excise sensitive material. Now a rare book.

Hanna, Colonel H B. *The Second Afghan War, 1878-80* (3 Volumes). Pub London, Constable and Co, 1899.

> Written at a time when adverse criticism of people and events by army officers was usually muted, this is a very forthright description of everything that went wrong, politically and militarily, by one who participated (but cautiously postponed picking up his pen and publicising until well into pensionable retirement). Now a rare book.

Hensman, Howard. *The Afghan War of 1879-80.* Pub London, W H Allen & Co, 1882.

> Hensman was Special Correspondent of the *Pioneer* (Allahabad), and *The Daily News* (London). He joined General Roberts' force when the war resumed in September 1879, and covers the period from the capture of Kabul to the end of the war, giving many first-hand descriptions. Now a rare book.

Roberts, Field Marshal Lord. *Forty-One Years in India.* Pub London, MacMillan & Co, 1897.

> Covering his time in India from a subaltern in 1852, through the Indian Mutiny and the 2nd Afghan War, in which he took a leading part, to C-in-C. Recently reprinted in paperback by the Naval and Military Press.

Robson, Brian. *The Road to Kabul.* Pub Arms and Armour Press Limited, London, 1986. ISBN 978-1-86227-416-7.

> A good and much-needed latter day description of the war using archive material, and covering the causes and the course of the campaign. Reprinted in paperback. Easily obtainable second-hand.

Chapter 11. The Zulu War, 1879

> This inglorious and nowadays over-glamourised six-month colonial war has become a modern industry for tourism and authors, spawning an excessive number of inevitably repetitive books. The list could be endless. At the risk of offending a multiplicity of authors, here is a limited selection.

Morris, Donald R. *The Washing of the Spears*. Pub 1966, London, Jonathan Cape.

> A well-researched and well-written account of the rise of the Zulu nation under Shaka, and its fall in the Zulu war. Subsequent books have little of substance to add to this overall description.

HMSO. *Narrative of the Field Operations connected with the Zulu War of 1879*, pub 1881.

> The contemporary official record of events. Not readily available, except in specialist libraries.

Chapter 12. The Expedition to Egypt, 1882

Maurice, Colonel J F. *The Campaign of 1882 in Egypt.*

> The Official History of the campaign, available in some military libraries.

Wright, William. *A Tidy Little War - the British Invasion of Egypt, 1882.* Pub 2009, Spellmount Press.

> A rather long-winded account of the campaign, and poorly edited, but gives much of the political background.

Chapter 13. The Expedition to Bechuanaland, 1884

> There are no particular books relevant to this expedition. It forms a small part of the ongoing friction between the Boers and the British, and is covered as part of any modern history of South Africa.

Chapter 14. The Nile Expedition, 1884-85

> A controversial expedition to rescue General Gordon from Khartoum, its failure caused principally by Gordon's wayward personality which unnecessarily led him into the situation, and further compounded by Gladstone's political procrastination.

Neillands, Robin. *The Dervish Wars - Gordon and Kitchener in the Sudan, 1880-1898.* Pub London, John Murray, 1996.

> A good overview of why Britain became involved in Egypt and the Sudan in the late 19th century, the history and the personalities, and the course of events.

Preston, Adrian. *In Relief of Gordon.*

> Mainly a reproduction of Wolseley's private journal, in which he fulminates about Gladstone and some of his military subordinates, giving a useful insight into the expedition, which proved to be the turning point in his successful career up to that point.

Chapter 15. Suakin, 1885.

> There are no particular books relevant to Suakin. It forms part of the wider Nile expedition histories.

Chapter 16. The Second Ashanti Expedition, 1895.

Baden-Powell, Major R S S. *The Downfall of Prempeh.* Pub London, Methuen, 1896.

> Perhaps better remembered these days as the founder of the Boy Scout movement, Baden-Powell participated in the expedition, and wrote of his experiences.

Chapter 17. The Reconquest of the Sudan, 1896-99.

Steevens, G W. *With Kitchener to Khartum.* Pub Wm Blackwood and Sons, 1898.

> Steevens was a war correspondent who joined the expeditionary force at Wadi Halfa in 1897, after the first phase of the reconquest that had seen the capture of Dongola the previous year. He accompanied them to Khartoum, describing many aspects of the advance.

Churchill, Winston. *The River War.*

> Churchill was a war correspondent during this expedition.

Marlowe, John. *Cromer in Egypt.* Pub London, Thomas Nelson Printers Ltd, 1970. ISBN 0-236-17656-0.

> A very detailed account of all the political background between the British government, Egypt, and the various countries involved with Egypt in the period 1876 to 2007, when Lord Cromer (Sir Evelyn Baring) was Consul General.

- Encouraging a culture of lifelong learning

- Recognising and rewarding excellence

- Supporting professional registration opportunities

- Professional networking with service counterparts and industry

Funded by the Royal Signals Charity as well as subscriptions from individual and corporate members, the Royal Signals Institution (RSI) exists to serve the interests of all ranks of Royal Signals from the Regular, Reserve and Retired communities.

We are probably best known for the awards and prizes we sponsor. These number over 150 per year and range from top student on Initial Trade Training courses, top apprentices in each trade group to recognition of outstanding performance and service by Corps personnel through the higher level RSI awards.

We also promote and support professional accreditation and recognition with prestigious external civilian organisations such as the IET, BCS, ITP and WCIT. Professional development also extends to RSI members receiving invitations to a series of regular presentations on relevant topics and updates. These range from the technological (such as the implications of 5G roll-out) to the more esoteric (such as the mind-set of a hacker).

Our strong links with industry allow us to act as a Corps 'think tank', bringing together professionals from the Defence and commercial arenas in briefings, workshops and seminars to analyse the challenges faced by the Corps and to provide impartial input from outside the chain of command to key Corps policy leads. We also use these same industry links to provide a jobs list – get your password from Alison.Brennan510@mod.gov.uk

Further details about the RSI can be obtained from Lieutenant Colonel PA Osment, Director, The Royal Signals Institution, contactable on rsi@royalsignals.org or Tel: 94371 2647 (Mil) 01258 482647 (Civil)

Roger So Far...

The first 100 years of the Royal Corps of Signals

ROGER SO FAR...

The Centenary book 'Roger So Far' is available from the Royal Signals Museum Shop at a cost of only £9.99, follow the link from our website www.royalsignalsmuseum.co.uk.

'Roger So Far' celebrates 100 years of Service, Innovation and Achievement with the Royal Corps of Signals. From laying telegraph line across the trenches of World War I to the cyber warfare of today, the goal of the Royal Corps of Signals to maintain the British Army's 'vital link' in communication has never wavered.

Now in its centenary year the Corps is celebrating the crucial service and innovative support their members have given – and continue to provide – across the globe in a new, fully illustrated book Roger So Far...

Written by multiple expert authors in an accessible and engaging format, and featuring many previously unpublished images, Roger So Far... is both a concise historical record and an archive of memories across all parts of life with the Corps.

Chapters include topics and themes such as:

- Changing technology, equipment and work, from using carrier pigeons and antennas to SIGINT and SOE work to cyber warfare.
- Humanitarian programmes and charity work carried out around the world.
- Sporting achievements, adventure training and family life within the Corps.
- Key military operations, including World War I, the Gulf War, Balkans, Iraq, and Afghanistan.

Roger So Far... hopes to inform those who are serving, or have served, as well as to educate those who have not served and those who may be thinking of doing so.

One hundred years on from its formation on 28 June 1920 the importance of the Corps is growing because conducting successful operations is becoming increasingly reliant upon being able to operate in a congested and contested electronic battlespace, where the cyber warfare skills of Royal Signals soldiers are vital.

Due to current circumsances distribution may be a little slower than normal. We would ask your patience in this respect.

Hard Back only, 288 pages full of photographs maps and illustrations

Tales of the Telegraph

Part 2

The Telegraph Battalion
in
the Anglo-Boer War, 1899 to 1902

by

Lieutenant Colonel David Mullineaux

The Telegraph Battalion

in the

Anglo-Boer War, 1899 to 1902

Contents

		Page
	Introduction	(ii)
Chapter		
1	The War starts	1
2	Operations in Natal, 1899	12
3	The Western Front, 1899	23
4	Communications at Colesberg	36
5	The Relief of Ladysmith	43
6	Kimberley	54
7	The Advance to Bloemfontein	67
8	At Bloemfontein	79
9	The Advance to Pretoria	88
10	Guerrilla Warfare	99
11	The War Ends	109
	Bibliography	116
	Glossary of Telegraph Terms	118
	Glossary of Afrikaans words	123

Introduction

The Boer War (1899 to 1902) is now largely forgotten. It remains controversial in many aspects; in retrospect, it was a war that better diplomacy by both sides should have avoided. As far as the army was concerned, it was quite different to the operations and expeditions experienced during the preceding forty or so years as the British Empire developed. It was against a modern enemy rather than ill-equipped native forces, and it was much more extensive in the number of troops and in the expanse of territory it covered. The British army suffered many setbacks, and perhaps the greatest benefit of the war was that it's many deficiencies in modern tactics, command structure, organisation, and tactical communications became apparent, serving as a 'wake up' call before the First World War.

The Boer War itself has been described in many books, mostly published soon afterwards, notably *The Times History of the War in South Africa* (six volumes) and the *History of the War in South Africa* (four volumes) - each with their defects in the light of subsequent revelations. Like most military histories they have little to say about communications. Yet for those involved in that department the Boer War was the biggest challenge so far - the number of troops involved, the volume of traffic, as well as the distances over which communications were required. It greatly exceeded anything experienced before. A particular advantage in this case, compared with previous expeditions in less developed countries, was that a telegraph system existed within the operational area. But the technology for field communications had not changed for several decades; it was still visual signalling at the tactical level, and electric telegraphy using landlines. Wireless had recently been invented, but was not yet practical and was not used. The telephone existed but its range was limited. The army's Telegraph Battalion, only some 300 strong at the start of the war, could not cope, although fortunately they were able to draw on many Post Office reservists, bringing their numbers up considerably as the war progressed.

In this situation the provision of communications for the army was to be an entirely new scenario. And as you read the story that follows, try and place yourself in the shoes of the relatively junior officers and NCOs responsible for providing the army's telegraph communications as they undertook the immense tasks that faced them, exceeding the level of responsibility they would be expected to undertake today.

Having myself spent many independent holidays in South Africa during retirement, I have been able to visit the operational areas and study the ground for myself. Indeed it was this that originally stimulated my interest in the war when, having read the history books, there appeared to be nothing readily available which described the communications, although of course that is a rather specialised topic not given to profitable publishing. Another interesting aspect was that the ground over which the operations were conducted, much of it away from the modern tourist trail and its associated building developments, remains mostly as it was at the time of the war, and thus it was possible to relate the events and the contemporary maps very accurately to the present-day scene.

My research has taken place in the Royal Signal archives (where there are a number of diaries of those who participated), the Royal Engineers library and archives, the Prince Consort's library in Aldershot, and various sources in South Africa itself. I am grateful for the help I have received from all of them.

The Boer War and its extensive communications requirements, despite what today might seem to be the limited technology of the time, were to be the trigger for developing better resourced and more soundly organised army communications for the future - all part of the 'wake up' call. As we now know, it was to lead from the Telegraph Battalion at the time of the Boer War to expansion into the RE Signal Service for the First World War, and after that, in 1920, the formation of the Royal Corps of Signals.

David Mullineaux

Laying cable Boer War circa 1900

Chapter 1

The Boer War Starts

Introduction

The Anglo-Boer War of 1899-1902, also sometimes described as the War in South Africa, was the greatest test of army communications so far. The various operations in the preceding decades had all presented their problems but the scale of this war brought things to a new level. [1]

Initially the Telegraph Battalion deployed its full resources, both 1st (regular) and 2nd (reserve) Divisions, amounting to just under 300 men. Reinforcements were soon needed; more regular troops were trained, and many more reservists from the Post Office volunteered, bringing their combined strength to about 2,400 by the end of the war. To this was added a locally employed workforce - some 900 unskilled natives who were taken on soon after arrival. As a consequence of this necessary expansion the Telegraph Battalion became known during the war as the Telegraph Division. Later, when the telegraph networks of the two Boer republics were taken over, some of their former employees, mostly of British extraction, were re-employed and worked under the direction of the Telegraph Division.

As for regimental signalling, although its separate organisation remained flawed in principle, it emerged slowly from many years of dubious quality in the British army. It had to! Although restricted to visual methods, signalling improved as the war progressed. After a faltering start in the first few months, commanders came to realise its necessity in a country where gallopers or messengers were no longer effective against an elusive and mobile enemy over great distances, and where the topography and fairly dependable sunshine suited it admirably .

The telephone was also used, but its range was limited so it was generally restricted to static defended areas; the invention of the thermionic valve, which could amplify the waveform of the human voice and enable the telephone to work over greater distances, was still about a decade away. Wireless, too, though invented a few years earlier, and trialled in South Africa at the early stage of the war, was not sufficiently developed to be of any practical use in the field. So communications still depended principally on the electric telegraph and visual signalling. Technology had not developed much in the previous three decades, although that was to change soon after the war. But the great advantage was that in South Africa, unlike previous operations conducted in undeveloped countries, there was a well-established permanent telegraph system.

Historical Summary

The war had its roots in many years of turbulence in South Africa. A somewhat convoluted story, it involved conflicts between black tribes, between white settlers and black tribes, and between white settlers of different nationalities in a continuously bubbling cauldron of disputes about land and power, Boer independence and British imperialism, devious politics and ephemeral policies, all spiced by the discovery of diamonds and gold. From the British point of view, the political vacillation and changing policies of successive governments owed much to the lack of communication with their administrators on the ground in South Africa - until eventually a submarine telegraph link to Durban was established in December 1879.

The history and the politics, however interesting they may be, are subjects that have been covered at length in many books, most notably *The Boer War* by Thomas Pakenham, researched in considerable depth and published in 1979 when many previously unpublished records were available, but are not part of this narrative. Suffice to say that Anglo-Boer antipathy had existed for a long time. Boers, literally farmers, the name reflecting their predominantly agricultural lifestyle, were mainly of Dutch extraction. They were also generally known as Afrikaners, and very proud of their Afrikaans language, a basic form of Dutch - language being another cause of dissension with the British.

After the discovery of gold in 1886 there was an influx of Uitlanders ('outlanders', mostly Europeans, a high proportion of them British) to the South African Republic (as the Transvaal was then known), bringing with them both mine engineering skills and the financial resources needed to fund mine development. But the restrictions placed on their enfranchisement by an Afrikaner government fearful for their independence caused yet more friction between the two white factions and ultimately contributed to the *casus belli*. Tension reached a new level with the botched Jameson Raid in 1895. The immediate

THE BOER WAR STARTS

cause of the war, though, was a failure of politicians and diplomacy, exacerbated by a clash of personalities, with blame able to be assigned to both sides - and well-described in Pakenham's *Boer War*. Even at the time many regarded it as an unnnecessary war, and with the benefit of hindsight this view is strengthened.

Following an abortive conference at Bloemfontein in June 1899, and trouble looming, Britain sent some military reinforcements to South Africa. President Kruger of the South African Republic, the arch-Boer, saw this as a threat and issued an ultimatum. When it expired, the Boers invaded the British colony of Natal, on 12 October 1899. The war had started; it was to prove costly, and ultimately futile.

The table below summarises the principal events in South Africa's history that led up to the war. As will be

South African History - Principal Events

Year	Event
1652	Dutch East India Company set up station at Cape Town.
1806	British gain control of the Cape.
1811-53	Various frontier wars with tribes in Eastern Cape.
1815	Cape Colony founded.
1820	4,000 British settlers arrive at the Eastern Cape.
1834	Slavery abolished at the Cape.
1835	Great Trek begins. Many Dutch settlers remain in the Cape Colony.
1838	Boer trekkers start concentrating in Natal.
1843	British annex Natal as a colony.
1844	Transorangia annexed as Orange River Sovereignty.
1852	Sand River Convention confirms independence of Transvaal, renamed as South African Republic in 1855.
1854	Bloemfontein Convention restores independence of Transorangia, now called Orange Free State.
1868	Basutoland (Lesotho) becomes a British Protectorate.
1870	Diamond rush to Kimberley.
1871	Annexation of Kimberley to Cape Colony.
1877	Annexation of South African Republic as the Transvaal.
1879	Britain invades Zululand and incorporates it in Natal.
1880-81	Transvaal rebellion. First Anglo-Boer War.
1881	Peace talks after British defeat at Majuba. Pretoria Convention restores limited independence to Transvaal.
1884	London Convention gives greater independence to Transvaal, which reverts to the South African Republic.
1884	Expedition to Bechuanaland (Botswana) which becomes a British Protectorate.
1886	Gold rush to the Witwatersrand, near Johannesburg.
1895	Jameson Raid.
1897	London inquiry into the Raid. Sir Alfred Milner becomes British High Commissioner at the Cape.
1899	Anglo-Boer War begins on 11 October.
1900	Annexation of Orange Free State and South African Republic to Orange River Colony and Transvaal Colony. Guerrilla war begins.
1902	Treaty of Vereeniging. War ends on 31 May.
1906	Self-government returned to the former Boer Republics, which become Orange Free State and Transvaal.
1910	Union of South Africa under Afrikaner control.

seen, the South African Republic and the Transvaal are synonymous, the correct name depending on the point in time. For consistency in this narrative it is referred to subsequently as the Transvaal - a name that eventually disappeared when apartheid was banished and black African majority rule was introduced with the Government of National Unity in 1994.

Hesitant Mobilisation

In Britain, with the war imminent, authorisation for army mobilisation was slow, partly to avoid inflaming the situation. In early September 1899, as a limited measure, the War Office set about drawing soldiers from India and stations in the Mediterranean to go to South Africa. Within two days of the order being given the first elements of the Indian contingent were on the quays at Bombay and Calcutta, ready for the passage to Durban, in Natal, and were despatched with remarkable rapidity, arriving there in late September. Within the next few weeks, some support troops from England were also sent to Durban; they included a headquarters element and one section of the Telegraph Battalion which sailed from Britain on 20 September.

It was not until 7 October, just a few days before war was declared by President Kruger on 11 October, that the general mobilisation order was belatedly issued and the main body of troops from Britain started to move. The one month delay in getting sufficient forces on the ground in South Africa in time, caused by political prevarication, was to lead to a situation that was to throw the army's operational plan into disarray, although to be fair to the politicians, lots of army shortcomings were going to contribute.

Once started, mobilisation continued apace, movement to South Africa being organised extremely efficiently by the Admiralty, using civilian shipping. The sea passage to Cape Town took just over three weeks. Embarkation of the original Army Corps in Britain, sailing mainly from Southampton, London, and Liverpool, started on 20 October and was largely complete by 17 November. The force consisted of three infantry divisions, a cavalry division, and supporting troops - some 47,000 men with their horses and equipment. They included the main body of the Telegraph Battalion which sailed from Southampton on 21 October.

Two more divisions were to be mobilised soon afterwards, virtually denuding Britain of all its regular

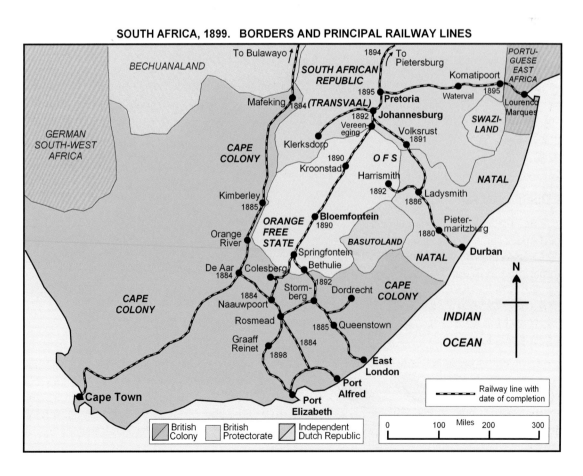

army. Subsequent reinforcements were to be reservists of various sorts – yeomanry and militia, many untrained and of poor quality, and many from urban backgrounds who were quite inexperienced in living in a country like South Africa. There were also contingents of volunteers from other parts of the empire, such as Australia, New Zealand, and Canada; their quality was high, and the conditions in South Africa not dissimilar to conditions at home. But to begin with, crucially, the troops were late arriving and were too thin on the ground.

The Strategic Plans

The British campaign plan had been drawn up in London in the preceding months. Only two lines of advance had been seriously considered - through Natal, or northwards from Cape Colony. Railways dominated the strategy; they were then the main transport artery, otherwise it was animal transport, with all the limitations that imposed, especially in a country where distances are great. The route through Natal offered a shorter distance with its base in a loyal colony, but the disadvantages were difficult topography between Natal and the Boer republics, more suited to Boer defenders, and secondly, the force would be supplied from a single port along a single railway line with low capacity due to steep gradients. From Cape Colony, once into the Orange Free State and then the Transvaal the ground was flat and more open, suitable for cavalry tactics, and there were three ports and and three railway routes converging on the southern border of the Orange Free State, but the disadvantage was a very long line of communication - from Cape Town over 500 miles to the border of the Orange Free State, and nearly 1,000 miles to Pretoria. In London, the army's Commander-in-Chief, Field-Marshal Lord Wolseley, well acquainted with South Africa, decided on the Cape option as the primary route. As its objective, the Army Corps was to advance to the capital, Pretoria, dealing with any resistance on the way, whereupon the Boer Republics would be defeated. The resistance expected from what was seen as a motley collection of ignorant farmers was considerably underestimated. The British troops were assigned as follows: three infantry divisions and the cavalry division were to advance from the Cape through Bloemfontein to Pretoria along what came to be known as the western front, and the fourth infantry division was assigned to Natal to provide protection against what the alarmed colonists there rightly saw as the prime threat.

The Boers, too, had a plan of campaign, and it was quite simple. A large force was to invade northern Natal, crush the small garrisons at Dundee and Ladysmith, and then rapidly overrun Natal down to Durban, thus gaining access to the seaport – a strategic asset which the landlocked Orange Free State and Transvaal were denied. Other Boer forces were to capture the towns strung along the Bechuanaland railway – Mafeking, Vryburg, and Kimberley – and would then advance southwards to help the armed uprising of the Dutch population in the Cape, which was to be expected after their initial successes. They would then hold on, and harass the British until the sympathetic European powers such as Germany intervened on their behalf or the British gave up the struggle. But like the British, the overoptimistic Boers got their calculations wrong and were also to see their plans upset – they never got to Durban, and there was neither an uprising of sympathisers in the Cape nor any practical support from Europe.

Both strategies fell apart, and many decisions in the opening stages of the war were made in response to unforeseen setbacks.

The Opening Phase

The Natal force, principally the troops from India who had arrived in Durban in late September, were placed under the command of Lieutenant-General Sir George White, who arrived at Durban from England on 7 October, only four days before war was declared. General Sir Redvers Buller, General Officer Commanding at Aldershot, had been appointed Commander of the Army Corps and the South Africa Field Force and, after a patriotic farewell from Southampton on 14 October, had arrived in Cape Town aboard the SS *Dunottar Castle* on 31 October - twenty days after the war started. Between his departure from England and his arrival in Cape Town the situation had changed, the only inkling during the sea voyage being snippets fortuitously exchanged on the high seas between passing ships.

After the declaration of war on 11 October the Boers had moved rapidly. In Cape Colony, Kimberley and the more remote Mafeking had been under siege since 14 October. In Natal, battles were fought at Talana (20 October) and Elandslaagte (21 October). Natal appeared defenceless, and Pietermaritzburg and the port of Durban were under threat. In early November Boers from the Orange Free State were threatening the undermanned garrisons at Naauwpoort and Stormberg on the Cape border, and they were ordered to withdraw.

THE BOER WAR STARTS

None of these setbacks had been foreseen in London and Buller, taking stock of the situation he found on arrival, was forced to change the plan. Ladysmith became besieged on 2 November, three days after his arrival. He had to send reinforcements to Natal, both to defend the colony and to relieve Ladysmith, as well as relieving Kimberley and Mafeking. To do this meant splitting his Army Corps, thus delaying the planned advance to Bloemfontein and onward to Pretoria. It wasn't going to be 'Home for Christmas' after all. Moreover, doubting the quality of the commanders in Natal, he decided to go there himself with a small personal staff. He left Cape Town in some secrecy on 22 November arriving at Durban three days later, but leaving a command vacuum behind in Cape Town. The wisdom of these decisions, and events in Natal that were to follow, have long been distilled by military historians.

First Telegraph Detachment to Natal

That summary leads into the task facing the Telegraph Battalion - a unit some 300 strong, expected now to provide communications for a force of five divisions deployed on two fronts some 300 miles apart, the western front advancing from Cape Colony and aiming to go to Pretoria, with the huge distances that involved, and the unexpected situation in Natal. The belated mobilisation, the initial setbacks, and the consequent change in plan were going to affect them quite considerably, as will be described. Clearly their numbers were hopelessly inadequate for the huge task ahead which was far in excess of anything they had previously undertaken, and steps were taken quickly to provide them with reinforcements.

At the time, the organisation of the 1st Division of the Telegraph Battalion (the regular element) was an administrative headquarters and four sections, each section consisting of two airline detachments, a cable wagon, and a number of telegraphists. As already mentioned, the first detachment - a headquarters element including the Battalion commander, Major W. F. Hawkins, and one section - were mobilised and sent with the first batch of troops from England to Natal where the initial threat was greatest. They left Aldershot on 20 September 1899, without horses, and embarked at Southampton on the SS *Jelunga* for passage to Durban, calling at Gibraltar, Malta, Crete, and Port Said on the way, the purpose of calling at these Mediterranean ports being to pick up other troops *en route*. When speed was of the essence it was frustratingly slow. Worse still, the ship broke down in Port Said for a few days.

After their lengthy passage on the *Jelunga* they reached Durban on 27 October, sixteen days after the war had started. By then Natal had been invaded, the battles of Talana Hill and Elandslaagte had been fought, and Ladysmith was threatened. On arrival they were moved quickly by rail from Durban to Ladysmith. Within days they were to be trapped in Ladysmith for the four-month siege. The ensuing events there will be described in the next chapter.

Main Party to Cape Town

Meanwhile mobilisation orders for the remainder of the Telegraph Battalion were received in early October, and on 8 October Major A. E. Wrottesley was posted in and appointed as Officer Commanding, to be Director of Army Telegraphs in South Africa. Others to be posted in were Captain E. G. Godfrey-Faussett as Second-in-Command, Lieutenant R. J. Jelf to command a section that was constituted to replace the section that had already been sent to Natal, and Lieutenant A. Bannerman to be in charge of the Clearing House (dealing with administrative and financial matters).

There were equipment problems. The Battalion was normally established with four sections. The section sent to Natal had been formed as an additional fifth section, but had taken the equipment of No 1 Section, which had not been replaced in the short intervening period. Equipment of Nos 2, 3, and 4 Sections was scattered and incomplete, some of it at Salisbury - all of which seems rather negligent considering that mobilisation was far from being unexpected. There was a last-minute rush to re-equip from stores held at Woolwich in the charge of the Ordnance Department, although these were earmarked for other roles and were not kept properly sorted out in storage. The outcome was that the main body departed with equipment deficiencies. Fortunately this seems to have been made good by stores available in South Africa from the civil Telegraph Department.

Wrottesley had for some years been in command of the 2nd Division of the Telegraph Battalion and was well-known in the Post Office's Southern District where he was Superintending Engineer. Due to his foresight, and in anticipation of using the permanent telegraph lines in South Africa, they took with them a small number of Wheatstone automatic telegraph equipments, used by the Post Office for many years but not by the army because it was not a robust equipment suitable for field conditions and needed permanent lines of much higher quality than the field telegraph lines the army used. The equipment's great advantage was its ability to deal with high volumes of traffic, and as things turned out this was going to prove vital.

This group, consisting of seven officers and 226 soldiers, departed from Southampton on 21 October aboard the Union liner SS *Gascon*. They included thirty-seven telegraphists from the Post Office reserves. Their destination was Cape Town, whence they were to support the main advance from Cape Colony through the Orange Free State to Pretoria, the western front, according to the operational plan already described.

The Telegraph Battalion Record shows them to be organised as follows:

	Major A. E. Wrottesley. In command, and Director of Army Telegraphs
	Captain E. G. Godfrey-Faussett, Second-in-Command
HQ Section	**No 1 Section**	**No 2 Section**	**No 3 Section**	**No 4 Section**
Lieut A.Bannerman	Lieut R.J.Jelf	Lieut E.O.Henrici	Lieut J.P.Moir	Lieut H.L.Mackworth
T.S.M. Foster	Sgt Hawker	Sgt Longley	Sgt Shergold	Sgt Cadwell
T.S.M. Kilburn	Sgt Salter			
T.Q.M.S. Low				

Also on board the *Gascon* was a detachment of six civilian engineers of the Marconi Company and some supporting Royal Engineer soldiers, under the command of Captain J. N. C. Kennedy RE, with five sets of experimental Marconi wireless apparatus. By coincidence they were going to South Africa on the same ship that had been used earlier that year for trials of wireless telegraphy in Algoa Bay near Port Elizabeth. Marconi's wireless had first been demonstrated to the British army on Salisbury Plain in 1896, and trials in England had been conducted subsequently. It had been decided to take the equipment to the Boer War, where its limited deployment will be mentioned later. To allay any premature excitement that may arise from the advent of this rather more modern method of communicating, it should be said now that the experiment was a failure as far as the army was concerned. Wireless did not contribute to the provision of any army communications during the war, and will not be mentioned further.

Post Office Reservists

The Post Office reservists who were part of this first group, and others who followed with subsequent reinforcements, were all civil servants in peacetime, forming 'L' Company of the 24th Middlesex (PO) Rifle Volunteers. Two publications give an insight into the work and thoughts of the Post Office telegraphists as they participated in by far their biggest call-out in support of the regular army. The first was the in-house newsletter of the Post Office, a publication known as *St. Martins-Le-Grand* (after the name of their headquarters office building in London), and secondly *The Telegraph Chronicle*, the Post Office telegraph service's regular fortnightly newsletter. An enterprising non-commissioned officer in I (Depot) Company of the 24th Middlesex RV, Colour Sergeant R. E. Kemp, encouraged those sent to South Africa to correspond with him, and he ran a news bureau for the duration of the war. This source of information gives a less formal record of events than either the published histories or the officers' diaries (which tended to confine themselves to operational matters), and provides an insight into the war from the point of view of the soldiers who served in the Telegraph Division. Extracts from some of their letters are quoted during the narrative.[2]

The reservists accompanying the Telegraph Division to Cape Town had been called up at short notice, medically examined, and ordered to report to Aldershot. They came from the Central Telegraph Office in London, other London offices, and provincial offices – places like Aberdeen, Birmingham, Edinburgh, Huddersfield, Liverpool, and Southampton. All the reservists were identified by the two-letter telegraphic address of their office (AB, BM, EH, etc). These reservists were the first of many of their ilk to follow.

In its issue published on 27 October *The Telegraph Chronicle*, under the heading 'TS and the War' (TS was

THE BOER WAR STARTS

the telegraphic address of the Central Telegraph Office, London), reported:

> The telegraph plays an increasingly prominent part in warfare with the advent of every fresh campaign, and that it is considered of first importance is evident from the number of operators required, and the fact that a strong Wheatstone staff has been selected from the service and a large quantity of apparatus for working and materials for erecting lines is on its way to the scene of hostilities.

The Telegraph Chronicle also describes the departure of this group from Aldershot:

> On Saturday morning, October 21st, we paraded at 4.50 a.m. In the cold, grey dawn, dressed in khaki full marching order, we looked like so many spectres in the thoroughly saturating mist. After roll-call we marched off at about 5.15 a.m. to the Ash Military Siding [near Aldershot] where we at once entrained, arriving at Southampton Docks at 7.30 a.m. We were enthusiastically cheered by the inhabitants of the wayside villages as we passed through. In the sheds at the docks carbines, with fixed swords, and valises were passed from hand to hand and were stowed away on board, after which we trooped on and took up our positions in our several messes – sixteen men to each mess. We were at once supplied with a meal of porridge, hash, coffee, and bread. Later on in the day the 2nd Coldstream Guards, to the tune of about 1,200 embarked, as well as 100 Medical Staff Corps. The Royal Engineers, numbering over 200 - with the Medical Staff Corps – were in the stern, the Guards in the forepart of the ship, and the officers amidships. At 5 p.m. we were ready to start. Owing to our being behind time, however, the tide had run out, and we were resting on the mud and could not move. By command the great crowd of men in the stern moved to the forepart of the vessel, which released the ship from her mud-bed, and finally we left the quay amid cheers, 'God Save the Queen', 'Rule Britannia', and 'The Soldiers of the Queen' from immense crowds assembled ashore, and whistles shrieking from all kinds of steamers. The sight when the ship commenced to move was magnificent. Many of the men had climbed the masts as far as it was possible to go and were frantically cheering and waving handkerchiefs in response to the enthusiastic send-off. Two large steamboats fairly packed with people accompanied us up the Southampton Water, one on each side. Then we all went down to tea, and when we came up again the Needles Lighthouse was just being passed, the last sight of old England being the revolving light of the Needles; and the men were happy at last to be *en route* for the seat of war.

As well as this rousing send-off from Southampton "the Duke of Connaught came on board to bid us farewell. The parting of husbands and wives was most heart-rending", wrote Sapper Stimpson, a regular correspondent with Kemp.

Two Fat Men

An amusing situation arose over two of the reservists who quite literally missed the boat – Messrs, or more accurately now, Sappers J. S. Tough from Aberdeen (AB) and J. F. Metcalfe from the Central Telegraph Office (TS) in London. When these two rotund men, straight from their sedentary civilian duties in telegraph offices, were being kitted out at Aldershot it was not possible to find uniforms large enough. They were classified as 'outsizes' and their measurements were sent urgently to the Pimlico Clothing Factory. The day of departure came and still no uniform, so they had to be left behind in Aldershot. Later investigation revealed that their uniforms had been sent straight from the factory to Southampton, where the *Gascon* departed with their uniforms but without them. Some days later, re-measured, and their second new uniforms having arrived, they departed on the next available ship, which happened to be the SS *Bavarian* sailing from Liverpool, two lonely telegraphists on a ship going next to Queenstown (now Cork) in Ireland to pick up the Dublin Fusiliers and the Connaught Rangers. Metcalfe described this in a letter sent from Las Palmas on 14 November:

> ... were seasick from Liverpool to Queenstown, where we stayed twenty-seven hours. Every type of Irish people came down to see us. We had a rattling send off, thousands of people cheering and burning coloured lights, while bands played on board. ... We have 2,700 on board ... and today we start physical drill [much needed in their case, it would seem]. Twice one of us has been a defaulter for being unshaven, our major being very strict in that respect. But how can a fellow shave when he's seasick and the boat is rolling so? We have got our sea legs now, and feel all right. There are concerts and bands every evening.

Having delivered the Irish troops to Durban for operations in Natal the ship ended up in East London where Metcalfe and Tough disembarked. They never rejoined the Telegraph Battalion. Metcalfe ended up on telegraph duties at Springfontein but died there on 20 May of dysentery, and Tough was invalided back to Britain on 31 May, also with dysentery. [3]

THE BOER WAR STARTS

Death of Major Wrottesley

During the Telegraph Division's passage to Cape Town an unfortunate incident occurred on the *Gascon*. Major Wrottesley, the Director of Telegraphs, was lost overboard from the ship somewhere near Tenerife. He had been suffering badly from seasickness and depression. Captain John Kennedy relates in his diary how Major Wrottesley had confided in him about his state of mind and Kennedy feared he was suicidal. [4] One of the reservist's letters reveals how "Major Wrottesley, who is in charge, is prostrated with seasickness ever since Sunday, and has had to have a guard to watch him." Last seen in his cabin with a cup of cocoa at 9:30 on the morning of 26 October, while the ship was battling with heavy weather and everything was battened down, he was later reported missing and was presumed to have fallen overboard, unseen, in the storm. Another telegraphist, Sapper Stimpson, thought otherwise, saying that Wrottesley was: " ... supposed to have jumped overboard, his health being very bad, and his mind was thought to be deranged by seeing the sergeant after his suicide at Aldershot." The exact circumstances of Major Wrottesley's demise remain a mystery. His death was reported back to England by telegraph when the ship reached Madeira.

Lt Col R. L. Hippisley.

A replacement was quickly appointed and sent to South Africa, Lieutenant-Colonel R. L. Hippisley, who sailed from Britain on 18 November 1899 on the SS *Guelph*, arriving in Cape Town on 10 December. Richard Hippisley, forty-six years old, had been commissioned in 1872 and during his service had completed several tours of duty with the Postal Telegraph Companies and their successor, the 2nd Division of the Telegraph Battalion, had participated in the expedition to Egypt in 1882, had been an instructor in telegraphy at Chatham, and had commanded the 1st Division of the Telegraph Battalion for six years, so was well-qualified for this unexpected appointment. He was to remain as Director of Army Telegraphs, South African Field Force, for the duration of the war. Meanwhile, Major F. C. Heath, already on the army engineering staff in Cape Town, was appointed Acting Director of Army Telegraphs from 3 November until Hippisley's arrival, and made various preliminary arrangements with the Postmaster General's staff in the Colony.

Back on the *Gascon*, some extracts from *Khaki Letters* give an idea of life aboard, the first dated 27 October, near Tenerife:

> The boys are bearing up well although the weather has been very windy and rainy, causing some discomfort. The Coldstreams do all the ship's duty, and we do a bit of physical drill every morning after which we are treated to a dose of 'Slingo' [contemporary slang, meaning instruction in technical telegraphy]. ... All the boys have had a machine crop, and are in the best of spirits.

> This is only about one-fifth of our journey. ... The nights are the most uncomfortable. ... It is sweltering down below and the hammocks are as close as they can possibly be hung, making the air very stuffy.

Having been fitted up, as far as possible in the confined space of the ship, and instructed in aspects of military telegraphy that they did not normally practise (reservists were principally operators; military telegraphists had to be able to set up a telegraph office involving correct line, instrument, and battery connections), they reached Cape Town on 12 November, and their story will continue in chapter 3.

* * * * * * *

The South African Permanent Telegraph Network

The provision of the army's communications in South Africa was going to depend heavily on the existing telegraph communications there. Unlike earlier campaigns undertaken by the Telegraph Battalion, South Africa was by now furnished with a telegraph network covering the two British colonies and the two Boer republics. Its extent will become apparent when the telegraph operations during the course of the war are described later.

The British colonies, Cape Colony and Natal, had postal and telegraph organisations similar to Britain - government-controlled under a Postmaster General, each with telegraph departments. They were manned mostly by staff of British extraction, many of them having emigrated from Britain (there was a 'passage paid' scheme to attract them) after working in the British telegraph system and, as in Britain, they were civil servants in status. The equipment and procedures were compatible with British practice - something that was going to prove to be a great advantage.

The two Boer republics, Transvaal and Orange Free State, also had telegraph systems. Apart from the concentration around Pretoria and the gold mining area of Johannesburg, their telegraph was rather thinner on the ground than the British colonies due to a combination of fewer and more widely spread townships, shortage of money until gold was discovered, and consequently later development. Their equipment was largely sourced from Germany, principally by the Siemens company. Nevertheless, by 1899 the separate telegraph networks in South Africa interoperated with a high degree of freedom and technical compatibility.

The Postmaster General in Cape Town, Sir Somerset French, had done much to prepare the Cape Colony telegraph system for the war, despatching men and stores to the forward base at De Aar, and ordering stocks of material. A group of his civilian telegraphists were besieged in Kimberley, where they were to play a part to be described later. Early liaison, before the arrival of the Telegraph Battalion at Cape Town, had been carried out by Major Heath. It was agreed that the civil Telegraph Department would repair all damaged line in the Cape - the Boers had, by the end of October, caused considerable damage to lines near the Orange Free State border - but that was moderated by agreement that in the loosely defined 'fighting area' the army would be responsible for temporary repair until the area was deemed 'safe'. Subsequent events in the Cape were to mean that the army did much repair work to civil lines during the operations to relieve Kimberley and along the Cape Colony border while these were in the fighting area. After the capture of Bloemfontein, although the area within the Cape border was safe, the need for communications was urgent and the Cape Telegraph Department did not have sufficient resources to repair quickly enough, and a joint 'rapid repair' programme was put in hand. Also, the civil department continued operations in what became 'fighting areas' around Stormberg and Dordrecht where there were simply no army resources at all, and in Kimberley, where they were trapped and had little choice in the matter. In the early stage of the war there was much cooperation between the army and the Cape Telegraph Department in this rather ill-defined arrangement. When matters stabilised after the capture of Bloemfontein, and the Boers were out of the Cape Colony, things settled down in the agreed way. This arrangement, however, meant that the colonial telegraph departments, with their civilian staff, had no responsibility at any stage for the telegraph networks in the two Boer republics; that remained an army responsibility, albeit one that was some months away. Financial arrangements were also needed for the passage of military and civil traffic, which was all charged for.

* * * * * * *

Boer Army Communications

What of the Boer Army communications? The Boer army, both in the Transvaal and the Orange Free State, was mainly a reserve army, the *burghers* (citizens) being called up as necessary. They were organised into units called 'commandos'. These were based on regional loyalties, and the Boers were organised into town commandos. Completely unlike the British, and in keeping with their republican upbringing, it was a very democratic arrangement. Leaders were elected by the men, and if the men lost confidence in their leader he was 'deselected' and another chosen. Alternatively, the individual soldier would go to another commando of his choice, or more likely he would drift off home. Decisions at all levels were reached by what might be called a committee meeting – a *kriegsraad* (council of war) at the higher levels. The obvious weaknesses in command and discipline that emanated from these arrangements in a military environment did much to counteract the Boer's greater skills in such things as shooting, horsemanship, scouting, use of ground, and local knowledge.

There were, however, two regular, professional elements of the Boer Army – the Staatsartillerie and the Field Telegraph Department. The Field Telegraph Department was placed under command of the Staatsartillerie, but this was mainly a peacetime administrative arrangement because the telegraph department operated independently of the artillery in the field. Both benefited in efficiency from the training and discipline that this arrangement imposed, and both gave a good account of themselves in the war - at least up to the start of the guerrilla war phase, by which time all the Boer artillery had been captured and the telegraph taken over. Heliography became their main method of communication during the guerrilla war phase. (5)

Signallers of the Transvaal Field Telegraph Section with a cable cart. Lieutenant Paff is seated on the left, and standing next to him, with the heliograph, is Scheepers.

The Boer army communications department started in 1890 when General Joubert (to be their commander in Natal in the opening stages of the war), having noted the great use of the heliograph made by the British in the first Anglo-Boer war in 1881-82 (unfortunately little of this has been recorded by anybody on the British side), formed a heliograph corps. Almost immediately this was expanded into a telegraph corps. The commander of the Field Telegraph Department was Paul Constant Paff, who had come to South Africa in 1888 from the Amsterdam Telegraph Department, in response to a request from President Kruger for the services of an experienced telegraphist. The Field Telegraph Department was formally established by the *Volksraad* (parliament) in May 1890, at which point Paff's contract changed and he was commissioned as a Lieutenant. He then trained fifteen men in Morse telegraphy, heliograph, lamp and flag signalling. After the Jameson Raid in 1895, the strength of the Telegraph Department was increased to thirty-one men.

The young Lance Corporal Scheepers.

One of those in the Field Telegraph Department trained by Paff was Gideon Scheepers, a Transvaal farmer's son, who joined as an apprentice telegraphist in about 1894 at the age of sixteen. After about four years with the Transvaal forces, Scheepers transferred to the Orange Free State staatsartillerie, which was structured similarly to the Transvaal, to organise their telegraph and heliographic corps – quite a job for a twenty-year-old. In the period immediately before the war, and realising that they were likely to be overrun and lose the use of their telegraph system, Scheepers reconnoitred heliograph sites. The tops of certain wooded hills were cleared of trees, and heliograph communications trials between selected sites in the Orange Free State, and across to the border to sites in the Transvaal, were carried out. When, as he had foreseen, the telegraph systems of both states were overrun, and operations turned into a very mobile 'hit and run' guerrilla war, these sites were used by the Boers for their heliograph communications.

During the early part of the war Scheepers fought in the operations on the western front, fighting against the 1st Division under Lord Methuen and serving the Boer commanders there, Cronje and De Wet. Scheepers continued in his signalling role with De Wet for some time, but as the war progressed he dropped this and took part in the guerrilla war, operating in the eastern area of Cape Colony, encouraging Cape Afrikaners to join them, wrecking trains, and attacking the British. He was eventually captured, tried for war crimes that included the murder and ill-treatment of black prisoners, was found guilty, and was executed by a British firing squad at Graaff Reinet in the Eastern Cape in January 1902. (6)

At the outbreak of war, Lieutenant Paff and the Transvaal Field Telegraph Department supported the operations in Natal. They connected the laagers surrounding Ladysmith, and used the permanent telegraph system to maintain communication with Pretoria, although this took time as some ignorant Boers had themselves damaged the lines they later wanted to use for their own purposes during their advance into Natal.

As the war progressed, and as Scheepers had foreseen, telegraphy became increasingly difficult for the Boers when they lost control of much of their permanent telegraph lines and offices. More and more they used the lightweight and mobile heliograph, well suited to their style of operations, increasing their stock of instruments with those captured from the British in the early stages of the war! A captured heliograph was a valuable prize.

Having set the scene, the main body of the Telegraph Battalion on the high seas and approaching Cape Town and destined for the western front, the next chapter turns to the detachment of the Telegraph Battalion initially sent to Natal.

Endnotes

1. Described by the author in *Tales of the Telegraph*, covering army communications from the mid-19th century up to the start of the Boer War.

2. *Khaki Letters from My Colleagues in South Africa*. Edited by Colour Sergeant R. E. Kemp of the 24th Middlesex (P.O.) Rifle Volunteers. Published during the war. A reprint is available from The Naval and Military Press Ltd, in association with The National Army Museum.

3. ibid. pp 166, 168, 210.

4. Kennedy's diary is held in the archives of the Royal Signals.

5. Detailed descriptions of the formation of Boer army signals units and their operations in the war will be found in *Army Signals in South Africa* by Walter V. Volker, pp 14-40. Pub 2010, Veritas Books, Pretoria. ISBN 978-0-620-45344-8.

6. More about Gideon Scheepers and his signalling and other exploits during the war will be found in the book *Commandant Gideon Scheepers and the search for his grave* written by Taffy and David Shearing, pub 2002, Mills Litho, Maitland, Cape Town. ISBN 0-6202-4535-4.

Chapter 2

Operations in Natal, 1899

Operational Situation

When Lieutenant-General Sir George White arrived in Durban on 7 October 1899 the greater part of the Indian contingent had arrived and the rest were close behind. Even so, it was no easy task to defend the colony. His force was inferior in numbers and mobility, and the topography of northern Natal made it difficult to take up advanced positions that could not easily be outflanked by an invading army. He decided to base his force on Ladysmith and to abandon Natal north of the Biggarsberg. It was morally difficult to abandon more, because of the message it would send both to the Boers and to the agitated inhabitants of Natal. The Natal Ministry urged a defence well forward. On the other hand Buller, who knew Natal well (he had served there with distinction in the Zulu War of 1879, winning the Victoria Cross), had, prior to his departure from England, recommended defending on the line of the Tugela river, a natural barrier south of Ladysmith, and not exposing the force any further north.

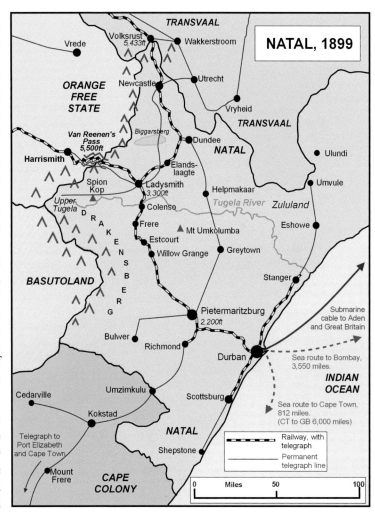

By 11 October, the day war broke out, General White was at Ladysmith. From the north the Transvaal forces under Joubert advanced on Dundee, and from the west the Free State forces entered Natal through Van Reenen's Pass. It soon became obvious that the Boers intended to encircle the British forces north of the Tugela. White, newly arrived and uncertain, wanted to withdraw the forward troops from Dundee; Major-General Penn-Symons, commander of the forces in Natal before the war, over-confident, and backed by the Natal Ministry, wanted to hold it. There were two battles against Transvaal Boer forces: firstly at Talana Hill, just north of Dundee, on 20 October, during which Penn-Symons was killed, and secondly at Elandslaagte on 21 October. There were tactical mistakes by both sides, and plenty of confusion, but generally the British came off best.

There were no military telegraph troops in Natal at this time. As a consequence of their delayed departure, on 20 September, for a war that was seen to be coming, and aboard a ship that created a record for slow passages from England to Durban via the Mediterranean, the army telegraph detachment did not arrive in Natal until 27 October. Communications up to that point were by means of the Natal civil telegraph system, operated by the telegraphists of the Postmaster-General's department. The contemporary sketch from the *Illustrated London News* on the next page shows Major-General French carrying out a reconnaissance beyond Elandslaagte in an armoured train, and reporting back by telegraph. Two civilian telegraphists accompanied General French on his reconnaissance and tapped into the telegraph line that ran beside the railway; the vital messages got through.

Continuing their advance, the Boer artillery started pounding Dundee. It was soon realised that the force at Dundee was isolated and that reinforcements could not be spared; Ladysmith itself was being threatened from the west by the Boer forces from the Orange Free State. Orders were given to withdraw from Dundee into Ladysmith. The town, and all the stores and wounded there, were abandoned. The withdrawal was made in appalling weather conditions - torrential, freezing rain. The exhausted soldiers and civilians seeking refuge arrived in Ladysmith on 26 October.

A correspondent wrote a letter to the *Telegraph Chronicle* describing these events as they affected the telegraph department:

> Natal has ample reason to be proud of her Telegraphists who are at present engaged in various arduous duties in different parts of the 'Garden Colony'. At most of the townships which have advisedly been evacuated – such as Charlestown, Newcastle, Dundee, and Colenso – the Telegraph officials have manfully and loyally remained at their posts until the very last moment, and were, in more than one instance, almost cut off by the Boers.. Mr A. E. Browning (Postmaster of Newcastle) and staff – Alderson (LV), McIsaac (GW), and three others – had an exciting time as they left in the last train, the Boers appearing on the hills as the engine steamed southwards towards Ladysmith. … [LV and GW were the telegraphic abbreviations for the offices the men had originally worked in before emigrating, Liverpool and Glasgow.]

20 October 1899. General French on reconnaissance towards Dundee by rail. Civilian telegraphists tapping the telegraph wire, enabling him to report back to Ladysmith. Sketch by war correspondent Melton Prior.

> The postmaster of Dundee (Mr. H. H. Paris) has already described in your columns the battle of Talana Hill and the death of General Sir Wm. Penn Symons. It was immediately after the battle that the Boers appeared on another prominent kopje accompanied by a forty-pounder. The British camp, which was then in range of the Boers' 'Long Tom' [artillery gun] had to be speedily evacuated. The town of Dundee being also under fire, the inhabitants were advised to move out to the new camp. Notwithstanding the danger, telegraph communication with Pietermaritzburg was still maintained, although it must have been performed with heroic self-sacrifice on the part of the Dundee staff, for upon each occasion the Telegraphist had to be accompanied by a mounted escort to the office.. At last the order came that all were to leave Dundee within ten minutes. … The writer remembers being at the PMB [Pietermaritzburg] end of the telegraph line sending to Dundee when the receiver would suddenly stop and say 'Oh, God, there's another shell just gone over the office!', but notwithstanding his imminent danger, he still kept at the instrument receiving important military messages for the General. A short time after we (PMB) received a regretful 'Goodbye'. …[1]

Boer forces followed the troops withdrawing from Dundee and on 30 October the battle of Lombard's Kop took place, to the east of Ladysmith. Soon afterwards, the confused and disastrous battle at Nicholson's Nek took place four miles north of Ladysmith (see map on next page). As a result 850 British troops surrendered to a small force of about 300 Boers, the infantry withdrew from Limit Hill, and the armoured train was also withdrawn into the town. Contributing to the confusion at Nicholson's Nek was the loss of the defending force's two heliographs, to be operated by regimental signallers, being carried on mules which panicked and stampeded as they advanced under cover of darkness to take up their defensive positions, leaving them without any means of communication and unable to report the situation or receive orders. A heliograph on Limit Hill failed to draw any response from Nicholson's Nek, and flag signals from Nicholson's Nek were not seen by anybody on Limit Hill or in Ladysmith. It seems strange that something as vital as the heliographs, portable and usually carried by the signallers, was entrusted to the temperament of mules - another example of the unprofessional bungling that was rife. As a result of the defeats at Lombard's Kop and Nicholson's Nek, the Boers were left in possession of the heights which dominated the north and east of Ladysmith.

OPERATIONS IN NATAL, 1899

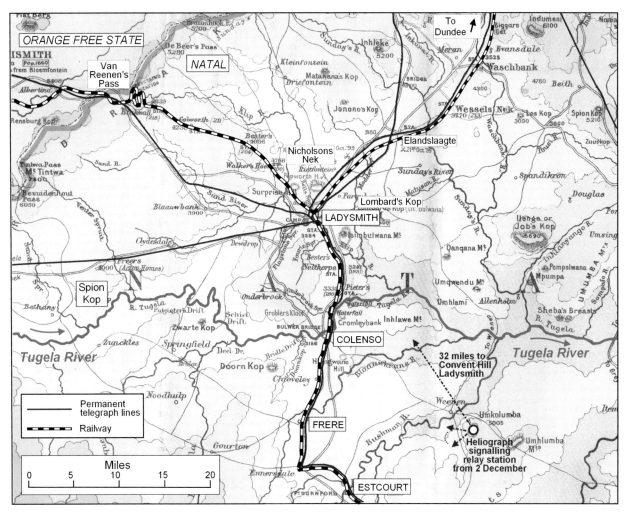

Ladysmith

Ladysmith, the third largest town in the colony in 1899, had been proclaimed a township in 1850, and named in honour of the Spanish wife of the Cape Governor, Sir Harry Smith. Sir Harry had gallantly rescued Juanita from his own pillaging troops when fighting at a previous siege in another continent – in 1812 at Badajoz, in the Peninsular War. Despite the pleasant image that the name might thus conjure up, Ladysmith was a hot, dusty, disease-prone town, surrounded by a circle of ridges and hills - a most unsuitable place to become entrapped. To stay there was a strategic error, but that is what happened. Buller had said to stay south of the Tugela, but he was now on the high seas en route to Cape Town, and thus out of communication. The Natal ministry and Penn-Symons wanted a defence further north but Penn-Symons had been killed at Talana. White was new to Natal. Amidst all this uncertainty and lack of firm command, it is not difficult to see how it all happened.

At the end of the 19th century Ladysmith owed its importance to the railway. The line from Durban reached there in 1886, and it was the junction for the line westwards through Van Reenen's Pass to Harrismith in the Orange Free State, and northwards over Laing's Nek (Pass) to the Transvaal at Volksrust, both lines climbing to an altitude of over 5,000 feet as they crossed the hilly Natal border to the tableland beyond. Up this line to Ladysmith from Durban and Pietermaritzburg were now being brought trainloads of troops and military stores of all sorts. General White had located his headquarters in the convent, on a prominent hill above the town. Convent Hill was, in a few weeks time, to become the site of their communications with the outside world. Churches and schools were requisitioned for military purposes; tents sprouted everywhere around the outskirts of the town. The soldiers gave names to some of the hills, replicating familiar names in the Aldershot training area such as Wagon Hill and Caesar's Camp.

Communications in Ladysmith

The first detachment of the Telegraph Battalion, which had left England on 20 September, arrived at Durban on 27 October. They left Durban by train that same day and reached Ladysmith on 29 October, joining the 13,000 troops assembling there, and were involved in the war immediately. The detachment consisted of Major W. F. Hawkins (then in command of the Telegraph Battalion), Captain R. H. H. Boys, attached to the Post Office but posted in on mobilisation, some senior NCOs of the battalion HQ, and No. 1 Section consisting of fifty-five men under the command of Lieutenant A. B. R. Hildebrand. Walter Hawkins had won the Sword of Honour at the Royal Military Academy, Woolwich, in 1875. He had subsequently served in the Nile expedition of 1884-85, had been an instructor at Chatham, and served a three-year tour with the Telegraph Battalion between 1886-89. On returning from a tour of duty in Singapore in 1896 he had taken over command of the 1st Division of the Telegraph Battalion. Reginald Boys, commissioned in 1886, served many tours of duty with the Telegraph Battalion, both before and after the Boer War. Arthur Hildebrand was commissioned in 1890, and had served in India with the Bombay Sappers and Miners for five years from 1893, before joining the Telegraph Battalion in 1898.

They brought with them fifteen miles of airline, eight miles of cable, and the equipment for three telegraph offices – a pitifully small amount it would seem. (Back in England there had been equipment problems, as described in the previous chapter.) Reserves of forty-five miles of airline, sixteen miles of cable and eight offices followed shortly after.

The telegraph section's first task was to provide communications for the defence of the town. On 30 October a telephone exchange was set up at the headquarters and the section started to connect the outposts by telephone. The existing telegraph line along the road from Ladysmith to the northwest to Dundee was used to run spurs off to the post office, the railway station, the armoured train, the observation balloon, and Limit Hill to the north east of the town, where the main body of infantry defending the town were located, so that they were all in communication. Working together with the army telegraph section was a small group of the Natal Telegraph Department's civilian staff who had stayed in Ladysmith. On 31 October, the telegraph section ran a line to Caesar's Camp about three miles to the south of the town, and on 2 November a line was run to King's Post, a sector headquarters two miles west of the town, and Cove Redoubt, an important defensive and artillery position nearby. After a few days all were connected by telephone, as shown in the diagram below. The defending troops also established numerous visual signalling links between their positions, operated by the regimental signallers – heliograph by day and lamp by night.

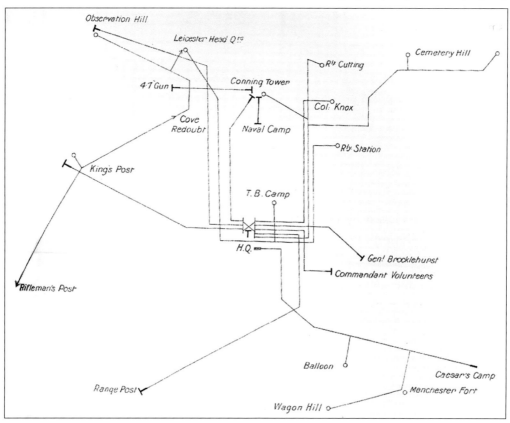

Local telephone communications in Ladysmith. Reproduced from the 'History of Telegraph Operations in the War in South Africa'.

The Conning Tower and Naval Guns

Of the outposts shown in the diagram one might ask, as an interesting little diversion, what was a naval observation post (locally known as 'the conning tower') and a naval gun position doing in Ladysmith? On 25 October General White had telegraphed Simonstown, the Royal Navy base near Cape Town, for assistance to reinforce his army artillery with heavier and longer-range naval guns. Within twenty-four hours the shore-based guns consisting of two 4.7-inch guns and four 12-pounders, together with a naval contingent of 280 men, had been loaded on to HMS *Powerful*. The ship sailed on the 26th, arriving at Durban on the 29th, unloaded straight away, and with great promptitude the guns reached Ladysmith the next day, the 30th, at 9:30 am. They arrived just as an attack on Ladysmith was taking place, with Boer artillery shelling the station. The guns were quickly unloaded from the train, and were in action immediately from the vicinity of the station. Their range, greater than the army's artillery, considerably surprised the Boers, who ceased fire.

Later on, the naval guns were directed by telephone communication with an observation balloon. The 2nd Balloon Section RE had reached Ladysmith just ahead of the naval guns. After the initial flurry, the two 4.7-inch naval guns were deployed in a number of emplacements and moved around, and used their superior range (about 8,000 yards) to counter-bombard the Boers. Targets were identified and the gunfire controlled by an observer in the captive balloon giving instructions to the gun position, improving considerably the accuracy and effectiveness of the guns, which of course now had limited ammunition available. In fact it was the gas for the balloons which ran out first, as well as the destruction by enemy fire of some of the balloons, which eventually curtailed this practice.

Telegraph Lines cut and Ladysmith Isolated

On 2 November, having already cut the telegraph lines to the north and west, the Boers cut the main civil telegraph line to the south. Ladymith was now isolated. No further telegraphic communications were possible until after the siege was lifted, but other means of communication with the besieged town were to be achieved.

Just before the telegraph line was cut General White had been able to telegraph General Buller, who arrived at Cape Town on 31 October, and during this exchange Buller acquiesced to White's plan to hang on in Ladysmith rather than abandon it. Nevertheless, it wrecked the plan of campaign for the advance of the main Army from Cape Town to Bloemfontein and Pretoria, and Buller had to divert troops to Natal. In accordance with Buller's telegraphed orders, General French and his staff left Ladysmith on 2 November in the last train before the railway line was torn up by the Boers later that same day. Although leaving his cavalry troops behind, against his wishes but overruled by General White, General French went on to conduct excellent operations near Colesberg while Ladysmith remained besieged. Another section of the Telegraph Battalion was to be involved with the Colesberg operation, an interesting diversion which will be described later (chapter 4).

Given the local communication facilities described above, the elderly General White directed the defence largely by telephone - in fact, he was much criticised by the units in Ladysmith for his 'bunker mentality', lack of aggressive action, and lack of personal visibility and leadership during the siege. He was also criticised for not sending out of Ladysmith before the siege began the non-combatant civil population and unwanted troops such as the cavalry and their horses who could perform no useful function during the siege; they were all to consume the dwindling resources of the besieged garrison and the horses contributed to the sanitary problems. Some of the horses were eventually themselves eaten for food, such did the situation become.

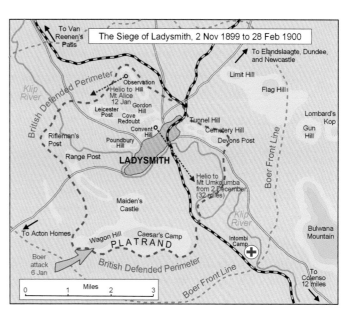

The Boers meanwhile had established their laagers (camps) around the town, some five miles distant. The Boer HQ was at Modderspruit, five miles north-east of Ladysmith, and all the laagers were connected by a telegraph system set up by the Boer field telegraph company, one of their few regular units, ably led by Lieutenant Paff. They were also in communication with the Orange Free State and Transvaal by telegraph using the permanent telegraph lines. Unfortunately, nobody on the British side seemed to think of going out and beating the Boers at their own game by wrecking these lines the Boers had taken over, although the lack of aggressive patrolling of any sort probably meant it was not even known about.

So it was that in early November 20,000 Boers surrounded the demoralised 13,000 British, but failed to exploit their immediate advantage. Speed and decisiveness were needed, but such characteristics and an unwillingness to attack and face losses were Boer weaknesses. Uninspiringly led by the aged and ill General Joubert, they were, like the elderly General White's conduct of the siege, too passive. During November and December they mounted a number of small attacks, but rather sporadically and unenthusiastically, providing nothing more than some temporary local excitement.

The Defensive Communications System

The telegraph section was kept constantly busy improving the network of defensive communications and repairing damage caused by Boer artillery gunfire and thunderstorms. The civilian Telegraph Department staff who had remained in Ladysmith worked jointly with the army telegraph section in providing the local communications. On one occasion two of them, Messrs. R. G. Honey and J. S. Hayburn, had an exciting experience. A number of shells fell near General White's house and one struck it, demolishing two rooms in the back. The telephone exchange, at which these men had been working until shortly before, was located in one of these rooms. The exchange was moved three times before what was considered to be a safe location was found.

The telegraphists of the Natal telegraph department taking part in the operations were named in a letter to the *Telegraph Chronicle* as follows.

> Besieged in Ladysmith: Messrs J. C. Burton (MR), W. McIsaac (GW), David Salmond (AB), J. O'Neill (West Australia), L. Bagley (BM), T. Alderson (LV), J. B. Craddock (CF), Acting Postmaster F. Hayburn (EH), and G. Joy (MR).

> With or near General Buller's relief column: A. R. Foster (YO), F. Banks (Cambridge), C. Neate (CF), W. Geraghty (CF), T. Phelan (DN), H. Rea (BM), Fitzgerald (West Australia), and R. T MacArthur (GW). [2]

These were mainly telegraphists who came to South Africa to join the Natal Post Office as civil servants. Their original telegraph offices are shown against their name using the two-letter telegraphic address – (MR) Manchester, (GW) Glasgow, (AB) Aberdeen, (BM) Birmingham, (LV) Liverpool, (CF) Cardiff, (EH) Edinburgh, (YO) York, and (DN) Dublin.

A high degree of vigilance was maintained throughout the siege to respond quickly to any Boer attacks, which were expected at any time. A signals officer always slept in the headquarters exchange, and linemen went out at night to repair faulty lines immediately, to keep the system in perfect working order.

Restoring Communications with Ladysmith

So much for the outline of the siege and communications within Ladysmith, but how were external communications to be established for what was to be seventeen weeks until relief arrived?

When the lines were cut there was an attempt to restore emergency telegraph communications. Captain Boys of the Telegraph Battalion records in his diary on Friday 1 December how a message was received, presumably carried by native runner, and was handed to Major Hawkins, the DAT, by the headquarters staff. It read:

> From C.R.E. Natal. Bunch all telegraph wires together and signal to us through them with Cardew Buzzer. We will try to communicate with you on phonopore at Frere. Also try wires separately. Make a beginning using number one and two railway telegraph wires. Connect them with departmental telegraph wires from one to four. Try every two hours commencing work at noon 1st December.

Captain Boys describes how the attempt to communicate was handed over to the civilian telegraph staff. The wires were cut at the railway station and they used a phonopore to try and communicate. A Cardew buzzer, the most sensitive instrument of all, able to work through the most unlikely conditions, was not available. (This is surprising as it was a standard equipment with the telegraph sections, but this section had come to Natal not well equipped.) The phonopore was an instrument used for railway signalling and combined a telephone and a telegraph sounder. As a number of these instruments were found in Ladysmith they were used in the defence communication system but were found to give trouble and often went out of working order. After a day or two without any success the experiment to restore rearward communications by telegraph was given up as a failure.

Meanwhile, a number of other methods were used to communicate: native runners, pigeon, searchlight, and eventually and most usefully, heliograph.

Pigeon

Shortly before the investment of Ladysmith General White had accepted sixteen carrier pigeons trained by a Mr Hirst of the Durban and Coast Poultry Club, who thought they might come in useful. Kept in Ladysmith, they were released from time to time, and carried messages describing conditions in Ladysmith back to their lofts in Durban. It was, of course, one-way traffic only. One of their early messages, on 9 November 1899, was congratulations from the beleaguered inhabitants to the Prince of Wales on the occasion of his birthday! Pleasant though such gestures may have been, something more effective in the way of communication was required.

Searchlight

The next attempt was to use searchlights. The origin of using a searchlight to communicate visually but without a direct line of sight appears to have first occurred many miles away from Natal in 1878. A Royal Engineers officer on board a Royal Navy flagship, describes how it happened:

> In 1878, the Admiral lay with four ships in the Dardanelles, the remainder of the fleet being in Xeros Bay on the other side of the peninsula of Gallipoli, which intervened and cut off all visual communication between the fleets. The distance between them was five or six miles.
>
> One cloudy evening, while experimenting with the lime light, a beam happened to be thrown upon a cloud which produced the effect of a lightning flash. At once the happy thought struck someone: why not signal by this means to the other ships? The 'preparative' was signalled, and repeated for about an hour, by which time the other vessel had got steam up and replied. Then was sent the message 'The Admiral will be pleased to see the Commodore at breakfast tomorrow'. The answer 'Thank you' was returned, and the Commodore duly appeared alongside for breakfast next morning.
>
> The moral of all this of course being that in cloudy weather, this means might be adopted for communicating between two forts or columns separated by a range of hills or other obstacle to visual signalling. [3]

The technique, in this initial case using the brilliancy of lime light, was thus discovered, but how was it done in Natal? The Royal Engineers, with considerable ingenuity, fitted up their apparatus in Fort Napier, Pietermaritzburg. It comprised a light projector, a steam engine, and a dynamo, which were obtained from a tug in Durban harbour, and a boiler from the public baths in Pietermaritzburg, the roof of the building having to be removed to extricate it. A shutter system was developed to go over the front of the projector, to enable the light beam to be interrupted in the manner of a heliostat and Morse code to be flashed, albeit slowly. The plant was mounted on two ox wagons for mobility. To steady the wagon carrying the projector, its wheels were run down sloping trenches until the body rested on the earth between them. A party of four civilians and a lance-corporal from the King's Royal Rifles who was a signaller were trained to flash signals into the clouds. The lower edge of the cloud was found to be the best place to direct the beam on. Trials were carried out in Pietermaritzburg and it was subsequently ascertained that these preliminary practices were seen and read at Estcourt and Ladysmith, distances of sixty and ninety-five miles respectively. The direction of Ladysmith had been accurately determined, and pickets set up the alignment. From 13 to 19 November the light was kept at work nightly. On 20 November it was ordered to be taken forward to Estcourt, then the most advanced post, to open up communication with Ladysmith.

The whole plant went up on two railway trucks, with a pilot railway engine in front, as the surrounding country was surrounded by Boers, and reached Estcourt at 9:30 pm. Everything was assembled, and the first message sent to Ladysmith thirty-two miles away that evening.

Estcourt was surrounded by Boers and cut off next day, so as a precaution the CRE asked the Royal Navy to make another searchlight. This second one was built in the railway workshops in Pietermaritzburg and was available on 30 November, crewed by two officers and five bluejackets. It was not as mobile as No 1 searchlight and was only able to operate from a railway wagon. By this time the battle of Willow Grange had been fought and won, and the Boers had retired from the Estcourt area and been driven back to Colenso. [4]

From then on, weather permitting, searchlight messages were sent to Ladysmith every night in Morse code. As a counter-measure, the Boers turned on their own searchlights (these were operated by the Boer telegraph corps), but despite their interference the signals were readable. Captain Arthur Crosby, of the Natal Carbineers, besieged in Ladysmith, kept a diary, and noted on

The searchlight, mounted on a train.

29 December that: "4 searchlights were working about us, 3 belonging to the Boers and one to the Relief Column signalling, the former playing on the latter in order to confuse the reading of the signals."

Within Ladysmith attempts were made to reply. A large parabolic mirror eight feet in diameter was constructed in a wooden framework lined by sheets of tin. Providing the source of illumination was the main problem. 360 battery cells were connected together, various lamps were tried including one used by the balloon section, and the signalling limelight was also tried. Unfortunately none of these methods produced sufficient illumination and the attempt was a failure. So searchlight communication with Ladysmith worked only in one direction, into Ladysmith but not out of it. In another besieged town, Kimberley, a similar method of communication was used and will be described in chapter 6.

South of the Tugela, both the British searchlights remained in the operational area and as well as signalling were used for work such as repairing damaged railway line at night. In the subsequent attempt to relieve Ladysmith No 1 searchlight was moved on its ox carts (two carts for the equipment, and two others with the coal and wood fuel for the boiler, together with the team of native drivers) up to the Spion Kop area during operations there, and was used to seek out enemy positions during night operations around Vaal Krantz. Later, after the Boers had been cleared from the area and the war had moved on, the plant was used in the Boer prisoner-of-war camp in Ladysmith to run electric light and boil water.

General Buller Arrives in Natal

As described in the previous chapter, General Buller, having arrived in Cape Town and assessed the situation, had decided to go to Natal. He arrived in Pietermaritzburg on the night of 25 November. During the three days that he had been in transit the war had progressed satisfactorily. Lieutenant-General Lord Methuen had won initial engagements and was advancing towards the Modder River; Major-General French, having left Ladysmith on 2 November and transferred to the Western Front, was pushing his patrols forward from Naauwpoort to protect the Cape border; and Major-General Gatacre was moving his troops up from Queenstown to protect the eastern Cape border. But the situation was soon to change.

In Natal, the invading Boers who had been around Estcourt and Mooi (pronounced 'Moy', Afrikaans for 'pretty') River, and were threatening Pietermaritzburg, had melted away in the face of increasing British troop numbers. The British forward headquarters moved on up the railway and telegraph line to Frere where a camp was set up for the continuing reinforcements. Buller himself reached Frere, only twenty-five miles from Ladysmith, on 5 December. It now seemed that it was only necessary with his superior forces to relieve Ladysmith, drive the Boers out of Natal, return to Cape Town, and resume the original plan. Such optimism was premature, for 'Black Week' was about to happen. In the eastern Cape Gatacre suffered a serious defeat at Stormberg on 10 December, and Methuen likewise at Magersfontein the next day - the first two disasters of Black Week. Little did Buller then know that it was he who was about to suffer the third. While that was incubating, a development occurred in communications - a heliograph link was established with Ladysmith.

Heliograph Communications Established

Ladysmith was screened from the main British reinforcement force located south of the Tugela river by the intervening range of hills known as the Tugela Heights, thus preventing direct heliograph communications. Although Ladysmith is mostly in a hollow, the area to the immediate north west of the town rises by just a few hundred feet, and two features were to become important for communications purposes, Convent Hill and Observation Hill (see map on page 16). Both were connected to the local defence telephone system, and both gave good long-range views out over the surrounding country, Convent Hill to the east and south-east, Observation Hill to the west. Nowadays modern housing covers these hills, and there is a water tower on Observation Hill, the highest point in the town, but the topography is otherwise unchanged and the limited scope for visual communications to the south of the Tugela river is still evident to anybody who cares to study it.

But there was one important way out of Ladysmith, to be exploited by Buller's divisional signalling officer, Captain John Cayzer, 7th Dragoon Guards. From Convent Hill to the south-east, between the two big features of Bulwana Mountain and the Platrand, there is a line of sight to a prominent high mountain to the south-east of Weenen - Mount Umkolumba, some thirty miles away. About 5,000 feet (1,500 metres) in altitude, and 1,700 feet higher than Convent Hill, Mount Umkolumba also dominates much of the ground south of the Tugela where Buller's forces were situated around places like Frere and Estcourt twenty miles away. It became the heliograph relay station into Ladysmith (see map on page 14). [5] Communications between Ladysmith and Umkolumba mountain were established on Saturday 2 December [6], one month after the telegraph lines had been cut. The heliograph flashes from Umkolumba were visible all along the ridge of Convent Hill in Ladysmith.

A photograph taken from Convent Hill in 2002, looking south-east through a gap in the hills towards Umkolumba in the distance - the line of sight for the heliograph link. The surrounding country is much as it used to be at the time of the war. The Platrand is rising on the right, and Bulwana on the left.

It was well-known that, even though the transmitted reflection formed a very narrow beam, interception of heliograph messages was possible, and care had to be taken. When the thirty-two mile heliograph link from Mount Umkolumba into Ladysmith was initially established, Captain Cayzer firstly wanted to ensure he was communicating with the right people. As it happened, he had a friend who was besieged in Ladysmith. He sent a message to the distant station, addressed to his friend, asking a personal question that only he would know the answer to - what was the name of his father's estate in Scotland? When the correct answer was promptly received as authentication, message traffic began!

General Buller was from then on in regular communication by heliograph with General White in Ladysmith, so each was aware of the other's situation. The heliograph interchanges could also of course be observed by any Boers who were in the path of the heliograph's narrow beam. However, both Buller and White had cipher code books and when appropriate the messages were in cipher. In those days encryption was a duty of the staff who originated the message. The signaller simply sent the message he was given. It would be a mixture of plain text in lower-case with the sensitive parts encrypted into four-letter cipher groups in upper-case, which he preceded and terminated with the signaller's abbreviated code <u>CC</u>, and transmitted without understanding what he was sending – hence the need for transmission accuracy. When receiving, he wrote the plain text in lower-case free hand and the cipher groups in upper-case. The staff at the receiving end were responsible for decryption of the upper-case.

The heliographic activity was watched with great daily interest by the anxious inhabitants of Ladysmith who no doubt, like the Boers, would also like to have read them. The heliograph link though Umkolumba became so solid that, when all military traffic had been passed, personal messages were permitted. Captain Arthur Crosby recorded in his diary on 14 December: "Put in a message to Nancy to be heliographed saying 'safe and well, acquaint generally.' This was through Brigade Office so hope it will go through."

The Battle of Colenso

On 15 December, Buller tried to breach the Boer defences, and the battle of Colenso took place. The details not relevant here, it was his first major battle against the Boers in his attempt to cross the Tugela river and relieve Ladysmith. Although it was difficult country to attack, the operation bore depressing hallmarks of other early Boer War battles – lack of proper reconnaissance, an unimaginative frontal attack, confusion arising from poor orders, poor tactics in the face of modern weaponry, and an underestimation of the enemy. Tactical signalling seems to have played little part in the battle, and rearward telegraph communications were provided by the colonial Telegraph Department from the permanent telegraph office at Frere.

Buller's force now had to reorganise and think again, and the dreary survival routine in Ladysmith was set to continue much longer than had been foreseen.

Wagon Hill

On 6 January 1900, after contemplating for over two months, during which many of the undisciplined *burghers* awarded themselves generous home leave for Christmas (some forgetting to return at all), the Boers launched their one and only major attack on Ladysmith. Their objective was the Platrand (flat ridge) feature to the south of Ladysmith. They attacked from the south of the ridge, firstly at Wagon Hill on the western end under cover of darkness at 2:45 am. The ridge was gallantly defended.

The Ladysmith telephone system came into its own and General White at his headquarters some two miles away was able to conduct some of the operation and order reinforcements to the scene. A further attack took place a little later at the eastern end of the ridge, near Caesar's Camp (where now stands a large microwave tower with many antenna dishes, but apart from that the Platrand area is largely undeveloped and remains much as it would have been then). The attack was pressed hard but the defenders resisted desperately – several VCs were won in the action. The sector commander, Major-General Ian Hamilton [7], organised the reinforcements and eventually the Boers were beaten back by the late afternoon, sixteen hours after launching the operation. It was a determined attack and a close run outcome. General White was to say afterwards that it was the telephone system which enabled rapid reinforcement to be brought to Wagon Hill and Caesar's Camp, and that perhaps saved the town – a turnaround of opinion about the telephone, considering the staff's usual aversion to the instrument. [8] As for the Boers, they became depressed and resolved not to attack again after that experience. Despite their many qualities as fighting men, close quarter combat against British bayonets was distinctly to be avoided; their well-rehearsed tactic was to fight then melt away into the veld before it came to bayonets, to fight again another day in another place – eminently sensible.

The Siege Continues

That attack having been defeated, the siege developed into a test of stamina. The defences were secure; Boer artillery bombardment continued (they fired some 20,000 shells into Ladysmith during the four-month siege); and the privations got worse - rations were reduced, sickness and disease increased, and morale declined while they waited. But it had its lighter moments. On the 101st day of the siege, the Boers on Bulwana Hill heliographed to the signallers at Caesar's Camp on the Platrand: "101 not out". To which the hard-bitten signallers of the Manchester Regiment – a regiment which had earlier, on 6 January, conducted a magnificent defence of their patch - replied: "Ladysmith still batting".

They carried on batting until 28th February, but before then many died. The records show that three members of the telegraph section died there from enteric fever. Many books and personal accounts have been written describing the appalling conditions. Nevertheless, relieved they eventually were, after yet another disastrous battle. A telegraph section, redeployed from the Western front, took part in these renewed operations and their story will be recounted in chapter 5. Meanwhile, the narrative returns to 12 November, with the main body of the Telegraph Battalion about to land at Cape Town and take part in operations on the Western front.

Endnotes

1. *The Telegraph Chronicle*, 16 Mar 1900, pp 123-124.
2. *The Telegraph Chronicle*, 2 February 1900, p 71.
3. *RE Journal*, 1 December 1880, p 272. Signalling between two 'invisible' stations.
4. *RE Journal*, 1 May 1901, p 90. The searchlight in Natal.
5. Technology and circumstances have moved on, and it is nowadays the site for an automatic VHF radio rebroadcast station for a local security network.
6. *The Times History of the War in South Africa* (Vol II, p 303) simply says 'the first few days of December'. Major G. F. Tatham of the Natal Carbineers, one of those besieged, kept a diary in which he notes on Saturday 2 December: 'Helio station fixed at Weenen on the Kolombo Mountain visible from Convent and all along that ridge'. *The Times History of the War* gives the date as 7 December (Vol II, p 547). Unfortunately research into the establishment and operation of the heliograph relay station has not yielded any further information. Captain Cayzer was a son of Sir Charles Cayzer, the shipping magnate.
7. Maj Gen Ian Hamilton was no stranger to Natal, being a survivor of the battle at Majuba Hill, a British defeat during the first Anglo-Boer war in 1881. He was to take a prominent part in the course of this war.
8. Captain Boys' diary, Saturday 6 January 1900. This diary is held in the Royal Signals archives.

Chapter 3

The Western Front

Arrival at Cape Town

The SS *Gascon*, with the main body of the Telegraph Battalion aboard, arrived in Cape Town at 2:00 pm on Sunday 12 November 1899. This group consisted of a headquarters and four sections, each section consisting of about fifty-five men commanded by a subaltern. They were equipped with sixty miles of airline, thirty-two miles of cable, and twelve offices. The Army Corps' reserve telegraph equipment – 195 miles of airline, 128 miles of cable, and thirty-two offices – followed from England soon afterwards, to augment the small scale they had brought with them.

They also brought with them three sets of Wheatstone automatic fast-speed telegraph equipment, and skilled operators of the Post Office were amongst the group of reservists who accompanied them. [1] The Wheatstone instruments were to prove invaluable; without them it would not have been possible to cope with the high level of telegraph traffic that was to be generated, even in the early stage of the war. Lieutenant W. C. Macfie RE, who had previously been employed on general duties in Cape Town, was now attached to the Telegraph Battalion and was put in charge of the Wheatstone equipment and its team of specialist operators.

Following Major Wrottesley's death during the voyage, they were now under the command of thirty-one year old Captain Edmund Godfrey-Faussett. Posted in as second-in-command on mobilisation, he was to rise admirably to the challenge ahead. Educated at Winchester College and trained at the Royal Military Academy, Woolwich, he had passed out top of his intake and was commissioned in February 1888. Between then and the start of the war he had served two tours with the Telegraph Battalion, firstly with the 1st Division from April 1891 to January 1894, and later, in 1898, with the 2nd Division. Promoted Captain in September 1898, he was an experienced and capable officer. He kept a diary throughout the war; anybody who reads it will soon see how it exudes his capability for level-headed thinking, organisation, and planning, and it has been used extensively in compiling this narrative. Like all formal diaries of the time it is 'politically correct'; it confines itself to a record of facts, and never a word of criticism about anything or anybody emerges, whatever his personal opinion might have been! [2]

Initial Deployment to De Aar

It was decided that the Telegraph Division would first be deployed to De Aar (see map on page 3), where the army was setting up a forward base. After disembarking and working through the night to unload the ship, the advance party set off on the afternoon of 13 November on the thirty-six hour, 500-mile rail journey to De Aar; the remainder followed on the 14th.

A letter dated 20 November from Post Office reservist Sapper Jenner, one of Colour Sergeant Kemp's regular correspondents, describes events:

> We landed at Cape Town on Sunday [12 November], feeling that we had all had enough of the sea after twenty-two days voyage ... and the latter part of the journey was very rough owing to the continuous strong wind. We got to work at once and were kept at it, with slight intervals for rest, till noon the next day, during which time I was employed in the lower hold, wheeling trucks and assisting in unloading the wagons – small items of 3½ tons. We then marched out to Green Point, at the foot of the mountains, where we encamped and remained till Tuesday night. ... The colonists are very enthusiastic over the soldiers; one man, a native of Chatham, who had made his pile, spent £4 9s 3d in whisky and beer for us seventeen men, so that we had a good damper before starting, which was very acceptable after our continued enforced abstinence, and then several Cape Telegraphists [many Cape telegraphists were former British Post Office telegraphists who had emigrated under a passage-paid scheme] were at the station, and they also 'turned it on', so that we were not without a drop on our long journey. We started at 10 p.m. with a fine send off, in a third-class carriage, four a side, which is not too much room considering that the railway is narrow gauge; but we were not going to consider this a hardship. ..... The next morning we were travelling through some lovely scenery, and early in the forenoon we commenced climbing a mountain range. We had two engines to draw us and reached the summit about noon, where we breakfasted. The scenery here was delightful, and the air splendid, with flowers in full bloom. We stopped at every big station for water, and reached De Aar after a 38 hours journey, which was a very enjoyable one. The inhabitants welcomed us enthusiastically as we passed by. [3]

Today De Aar is described in a guide book as 'an historic railway town, in the middle of nowhere, the most important junction in the whole of the South African railway system ..... well known for its extreme winter and summer temperatures ..... a deadly dull place in which to live'. De Aar, meaning 'the vein', was originally the name of a farm settlement established there in 1870 and possessing the scarce resource in that arid area of an underground watercourse, vital not only for farms but also for steam engines, and thus the railway that was now the transport artery for the advance.

As De Aar was likely to be the base for some time it was decided to form a store of all reserve equipment and to establish the clearing house (dealing with administration and money matters) near the store. Under the arrangements explained in chapter 1, it was the handover point for communications between the Cape Telegraph Department and the army - beyond De Aar was for the moment the 'fighting area', and so for the time being the army's responsibility. Captain Godfrey-Faussett, in command, obtained the use of an empty cottage belonging to the Cape Government Railways (CGR) for offices, and pitched four marquees for storing equipment. The CGR laid a railway siding into the storage area to facilitate loading and despatch of equipment by rail. Lieutenant Alexander Bannerman had been instructed to purchase a corrugated iron building of suitable size for the clearing house, and this duly arrived from Cape Town and was pitched alongside the cottage.

Next was a visit to the Remount and Transport Department, also based at De Aar. Mules and oxen were incompatible in terms of their daily cycle of work and grazing, and in their speed. Oxen were generally provided for wagon transport, being strong, but slow and plodding. They were unsuitable for the Telegraph Division's wagons which required horses or mules, despite the latter's temperamental character. Mules were eventually arranged.

The Field Force's advanced base at De Aar was some seventy-five miles from the Orange Free State border and, due to shortage of troops, was only weakly protected – described as "£1 million pound's worth of stores detrained on the open veld and protected by 900 rifles". [4] It was tactically difficult to defend and, being so close to the Orange Free State's border, was vulnerable to Boer attack, but fortunately none was made. It was at that time protected by a battalion of the Yorkshire Light Infantry, rushed from Mauritius. A Boer commando, advancing from Naauwpoort, came within sight of the base at De Aar on 20 November but did not attack.

Communications at De Aar

The first technical work undertaken by the Telegraph troops was to set up local communications at De Aar, and this was carried out by Lieutenant Dickie Jelf's section. A telephone switchboard was placed in a room at the railway station adjoining the postal and telegraph office. This room was later developed into the military telegraph office and was to become a large and busy centre for the operations in the area that followed. The

The voucher for wagons, men, and animals issued to the Telegraph Division at De Aar, date stamped Nov 13 1899. The original is held in the Royal Signals archives.

switchboard was connected by telephone lines to outposts and the railway signal box. Captain Godfrey-Faussett noted in his diary on Friday 17 November: "Telephones borrowed from Post Office until our reserve stores arrive. Very difficult working in sandy soil, and bad earths".

Continuing his letter of 20 November, Sapper Jenner wrote that:

> I am in charge of the camp telephone exchange, which is good 'biz'. Ruffell is in charge of the military telegraphs at the station ..... They are just through to Orange River [Orange River Station, about seventy miles further north]; all lines beyond that the Boers pulled completely down with teams of oxen. ..... The climate is lovely and warm, but at night when the sun goes down we get sandstorms, which are very unpleasant. Whirlwinds of sand, which go straight up to the clouds, travel along at a great rate. We are using teams of ten mules and twelve oxen for the wagons. It is a very picturesque sight to see the Cape boys driving the teams. ..... My health is better for the change, but I would not like to live here.

Another regular correspondent, Sapper Stimpson, also wrote from De Aar on 20 November:

> We have twelve airline wagons and four cable carts, with three sets of Wheatstone apparatus with us, but they do not think it will be a success on account of the lines. ..... De Aar is 500 miles from the Cape, and is right in the great Karoo desert, so you can guess 'it ain't no catch'. ..... The sand is simply choking and blinding; shall not be sorry to get away. All 24th [Middlesex RV] men are retained for the Headquarters Staff, so some of us will go up with Buller as soon as some of the troops get up country. The Scotch [sic] Brigade is mobilising here, consisting of the H.L.I., Seaforths, Black Watch, and the A. and S.H., and one could not wish to see a finer body of men – the blacks say it is wonderful to see women fight; but it is rumoured they are to leave their kilts behind as they make such a good target. General Wauchope is in command here and has daily [telegraph] conversation over the wire with Buller [then in Cape Town].

Preparing for the Advance to Kimberley

While the Telegraph Division were getting themselves established in dusty De Aar, the 1st Infantry Division under Lieutenant-General Lord Methuen was preparing to advance northwards. The troops arriving in Cape Town during October and November and were being rapidly despatched by rail northwards. Lord Methuen had set up his Divisional HQ at Orange River Station, seventy miles further up the railway from De Aar and close to the border with the Orange Free State. Lieutenant James Moir went forward from De Aar with his section, with instructions to accompany the 1st Division and to make such temporary repairs to the wires as time permitted during the advance, and to open offices as required. He arrived at the Orange River Station on the night of 18 November, and the following day opened the telegraph office there using one of the permanent lines back to De Aar.

On 20 November the 1st Division under Lord Methuen advanced with a force of two brigades and supporting troops, some 8,000 men. Their objective was the relief of Kimberley, seventy-seven miles further on from Orange River Station across mostly flat, featureless land, in some places interspersed with small kopjes or low-lying hills. The plan was to advance along the line of the railway for transport and communications reasons, although this was quite predictable to the Boers. The ineffectiveness of foot-slogging infantry in the vast spaces of the veld then dawned; the need was for mounted men, like the Boers. This basic problem, in a vast and then mostly undeveloped country like South Africa, was to be responsible for much apparent timidity of British strategy and lack of tactical success during the war. Mounted infantry were needed.

Moir, now that he was aware of the local conditions, realised that his mobility during the advance was going to be impaired by lack of water for the mules; there was none along the railway line. Apart from the animals, this lack of water in the arid area for the numbers of troops now deployed was to be a serious problem for all of them. The Orange River water was not fit for drinking, other sources of supply were few, and transportation was difficult and unhygienic, as was the design of the army water bottle. All this led increasingly to outbreaks of dysentery and death from enteric fever. Consequently Moir had to load twenty miles of airline equipment into a railway truck and arrange to move it with the RE parties repairing the railway line. All stores had to be carried forward from the truck by hand. By great physical exertion they managed to keep up with the advancing force and invariably opened an office for the HQ at night and cleared message traffic.

Advance to the Modder River

There was a small and successful battle at Belmont on 23 November, the first battle on that front. Captain

WESTERN FRONT, 1899

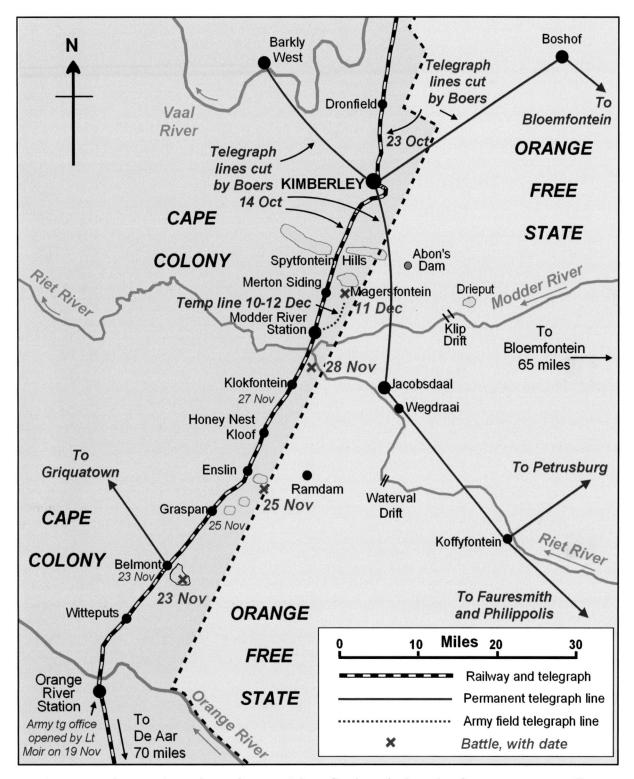

The area of operations from Orange River Station during the first attempt to relieve Kimberley, 14 October to 12 December 1899.

Godfrey-Faussett's diary notes that Lieutenant Moir opened the Belmont telegraph office at 10:00 that day, and they transmitted Lord Methuen's report of the battle to the GOC at 16:35, so they were well up with events. There followed another battle, at Graspan on 25 November, although the Boers, occupying the slopes of the kopjes, slipped away on their horses before they could be finally engaged and destroyed - a rearguard tactic at which they excelled throughout the war. Casualties were taken on both sides in each of these battles. The Boers, dispirited at the ease with which they had been dislodged, but inspired by a new leader, General De la Rey, decided to make a stand at the Modder river, and to employ new tactics.

WESTERN FRONT, 1899

Meanwhile, back at De Aar, Godfrey-Faussett was working up and exercising the other sections, many of the troops being reservists who needed this training. The usual method of keeping up with the advancing Divisional HQ was for a small detachment of a few men to lay cable from drums in the mule-drawn cable cart straight on to the ground; thus it was possible for the advancing HQ to keep in contact. That was quick, but of course the cable was vulnerable to damage from following wagons and supporting troops, and this frequently occurred. Line parties would follow behind erecting airline, and when this was done the cable was recovered and re-used later. While this was in progress the line would be part cable and part airline. When the location for the new telegraph office was reached, somewhere near the HQ, the office would be set up. As he exercised his sections, Godfrey-Faussett's diary for 25 November recorded a simulated six-mile advance:

> Used 1 cable cart for cable, this [6 miles] in 1 hour. 3 detachments had airline through in 4 hours (rather slow). Wheatstone took 10 minutes to be ready after waggon halted, working in bell tent with punchers [telegraphists preparing messages on punched tape] in the open. This time can be improved on, but marquee should be used, which takes all hands 12 minutes to pitch. Worked duplex at 190 [wpm] through 4 miles of airline and 2 [miles of] cable, and 300 [wpm] through 6 miles airline.

Sapper Jenner wrote on the same topic from De Aar on 22 November:

> At length, the advanced, or fighting, section has been detailed, causing extreme satisfaction or disappointment accordingly as the men have been selected for the advanced or relegated to the base section. We now expect to see some excitement and fighting, and it was with great pleasure that we who had staff or 'cushy' jobs gave them up to our more unfortunate comrades. We now practise pitching and striking the marquee which is to serve as an office at the front. We have a very efficient and able officer in Lieut. McPhee [Macfie], and under his tuition we are already able to pitch the office and get wires and instruments (Wheatstone duplex) fitted up in fifteen minutes, and strike and pack in wagon in eleven minutes. We expect to leave here Saturday, November 26th, and join the main Western Army at the Orange River. We shall, of course, go by train so far, after which we march with the army, all our equipment being loaded in general service wagons, drawn by a team of mules. … Merry Christmas and a Happy New Year to all. (5)

Sapper Pallet wrote in similar vein from De Aar on 25 November:

> On our journey here we saw a few ostriches and kraals [native villages]. ….. We stopped for food at large stations. A large station comprises a brick platform on one side only and two houses; the majority of stations seem to consist of a disused cattle truck for a signal box, and a signboard. De Aar is a place – a desert of sand surrounded by a few low hills. We eat the sand at every meal. ….. One good thing, there is a Soldiers' Home here, where one can get a good cup of tea for a penny and cakes at a like price. Another good thing, we get fresh meat and vegetables and other things, absolutely first class. Nobody has got drunk yet, for the simple reason that there is very little beer. ….. All 24th [Middlesex RV] men have been attached to the H.Q. Section – in fact, it comprises few others but them. Two days ago we heard that the H.Q. Section was to send some to the front – nearly everybody wanted to go. ….. The advance party were yesterday examined in punching and slipwriting [Wheatstone techniques]; all passed. The instruments were worked through a loop at Orange River, so we had a real wire, and attained a speed of 280 words per minute duplex. All our stuff is packed in a G.S. wagon with a team of twelve mules, and today it went out six or seven miles, and fitted up and worked a line to headquarters. We constantly expect attacks as the Boers are only fifteen miles away, and 6,000 of them reported. Yesterday forty Boer prisoners arrived at the station. They are a terribly dirty-looking lot. (6)

Lance-Corporal Nelson wrote from De Aar on 1 December, and after describing the train journey from Cape Town, a few promotions for soldiers of the 24th Middlesex RV, and domestic matters at the camp, he continues with a description of their preparations:

> The 24th [Middlesex RV] party has been divided into two 'Wheatstone' detachments – one for the front and the other for the base – a moving base it is believed will be the case. In addition to Sergeant Way, of EX [Exeter], Davis and myself are the non-coms. of the advanced party with Lieut. McFie in command. The sappers comprise Rory, Stimson, Horton, Pallet, Quinn, Alliston, Lang, Greenfield, Roberts, O'Sullivan, R. Poole, Woodrow, Jenner, Urquhart, and Weir. The remainder will form the base staff, the whole being termed H.Q. Section. Three out of the four working (airline and cable) sections have already moved away to the front, and the advance party of the H.Q. section are expecting to be moved forward at an hour's notice. We have had several rehearsals of our duties and movements. Troops for the Kimberley front have been passing through here in large numbers, and so also, I regret to say, have the wounded on the return journey to Cape Town. Reports of the fighting are very confused, even here, but the large numbers of wounded [from the battle at the Modder River] speak quite definitely enough of the severity of the conflicts. (7)

The experiments with the Wheatstone automatic were important. The volume of traffic had already built

> ### The Wheatstone Automatic System
>
> The highest speed at which a good telegraphist could send using a Morse key was about 35 words per minute, but he soon got tired and made mistakes, and time was then wasted in making corrections. With the volume of message traffic being experienced and the limited number of lines available, manual telegraphy could not cope. The Wheatstone was essential.
>
> The Wheatstone apparatus consisted of three parts - a perforator, a sender, and a receiver. It could send at between 200 and 400 words per minutes, depending on the line conditions. Before transmission the messages were prepared off-line by punching holes in a paper tape using a perforator, the pattern of holes representing letters of the alphabet. A number of telegraphists, or 'punchers', each with a perforator, could do this simultaneously. This was often done when lines were 'down', so that the messages were ready to be sent as soon as the lines were restored.. The prepared tapes were then fed into the transmitting part of the apparatus. The receiving part of the apparatus was an instrument similar to a Morse recorder, or 'inker'. The Wheatstone could be used in simplex or duplex circuits. It could thus send traffic at about six or seven times the rate of a hand speed device, but it was a heavy and rather delicate instrument, generally unsuitable for field work. Setting it up and operating it required special skills.

up quickly, described in Godfrey Faussett's diary as "work very heavy", and the high speed capability of this instrument was badly needed.

The Battle of Modder River

The advance of the 1st Division under Lord Methuen continued. The country in the northern part of the Cape where the troops were advancing was dry, barren, and rocky. There were a few natural obstacles ahead, particularly the Riet (reed) river and the Modder (muddy), a meandering river of no great size but deep and in parts with steep, unscaleable banks, and exactly as its Afrikaans name describes it. The water in both rivers was not potable. The two rivers joined near the Modder River railway station, where there was also a hotel, the *Crown and Royal*, which later became Methuen's headquarters, and some other buildings; it had been used as a weekend retreat for the workers in Kimberley. The Modder river was a deceptive position, and the Boers dug in. Conflicting intelligence from Kimberley told of the Boer intention to make a stand at Spytfontein, nearer to Kimberley. British scouting and reconnaissance was poor, as it was to be throughout the war, nearly always inferior to the Boers. Methuen was unaware of the nature of the ground at the river and of the Boer dispositions.

As the British approached the river, the well-concealed and protected Boers held their fire. The advancing British force was completely surprised by the ambush they walked into at 8:30 am on 28 November. In the ensuing battle, which lasted all day, tactical communications and command and control were poor. Whatever the shortcomings of the signalling arrangements may have been, the topography there is flat and featureless scrub for miles around, useless for visual signalling, and any flag-waving signaller who made himself prominent would have been quickly despatched by the accuracy of the Boer marksmen.

Eventually, by late afternoon, the left flank managed to cross the river at Rosmead drift (Rosmead is now called Ritchie) and turned the Boer positions. Had the British been better informed and better organised, the situation could have been exploited with greater advantage at the end of the day. But the Boers were beaten and retreated to Jacobsdal during the night. By the night of 28 November the Modder River Station was in British hands. Lord Methuen himself had been wounded late in the afternoon and was taken to the field hospital; his principal staff officer, Colonel Northcote, had been killed.

Telegraph to the Modder River

During this phase of the operation the telegraph section followed the advancing troops. The permanent telegraph line had been wrecked by the Boers in many places but they bridged the gaps with airline. This would usually have been a simple process, but in this hard and rocky area, typical of the karoo, it was difficult, with very little top soil and rock only a few inches below. 'Jumpering' – that is, making holes for the wooden poles by first driving in a strong metal spike with a sledgehammer - was difficult, and the jumpers were often damaged. Wooden stays were out of the question. Nevertheless, they progressed, and offices were opened at Witteputs on 21 November, Belmont on the 23rd, Graspan on the 25th, Enslin on the 26th, Klokfontein on the 27th, and at Modder River on the 28th. Godfrey-Faussett, still at De Aar, noted in his diary on Wednesday 29 November that: "Signals good to Modder River which opened today".

All these offices were to remain open for some months because Kimberley was going to take longer to relieve than was then imagined. Telephone exchanges were installed at Orange River and Modder River, each with six branches, connecting outposts.

It was soon realised that separate lines for army telegraph traffic and railway telegraph traffic was essential; the vibrator, despite its ability to use circuits of poor electrical quality, and thus often used by the field telegraph, interfered with the equipment used by the railway telegraph. Planning throughout the war always had to allow for a separate railway telegraph line. The Wheatstone also required its own telegraph line with no intermediate stations. [8]

In order to relieve Lieutenant Moir of responsibility for the lines and offices to his rear, Lieutenant Ernest Henrici and his section was detailed, on 21 November, once the advance began, to take over and maintain the lines and offices back to De Aar behind the advancing troops. On 7 December the Boers raided and wrecked the line near Enslin, and communications to Modder River were disrupted for a day-and-a-half.

Sapper Jenner wrote again in a letter describing the move of the advance Wheatstone party from De Aar to the front:

> The advance section paraded at De Aar at 8.15 a.m. on 4th December to go to the front. After loading our general service wagon and an airline wagon, we boarded the train at 9.30 a.m. After some shunting we left De Aar at 11 a.m. for the railhead; arrived at Kran Knil we began to feel the effects of the congested traffic on this single line system. We waited there an hour, and then in a siding four miles from Orange River we remained another two hours, finally arriving at Orange River about 7.00 p.m. (eight hours for seventy miles). Here I was appointed cook to the section, on my Aldershot reputation, with Quinn as assistant. We prepared the first meal, which was a great success, in the dark, and then slept comfortably on board (second class saloon) till 3.00 a.m., when we kindled a fire by the railside and prepared the matutinal meal, expecting to move off every moment. Anyway, we stayed on and on till dinner [lunch!] time, and then tea time, each meal being cooked in feverish haste, but we are still here – December 5th., 5 p.m. They are now making up the train. We are anxious to get off and see the battlefields at Belmont, Witteputs, and Graspan. We expect to get a bar [to the campaign medal] for the relief of Kimberley [rather anticipating the pace of events!] The scenery en route was fairly good, but the flat-topped mountains are rather monotonously numerous. We passed a great many ostrich farms on the way, and they, with horses and goats, seem to be the only source of revenue about here. [Sapper] Arundel came with us as far as Orange River, where he will be stationed till the Army has got right away. A lot of wounded English and Boers are here; the latter seem a very motley crew, and are composed of ragged, immature youths and old men – in fact they resemble a mob of tramps. I feel sorry for them sometimes, they seem so ignorant and misguided. The water taken from the river here (Orange River) is bad – quite yellow. The clear water has been condemned. Great slaughter took place near here and we can already smell the stench caused by the partly-buried bodies. We left Orange River at 8.30 p.m., and passed the night on the train under arms, as Boers had been seen in the vicinity. Next morning we detrained at Klokfontein as the train could get no further.

Lance-Corporal Nelson, a reserve telegraphist also from the Central London Telegraph Office and a member of the same group, elaborated on the scenes after they got off the train and marched the last stretch to the Modder River:

> It was somewhat trying, as the heat was great and we were in full marching order. The road was heavy with soft sand and dust. ... For several miles we travelled over the most recent battlefield [Modder River], and there was plenty of evidence about denoting the fact of the struggle; ruined farmhouses perforated with shot holes, dead horses and mules lying putrefying here and there, empty shell cases, and rude graves of the killed. ... One or two of our fellows were pretty well knocked up by the time Modder River was reached, and very glad all were when we had crossed the river by the pontoon bridge and halted at the railway station where we are presently quartered. ..... About 1.30 p.m. we were busily working to De Aar, endeavouring to dispose of a great accumulation of work, some of it dating back to December 2nd, and which had been brought safely past the Boers by special dispatch riders from Kimberley. ..... On Thursday the first train came safely across the new R.E. bridge, amidst cheers. The naval gun presents a fine sight with the thirty-two oxen necessary to get it along.

The telegraph traffic along the line between Modder River Station and De Aar was heavy. In addition to the army traffic, large batches of messages arrived every few days by messengers from Kimberley who managed to evade the Boer cordon - the Boers had cut all the telegraph lines from Kimberley as soon as the siege began. The cordon around Kimberley was rather loose, and messengers on foot were able to get in and out under cover of darknesss without too much difficulty, although any natives who were caught by the Boers were summarily killed, so they were well paid for the high risk they took.

The Wheatstone automatic was installed at Modder River Station on a through-line to De Aar. The foresight and initiative of Major Wrottesley, the unfortunate initial Director of Telegraphs, in getting this apparatus together with specially trained operators from the Post Office Reserve, was to pay off handsomely. Here, on 6 December, between the commandeered first-class ladies' waiting rooms at both Modder River station and De Aar where it had been installed, the Wheatstone was used for the first time in war.

Sapper Alliston one of the party, wrote on 11 December, saying:

> The authorities are very pleased with the Wheatstone experiment - it is a regular take down for the Colony [Cape Telegraph Department] as they were most certain we would not be able to do anything, they having failed on several occasions. (9)

The country in this area is today still much as it would have been then. The scrub-covered ground between the Modder and Riet rivers, and the surrounding area, remains flat and featureless – easy to get disoriented and lost, and abominable for visual signalling so it is not surprising that tactical communications were poor. The wartime Modder River bridge is still there but not used, the railway line having been slightly realigned and going over a new bridge alongside. There is an old blockhouse which later during the war guarded the strategically important bridge, there are some war memorials nearby, and there is still a small township around the Modder River station, including the rather delapidated *Crown and Royal* hotel, but the first-class ladies waiting room, where the all-important Wheatstone automatic burst into military operational life, is no more.

The Modder River railway bridge.

This photograph was taken in 2002. The modern bridge shown has been built alongside the old bridge. The blockhouse beyond the railway line is the remains of the original structure built to guard this vital strategic point. Modder River Station, where the Wheatstone telegraph equipment was located in the commandeered Ladies Waiting Room and first used in war on 6 December 1899, is a few hundred yards over the bridge. The station itself has been rebuilt - but the nearby Rose and Crown pub, looking a bit dilapidated, is still there.

Mackworth's Section Detached

Meanwhile, although Kimberley was still the objective, Boer incursions were taking place in the eastern Cape Colony, and Lieutenant Harry Mackworth's section, still at De Aar, was detached on 29 November, the day after the Modder River battle, to provide telegraph support to operations being conducted there by Major-General French who it will be recalled had escaped from Ladysmith at the last moment, to be redeployed for this task. The next chapter describes this interesting diversion.

Thus at the end of November the four sections were deployed, one (Jelf) at De Aar setting up and operating local communications, two (Moir and Henrici) supporting Lord Methuen's operations in the advance to

Kimberley and spread out between Orange River Station and the Modder River, and the remaining one (Mackworth) detached and supporting General French. A fifth section, the one sent out first in September to Natal, remained locked up in Ladysmith. The Telegraph Battalion's manpower was insufficient – it was a ridiculously small number in proportion to the total number of troops already deployed in South Africa and the vital task they did. It was already realised that there were not enough of them, and reinforcements were on the way.

Communications with Kimberley

All telegraph communication with Kimberley had been cut by the Boers at the start of the war. (The situation in Kimberley will be described in chapter 6.) Back in the area south of the Modder River, initial searchlight communications with Kimberley began on the night of 26 November when the advancing troops reached Enslin. Unlike Ladysmith, this was two-way communication. The advancing troops used a searchlight mounted on a train, and the Kimberley garrison had numerous searchlights which had been used before the war for mining purposes, now used as part of their defence system. As at Ladysmith, the light was reflected off the cloud base. It was dependent on weather, and even at best was slow. A coded searchlight signal on 4 December from the commander at Kimberley (Lieutenant-Colonel Kekewich), in reply to a signal from Lord Methuen, told that Kimberley had enough food to be able to hold on for another forty days.

The searchlight, and a 12-pounder naval gun, mounted on a train, used to signal to Kimberley.

The Battle of Magersfontein

While the British reinforced and consolidated after the Modder River battle, the defeated Boers entrenched at the next defensive barrier on the way to Kimberley, Magersfontein Kopje. Throughout this period of build up, telegraph traffic levels at Modder River Station and the intermediate telegraph offices were high.

Methuen, recovered from his wound, unimaginatively decided on another frontal assault on the Boer defences at Magersfontein. Again, unbelievably, because of lack of reconnaissance, he did not know the enemy disposition or strength. It was assumed, as before at Belmont and Graspan, they would be on the high ground. In fact the Boers were well entrenched and camouflaged at the base of the hill; they had already decided that occupying the high ground only made them conspicuous targets for accurate British artillery bombardment. A preliminary heavy bombardment of the hill the previous evening, 10 December, made an impressive noise but did little actual damage, and served to warn of the impending attack. On the same day, a temporary telegraph line had been run from Modder River towards the Magersfontein position - another 'give away'. The night approach march in bad weather failed to keep up to time, and the soldiers were not deployed into battle formation soon enough, advancing over the last stretch in daylight. The Boers waited until they were some 400 yards away, and then released a murderous fire from their well-protected positions. On the utterly flat and featureless veld, covered by low scrub, the British soldiers of the Highland Brigade were pinned to the ground all day, protected only by the guns of the artillery. Their commander, Major-General A. J. Wauchope, was killed early in the battle. Late in the day, 11 December, the soldiers retreated in some disarray. There had been heavy British casualties – 23 officers and 182 soldiers killed, 45 officers and 645 soldiers wounded, 76 men missing or captured. Next day the Royal Engineers reconnaissance balloon rose in the sky, and telephoned down the news that the Boers were still in occupation of Magersfontein Kopje.

Lord Methuen's headquarters during the battle was at Headquarter Hill, something of a misnomer for it is a scarcely perceptible knoll rising no more than a few feet above the otherwise flat and open landscape, some two miles behind the front line. The telegraph line was in use until 12 December, and was used during the battle, but presumably only for administrative purposes such as casualty evacuation.

After the battle, telegraph traffic with casualty lists was extremely high and was cleared efficiently from Modder River Station with the Wheatstone automatic. The casualty list of some 6,000 words was sent within 1¾ hours of its receipt.

Lance-Corporal Nelson, at the Modder River Station telegraph office, wrote on 15 December:

> ... The work is pretty hard when a heavy batch arrives by special dispatch rider [from Kimberley]. The disastrous battle of Magersfontein caused a big rush of both Press and commercial telegrams and the twenty-seven long pages of casualties which we sent on via De Aar ended a long and arduous day on the 13th. ..... The advance party is divided into two 'watches', one under myself and one under Lance-Corporal Davis. Our hours of duty run as follows: Nelson's staff, midnight to 8 a.m. ; Davis's, 8 a.m. to noon; Nelson's, noon to midnight; Davis's midnight to 8 a.m. ; and so on.
>
> The incident of the casualty list is worth referring to, as it has opened the eyes of the military authorities to the value of Wheatstone working, and caused great satisfaction. ..... The character of the work being so important, all the sad particulars were treated with extreme caution, so as to be accurately rendered; but by 11.16 p.m. the whole of the 6,126 words – counting groups of figures as one, for it was a military commercial message – had been run through to De Aar, to the great credit of the six punchers engaged. Only one repetition was required, and that was caused by a broken slip. It was subsequently ascertained that the whole message was written up by our colleagues at De Aar and handed over to the Civil Telegraph for transmission to Cape Town by midnight.
>
> On the evening of the 12th and all day on the 13th the wounded and dead were coming in on the ambulance wagons, and the horrors of war time were and are still painfully in evidence. As far as office duties permitted our own fellows put in a lot of time, voluntarily rendering as much assistance as they could to the poor wounded men. [10]

After the serious reverse at Magersfontein on 11 December operations on that front came to a halt for some time. The objective, the relief of Kimberley, was not to be achieved until the middle of February. There followed a period of 'phoney war' at Modder River while reinforcements were assembled. The telegraphists were not overwhelmed with traffic as they had been, except when batches arrived from Kimberley.

Changes in the Higher Command

When the war was seen to be faltering, as it now certainly was, the Government at home became uneasy. Field-Marshal Lord Roberts, veteran of campaigns over many years with the Indian Army culminating as its Commander-in-Chief, and presently in Ireland having been brought home and 'put out to grass' prior to retirement, was brought to a meeting of Ministers at Lansdowne House, London on Sunday 17 December. He had for some time been itching to go to the scene of operations in South Africa and sort things out. Before the meeting started, he heard of the death of his only son at the battle of Colenso, near Ladysmith, on 15 December. At the meeting he was asked if he would accept command. After a few minutes consideration he did, and that same evening was appointed Commander-in-Chief of the South African Field Force, superseding General Buller who was to remain in Natal and give that front his undivided attention. Lord Roberts then selected Major-General Lord Kitchener of Khartoum as his Chief-of-Staff, thus ensuring the upper house was to be well represented in the forefront of war. Under a week later, on 23 December, after selecting other staff for his new headquarters, Roberts sailed from Southampton, to arrive in Cape Town on 10 January 1900, just under three months after the war had started.

Fd Marshal Lord Roberts.   *Maj Gen Lord Kitchener.*

Kitchener was in Khartoum when his appointment was telegraphed there on 17 December. Like most well-trained soldiers, he knew how to pack his bags quickly. He left Khartoum the next day, first by gunboat to railhead at Halfaya, then by train to Alexandria. There he went aboard HMS *Isis*; at Malta he transshipped to HMS *Dido* for a rough passage to Gibraltar; and there on 27 December he joined up with Lord Roberts on the *Dunottar Castle* on its way to Cape Town.

Along with these changes, further troop reinforcements were sent, many of them of poor quality as resources were now drained.

One Telegraph Section transferred to Natal

General Sir Redvers Buller, in Natal, and now appointed the commander there, was drawing up fresh plans for the renewed efforts to relieve Ladysmith. With the original section of the Telegraph Battalion still besieged in Ladysmith, he desperately needed field telegraph communications. On 20 December orders were received to prepare a telegraph section to transfer to Natal, in support of Buller's renewed offensive. Lieutenant Jelf's section, still at De Aar, was assigned to this new task, and on 23 December they departed for Natal. Their story in support of the tortuous operations to relieve Ladysmith will be recounted in chapter 5. This left only three sections at the Western Front, with one of those, Mackworth's, still detached at Colesberg.

Christmas at the Modder River

Christmas at the Modder River was described by Sergeant Nelson, one of the telegraphists from the 24th Middlesex RV:

> Neither carols, waits, nor peals of bells broke the stillness of the camp to usher in the Christmas of 1899, the only sounds to break the utter silence being the occasional challenges of the sentries and the sounder clicks and the buzz of the vibrators. ... It was decided to combine our forces with those of No 3 section in the formation of a grand Christmas dinner party. At 6 p.m. prompt our office was closed, and the party of twenty trooped across to No 3's quarters. Here we found all ready for the repast, and down we sat, shoulder to shoulder, and soon the rattling of dining utensils and the clashing of elbows evidenced the commencement of operations. Unfortunately the Christmas pudding gifts from England for the troops had been miscirculated to Natal, and it was only due to the kindness of Lieut. Moir – who supplied not only a private pudding, but also the spirit illuminant – that our Christmas feast was blessed with a taste of the season's delicacy at all. ... A rough and ready smoking concert followed and although the ever-present dust and lack of lubricants made the top notes somewhat difficult of attainment, general enjoyment prevailed until 9 p.m., when song and jest were abruptly terminated by camp regulations. [11]

Sapper Jenner wrote to Colour Sergeant Kemp, thanking him for all the Christmas cards he had sent, saying that they were all going to send them home "so that they might be preserved as a pleasing memento of the campaign". He also described the working of the telegraph:

> We have buzzer circuits: De Aar to Orange River, and Orange River to Modder - with intermediate stations at Belmont, Graspan, Witteputs, and Enslin. The Wheatstone is a great success, but we find we cannot tee a buzzer on to a line connecting WX [Wheatstone] apparatus, as is done with the Morse [sounder]. Terribly hot here ... [12]

The Wheatstone always required its own dedicated 'through' wire to work, and this later became a problem during the advance to Pretoria because of shortage of wires, the need also to communicate with intermediate stations, and the need for separate railway operating wires along the telegraph lines that ran beside the railway. The more common instruments, the Morse sounder and the buzzer (vibrating sounder) could co-exist on a single line provided that the buzzer was fitted with a 'separator' - essentially a condenser as it was then called (capacitor) - which acted as a block to the direct current of the sounder and thus prevented interference caused by the buzzer. One advantage of the Wheatstone was that its traffic, passed at very high speed, could not be be read by any intercepting Boer operators 'tee-ing in' to the line, who had nothing to match it. There was in fact no evidence of any organised Boer telegraph intercept operation during the war, although the occasional random message may have been read.

Telegraph Reinforcements

The scale of the telegraph undermanning problem had been realised in November and back in England further reinforcements had been mustered. *The Telegraph Chronicle* reported:

> ... a much larger number [than expected, of Post Office volunteers] received the summons on Friday, November 24th. Next day they were medically examined, and on Monday November 27th, they proceeded to Aldershot. ... Most of the provincial men journeyed to the 'Shot' independently, but the C.T.O. [Central Telegraph Office, London] men and a few others left Waterloo by the 2.45 p.m. train, where a very large gathering of friends and relatives assembled to give them a hearty send off. Special compartments were procured for them, and as the train slowly steamed away more than one of the ladies present were compelled to give way to their feelings. [13]

This party of two officers and 106 soldiers (consisting of about fifty linemen and fifty telegraphists, including thirty-six Post Office Telegraph reservists from the Central Telegraph Office and provincial offices) left Aldershot, where they had assembled, and travelled to Liverpool, where they embarked on the SS *Canada* at Liverpool on 1 December 1899. The party was headed up by Captain J. S. Fowler, DSO [14] and Second Lieutenant A. W. Hepper. Amongst the reservists of the 24th Middlesex was CSM W. G. Tee from the Central Telegraph Office, London. Private Tee had served in the Army Postal Corps in Egypt in the campaign of 1882 (when 'C' Troop, forerunner of the Telegraph Battalion, had been present); Corporal Tee had again served in Egypt as a telegraph reservist in 1884 (the Gordon Relief Expedition), and in September of that year he was Mentioned in Despatches. Now, as CSM Tee, this veteran reservist was again answering the call.

Tee wrote a letter just before their departure:

> ... We go on board the 'Canada' about 2 p.m. There are about 70 T.B.'s [soldiers of the Telegraph Battalion] besides us, with Captain Fowler of Afridi fame, in command. ... All are fit and well and eager for foreign service. Our photos were taken ... I think we are ready now, and hope again to do our duty for the Old Country. [15]

Sapper G. W. Bannister, one of the reservists, also wrote an informative letter just before departure:

> ... It will be interesting for friends to know the kit we have received. It is really a new thing, and is different to that issued to the first detachment. A suit of brown serge (like khaki in colour) for warmth, scarlet serge, trousers and field service cap, two pairs of boots (to be kept brown), two pairs pants, two good quality flannel shirts, two pairs of socks, braces, Jersey, hair brush, clothes brush, towel, razor, jack-knife and lanyard, hold-all, housewife (needles, cotton, worsted, etc) and a waterproof kitbag. Today we received our khaki and sea kit, including an additional bag, two cholera belts, Balaclava, night cap, canvas shoes, pugaree, puttees and foreign service helmet, also a Slade-Wallace equipment, great coat, carbine and sword. ... The 'Canada' is a fine boat belonging to the Dominion Line. .. We have no cleaning to do – all straps are buff, and buttons are to be kept dull. ... Look out, MR! [Manchester Telegraph Office, where he came from]. [16]

This group arrived in Cape Town on 21 December, to form what was called the 'permanent line party'. Supplemented by local native labour, their role was to repair and maintain the permanent telegraph lines behind the army – their vital line of communication - as it advanced through Cape Colony and later the Orange Free State to Pretoria. Most of the lines were destroyed by the Boers as they retreated, so they were kept busy. More about all that later.

Their arrival brought the Telegraph Division's strength by the end of December 1899 up to thirteen officers and some 390 soldiers, of whom three officers and fifty-five soldiers were still locked up in Ladysmith. It appears that they did not arrive in the operational area until Saturday 20 January. [17]

Godfrey-Faussett's diary notes that on 20 January Lieutenant-Colonel Hippisley, Captain Fowler, and he discussed and agreed the arrangements for handing over lines and future responsibilities. Hippisley, the Director of Army Telegraphs, was still based in Cape Town, where he was responsible for staffwork involving liaison, strategic planning, and obtaining the necessary resources, leaving Captain Godfrey-Faussett responsible for liaison and close support to the operational commander and day-to-day executive control of the sections. Fowler became responsible for the permanent lines and terminal offices, and his permanent line party took over the lines and telegraph offices between De Aar and Modder River, and between Naauwpoort and Rensburg, thus freeing the telegraph sections for field work more directly in support of the troops as they prepared for the advance into the Orange Free State. The successful rehabilitation and use of these permanent lines was to be vital, as they provided greater capacity and were far superior in robustness and transmission quality than the tactical airlines or field cable hastily constructed by the sections of the Telegraph Division.

Wireless

The move to South Africa of the Marconi experimental wireless detachment was mentioned earlier (chapter 1). They were deployed to De Aar in early December, and from there sets were sent to Orange River, Belmont, Enslin, and Modder River, a total distance of some seventy miles, with the intention of establishing wireless communication between them. For six weeks they attempted to communicate, but without much success. The range of these early experimental wireless sets was generally too small for South African distances. Using groundwave propagation, they needed to use long vertical wires as antennas, and there were problems with masts and earthing in the semi-desert of the karoo. (This problem also affected telegraphy, cable and airline both using earth return circuits.) Another problem affecting the wireless was the fierce electrical storms and accompanying lightning that was locally prevalent, exceeding anything experienced in England, causing damage to the equipment. The wireless experiment failed and on 12 February 1900 Lieutenant-Colonel Hippisley ordered the wireless equipment to be returned to Cape Town. Captain John Kennedy, the officer in charge of the experimental team, was redeployed to become the ADAT in Natal, his presence in that role being urgently needed for reasons that will be explained later. The wireless equipment was transferred to the Royal Navy where it proved to be much more successful, being used for some time by ships around Delagoa Bay. Range was much greater, due to propagation over sea rather than very dry land. The equipment stayed in service with the Royal Navy until November, and was then withdrawn. It is somewhat academic to say that this was the first occasion that wireless was used in war. It actually made no practical contribution to army communications. [18]

Endnotes

1. *The History of the Telegraph Operations during the War in South Africa* states that two Wheatstone equipments were brought. A letter by Sapper Stimpson, one of the Post Office Wheatstone operators, written on 20 November and reproduced in *The Telegraph Chronicle,* states there were three. The latter, closer to the subject, is probably correct, and it makes more sense (two to work a circuit, plus one in reserve or for movement). Whatever the correct number initially, more Wheatstones arrived later after it proved to be so successful.

2. The diary is held in the archives of the Royal Signals, along with various other documents he wrote. There are diaries by several other officers in the same archives, in similar vein. It would appear that officer cadets at the R.M.A. Woolwich were trained to record events in that way.

3. *The Telegraph Chronicle*, 22 December 1899, p 3.

4. *The Times History of the War in South Africa*, Vol I, p 136.

5. *The Telegraph Chronicle,* 5 January 1900. p 30.

6. *The Telegraph Chronicle,* 5 January 1900, p 31.

7. *The Telegraph Chronicle,* 5 January 1900, p 32.

8. Railway telegraphs during the war were run by Captain M.G.M. Bowman-Manifold RE under the Director of Railways, Major Percy Girouard RE. Graham Bowman-Manifold had been Director of Telegraphs in the operation to reconquer the Sudan between 1986-98, so was well experienced. The Sudan expedition had been commanded by Kitchener, and Percy Girouard had been Director of Railways.

9. *The Telegraph Chronicle*, 19 January 1900, p46.

10. *The Telegraph Chronicle,* 19 January 1900, p 45.

11. *The Telegraph Chronicle,* 16 February 1900, p 86.

12. *The Telegraph Chronicle,* 16 February 1900, p 86.

13. *The Telegraph Chronicle,* 8 December 1899, p 202. Names and telegraph offices are shown.

14. Already the holder of the DSO as a result of service in India in the Chitral Expedition - a very interesting and well-deserved award - and later to be Lieutenant-General Sir John Fowler, KCB, KCMG, DSO, the first Colonel Commandant of the Royal Corps of Signals, from 1923 to 1934.

15. *The Telegraph Chronicle,* 8 December 1899, p 202.

16. Ibid. p 202.

17. There is some discrepancy between *The History of the Telegraph Operations* and Godfrey-Faussett's diary. The former gives details of arrival date and number as given above; the latter states that 72 men arrived in the operational area on 20th January. It is possible, though inexplicable, that they stayed in Cape Town for a month, and some of the men may have been diverted elsewhere, perhaps to reinforce Jelf's Section although no mention is made of this and it seems unlikely.

18. *Wireless Telegraphy during the Anglo-Boer War of 1899-1902*. Military History Journal, Dec 1998, Vol 11, No2. See also *Wireless in the Boer War*, Journal of the Royal Signals Institution, Spring 2004.

Chapter 4

Communications at Colesberg

The Situation in Eastern Cape Colony

The eastern Cape Colony, from De Aar eastwards through Naauwpoort, Stormberg, and Dordrecht to the Basutoland (Lesotho) border, adjoined the southern part of the Orange Free State (see map below). The majority of the white population in that part of the colony were of Dutch extraction and sympathetic to the Boers. Indeed there were fears that they, too, would rise up and fight the British. Although this never happened it was always a threat, and they supported the Boer commandos from the Orange Free State in every way they practically could, short of actually fighting with them.

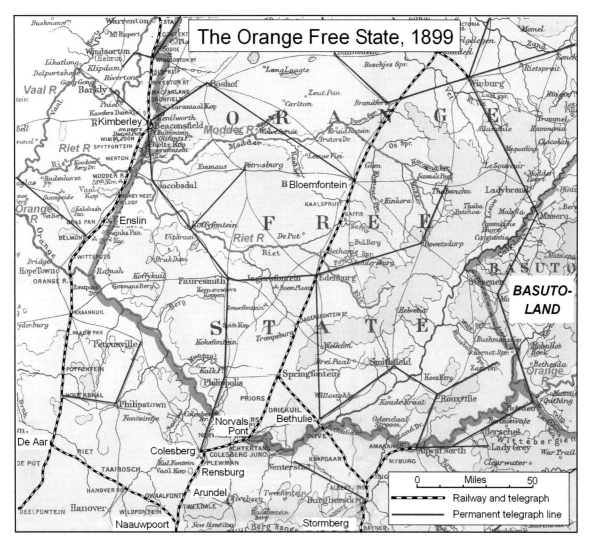

From early October, small groups of British soldiers - all that were available - were sent as quite inadequate garrisons protecting the strategic towns along the railway line. For three weeks after the war started nothing happened, until on 1 November the Free State Boers seized the important railway bridge at Norval's Pont and others crossed the river at Bethulie and threatened the lightly defended towns of Naauwpoort and Stormberg. Fortunately telegraph lines had not been destroyed and communications over the civil system were maintained by the Cape Telegraph Department.

On 3 November, fearful for the security of those manning these inadequate defences, General Buller, who had arrived in Cape Town on 31 October, ordered the thinly dispersed troops to retire to De Aar and Queenstown, leaving the border area undefended. Fortunately the hesitant Free State Boers did not take immediate advantage of this opportunity. Over the next two weeks they wrecked a series of bridges and railways, while the British also made a number of protective demolitions. Then, under General Schoeman, a force of 2,000 Boers headed unopposed towards Naauwpoort but failed fully to exploit the situation caused by the withdrawal of the garrisons. After a leisurely advance they arrived in the general area of Colesberg, remaining there in a state of inactivity for some time.

To deal with this situation, a British force was being hastily assembled from various reinforcements as they arrived. Placed in command of part of this force, and assigned to an area around Colesberg and Naauwpoort, was Major-General French, who it will be recalled had left Ladysmith in Natal on 2 November with a small staff on the last train out before the siege began (chapter 2). Further east, Major-General Sir William Gatacre was given a similar role around Queenstown. On 20 November French arrived at Naauwpoort, and next day carried out a reconnaissance along the railway line from Naauwpoort to Arundel and on to Rensburg station. Having seen the ground, he formed his plan. This was the start of a three-month period of skilful operations, deceiving the Boers about the still weak state of British forces in the area, keeping the lid on a threatened Boer rebellion in the Eastern Cape, protecting the strategically vital railway communications, and generally getting control of the area in preparation for the main advance to Bloemfontein - although it was already clear that this was going to be much delayed.

Mackworth's Telegraph Section detached to Colesberg

In forming his plan General French quickly recognised that he needed good communications. On 26 November, he telegraphed to Cape Town asking that some of the Telegraph Battalion's resources should be sent to assist with his operations in the area. Consequently, Captain Godfrey-Faussett was summoned from De Aar to Naauwpoort, where he arrived late on 27 November. There, on the 28th, he was briefed about the operational requirement by French, and by 7:30 am next day, No. 4 Section of the Telegraph Battalion, commanded by 2nd Lieutenant Harry Mackworth, was on the train from De Aar to Naauwpoort. The section consisted of thirty-three NCOs and men, fourteen Cape boys, eleven riding ponies, thirty-nine mules, twenty-four oxen, three air line wagons, one cable cart, one GS wagon, and one scotch cart (so-called because wagons in South Africa were taxed according to the number of wheels. The normal wagon had four wheels; to save tax, a 'scotch' cart had only two wheels). The section carried fifteen miles of airline and eight miles of cable, and was generally self-contained and independent - a fine command for so junior an officer, and deployed on this new mission with commendable speed.

2nd Lt Harry Mackworth.

Young Harry Mackworth had followed in his father's footsteps. Sir Arthur Mackworth had commanded 'C' Telegraph Troop from November 1881 to August 1883, and during his tenure the Troop had participated in the successful expedition to Egypt in August 1882. Harry was Sir Arthur's fourth son, out of a rather prolific family of seven sons and six daughters, and had been commissioned from the RMA Woolwich into the Royal Engineers in March 1898. After the usual young officer course at Chatham he was posted to the 1st Division of the Telegraph Battalion (as his father's previous command, 'C' Telegraph Troop, had become) on 5 September 1899, taking charge of No 4 Section. The war had started some five weeks later.

Harry Mackworth arrived in Arundel and worked closely with General French and his chief staff officer, Major Douglas Haig. He obviously struck up such a rapport with General French during this period that he was chosen to provide the telegraph communications for him on many of French's subsequent operations during the war.

Integrating the Communications

For the next few weeks French, with his gradually accumulating force, carried out small patrolling forays intended to keep the Boers guessing while gathering information and intelligence but avoiding set piece battles. The telegraph section participated in these, sometimes providing communications from the cable cart, sometimes tapping into and using the railway telegraph line which ran between Arundel and Colesberg.

Even when his reinforcements had arrived French was numerically inferior to the Boers. He had to appear stronger than he really was, and he had to eke his rather thin resources across a wider frontage than he could

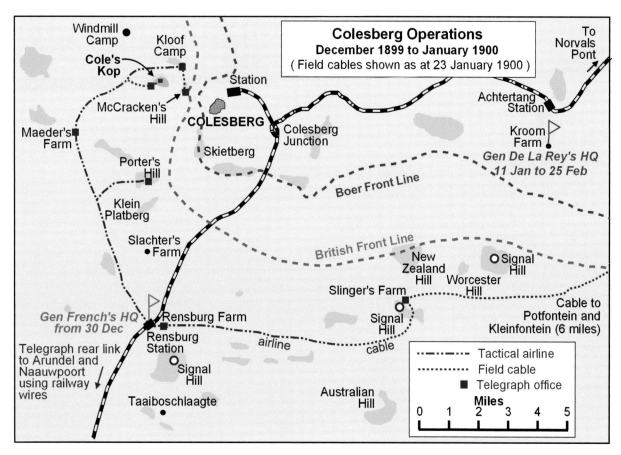

normally control. As well as aggressive patrolling another tactic, and the one of interest here, was to exercise command and control of his force in a manner which at that time was not practised - he integrated his communications system. What that means in the context of those days was that he brought together the visual signalling of the regimental signallers - essentially heliograph, and habitually planned and operated by regimental signallers in isolation - with the electric telegraph, which normally did not extend to lower formation level but which French was using for the first time in a tactical role. This enhanced system of communication, visual and telegraph working together, was the key to exercising greater control of his force over a larger area.

By 7 December French was in occupation of Arundel. The main Boer force, which had been in the area for a month in far superior numbers to the British, but had squandered their opportunity by inactivity, was itself being reinforced as a result of the perceived threat that French had induced. It was 4,000 strong, with the more effective General De la Rey now in command. Some of these Boer reinforcements had been withdrawn from Natal, relieving the pressure there.

Defeat at Stormberg

Meanwhile, further east along the threatened border with the Orange Free State, reinforcements being brought by train from Port Elizabeth were forming up under General Gatacre's command at Queenstown. Gatacre decided his first task was to re-occupy Stormberg. His plan was to surprise the Boers by advancing by train from Queenstown to Molteno in the late afternoon, detrain, march ten miles overnight, arrive at the right point at first light, and attack. The plan was too ambitious; it asked a lot of the troops and allowed no margin for errors, of which there were plenty. A civilian telegraph clerk forgot to send a vital message calling up part of the attacking force, entraining was delayed, native guides lost the way, and Gatacre did not know where he was. There had been no reconnaissance either of the route or the enemy positions, so nobody knew how to get there and where to attack.

The detail of the battle, which took place on 10 December, is not part of this story, and so short and disorganised was it that no communications of any sort ever seem to have been established. The telegraph Division was not involved, communications such as there were being either by the Cape Colony's civilian telegraph offices or regimental signallers. Despite a brave performance from the mounted infantry and artillery, it was another shambles. 634 British infantrymen were captured by the Boers, uninjured. British casualties were 28 killed and 61 wounded. Suffice to say that, on 10 December, it was the first of three disasters that contributed to what became known as 'Black Week'. The second was Magersfontein on the following day, 11 December, and the third was on 15 December at Colenso in Natal, both of these already recounted.

Operations at Colesberg

Back at Colesberg, General French's incessant skirmishing, enthusiastically carried out by his troops, had by 29 December forced the Boers to withdraw to the strong defensive position generally running to the north and to the east of Colesberg, along a line of hills known as the Skietberg.

General French moved his headquarters forward from Arundel to Rensburg Station, nine miles, on 30 December. The telegraph section now maintained and worked the railway telegraph wire from Rensburg to Naauwpoort, thirty miles away, as part of the rearward communications. At Nauuwpoort the responsibility was handed over to the Cape Colony Telegraph Department.

On 31 December an offensive movement against the enemy's right resulted, early next morning, in the capture of McCracken's Hill, which was held as an advance post throughout the operations. Working closely with French, Mackworth's men ran a cable from Rensburg to Porter's Hill to connect up the post there.

Coles Kop, the dominant hill, seen here from the south. Apart from the various other small hills, the open veld gives no cover to daylight operations.

Attack and counter-attack went on through 1 January and by evening the British were well-established from Kloof Camp and McCracken's Hill on the north-west flank to Jasfontein on the south-east, a front of some ten miles, with a supporting post at Maeder's Farm.

There was a pause in the fighting on 2 and 3 January and, using this opportunity, the cable from Porter's Hill was extended to Maeder's Farm on the 3rd, and a telegraph office opened there. That same evening the line was further extended to Coles Kop, a strikingly steep and symmetrical conical hill, guarded by a small escarpment around its slopes about half-way up (a geological feature of a number of hills in South Africa), which dominates the country for miles around. Telegraph offices were opened at the foot and at the top of

A 15-pounder artillery gun of 4 Field Battery on top of Coles Kop.

Coles Kop was also used as a heliograph signalling station.

Both these contemporary photographs were originally published in 'With the Flag to Pretoria'.

Coles Kop. This was a frantic day's work for one cable cart, but it was fortunately achieved in time to be of great importance, because early the next morning the Boers, reinforced, made a strong attack on the extreme British left flank at Kloof. The detachment of the Suffolks there was outflanked and was in serious danger for some time, but General French, who at the time reached Porter's Hill, was able to use the telegraph to order reinforcement from the area of Maeder's Farm. The Boers were then driven back, with a loss of about 130 men. On 5 January the line was rapidly extended from Maeder's Farm to Kloof by bare wire being laid on the ground, the cable having all been used up.

In early January Captain Godfrey-Faussett and the Director of Army Telegraphs, Lieutenant-Colonel Richard Hippisley, who was based in Cape Town for the first phase of the war, visited French's Headquarters at Rensburg station. (Interestingly, Hippisley had served as a subaltern in 'C' Troop in the 1882 expedition to Egypt, under the command of Sir Arthur Mackworth, Harry's father.)

Sadly, while Harry Mackworth was doing such good work for General French at Colesberg, his eldest brother, Digby, heir to the baronetcy, was killed at Ladysmith, some 300 miles away in Natal. Digby was a Brevet Major in the Royal West Surrey Regiment, but was attached to the King's Royal Rifle Corps, and was killed, shot in the head, while gallantly leading a charge during the serious Boer attack on the Ladysmith defences at Wagon Hill on 6 January 1900 (see chapter 2). [1]

On 7 January a reconnaissance had shown a good position at Slingersfontein Farm, eleven miles east of Rensburg, and this was occupied two days later by cavalry and guns under Colonel Porter. General French intended to use it as a base for a mobile force on his right flank. The cable cart accompanied this force, laying thirteen-and-a-half miles and opening a telegraph office at Slingersfontein. During this march, General French who was with the column, was in direct communication through twenty miles of cable with the guns on Coles Kop, and was thus able to issue instructions for an artillery demonstration on the enemy positions which distracted their attention while the move was taking place.

Next day some of this cable was replaced by airline, the party laying six miles in four hours despite dry and stony ground. Mackworth also pre-positioned more cable at Slingersfontein, ready to support a reconnaissance on the 11th. It was not used, however, for the Boers were in too great a strength for any further advance.

On 11 January with the help of a field troop of the Royal Engineers, a 15-pounder artillery gun of 4th Battery Royal Field Artillery was raised to the top of Coles Kop – a great feat, as the position commanded the entire Boer defences. (The Battery, to this day, keeps Coles Kop in its title.) A wire rope was also installed to enable the supplies and ammunition to be hauled up. This gun, and another installed a few days later, fired straight into several of the Boer laagers, to their immense astonishment!

The next few days were spent improving the lines by replacing cable by airline. Morse sounder was then substituted for buzzer at Kloof, and cable was laid to McCracken's Hill. Captain Godfrey-Faussett had come over to Rensburg to assist Lieutenant Mackworth for a few days, and his diary over this period includes diagrams of the telegraph arrangements.

On the 15 and 21 January further reinforcements joined under Major-General Clements, including two howitzers which shelled Grassy Hill on the 19th, their fire being directed by telegraph from an observation post on the top of Coles Kop. The cable on the right flank was extended to Kleinfontein. On the 20th more cable was obtained and the cable detachment of No 2 Section was also brought to Colesberg as it was not presently being used, the rest of that section being engaged in maintenance duties at Orange River.

The cable from Porter's Hill to Coles Kop was still lying out, though no longer in use. It was reeled up on the night 23rd/24th, the experience being rather exciting for the line party as it involved approaching the sentries at Porter's Hill from the enemy's direction. An advanced guard was detailed who, to avoid 'friendly fire', whistled popular British airs as loudly as possible when approaching the outpost line!

Of course the communications were not only telegraph. A good visual communications network was established by heliograph, in what was ideal country for such signalling – crisp, clear air, numerous kopjes rising above the generally flat karoo, dominated of course by Coles Kop on which signalling stations were sited. Details of the visual signalling operations are not available, but 'Signal Hills' are marked on contemporary tactical maps near Rensburg Station, near Slingersfontein, and near Worcester Hill, and doubtless heliographs were deployed on numerous other kopjes. As has been described, some of these sites were colocated with telegraph offices, so that the signalling network was well integrated. It was certainly an interesting experience for Lieutenant Harry Mackworth of the Telegraph Battalion, and more about him later.

The work around Colesberg carried on until the end of January when General French was withdrawn in secrecy, as part of a deception strategy, to command the Cavalry Division being formed up under the new Commander-in-Chief, Field-Marshal Lord Roberts. Harry Mackworth's section was also withdrawn, and rejoined the Telegraph Battalion at Enslin ready for the renewed operation to relieve Kimberley and advance to Bloemfontein. The work of his section over the period at Colesberg had been exemplary, reflecting great credit on this relatively junior officer.

The Success of the Colesberg Operation

The Colesberg threat had been contained. General French's handling of his ad hoc force won much praise. The operations he conducted were extremely successful – a refreshing change at this otherwise dismal point in the war. His success was attributed to the way in which he deployed and organised his troops, inferior in numbers to the Boers, into a mutually supporting all-arms force. Into this plan he integrated his communications system, visual and telegraph, operating together tactically. It was to be the first time that telegraph was used in direct support of operations at tactical levels lower than divisional HQ. Nothing new in that in the years to come, but in the stultified tactical thinking that prevailed in the late-Victorian army, all of this was extremely innovative.

General French himself said that he would have been unable to hold his extended positions without such good communications. Valuable though it was, it should be remembered that it was a fairly static operation in a limited area. Telegraph operations for more mobile operations, as the war later turned to, were another thing. While his HQ remained at Rensburg throughout, General French moved from one telegraph office to another as appropriate during the course of the operations - perhaps the first practical concept of a tactical HQ, made possible because of the communication facilities.

Major-General John French

French was a man with a colourful private life, a penchant for financial irresponsibility, and was surrounded in later years by personal and professional controversy. That apart, he had as a cavalryman, studied his profession well and, commanding the Cavalry Division, was one of the few successful Generals of the Boer War. He later rose to be Field Marshal Sir John French, 1st Earl of Ypres, KP GCB OM GCVO KCMG PC, and held the posts of CIGS 1912-14 and Commander BEF 1914-15. From 1918-21 he was Lord Lieutenant of Ireland. An interesting biography is *The Little Field Marshal* (the title reflecting his physical stature) by Richard Holmes, who had full access to French's private papers, published in 1981.

Major Douglas Haig, chief staff officer.

Haig, a scion of the whisky distilling family, was also to become famous. He rose to be Field Marshal Earl Haig of Bemersyde, KT GCB OM GCVO KCIE ADC. During the First World War tensions arose between Haig and French, and Haig replaced French as Commander BEF in 1915 - all a long way from earlier days in South Africa when they worked well together. Earl Haig is perhaps best remembered today for his work in creating the British Legion, and its eponymous poppy fund.

Nowadays the ground over which these operations were fought remains much as it was at the time of the war. Colesberg has of course become a bigger place, and an important road junction. Being about half-way between Johannesberg and Cape Town, a distance of nearly 900 miles, it is a busy stopover, and the road junction to Port Elizabeth divides here. This road, passing through the former battlefield area, carries only light levels of traffic by modern standards, and does not detract from its earlier wartime situation. It passes just to the west of Rensburg station, still there but nothing more than a platform, and one might have to wait a long time to catch a train. Away from the town and modern roads, the country remains undeveloped and it is easy to reconstruct the operations. Exploring around the minor roads and farm tracks, it is easy to identify the features: Coles Kop is unmistakeably prominent; the various Signal Hills are obvious small kopjes rising perhaps fifty to a hundred feet above the surrounding country, and there are even the remains of some trenches in the area of Maeder's Farm. There is no natural cover, and concealment in the arid karoo is obtained by using the features of the ground. Colesberg is a long way from the modern tourist trail, but it repays a visit for those interested in the detail of its military history.

The Colesberg operation over, Lord Roberts and Lord Kitchener now in South Africa to bolster command arrangements, and consolidation taking place in the area of the Modder River in preparation for renewed operations there, it is now time to skip back to Natal, where Lieutenant Dickie Jelf's section had been sent, to catch up with General Sir Redvers Buller's attempt to relieve Ladysmith.

Endnotes

1. Although he was the fourth son, Harry Mackworth eventually succeeded to the baronetcy. On Sir Arthur's death in 1914, and as a consequence of Digby's death, the title passed to the second son, Humphrey. Humphrey died in 1948. A third son, Francis, who might then have succeeded, had sadly also been killed in action in France in 1914 while serving with the Royal Field Artillery, and so in 1948 the title, as the 8th baronet, passed to Colonel Harry Mackworth, as he had become. There were no heirs from Harry's marriage, and on his death in 1952 the title passed to a nephew.

Chapter 5

The Relief of Ladysmith

Introduction

It will be recalled that things had gone badly wrong in Natal (see chapter 2). Ladysmith had been besieged on 2 November, General Sir Redvers Buller had gone to Natal at the end of November, and his attempt to breach the Boer positions and relieve Ladysmith had been badly defeated at Colenso on 15 December. The Natal Field Force, as it was known, had withdrawn south to the area around Estcourt and Frere, and was being reinforced in preparation for a renewed attempt to relieve Ladysmith. As part of his preparation, Buller had ordered another telegraph section to be deployed from the western front to Natal with him. This task fell on No 1 Section of the Telegraph Battalion commanded by Lieutenant Dickie Jelf. His reaction on being detached from the Battalion at De Aar for the operation in Natal was, as he said in a letter to his parents, one of "unspeakable joy". Sadly, it was not to end that way. [1]

Reinforcement Telegraph Section to Natal

Jelf and his section left De Aar on 23 December and, after some slightly over-exuberant celebration of Christmas in Cape Town, travelled by ship from there to Durban, which they reached on 3 January 1900. Dickie Jelf, like Harry Mackworth, was another who was following in paternal footsteps, for his father, Colonel R. H. Jelf, had served in 'C' Telegraph Troop in 1871-72 and commanded the Telegraph Battalion from 1885 to 1889. Young Dickie had been born in what at the time was descriptively called the 'Adjutant's Tin Hut' in Aldershot, when his father, then a Captain, held that appointment in 1872. After that unpretentious arrival matters had improved, and he had been educated at Cheam and Eton. He had several brothers serving in other regiments and, as it happened, one of them, Rudolf Jelf, in the King's Royal Rifle Corps (to which Harry Mackworth's brother was attached), was besieged in Ladysmith – so there was a certain incentive for Dickie to assist in its relief. Dickie Jelf himself had got married in England in September, shortly before war broke out, and was about to return with his new wife to his post in a RE Company in Gibraltar a few weeks later when he was recalled for active service with the Telegraph Battalion.

Dickie Jelf, as a 2nd Lieutenant. Photo: RE library.

After landing at Durban Jelf and his section immediately moved up by rail to Estcourt to join the relief force. There he reported to Lieutenant-General Sir Charles Warren, Commander 5th Infantry Division. In 1885, the then Lieutenant-Colonel Sir Charles Warren had commanded the Bechuanaland Expedition in which Dickie Jelf's father, then Captain Richard Jelf, had provided the communications by constructing a 350-mile telegraph line from Barkly West (near Kimberley, and at that time the limit of the colonial telegraph system) through Vryburg and Mafeking to Molepolole. After this meeting with Warren at Estcourt, Jelf and the section moved on to Frere, the forward headquarters, served at the time by a civilian telegraph office. [2]

Here preliminary discussions took place with the colonial Telegraph Department as to the working of the circuits to be established for the army and the civil-military interface. It was decided for the present to make Frere the point of demarcation between the army and civil telegraphs, and the army would open a military telegraph office there alongside the civil one. North of Frere the telegraph lines were cut and the country was in the hands of the Boers. But otherwise, rearwards, the Natal system was connected through Pietermaritzburg to the Cape Colony and internationally by the submarine cable from Durban. General Buller was thus in communication with Lord Roberts from the moment of Roberts' arrival in Cape Town on 10 January, and with London throughout the entire operation. That there may have been some difficulty in communication between Roberts and Buller, as some historians describe, was a matter of personalities; it was no fault of the telegraph system.

The task facing Jelf's telegraph section was enormous. The resources of only one section, consisting in this case of one subaltern, one sergeant, two corporals, four lance-corporals and fifty-three men, were tasked with providing communications for an army which by early January had reached a strength of some 30,000 men, including nine officers of General rank. The section was quite insufficient for the task and was to be heavily overworked in the stressful weeks ahead as Buller's army tried to cross the Tugela river and relieve Ladysmith.

Lieutenant Jelf kept a very detailed diary describing his section's work, and it is possible to relate their activity closely to the operations that were about to be conducted in the Upper Tugela at the battles of Spion Kop and Vaal Krantz, and after those failures, back downstream again in the area of Chieveley at Hlangwane, leading eventually to the crossing of the river, the capture of the numerous hills on the opposite side generally described as the Tugela Heights, and then at last the relief of Ladysmith. [3]

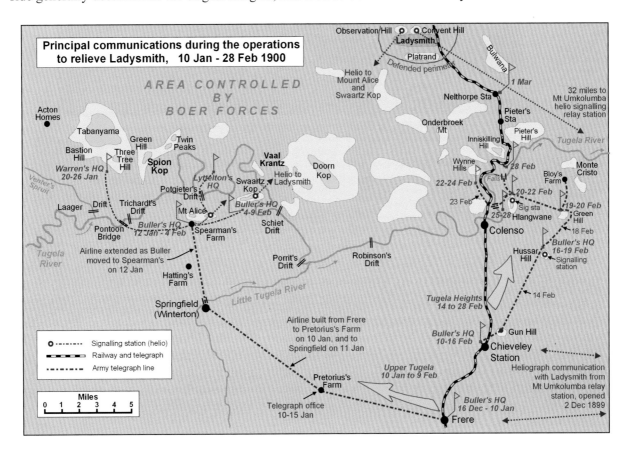

The Advance to the Upper Tugela

General Buller had decided to try to cross the Tugela further upstream than the normal approach through Colenso, although several of his Generals had argued that the best route to Ladysmith was via the Hlangwane and Monte Cristo features just downstream from Colenso. His plan was to establish a forward headquarters at Springfield (now called Winterton), from which to launch another attack. This strategy, and the conduct of the tactical operations in the Upper Tugela, have attracted much criticism from historians.

Sir George White in Ladysmith was appraised by heliograph signals of Buller's intentions. (The heliograph link to Ladysmith through the relay station on Mount Umkolumba had been established in early December, as previously described in chapter 2.) Captain Walker, the signalling officer in Ladysmith, was warned to look out for new heliograph signals from the south-west. [4]

On 9 January a cumbersome and overloaded column started the advance from Frere - despite the problems of transport Sir Redvers never travelled lightly. The area into which they were advancing was not served by any permanent telegraph lines, so all communications were going to have to be found from their own resources. Heavy rain had begun to fall on the previous day; wagons sank up to their axles, progress was laboriously slow, and the Tugela river was in flood. As the troops found out, summer in Natal is the rainy season, and the rain there can be very heavy. Eventually 23,000 troops moved to the Upper Tugela, leaving 7,000 at Frere, their base camp.

An intermediate depot was established at Pretorius's Farm, about seven miles from Frere, defended by a small number of troops. On 10 January the telegraph section built an airline to it and a telegraph office was opened there. This line, for its preservation, was kept about a mile away from the road along which the transport marched, and it worked well. It had been preceded, as was the practice for rapid laying, by a cable laid from a cable cart following a more direct route, but the cable was considerably damaged by the baggage wagons.

Springfield Bridge - a pivotal point for the advance.

Also on 10 January the leading troops, the Mounted Brigade under the command of Colonel the Earl of Dundonald, found the bridge across the Little Tugela river at Springfield undefended and intact. The Boers, fearful of being cut off by the rising river, had withdrawn. Seizing this unexpected opportunity, Dundonald's force pushed on and established themselves on Spearman's Hill. This hill is a dominant feature with two crests, Mount Alice on the western end, and at the eastern end, where naval guns were later emplaced, what was to become known descriptively as Signal Hill. From there the important Potgieter's Drift across the Tugela river was in view about a mile ahead. From Mount Alice, on 11 January, the signallers with Dundonald's force established heliographic communications with Ladysmith.

Also on 11 January, the airline was continued by the telegraph section to Springfield, which was to be General Buller's headquarters that night, and an office opened there working to Frere, with Pretorius's Farm as an intermediate office. On the morning of 12 January General Buller pushed his headquarters further forward, to Spearman's Farm just at the base of the hill. The airline followed him there and the telegraph office at Spearman's Farm opened on 12 January, working back to Frere just over twenty miles away. Buller himself rode up to Mount Alice, 1,000 feet above the surrounding country, and enjoyed the splendid panorama it afforded northwards and eastwards over the intervening range of hills to his now visible objective - Ladysmith.

Due to the bad weather and condition of the roads it took six laborious days, until 15 January, to move all the troops and the supplies up to their positions at Springfield and Spearman's Farm, ready for the operations to follow. This done, the intermediate depot at Pretorius's Farm was no longer needed, and the telegraph office there was closed. The insufficient number of men in the telegraph section meant that they constantly had to be redeployed to where the workload fell heaviest.

Meanwhile, Buller was considering his next move. He concluded that a direct advance from Potgieter's Drift was not possible, and inclined to the view that he should move yet further upstream, with a small force holding Potgieter's. He instructed General Warren to reconnoitre Trichardt's Drift about five miles upstream from Potgieter's; Warren pronounced the crossing there feasible. On the 13th Buller telegraphed his new intention to the Secretary of State for War in London and, on the 14th by heliograph to Sir George White in Ladysmith.

On the evening of 16 January Warren's troops moved forward to Trichardt's Drift. A pontoon bridge had been built just upstream of the drift, but even so the crossing was beset by all sorts of difficulties. At much the same time General Lyttelton's troops advanced to Potgieter's Drift, and that crossing was not easy either. By the evening of the 17th both columns were in occupation of bridgeheads on the enemy's side of the river, but it took until 19th before Warren had all his supplies across and, after much confusion due as much to poor organisation as to the difficult conditions, laagered near Venter's Spruit (stream).

The Boers, from their observation posts on the hills north of the river, had seen and reported (by heliograph from Doorn Kop to their laagers, and from the laagers by telegraph to Kruger in Pretoria) all the movement to Springfield, and the two crossings of the Tugela, so there was absolutely no element of surprise. The ponderous progress gave them ample time to determine Buller's intentions, to identify where the attack was to be made, to deploy their forces, and to prepare defences.

Telegraph Communications, 16 to 19 January

Soon after the troops left Frere, Captain Kennedy was attached to the force as the Assistant Director of Army Telegraphs (ADAT). John Kennedy, it will be recalled, had been the officer leading the experimental wireless detachment, mentioned earlier (chapter 3). After it had been decided that the wireless was unable to contribute effectively to the operations and the trial had ended, Kennedy was diverted to assist the overstretched telegraph section in Natal. As ADAT, based at Frere, he was now principally acting as a coordinating officer between the army telegraph section and the colonial Telegraph Department, and dealing with such things as press traffic. Press traffic was something the army could well have done without at that time. It involved censorship, collection and accounting for money, and the verbose use of telegraph lines of limited traffic capacity.

Buller's Chief-of-Staff at this time was Colonel A. S. Wynne - the same officer who in 1878-79, as Captain Arthur Wynne, had been the signalling officer to the then Major-General Roberts in the first part of the 2nd Afghan War. He had also participated as a signalling officer in the 1st Anglo-Boer War of 1881, and was responsible for establishing a heliograph chain from Natal to Pretoria, the Boers having destroyed the electric telegraph route. An experienced signaller, it was he who gave the operational plan and communications requirement to Lieutenant Jelf. (5) It becomes apparent from Jelf's diary, that Wynne's experience ensured that he was well-briefed about operational plans and future communication requirements. For example, Jelf's diary relates how he was told on the 15th about the intended move westwards and the crossing at Trichardt's Drift when for security reasons the plan was restricted to but a few. His task was to run out a line to keep General Warren in communication as he made his movement westwards to Trichardt's Drift, enabling him to keep in communication with Buller at Spearman's, and also to run a line to General Lyttelton who was tasked with moving to Potgieter's Drift and attempt a crossing of the river there.

A signalling station had been established on the eastern crest of Spearman's Hill and heliograph communications opened with Ladysmith, over fifteen miles away. The heliograph in Ladysmith was located on Observation Hill.

On 16 January a cable was laid from the headquarters at Spearman's Farm to the signalling station; it was initially a telephone circuit but on the 17th this was changed to a vibrator (presumably the telephone circuit was unsatisfactory), and one of Jelf's telegraphists was attached to the signalling station. That evening Warren's force advanced from the holding area at Springfield towards Trichardt's Drift, and early on the 17th Sergeant Hawker

The heliograph signalling station on Observation Hill in Ladysmith. A photograph dated 24 January 1900 from 'With the Flag to Pretoria'.

began to construct the telegraph line from Spearman's Farm to Trichardt's Drift. It was hard work in the rough and rocky ground and progress was slow, so the second half of the route to Warren's headquarters was initially run out with cable to save time, some of it being replaced later by airline. Warren crossed the river on the 18th and, accompanied by Sergeant Hawker and the cable cart, the line was continued to his temporary headquarters near Venter's Spruit.

Meanwhile, on the 17th, a cable had been run down the steep slope from the signalling station to extend that line on towards Potgieter's Drift for General Lyttelton's headquarters, but the cable cart had to remain on the southern side of the river while Lyttelton's troops crossed on the pontoon. Communications were maintained without interruption and traffic was passed all through that night.

On 18 January, south of the river, a section of airline was built to replace part of the cable laid the previous day on the route to Warren's headquarters, so that the cable could be reeled in and be ready for rapid laying during the forthcoming advance. Buller's headquarters changed its location that day, moving one-and-a-half miles further north, closer to Mount Alice, and the lines had to be adjusted accordingly. [6] On the right, near Potgieter's Drift, the situation was unchanged, the cable cart remaining south of the river. The forward communications from Buller's headquarters to Warren remained good all through the day and that night, but there was a fault for one-and-a-half hours on the rear link to Frere. The forward communications and the line to Frere were heavily used, Jelf noting in his diary that "immense amount of work at front and main office, and Frere, but hardly any at Springfield".

He was, in fact, very short of telegraphists for the amount of traffic being passed and the number of offices being manned. At some offices, there was only one telegraphist; they were rapidly becoming exhausted, and unable to stay awake. To relieve this problem Captain Kennedy back at Frere arranged for the line there to be connected directly with Pietermaritzburg. This had the effect of reducing the staff needed at the handover point at Frere, where there had been two full time offices, one civil and one army. Buller's headquarters at Spearman's Hill could now communicate directly with Pietermaritzburg, and only local traffic for Frere needed an operator. The army telegraphists thus saved were redeployed to Jelf at the sharp end. A number of the civilian telegraphists from Frere also volunteered to come forward and help. A Mr Rea was working at the headquarters near Spearman's Farm, later to be joined by Messrs McArthur and Anderson, and a Mr Banks assisted the lone army telegraphist at the telegraph office at Springfield. Banks had been assisting in the military telegraph office at Frere for some time.

There were other more mundane administrative problems. The telegraph section was not an integral part of any larger organisation. The section was issued with the rations for their men, and left to get on with it. The distribution of these rations to his scattered section, and cooking them, were all additional problems for Jelf and his men.

On the 19th, on the left flank, Sergeant Hawker and the cable cart continued the advance to Venter's Spruit. They were to follow Warren closely for the next few days. On the right flank, where Corporal Wake was in charge, an airline was now extended across the river for a short distance and the detachment was ready to follow any advance that General Lyttelton might make, although for the time being Lyttelton himself remained on the south side of the river. At Buller's headquarters Corporal Adams was in charge.

The line configuration was now as shown in the map above. Buller's headquarters was connected across the pontoon bridge over the Tugela to Warren on the left flank by a mix of cable and airline. With Warren was Sergeant Hawker, who with his cable cart was to follow Warren in the forthcoming advance of his headquarters, to a distance of about seven miles from Buller's headquarters. The line continued to the right flank with another mix of airline and cable connecting General Lyttelton, this line being extended across the Tugela when Lyttelton crossed at Potgieter's Drift. Thus, apart from the occasional fault or break in the line, Buller, Warren, and Lyttelton were in almost continuous telegraphic communication between their respective headquarters over the forthcoming period. In addition, there was the line to the signalling station which was in heliograph communication with Ladysmith, and the rear link through Springfield with Frere, which again suffered the occasional fault but nothing too lengthy. On 21 January, the section's linemen built an additional airline from Buller's headquarters to Potgieter's Drift in anticipation of them moving north across the river on their way to Ladysmith - something that was soon to prove overoptimistic.

The Battle of Spion Kop

On the evening of 19 January Warren issued orders for an advance to Bastion Hill and Three Tree Hill. He moved his headquarters on 20 January to a point about a mile north of the Venter's Spruit laager, and again next day to a position about half a mile south of the peak of Three Tree Hill and some three miles north of Trichardt's Drift, Sergeant Hawker moving with him and maintaining communications with the cable cart.

On 22 January Warren issued orders. "As a temporary measure the following arrangements for staff duties are being made: ..... Signalling Officer - Captain MacHardy." [7] Characteristic of so many failures of staffwork and organisation, it was a ridiculously late stage in the proceedings to appoint a divisional signalling officer, who would have to assemble an ad hoc team of regimental signallers. This appointment should have been foreseen and put in place when the force assembled at Frere. The signalling deficiencies that followed are probably due to this belated arrangement.

Warren called a meeting of his Generals that evening, and it was decided that the best option now was to attack and capture Spion Kop. The first white men to stand on the summit of that great mass were Boer Voortrekkers, approaching from the west in 1838 as part of the so-called 'Great Trek' out of the Cape Colony to rid themselves of the British. From its lofty summit, at an altitude of nearly 6,000 feet (1,830 metres), affording a magnificent view, they saw the fertile land of Natal stretching below them to the east, and they named it Spion Kop – Lookout Hill.

Spion Kop, about ten miles away, seen from near Observation Hill in Ladysmith. Spearman's Hill, not visible, is to the left of the picture.

On 23 January, Warren carried out a reconnaissance of the approaches to Spion Kop as best he could, and on his return met General Buller. It was decided to attack that evening, and orders were issued by Warren to his attacking force. Paragraph 14 of the orders states that: "Signalling communication must be established as soon as secrecy is no longer possible", and that is all it says on the subject. [8] An attack was about to be launched at short notice on an objective whose selection had depended more on running out of other options rather than a serious consideration of its strategic value in fulfilling the main aim of relieving Ladysmith. There had been inadequate reconnaissance, despite the days already spent in the area, and insufficient time had been spent on detailed planning and preparation, these inadequacies compounded by poor staff work and poor generalship. From these factors stemmed the subsequent chaos in command, control, and communications. It was a scenario already depressingly familiar, and so was the inevitable result.

The battle that subsequently took place was a disaster. The bare-topped hill itself was lightly defended and quickly captured at first light on the misty morning of 24 January, but when the mist lifted about three hours later the deceptive ground and vulnerability of their position was revealed. The Boers reacted quickly and throughout the day raked the hilltop from several directions with well-aimed artillery fire, as well as Mauser rifle fire. Entrenching in the rocky ground was well nigh impossible; there was no refuge anywhere. Casualties mounted, while Buller watched through his telescope from Mount Alice. Confusion reigned. Many descriptions have been published and it is unnecessary to add to them here. Suffice to say that the attack failed and the hill was evacuated that night. The British suffered thirty-two officers and about 290 men killed, and there were some 560 wounded. The many graves and memorials now on the hill record the disaster. The next morning the Boers, who were on the point of withdrawing, discovered that the hill had been abandoned and re-occupied it.

Jelf's diary on the 24th, the day of the battle, describes events:

> All working well. Fighting has been going on for several days and Sir C Warren is especially engaged. Sergeant Hawker managing left [flank] cable very well indeed.
>
> About 10.00 am firing got very heavy on the left and pressure on the local lines was very heavy in consequence.
>
> [Warren sent a telegram to Lyttelton at 9.53, received 9.55. 'Give every assistance you can on your side; this side is clear but the enemy are too strong on your side … if assistance is not given at once, all is lost. I am sending up two battalions but they will take some time to get up.' [9]]
>
> At 12 noon 'Clear Line' messages occupied local circuit for considerable time. [Clear the Line messages are authorised only by the senior commanders and clear the line of all other traffic ready for operational messages of very high precedence.]
>
> 12.30. Hawker, who had stuck to Sir C Warren throughout, reported 5 shells falling within a few yards of the cable cart. No casualties among his party DG [thank God], but one orderly killed and one horse, and one man badly wounded. Cart went dis [disconnected] for ¾ hour and took up safer position.

> An appeal from LF [the telegraphic address for the intermediate telegraph office on the Left Flank, on the way to Warren's headquarters] for relief of clerk but can't spare one. Have tried to arrange for a sapper to listen there and wake clerk, but seems impossible to teach them .....

So much for the telegraphic communications on the day of the battle of Spion Kop. Although there were short periods of line failure, and apparently a delay due to exhausted telegraphists, the telegraph generally worked and the important messages were passed quickly.

Efforts to reconstruct the signalling forward of Warren's headquarters are more difficult. Tactical communications had been haphazard. For a start, with early morning mist prevalent on hills in the area at that time of year, visual signalling, even if arranged properly, would have been impossible. Imagination does not seem to have stretched to the idea of running a cable up the hill to provide telegraph or telephone communications instead. The organisation of visual signalling was left to the last minute, was inadequately planned, and along with the command arrangements became chaotic. Nobody at the bottom knew what was going on at the top, and orders to react to events could not be passed.

There was a signalling station at Warren's headquarters, and two other signalling stations were sent to the top. The signallers were a prime target. The heliograph of one was destroyed by Boer gunfire, and the other station had to be relocated on the reverse slope due to the heavy fire sweeping the summit. Flag signalling attracted heavy Boer fire. Begbie signal lamps for use at night had been sent up without any paraffin, so were useless. Most of the messages that passed from and to the summit were hand-carried messages, but were so late and so confused that they had little value. One was even taken down by Lieutenant Winston Churchill.

General Lyttelton, just north of Potgieter's Drift, was in communication with Warren's headquarters throughout, by means of the telegraph line, and ready to cooperate as required. His two 4.7-inch naval guns were on Mount Alice and during Warren's attacks on Three Tree Hill and Bastion Hill he used his troops to demonstrate and make excursions into enemy territory with the intention of diverting the Boers' attention away from Warren's force. While the battle of Spion Kop was in progress, and in response to a heliograph message he had received from the top of Spion Kop [10], Lyttelton sent troops up, some to reinforce Spion Kop and some to attack the Twin Peaks where Boer fire was coming from. In the latter they succeeded brilliantly, and in doing so relieved much of the fire on Spion Kop. But it was all too late to save the day; that they had not attacked earlier as part of a co-ordinated plan was another example of the senior commanders' tactical ineptitude. Warren, with no previous fighting experience, was out of his depth.

The military graveyard at Spearman's Farm, where the wounded from Spion Kop who did not survive were buried.

Buller and Warren, the two architects of this disaster, met, and Buller decided to withdraw. The detachment that came back with Warren were, in Jelf's words, "quite done" after their exertions. The 26th and 27th saw the lines being dismantled and reeled in as the force withdrew back across the Tugela.

Meanwhile work continued for the telegraph section, repositioning lines and offices, traffic still very heavy. More mundane work came Jelf's way with having to settle his financial accounts - such things as payment for officers' private telegrams. For three days it rained heavily; the section was becoming exhausted.

Buller next attempted to cross at Vaal Krantz, further downstream. The signalling station to Ladysmith was repositioned on Swaartz Kop. Jelf was kept very busy again laying further lines to the signalling station and to new positions downstream. This was provided by cable as the ground was too rugged for the construction of airline, and they were under observation and sporadic artillery fire from Boers on high ground north of the river. Kennedy, meanwhile, had been called to the headquarters near Spearman's Farm to run things there while Jelf was busy with the new work.

THE RELIEF OF LADYSMITH

After unsuccessful operations, it all came to nought at Vaal Krantz, and after two days, on 8 February, the newly laid cable started to be recovered. It all involved considerable work. As the withdrawal continued, more lines were dismantled and recovered. The office at Spearman's Farm was closed at 2.15 pm on 10 February. By 11 February lines had been recovered as far as Springfield. There, the office and the airline rear link back to Frere were left working, to provide communications for a Cavalry Brigade being left there. Having spent the night at Pretorius's Farm, the main body of the telegraph section marched to Chieveley, on the main railway line and just north of Frere rejoined Buller's headquarters, pitched their tents, and rested overnight.

It was a month since Buller's force had left Frere. In that time they had suffered more than 1,750 casualties, and were no nearer to attaining their objective – the relief of Ladysmith.

Operations at the Tugela Heights

At Chieveley, Buller prepared a new plan (see map on page 44). This time it was to advance north from Chieveley and establish a base on Hussar Hill; to capture the Boer positions south of the Tugela at Hlangwane Hill, Green Hill, and Monte Cristo; to cross the river downstream from Colenso; and then to advance generally along the line of the railway as it cut its way through the difficult and hilly country known as the Tugela Heights to Pieter's station, and once past there across the plain to Ladysmith some fifteen miles away. (This was what several of his Generals had recommended before the futile Upper Tugela operation was decided.) Buller carried out his reconnaissance on 10 and 11 February.

The operation started on 12 February, and on the 14th, after some skirmishing, Hussar Hill was taken. Unfortunately all this had been done rather slowly, the intended surprise was lost, and the rather more agile Boers had redeployed to meet the obvious new threat. By 18 February the various hills south of the Tugela had been captured, Hlangwane having been found unoccupied, but the Boers had escaped across the river destroying the ponts behind them, and were seen withdrawing towards Pieter's station. After more inexplicable delay, when a quick follow up against the retreating and demoralised Boers (they had heard of the relief of Kimberley on 15 February - see next chapter) would have been very effective, the river was crossed on 21 February, over a pontoon bridge built by the engineers to the west of Hlangwane. The bridge used by the Boers near the Colenso Falls had also been repaired sufficiently for infantry to cross. On the 22nd Buller transferred his own headquarters across the pontoon bridge and established himself north of the river. There followed a week of desperate and brave fighting around the areas which became known as Hart's Hollow, Inniskilling Hill, and Railway Hill, the British having to fight hard every inch of the way, but determined this time that they were going to succeed. It was punctuated by an agreed armistice on Sunday 25th to recover the dead and wounded on both sides left lying on the battlefield, so while stretcher parties went about their task the troops fraternised. On the 27th the pontoon bridge was moved to a better position further downstream near the Colenso Falls, and more troops poured across. Coincidentally, on the 27th, on the western front, Cronjé surrendered to Lord Roberts at Paardeberg (yet to be described - see chapter 7), a crushing defeat. Both the British and the Boers on the Tugela heard about it through their respective telegraph systems. The Boers, whose morale their leaders had found hard to sustain during the irrepressible force against them, could take it no more; they cracked, and ran. At about 4:00 pm on 28 February the leading elements of Lord Dundonald's brigade rode unopposed into Ladysmith. The siege was over. Many regimental histories have much to recount of those days on the Tugela.

Communications in support of the Operation

What part did the telegraph section play in this final and eventually successful operation? On 13 February telegraph office arrangements at Chieveley were set up, so that military and civil lines were coordinated, as had been done previously at Frere. Diagrams in Jelf's diary show the detail. That evening Jelf was briefed by the chief-of-staff for the forthcoming operation. On the 14th a cable, subsequently replaced by airline, was run from Gun Hill to Hussar Hill four miles further on, and the office located near the headquarters and the signalling station. Another line was laid for a telephone to the 5th Brigade at Schutter's Hill, a few miles north-west of Chieveley. 16 February arrived with a rude awakening; just after breakfast at first light a Boer shell burst about fifteen yards from the telegraph office at Chieveley camp, fortunately with no serious casualties. More shells arrived, and the office was moved to a more protected position in a donga (dry river bed). Civilian telegraphists helped the undermanned army telegraphists at the offices at Frere, Springfield, and Gun Hill.

During this period Jelf was impeded by numerous administrative problems. The system of orderlies, provided by units which had been briefed about the requirement and who were needed to deliver and collect messages from the telegraph offices did not work. Frustratingly, many messages, having been sent and received, piled up uncollected at the telegraph office. "Whole system of telegraph orderlies needs to be on a much more organised and important footing", noted Jelf in his diary. Rations, mentioned previously, and now also drinking water, had to be drawn from a central point at Chieveley and then redistributed to the section scattered at various telegraph offices and laying line all over the place. For several days the temperamental Natal weather had been very hot, so the water was essential.

On 18 February the Hussar Hill line was extended about three miles to Green Hill, where General Buller established his headquarters on the 19th and 20th, and later on in the day it was extended another two miles to Bloy's Farm. The line was heavily used by the headquarters to the guns at Gun Hill, calling for and directing artillery fire. Various other lines were laid locally at the time. On the following morning the line was extended to Hlangwane Hill, where General Buller temporarily relocated his headquarters. On 21 February Jelf received further instructions, his diary noting: "Sir R. Buller sent for me before daylight this morning, and said he wished airline brought direct from Hussar Hill etc. ..... Was kind in what he said when I told him how short-handed we were". Later that day the line was re-routed direct from Hussar Hill to Hlangwane, and extended to the pontoon bridge over the Tugela. Buller moved his headquarters across the river on the 22nd.

On 23 February the wire was pushed across the river at the pontoon bridge to the left bank, where an office was established. The telegraph detachment was exposed to severe rifle fire from the overlooking heights, and had to seek refuge in a donga, where the office instruments were brought. Heavy rains fell suddenly, turning the donga into a rushing torrent which carried the instruments away, and they were only just saved before they were swept into the Tugela river.

On the morning of the 24th Jelf got all his wagons and men across the Tugela. Finding the new headquarters position too exposed to enemy fire, and reminiscent of the Grand Old Duke of York, Buller that evening moved his headquarters back across the river to Hlangwane Hill! The heavy rain continuing, the lines and offices had to be rearranged in the darkness. Jelf was soaked and so exhausted that he fell off his horse, fortunately without injury. The 25th was the day of the temporary cease fire. The headquarters now occupied two positions, during the daytime on top of the hill, and at night just to the west of the hill, each requiring a telegraph office.

On the night of the 26th the pontoon bridge was removed to be set up further downstream as already described, but without any warning being given to Lieutenant Jelf whose cable across the river had been supported on the bridge. The cable was, of course, carried downstream and had to be cut and replaced by another piece, heaved with difficulty across the river where it narrowed into rapids. General Lyttelton advanced down the left bank of the river and close to the railway wires, which were used to keep him in communication.

On 27 February Jelf was at the headquarters at the northern tip of Hlangwane Hill when an important telegram reached them. His diary records events: "Just as the troops were going to attack, the news of Kronje's [sic] surrender to Ld Roberts on the anniversary of Majuba was flashed through the wire to this telegraph office, and telegram handed to Military Secretary and to Sir R. Buller, who had it immediately signalled to all the Brigades and Divisions not in touch already by wire, and to Ladysmith. The Military Secretary said: 'This news is worth 100,000 men to us today'. "

On 28 February, the headquarters moved across the river again, this time to Kitchener's Hill, not far from the railway. (The hill was named after Colonel Walter Kitchener, commander of the West Yorkshire regiment, and not to be confused with his brother, Lord Kitchener, who was at that time at the western front with Lord Roberts). Unsure of the state of the railway telegraph line, Jelf sent one of his men across the river to test and repair it, while starting to lay a new line across the river to the headquarters' new position. Minor repairs were made to the railway telegraph line, and it was brought to working order, a cable completing the short distance to the headquarters and thus putting them in direct communication with Chieveley and Frere. Later that day the first troops rode into Ladysmith.

On 1 March General Buller moved on to Nelthorpe Station, where a telegraph office was opened . A cable was pushed on towards Ladysmith, and at about 3.30 pm the line party met a line party from Lieutenant

THE RELIEF OF LADYSMITH

Hidebrand's previously besieged section running a line out from Ladysmith, and so telegraph communication with the town was restored. Thus, at last, ended the siege of Ladysmith. After the relief, the two telegraph sections, the one that originally went and had been besieged, and the other which had come from De Aar with Jelf, joined forces. They remained in Natal under Major Hawkins to continue operating there with General Buller when he later advanced into the Eastern Transvaal.

The telegraph traffic passed during the operations on the Tugela river from early January until the end of February was immense, covering as it did the build-up, movement and logistics for 30,000 troops, the operations firstly to the Upper Tugela and then to the Tugela Heights, the battles and the traffic generated in their aftermath – casualty lists, and so on - handled mostly by one telegraph section with some civilian assistance. The line laying and maintenance throughout the movement and various redeployments of headquarters had also been immense.

Dickie Jelf was known personally to General Buller, who mentioned him in his despatches from Ladysmith to Field-Marshal Lord Roberts:

> From: The General Officer Commanding, Natal
>
> To: The Secretary of State for War
>
> (Through the Field Marshal Commanding the Forces in South Africa)
>
> The Convent, Ladysmith
>
> 30th March, 1900
>
> Sir,
>
> Ladysmith having been relieved on the 28th February, this seems to be a period in the South African campaign at which I may suitably bring to your notice the names, not previously mentioned, of officers, non-commissioned officers, and men whose services, in contributing to that result, deserve special mention.
>
> [amongst a number of others]
>
> Lieutenant R. J. Jelf, Royal Engineers, has been indefatigable in charge of the Field Telegraph, and has constantly had to work day and night. No difficulty was too great for him.
>
> ... *etc.*

Jelf describes in his diary how, at Nelthorpe Station on 3 March, he met up with his brother "Ru" (Rudolf), who rode out to meet him. Dickie described him as "thin, and suffering for a month from an upset inside". They met again the following day.

Unfortunately the strain of these operations, lasting over two months, took their toll on the unfortunate Lieutenant Dickie Jelf, aged 28. The work he had to do in Natal with a small number of men and no supporting administrative staff was overwhelming – communications ever changing, the telegraph system heavily worked, desperate fighting at the Upper Tugela and then the Tugela Heights. Once Ladysmith was relieved, and suffering from dysentery, he collapsed. He suffered mental and physical exhaustion and what would appear to have been a severe breakdown. He was taken to Pietermaritzburg hospital, kept under observation for some time, and then sent back to England to recuperate. He died on 2 June during the voyage home, and after his short marriage, leaving a young widow.

A memoir appeared later in the RE Journal, written by his father, Colonel R H Jelf CMG. It discreetly omitted the full story. In fact, as a sad ending to this chapter, Dickie Jelf committed suicide on the ship on the way home. In *Khaki Letters*, a collection of letters written by civilian telegraphists from the Post Office who were sent to South Africa as reservists, the detail is exposed:

> It is with much regret that we record the death of Lieut R. J. Jelf, R.E., an officer well-known to all the men of the T.Bs., not only in South Africa, but at home. He had been invalided home suffering from dysentery

which had followed enteric fever. He embarked at Cape Town on board the transport "Dilwara". All the way home he was in very low spirits and generally despondent. About 7 o'clock in the evening of 2nd June, just before dinner time, he left his attendant and went right aft on the starboard side of the promenade deck. Without a moment's hesitation he took out a revolver from his pocket, and before he could be prevented, placed it to his temple, and pulled the trigger. .... He was buried at sea next morning. [11]

Endnotes

1. *Royal Engineers Journal*, 2 July 1900, p 157.

2. Warren had a varied career. Commissioned into the Royal Engineers in 1857, he subsequently worked for the Palestine Exploration Fund and carried out excavations in Jerusalem, conducted boundary commissions in South Africa, stood as a Liberal MP (which brought his army career temporarily to an end), became Chief Commissioner of the Metropolitan Police in 1886, and rejoined the army in 1889. In 1897 he was promoted to Lieutenant-General. At the start of the Boer War he was on the retired list, but was recalled to command the 5th Division in South Africa. He was quite unsuited to such an appointment, with no experience of field command or the conduct of tactical operations, and personal relations with Buller were strained.

3. Lieutenant Jelf's diary is held in the archives of the Royal Signals. It describes the daily work in great detail, including the names of the soldiers and their duties, as well as diagrams of the line layout (including links to signalling stations) and telegraph offices to support each stage of the operations.

4. *South Africa and the Transvaal War* by L Creswicke, vol III, p 95.

5. After the battle of Spion Kop, when General Woodgate was killed, Wynne took over his command of 11 Brigade. He was in command for the ensuing operations around the Tugela Heights in February, and gave his name to Wynne Hill, about three miles north of Colenso. He carried out further operations during the war and ended as the commander of the forces in Cape Colony. He was the brother of Captain Warren Wynne who served, and died, in the Zulu War of 1879.

6. Jelf's diary conflicts with the account of this move in the *Times History of the War*, vol II. Jelf's records how the headquarters camp moved on 18 January, causing him to rearrange lines. Line diagrams in his diary show this.

7. Force Orders by Sir C Warren, 22nd January 1900. Reproduced in the *History of the War in South Africa*, vol II, Appx 9 (L), p 635.

8. Orders by Sir C Warren for the Occupation of Spion Kop, 23rd January 1900. Reproduced in the *History of the War in South Africa*, vol II, Appx 9 (M), p 636.

9. *History of the War in South Africa*, vol II, Appendix 10, p 639.

10. Heliogram from someone unknown on Spion Kop to General Lyttelton, received 10.15 am, 24 Jan 1900: "We occupy all the crest on top of hill, being heavily attacked from your side. Help us. Spion Kop." *History of the War in South Africa*, vol II, Appx10(c), p639.

11. *Khaki Letters from My Colleagues in South Africa*, p195. Reported by Colour Sergeant R E Kemp.

Chapter 6

The Siege of Kimberley

Following the relief of Ladysmith we now step back in time to the Western Front and describe, with particular reference to communications aspects, the similar set of events in Kimberley - the start of the siege on 14 October, the initial attempt to relieve the town prevented by the disastrous battle of Magersfontein on 11 December (chapter 3), an interval of several months while reinforcements arrived and a new plan was evolved, the situation within the besieged town meanwhile, and eventually its relief. Kimberley, of course, had to be relieved before the main advance to Bloemfontein could be resumed.

The Early History of Kimberley

The diamonds in the area of Kimberley originated during a phase of volcanic activity which occurred about sixty million years ago. They lay there undiscovered until 1871.

At the end of the 18th and the start of the 19th century, the Griquas, a race of mixed European and Hottentot origin, settled in the area north of the Orange River, extending to the present day Botswana border. Their main settlement was Klaarwater, later renamed Griquatown, and their territory embraced the area where Kimberley is today.

Alluvial diamonds (found in river beds) were discovered at Barkly West on the Vaal River early in 1870, and Griqualand West suddenly became politically important. The main diamond rush to what is now the town of Kimberley started in 1871 when there was a frantic stampede to the mining area of what is today the 'Big Hole of Kimberley'. One of the biggest finds was at the farm *Vooruitsicht* (forward lookout), owned by the brothers Diederick and Nicolaas De Beer. The name of these two humble farmers has been associated with diamonds ever since.

Three countries - the Cape Colony, the Orange Free State, and the South African Republic (Transvaal) - became involved in a dispute of ownership with Nikolaas Waterboer, at that time chief of the Griquas. The Lieutenant-Governor of Natal, R.W. Keate, a barrister of Lincoln's Inn, was called in to arbitrate and in 1871, in what became known as the 'Keate Award', the matter of ownership went in favour of the Griquas. The diggers themselves were not interested in rightful ownership, took to arms and declared themselves independent. Their independence was short lived when Britain, in her imperial manner, simply took cession of Griqualand West from the Griquas. In 1873 the British proclaimed it a separate colony. The matter went to the law court, and Judge Andries Stockenstrom caused an upheaval when he announced in 1876 that Britain's possession of the Diamond Fields was illegal. Britain was forced to compensate the Orange Free State by paying the sum of £90,000 – good value, taking into account the staggering amount of money that has been made from the precious stones dug out of the ground since then. Griqualand West was annexed to the Cape Colony in 1880.

Two townships grew around the mines, Kimberley (named after the Earl of Kimberley who became the Secretary of State for the Colonies in Gladstone's Liberal government in 1880), and Beaconsfield, embracing the Bultfontein, Wesselton, and Du Toits Pan mines (named after Britain's Conservative Prime Minister from 1874 to 1880, Benjamin Disraeli, Earl of Beaconsfield). Kimberley became a municipality in 1877; its twin town, Beaconsfield, rose to that status in 1884. The two towns were merged into the city of Kimberley in 1912.

The year 1871, at the start of the diamond rush, also saw the arrival on the scene of a young man destined to make his fortune and play a leading part in shaping the history of southern Africa - Cecil John Rhodes. Rhodes, at the age of nineteen, having left England to seek sunny climes for the sake of his health, and after a failed cotton-growing venture in Natal, started out in the mining camp by manufacturing ice and pumping water from the claims. With an eye to business, he bought up claims. He then met up with C.D. Rudd, a man ten years his senior, and they each acquired 25% of the claims of the De Beers farm, and this led to the forming of the De Beers Mining Company. In 1880 Rhodes became the Member for Barkly West in the Cape Legislative Assembly, and ten years later he was Prime Minister. One of his great accomplishments was the amalgamation of the interests of the De Beers mine with that of the Kimberley mine in 1888, to become the new consortium, De Beers Consolidated Mines Limited, with Rhodes its chief executive.

The Town of Kimberley

The town of Kimberley, with its complex early history described on the previous page, became established mainly for the purpose of diamond mining, which began in about 1870. The majority of its inhabitants worked for or in some way supported the De Beers mining concern which was headed by Cecil Rhodes; it held a virtual monopoly of the town. Just before the outbreak of the war Rhodes himself went to Kimberley, arriving on 10 October, and was besieged there along with his employees. His motive in going to Kimberley is understandable, but his failure to get out of such an obvious Boer target when the war started was seen by most as an error of judgement, and his presence there was to become an unwarranted trial for the relatively junior military commander of the town, Lieutenant-Colonel Robert Kekewich.

The Boers decided to make Kimberley an objective for a combination of reasons - its reputed wealth, its position on the railway line, the lingering grievance of the boundary dispute with the Orange Free State when the diamonds were discovered, and their hatred of Cecil Rhodes, seen as the arch villain of money and power, British imperialism, and the principal perpetrator of the failed Jameson raid. Also, it appeared a soft target, near their border, isolated and weakly defended. They arrived on 14 October 1899, three days after the war started, and the siege lasted from then until 15 February 1900.

The siege of Kimberley differed from those at Ladysmith and Mafeking in that its larger population, some 50,000 people, were predominantly civilian, a mixture of white and natives, with only a relatively small military contingent. Due to the diamond mining, it also possessed better infrastructure and more skilled people.

Pre-War Permanent Communications

By 1899 the permanent telegraph system was well established and at the start of the war Kimberley was a telegraph hub for the area, connected by the routes shown on the map on page 36. The first telegraph had been constructed from Cape Town to Kimberley in 1876, when the diamond mines were being developed; that was the line running by the most direct route across the karoo through Colesberg, Philippolis, Fauresmith, Koffiefontein, and Jacobsdal. Although it ran across the Orange Free State, it had been paid for and was operated by the Cape government. The Orange Free State bought this line in 1891. The railway from Cape Town came later, in 1885, by a different route, reaching Kimberley through De Aar, Orange River Station and Modder River Station, and another telegraph route from the south was built alongside it, this becoming the principal route. North of Kimberley this main route divided. One branch continued along the railway through Taungs, Vryburg and Mafeking, serving intermediate telegraph offices in Cape Colony, and thence on northwards to Bulawayo in Rhodesia. The other branch, a busier route with more lines, headed north-east into the Transvaal to Johannesburg and Pretoria via Bloemhof, Wolmaransstad, Klerksdorp, Potchefstroom, and Vereeniging, which was also a main telegraph hub with other routes from the Eastern Transvaal and Natal. From Kimberley there was also a telegraph line into the Orange Free State to Bloemfontein through Boshof; as will be described in chapter 8, this line became important for a short period in keeping Lord Roberts at Bloemfontein in communication when other lines had been cut by the Boers. Kimberley was thus an important telegraph junction and the Cape Telegraph Department maintained a section of men there, some forty strong, for the operation and maintenance of the telegraph system within the area. At the start of the war they were amongst those besieged within the town.

Planning the Defence

Prior to the war, at the end of July 1899, three Special Service Officers had been sent to Kimberley as a precautionary measure: Major (Local Lieutenant-Colonel) H. Scott-Turner of the Black Watch who later commanded the troops and volunteer units, Captain (later Major) W. A. J. O'Meara RE acting as Intelligence Officer and Chief of Staff, as well as being responsible for communications, and Lieutenant (later Captain) D. S. MacInnes RE. Duncan MacInnes had served with the detachment of the Telegraph Battalion in the Ashanti War of 1895-96, and was afterwards placed in charge of planning and building the fortification of Coomassie for the defence of the garrison there, so he came to Kimberley with all the right experience. In Kimberley he again planned the fortifications, and then joined Kekewich as a staff officer.

Captain Walter O'Meara RE had so far had an interesting career. Born in Calcutta in January 1863, he passed out of the RMA Woolwich in February 1883 and in March 1885 was posted to the Bengal Sappers and Miners. He took part in operations in Burma in 1886 where he was seriously wounded and invalided to

England. In 1889 he joined the 2nd Division of the Telegraph Battalion, which at that time was responsible for the maintenance of the telegraph system in southern England. In 1890, he surveyed and constructed the first telephone trunk line through Kent, part of the route then being developed between London and Paris. In 1894 he went to Cape Town to command a RE Company, and in early 1896, after the Jameson Raid, had been sent to Pitsani, north of Mafeking, where the raiders had set out from, to take over their base camp. He later attended staff college in England before returning to South Africa on special service in June 1899. On arrival in Kimberley he was tasked with surveying and providing intelligence for a route for the intended advance from Colesberg to Bloemfontein, and when war became imminent he was ordered to stay in Kimberley. [1]

Early in September, Lieutenant-Colonel Robert Kekewich, the commanding officer of the Loyal North Lancashire Regiment, at that time stationed in Cape Town, was sent to Kimberley on a reconnaissance and reported the vulnerability of the town. As a result, four companies of his battalion including some regimental signallers (the rest of the battalion being needed elsewhere), detachments of Royal Artillery and Royal Engineers, and a small staff, some 500 regular soldiers in all, were sent to Kimberley. They arrived by train on 21 September.

The residents of Kimberley were reassured by their presence. The existing volunteer units in Kimberley, formed from the civilian population (some Field Artillery, the Diamond Fields Horse, the Kimberley Light Horse, the Kimberley Regiment, as well as the Cape Police) could now be strengthened and better trained by the regular soldiers, and a Town Guard was also raised and trained. With these units the military strength of the garrison was eventually about 3,700 men.

During the subsequent siege, Kekewich and the three Special Service Officers - Scott-Turner, O'Meara, and MacInnes - became known as 'The Inner Circle' and effectively ran the military operation. In doing so Kekewich was to come into conflict with the egocentric and arrogant Cecil Rhodes who had different ideas on how the siege should be conducted in 'his' town.

'The Inner Circle'. L to R: Lt Duncan MacInnes, Lt Col Robert Kekewich, Capt Walter O'Meara, Lt Col Henry Scott-Turner.

Cecil Rhodes

Rhodes had been prime minister of the Cape Colony from 1890-96. Because of his position in Kimberley, he did much to support the defence and look after the civilians, but he was at odds with military concepts of operations, discipline, organisation, and security, and was not prepared to do other than what he alone thought. Given confidential information on several occasions, he breached trust and immediately publicly announced it. Tension arose between him and the relatively junior but effective Kekewich, who probably had more to endure from Cecil Rhodes than from the Boers. Rhodes also agitated about the need for rapid relief of Kimberley in messages sent by despatch riders without Kekewich's knowledge. The first was to the High Commissioner within two days of the telegraph being cut, then to Sir Redvers Buller, and later to Lord Roberts, all exaggerating the situation and threatening to surrender the town to the Boers if relief was not imminent. Despite Kekewich's more measured assessment, Rhodes' exaggerated clamouring caused General Roberts to change his plans to the detriment of future British operations. Such was Rhodes – a controversial and complex character.

A thirteen-mile perimeter fence of barbed wire was built around the town, redoubts and other defensive fortifications were constructed, in many cases using the mine spoilings that were dotted around forming small hills, five searchlights from the mining operations were repositioned and used as part of the defences (they were known as 'Rhodes eyes'), and a telephone system to the outposts was set up, connected to an exchange. The telephones were supplied both from military sources and by De Beers, who already had a telephone system in the town (incompatible with the military telephones, but the technical problem overcome by the addition of a third wire).

On 10 October, the day before war was declared, Kekewich 'talked' by telegraph to Colonel Baden-Powell, commander of the Mafeking garrison. [2] They discussed the situation and agreed some plans. Again, on the 13th, he similarly 'talked' to the High Commissioner in Cape Town, Sir Alfred Milner, and they discussed his legal and constitutional position in the event of communications being cut. Amongst Kekewich's earlier decisions was the enforcement of a censorship on telegrams out of Kimberley. The Postmaster-General in Cape Town had protested at the presence of the Press Censor in the telegraph office and had sent a message to the Postmaster at Kimberley giving instructions for the Censor to be excluded, but the Censor resisted. The High Commissioner agreed Kekewich's orders and the matter was put right in Cape Town.

Just before 10:00 pm on 14 October Kekewich was again 'talking' to Cape Town, this time to the ADC to the GOC. By the time the GOC (Lt Gen Sir F. Forestier-Walker) was called to the office, the telegraph line was cut. "At that moment, the telegraph instruments in the office, which had been clicking away merrily like a lot of crickets, one by one, in rapid succession, ceased their chattering and within a few seconds a dead silence reigned in the room." [3] The Boers had cut the telegraph lines south of Kimberley. They also cut the railway line near Modder River and the town's water supply, which came from a source a long way outside the defended perimeter. The De Beers chief engineer, an American, George Labram, solved the water problem by connecting deep springs within the defended area to the town's supply system. But the lack of communications by railway and telegraph, though not unexpected, were not so easily overcome.

> **'Talking' by telegraph**
>
> An amusing story of 'talking' by telegraph between senior figures is related by Field-Marshal Lord Birdwood in his autobiography *Khaki and Gown*. Later during the course of the war, as Major Birdwood, he was on Kitchener's staff in Pretoria as Deputy Assistant Adjutant-General. (Kitchener had by then assumed the appointment of Commander-in-Chief.) In early 1901 General Tucker took over command of the troops in the Orange River Colony, as the Orange Free State came to be called. Birdwood tells the story:
>
> "Tucker was probably the best-known and most popular of the generals with the troops, owing to – rather than in spite of – his wonderful and ever-ready flow of strong and picturesque language. It had been Kitchener's custom, while the situation in the Free State was so critical, to have a direct 'talk' over the telegraph wires with Hunter [Tucker's predecessor], there being no telephones to such a distance. Similarly, he got in touch with Tucker on his first morning. After the telegraphist had tapped through his report, Tucker asked: 'Do you wish me to talk with you like this every morning?' To which Kitchener got his telegraphist to reply: 'Heaven forbid! The wires would fuse!' "

The Siege Begins

Boers surrounded the town, but surprisingly they did not cut the telegraph line from Kimberley to the north until the night of 23/24 October. Kekewich had until then been able to communicate with his police detachments at places such as Fourteen Streams, Taungs, and Vryburg, and these detachments were safely brought in to Kimberley. A mounted force under Scott-Turner, sent out to investigate on 25 October, discovered that about ten miles outside Kimberley the Boers had broken nearly all the telegraph poles along a four-mile stretch beside the road.

Within two weeks of the start of the siege the Boers had an estimated 5,000 men surrounding the town. The detailed conduct of operations need not be related, for many descriptions have been published. Suffice to say that the Boers, laagered up with their wagons (and their wives!), were relatively passive. Starting on 6 November, after the expiry of an ultimatum to surrender which Kekewich ignored, there was rather ineffective artillery fire, but no determined assault. It was typical of Boer operations elsewhere - somewhat unwilling to take close-quarter offensive action against a defended position.

In Kimberley, an observation post, known as 'the conning tower', was at the top of a 150-feet high scaffold tower over the winding gear of one of the mine shafts, built for the purpose by the De Beers chief engineer, George Labram. Kekewich climbed the tower daily to obtain a view far out into the veld in all directions as well as complete coverage of the town and its local defences. The conning tower was connected to the telephone exchange and was also used as a heliograph signalling station.

The artillery duel, initially ineffective on both sides, was later intensified. In Kimberley, George Labram designed and built a gun, known as Long Cecil, and also made ammunition for it - a remarkable achievement, but Labram was nothing if not resourceful. It came into action on 19 January, its 8,000-yard range having a considerable effect on the Boers. The Boers responded by bringing in a Creusot 6-inch gun, commonly known at the time as a Long Tom (to which, of course, Long Cecil was the obvious riposte), which opened fire on 7 February. It was the very same gun that had been damaged at Ladysmith by a British fighting patrol sent to destroy it. It had been taken from Ladysmith, repaired in Pretoria,

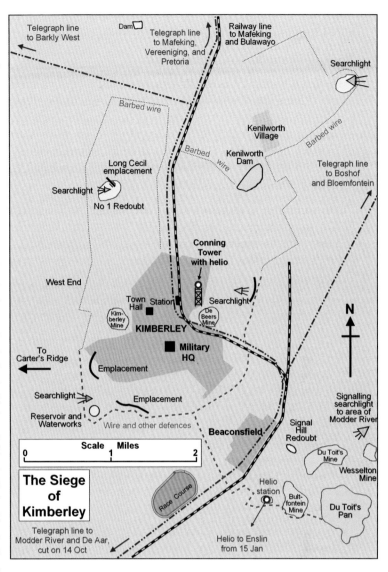

and was back in service, albeit with a shorter barrel than before. The Kimberley inhabitants called this particular gun 'Oom Paul' – Uncle Paul, after Kruger. It could penetrate into the heart of Kimberley and had a very demoralising effect on the inhabitants as it went about its business.

During the siege the Boers made great efforts with their Long Tom to destroy the conning tower, and came close to succeeding on several occasions. One shell passed under the canvas awning of the tower, another between the bars, lower down (one of these while Kekewich was there). A great splinter struck about five feet below the signalling platform and made the whole structure quiver. Splinters and bullets from shrapnel were found all around and under the tower. One shrapnel was timed about half-a-second too late, and exploded with a terrific bang over the open mine just behind.

A very basic form of signalling was introduced the day after Long Tom's appearance. A signaller was posted on the conning tower and kept observation on the Long Tom gun emplacement. As soon as he observed the puff of smoke indicating that the gun had been fired, he waved a flag. On this signal, buglers who were stationed around the town

The 'Conning Tower' at Kimberley.

sounded a 'G' (take cover). The inhabitants then had about fifteen seconds to dive for cover, the time it took for the shell to travel along its four-mile trajectory.

Communications During the Siege

With that outline of the general situation in Kimberley, it is now time to turn to communications during the siege. Rather like Ladysmith, there was an internal defence communications system from the command post to the outposts by telephone and heliograph. Similarly, after the telegraph lines were cut, external communications to the force attempting relief from the south were by messenger, searchlight, and eventually heliograph. There was a small detachment of signallers of Kekewich's regiment, and two regimental signalling officers were appointed, Lieutenants Woodward and de Putron, who conducted visual signalling within the defended perimeter.

There were also members of the Cape Telegraph Department in Kimberley, now without any telegraph system to operate, and they became part of the local defence force. An article written anonymously by one of them after the siege for a commemorative issue of the local Kimberley newspaper, the Diamond Fields Advertiser, describes their contribution :

> The Premier's instructions to the effect that civil servants were to take no part in any operations, defensive or offensive, applied to the Postal-Telegraph Department as well as to other branches of the Civil Service. It is worthy of note, however, that they were very early in the field. In fact the telegraph wires had not been cut for more than an hour, when arrangements were completed to arm 40 men, and at 11 p.m. (the wires were cut at 9.45 p.m.) on the 14th October, 1899, several telegraphists were making their way round the town, warning men to turn out, in obedience to instructions from the military authorities. Early the next morning - the notable 15th October - 40 men reported themselves at the De Beers Rock Shaft, where they fell in under Capt. Bowen (then of the Town Guard) who appointed Mr. J. E. Symons as Captain of the 'Telegraph Squad,' and assigned a position to them in the trenches to the left of No. 7 Fort.

> The following are the names of the Squad which included men from all the branches of the Postal-Telegraph Department here:-Symons, Gardiner, Darcy, Black, Whyte, Grey, Brown [a colourful lot!], Heads, Foot, Hills, Daly, Craigie, Richardson, Upfold, Bevan, Rankin, Kay, Cane, Behrends, Mortimer, Lunnon, Lambert, Lloyds, Simpson, Barnes, Pearce, Gatland, Goddard, Baxter, Henman, Backmann, Jubber, Ledger, Walt, Donoghue, Foster, Ward, Irving, and Wilson. [The names indicate they were almost exclusively of British origin.]

The civilian Telegraph Squad were soon afterwards taken off routine military defensive duties and re-deployed to communications tasks, as the article in the Diamond Fields Advertiser explains:

> The armoured train went out early on the morning of the 15th, and a member of the Maintenance Staff, Cruickshank, went with it for the purpose of ascertaining if there was any chance of being able to repair the telegraph lines. The Boers, however, brought a big gun to bear on the train and it had to return.

> Exceedingly good work was done by the Telegraph Engineering Staff under Inspector Gilbert, who erected telephone lines in a remarkably smart manner to many different points, including the redoubts. These lines have since been maintained in such a manner that complaints as to bad working have been conspicuously absent. A telephone was fitted up in the armoured train, and whenever it went out, which was often during the early days of the siege, telephonic communication was established with the Conning Tower.

> The number of telephone wires erected and the urgent need for skilled attention to the telephones at the Headquarter Office, Lennox Street, the Drill Hall, Conning Tower, Mounted Camp, Reservoir, Exchange, &c., and the necessity for relieving many of the officers, whose duties were interrupted by the numerous telephone calls, prompted the military authorities to recall the 'Telegraph Squad' from the trenches and caused them to take up duties at the various points mentioned. [4]

The more effective bombardment of the town by the Boers when Oom Paul came into action also had an effect on the operation of the telephone exchange which interconnected the defensive outposts and the conning tower. Until then it had been operated by the civilian telegraphists. When the bombardment intensified the civilian telegraphists refused to continue working there, despite its fundamental importance to the defence of the town. Royal Engineers sappers were quickly trained how to operate the exchange, and replaced them. [5]

Due to their heavy workload the small detachment of signallers from the North Lancashire Regiment, operating heliographs, requested assistance from the civil telegraphists. A number of telegraphists were

The civilian 'Telegraph Signallers' in Kimberley, having quickly learnt to use the heliograph in support of the regimental signallers.

detached and were trained for three days after which they became competent heliographers, helped of course by the fact that the heliograph, like the electric telegraph, used the Morse code, so they only had to learn to use the instrument and not the signalling code.

The correspondent in the Diamond Fields Advertiser describes how the civilian telegraphists were trained in the heliograph:

> The 'telegraph signallers' were Messrs. Symons, Osman, Lunnon, Rankin, Bevan, Upfold, Simpson, Backmann, Henman, Stephen, Foster, Whyte, Jubber, Wilson and Dennison. ..... The 'Telegraph Signallers' manned the Conning Tower, the Reservoir and No. 1 Searchlight, assisting the hard worked North Lancashire Regimental signallers. Everybody knows how difficult it is to get a 'show in' where the military are concerned, and although telegraphists are essentially signallers, they did not get a chance of distinguishing themselves until the work was too heavy for the Regulars to cope with. There was signalling to be done of course, but nothing very great. During the last week of the siege, however, and especially after Gen. French arrived, there was a tremendous lot of signalling work, and the Telegraph Signallers rose to the occasion. On Friday morning, the 16th February, helios were flashing all over the country. [6]

One of the Loyals' regimental signallers, J. Chambers, fell from the signalling platform on the conning tower one night and was killed outright. "He was undoubtedly dozing, and slipped through the bars", stated Mr J. E. Symons, the head of the civilian 'Telegraph Squad'.

Despatch Riders

As soon as the telegraph lines to the south were cut the Intelligence Department, under O'Meara, organised a communication service, and so-called 'despatch riders' operated the service between Kimberley and the nearest British headquarters south of the town. Some were men of the Cape Police, some were native runners, and some were white volunteers. Although called despatch riders, few of them rode horses. It was a dangerous task, and they mostly preferred to go on foot, relying on stealth and easy concealment to avoid the Boers, but taking longer.

A considerable number of messages were sent into and out of Kimberley by these despatch riders, as they broke through the siege cordon, which was not enthusiastically policed by the Boers during their sleeping hours, so the town was not totally *incommunicado*. These bulk deliveries would arrive at Orange River Station, or latterly at Modder River Station once it was occupied by the British forces at the end of November. On arrival at the telegraph office it would take quite some time to transmit the messages by the telegraph system. Similarly, replies for Kimberley were bundled up and sent back to Kimberley by messenger. Private correspondence and even newspapers were received in this way. All military despatches of a sensitive nature were encoded, in case the messenger was captured. The despatch riders also brought back information on Boer strengths and dispositions, and were debriefed on their return.

Rhodes insisted on using his own messengers and scouts who reported on Boer movements, preventing effective co-operation in communications and intelligence matters, causing potential security hazards, and increasing the friction between himself and the military. Further, he refused to coordinate with the military authorities about what was said, when it was said, and who it was sent to.

Searchlights

Kimberley's searchlights had been installed in the town before the war by De Beers for mining purposes and were now used for defence illuminations as well as communicating with the British force to the south, using the same technique as that employed at Ladysmith - by reflecting off the clouds. The searchlight at the Wesselton mine was mounted on a superstructure and used for communications. The signalling was carried out by signallers of the North Lancashire Regiment under Sergeant Herbert, and started during November, flashing 'MD' repeatedly in the direction of the Modder River. Eventually, on 27 November, this elicited the reply 'KB' from the area of the Modder River, and Kimberley was in tenuous communication with the relieving force.

The searchlight at Wesselton mine, used to signal from Kimberley.

The Royal Engineers with Lord Methuen's force near the Modder River had searchlights as well, mounted on a railway truck. So, unlike Ladysmith where the besieged town had no searchlight and efforts to build one had failed, no improvisation was needed in Kimberley and two-way communication with the relieving force was possible. However, it was weather-dependent, unreliable, limited to the few hours of darkness (it was mid-summer there), and even when working, was exceedingly slow.

Attempt to Relieve Kimberley

The task of relieving Kimberley was given to Lieutenant-General Lord Methuen and the 1st Division. As already described (chapter 3) they assembled at Orange River Station and started their advance towards Kimberley on 21 November, along the line of the railway to ease transport problems. They expected to relieve Kimberley within the week.

A despatch rider arrived in Kimberley early on the morning of 23 November with a message in cipher. When it was decoded it read: "18th November. No. R 98. General [Methuen] leaves here with small force on 21st and will arrive Kimberley on 26th, unless detained at Modder River. Look out for signals by searchlight from us, they will be in cipher." The anticipated arrival date of 26 November proved hugely optimistic. The relieving force underestimated the problems that lay ahead of them, despite being informed reasonably accurately by the intelligence staff at Kimberley of Boer intentions and dispositions, and the problems of topography in the Spytfontein and Magersfontein area.

On the evening of 27 November, gunfire having been heard in Kimberley during the day, a searchlight beam could be seen in Kimberley, coming from the south. Slowly and with difficulty the word KLOKFONTEIN was read, the name of a farm south of the Modder River, so it was presumed the relief column had reached there. It was imagined that relief might come on the morrow. Kekewich was able to signal that he had forty days provisions and plenty of water.

Kekewich decided to attack the Boers next day, 28 November, as a demonstration to divert their attention and in order to tie them down and prevent them attacking the relieving column. Unfortunately Lieutenant-Colonel Scott-Turner, one of his 'Inner Circle', was killed during the engagement at Carter's Ridge, about three miles west of the town. One other officer and twenty soldiers were also killed.

In fact the relief force was held up at the battle of Modder River, described in chapter 3. It endeavoured to communicate by searchlight with Kimberley that evening and also the following night, but did not succeed in getting a message through. On 29 November a despatch rider arrived with a message for Kekewich from Lord Methuen, and Kekewich replied giving details of the Boer forces and again reassuring him that the situation in Kimberley was not critical. Another message reached him on 1 December giving hitherto unknown details of the relief column – only a small division, due to the pressing need for troops elsewhere.

In Kimberley on 11 December the noise of another battle was heard in the area of the Spytfontein hills – it was the disaster at Magersfontein. On the evening of that day a short message was flashed by searchlight from Methuen to Kimberley: "I am checked". The next day a Boer heliographer flashed a message to the defenders: "We have smashed your column". On the 13th three enciphered messages arrived by despatch riders. They were from Lord Methuen. When decoded, they revealed the true and discouraging situation. Kekewich was advised not to expect relief until the middle of February, and to conserve his supplies accordingly. On 17 December, despatch riders arrived with Cape newspapers, and the secret was out – not only Magersfontein, but Stormberg as well. Morale in Kimberley sank.

Heliograph Link Established

Communications were still only by slow and unreliable searchlight at night, or by despatch rider taking days. At long last, on 15 January, three months after the siege began, a heliograph was brought into use. In Kimberley it was located, not on the conning tower, but on top of the headgear at the Bultfontein mine, the one spot in Kimberley that offered a line of sight through a small depression in the Spytfontein hills to an elevated signalling station on a kopje near Enslin, forty-one miles south of the Modder River. O'Meara described how it happened:

> At this time we were in signal communication with the Relief Column only at night, and this was not altogether satisfactory; owing to the difficulty experienced in working the shutters fitted to the searchlight at the Premier Mine Redoubt for signalling purposes, it was very slow work getting the messages off, and, consequently, only a small number could be sent during the few hours of darkness available at this time of year. Kekewich now made a suggestion to Methuen that an attempt should be made to establish heliographic communication by day between Kimberley and a post within the lines of the Relief Column. The proposal was accepted and a heliograph station was at once established for this purpose at the Conning Tower, the highest spot in Kimberley. The signallers began to call up the Relief Column, but no response could be got; the Spytfontein Hills proved to be a serious physical obstacle in the way. Fortunately, the Relief Column continued its endeavours to signal to us and, by mere accident, on the morning of 15th January, the beam of light from one of its heliograph stations penetrated over an almost imperceptible depression in the Spytfontein Hills. Keen-sighted Fraser [Major J R Fraser, Beaconsfield Town Guard, formerly of the North Lancashire Regiment] was at the time in the headgear of the Bultfontein Mine and seeing the flash immediately asked for a heliograph instrument. His request was satisfied without delay; he directed a beam of light on the spot where he had seen the flash a little earlier in the day; to his delight his signal was at once acknowledged, and the result was that henceforward we had signal communication with the Relief Column by day [heliograph] as well as night [searchlight]. It was later learnt that the signal which Fraser had seen came from a station on a kopje at Enslin. [7]

We need to go to the regimental history of the Gordon Highlanders to find out what happened at Enslin:

> Throughout the month signallers were trying to get in touch with Kimberley, communication with which was only clumsily maintained by flashlight [searchlight] signalling on the night sky. All efforts failed till one morning a piquet commander reported from his kopje [hill] that he could see the Kimberley searchlight. A Gordon heliograph was laid on, was answered, and by three days incessant work the great accumulation of public and private messages was cleared off. [8]

The kopjes to the east of the railway line at Enslin are the last high ground before the flat and featureless plain northwards to the Modder River. Unfortunately, like most visual signalling operations in the war, precise details of the heliograph operation at Enslin have not been found. Once the link to Kimberley was established it probably acted as a relay station, relaying the messages between Methuen's HQ at Modder River Station (about twenty miles from Enslin, and in the shadow of the Spytfontein hills) and Kimberley (about forty-one miles from Enslin, back over the top of the Modder River station and the gap in the Spytfontein hills into Kimberley). It was realised that the heliograph signals were being read by the Boers, so all confidential information was transmitted in cipher, encryption and decryption being the responsibility of the staff, not the signallers.

It is surprising that heliograph communications had not been set up earlier, on the initiative of the relieving force, Kimberley being their objective. The heliograph station at Enslin could have been used from 25 November, when the kopje on which it was sited was captured, but it took until the middle of January, and then only at Kekewich's prompt, for heliograph communications to be established. It again suggests that

THE SIEGE OF KIMBERLEY

not enough thought or effort went into the planning of signalling operations, the root cause being that it was not properly staffed and organised. At that stage of the war, before the Army had shaken itself out of its Victorian peacetime torpor, the same applied to many other facets of its operations.

The heliograph link with Kimberley now working, Mr J. E. Symons of the civilian 'Telegraph Signallers' relates a little more about signalling activity:

> Wednesday 17th January. I went down to Bultfontein and took a short turn at signalling to the Gordon Highlanders at Enslin, 41 miles from here. The light from their helio just cleared the ridge which intervenes between them and us; and, in spite of the distance, I had no difficulty reading.
>
> Thursday 18th January. Boer helio said: 'We are the Royal Artillery at Modder Rivier' – note the <u>Rivier</u> – we didn't fall for that one.
>
> Saturday 20 January. Drove down to the Premier Mine this evening with Lieutenant de Putron – one of the signals officers of the North Lancashire Regiment – as I wanted to get practical knowledge of the working of the electric searchlights, and thus complete my knowledge of the signalling methods in vogue here. (9)

Young Rhodes signals to Elder Brother

The heliograph link established, one interesting little story emerges. Lord Roberts' Director of Signalling at this stage was Local Lieutenant-Colonel Elmhirst Rhodes DSO. Elmhirst had been commissioned in the Royal Berkshire Regiment in 1878, and had fought in a number of expeditions in Egypt and the Sudan. He was the signalling officer of 2nd Brigade in Suakin in 1885 and participated in the battle of the Tofrek zariba. He was also the signalling officer with the Egyptian Frontier Force in 1885-86 and had his horse shot dead beneath him at the battle of Ginnis. For his part in this action he was awarded the DSO, one of the first officers to receive the award of that medal after it was instituted in September 1886. He next served with his battalion in Malta from 1888-93, being best remembered for his love of polo and the string of thirteen polo ponies he kept there. In January 1895, by then a Major, he had been appointed as the Superintendent of the Army School of Signalling at Aldershot where one of his students was Lieutenant de Putron of the Loyal North Lancashire Regiment, now serving as a regimental signalling officer in Kimberley. De Putron's certificate of signalling, awarded on successful completion of the course, dated 26 May 1999, and signed by Elmhirst Rhodes, is shown here. (10)

Besieged in Kimberley, as well as Lieutenant de Putron, and rather more importantly, was Elmhirst's elder brother - none other than Cecil Rhodes himself! (There were nine Rhodes brothers out of a family of eleven children, their father being the Reverend F W Rhodes, rector of Bishop's Stortford.) Elmhirst decided that he wanted to send a personal message to older sibling Cecil, and climbed on to the hill at Enslin to transmit it. The photograph (see next page) shows Elmhirst at work with the heliograph, and the message, reproduced below, shows how Elmhirst, like most others, thought his elder brother was a bit of an idiot to get himself besieged there! (Frank, mentioned in the message, was another brother, Colonel Frank Rhodes, also serving in South Africa, and concurrently besieged in Mafeking.)

The text of the message reads as follows:

> Have been staying at your house [in Cape Town] for few days. All are well. No letters from you lately. Both you and Frank are very foolish in getting yourselves shut up. You would have done much more good at large.

Presumably this message was signalled by Elmhirst himself. It is dated 17 January 1900, two days after the heliograph link was established. How Elmhirst Rhodes had stayed in Cecil's large mansion near Cape Town is somewhat of a mystery. Perhaps he had only just arrived in South Africa and spent a few days there before moving up to the front to become Roberts' signalling officer.

Elmhirst was replaced as Director of Signalling after Bloemfontein was captured and there was time for some reorganisation, the appointment being taken over by Lieutenant-Colonel Tom O'Leary of the Royal Irish Fusiliers, and Elmhirst reverted to his substantive rank of Major. It appears that Lord Roberts was not pleased with the standard of visual signalling in the early part of the war. Elmhirst was sent to a mundane posting as garrison commander at Dalmanutha in the Eastern Transvaal. He retired in 1903 and died in 1931. Cecil Rhodes, not in good health, died before the war ended, near Cape Town in 1902.

Kimberley Relieved

Since the failed attempts to relieve Kimberley in late November and early December, Field-Marshal Lord Roberts had arrived as Commander-in-Chief, with General Lord Kitchener as his chief of staff, and reorganisation and reinforcement of the British forces on the western front had taken place. It was now time to try again to relieve Kimberley, an essential precursor to resuming the planned advance to Bloemfontein.

On 10 February 1900, in a heliogram from Lord Roberts, Kekewich was advised that relief was imminent. This seemed to indicate that the route which had been recommended to the relief force at the outset, when it was still at Orange River, was to be adopted, and he was able to give information about topography and water supply on the intended route as a result of reconnaissances made before the siege began. Kekewich was told to look out for heliograph signals in four or five days time in a direction south-east and south-south-east of Kimberley.

So that Roberts could be forewarned of the circumstances of his strained relationship with Rhodes, Kekewich sent this telegram to Lord Roberts on the morning of 11 February:

Lt Col Elmhirst Rhodes signalling from Enslin. The photograph was published in 'With the Flag to Pretoria', Vol 1, and entitled 'Colonel Rhodes heliographing to his brother Cecil Rhodes, besieged in Kimberley.'

The message from 'Col. Rhodes' to 'Rhodes', dated '17-1-1900'. The original message form is held in the Royal Signals archives.

> Rhodes during siege has done excellent work; and also, when his views on military questions have coincided with mine, he has readily assisted me, but he desires to control the military situation. I have refused to be dictated to by him. On such occasions, he has been grossly insulting to me, and in his remarks on the British Army. More can be explained when we meet. I have put up with insults so as not to risk the safety of the defence: the key to the military situation here in one sense is Rhodes, for a large majority of the Town Guardsmen, Kimberley Light Horse and Volunteers are De Beers' employees. I fully realize the powers conferred on me by the existence of Martial Law, but have not sufficient military force to compel obedience. Conflict between the few Imperial troops here and the local levies has been, and must continue to be, avoided at all costs. [11]

Roberts' plan was to make a rapid advance using a cavalry Division commanded by Major-General French, and water supply for the horses along his planned route was essential. The route to Kimberley was in fact a diversionary part of his main advance to Bloemfontein, and will be covered more fully in the next chapter.

French's right-flanking route to Kimberley started from Enslin, through Ramdam with its abundant wells, across the Riet river at Waterval Drift, then across the Modder river at Klip Drift, past Abon's Dam, and on to Kimberley. This route of some seventy miles can be followed from the map at page 26. It bypassed the dithering Cronje and his force still laagered passively at Magersfontein.

From Kimberley, during 13 February, signs of battle could be detected in the direction of Jacobsdal. It was also evident that the Boers besieging Kimberley were unsettled, and they were seen from the outpost at Otto's Kopje to be making adjustments to their telegraph lines. O'Meara relates that: "They apparently expected that the direct telegraph wire between Bloemfontein and their headquarters at Spytfontein would be cut." (It probably already had been, for a patrol of the 10th Hussars, advancing from the Riet River on reconnaissance, had cut the telegraph line between Koffiefontein and Jacobsdal to prevent its continued use by the Boers.) They were busy building a new line to the west of Carter's Ridge in order to connect the Intermediate Pumping Station with Spytfontein, the Pumping Station being connected to Pretoria by the permanent line that ran via Fourteen Streams and Potchefstroom.

In Kimberley it was decided to make an attempt that night to cut the telegraph line north of Kimberley, to disrupt the Boer's alternative telegraph route. A native was sent north as soon as it was dark, with instructions to cut and remove as great a length as possible of the telegraph wires at some point north of Dronfield Siding. This task was succesfully accomplished. The subsequent capture of some Boers during a skirmish on 14 February elicited the information that both their telegraph line to Bloemfontein and the line north to Pretoria had been cut during the night, causing dismay to their leaders who imagined that British forces had got through under cover of darkness and were threatening their lines of retreat. Their unease at this news, reinforced by a report brought in by a Boer galloper of strong British cavalry across the Riet River, was soon to lead to a general Boer retreat. It was indeed the cavalry force under Major-General French making a direct thrust at Kimberley that had been spotted.

On the day that Kimberley was relieved, 15 February, the advancing cavalry force signalled to Kimberley by heliograph from their approach route to the southeast of the town, as had been indicated by Lord Roberts in his message on the 10th. The first signals were seen at about 3:00 pm, at a distance of about fifteen miles. There are stories that the garrison suspected they were being 'spoofed' by a Boer deception, and took a long time to authenticate them. This seems unlikely, as they had been well warned. The advance and the heliograph signals were reported immediately from the outpost to Kekewich in the conning tower, and they were at the time and from the direction expected, so it is difficult to imagine that much time was wasted getting correct authentication. At 4:00 pm Kekewich received a heliogram from French himself, confirming his advance and asking if he was on the right road, and requesting the preparation of facilities and water for 5,000 men and 10,000 animals. Kekewich replied at once with the information. By the evening the troops were in the outer areas of the town.

The relieving force was followed all the way to Kimberley by a cable cart commanded by Lieutenant James Moir of the Telegraph Battalion. It was something of an epic that the cable cart could keep up with the cavalry over that distance, but the detail of that forms part of the next chapter.

Permanent Communications Restored

After the relief of the town, and the area cleared of Boers, the recent army telegraph reinforcement, the permanent line party under Captain Fowler and Lieutenant Hepper, assisted by men of the colonial Telegraph Department, quickly restored the permanent telegraph facilities between Modder River Station and Kimberley. By 20 February three lines were in service and the Kimberley telegraph office remained open thereafter. The army's Modder River Station telegraph office was closed as the permanent line from Jacobsdal to Kimberley was also restored, enabling Lord Roberts rear-link communications during his advance to Bloemfontein to be re-routed to Kimberley.

Kekewich Superseded

When Major-General French reached Kimberley, the first person of note to gain his ear was the vindictive Rhodes, who wasted little time in telling him scurrilous stories about Kekewich. Some thirty-six hours after the relief of the town, on 17 February, Kekewich arrived at his office to find it taken over by Colonel Porter, the Commander of the 1st Cavalry Brigade, part of French's force. Rhodes' poison had permeated even a hardened soldier like French. Kekewich had been superseded in command of Kimberley, without the courtesy of even being told by French, who had already departed to hunt down Cronje.

In his post-Kimberley despatch, published in the London Gazette of 8 May 1900, Kekewich mentions Captain Walter O'Meara and Lieutenant Duncan MacInnes in glowing terms. O'Meara went on to other duties in South Africa, and MacInnes joined the hard pressed Telegraph Battalion as one of a number of officer reinforcements for the continuing operations. Also mentioned in Kekewich's Despatch are Lieutenant Woodward of the 1st Battalion, Loyal North Lancashire Regiment, who acted as signalling officer, and Mr J E Symons and Mr J Gilbert, Superintending Engineer, Cape Government Telegraphs, who "did good work in connection with the telephone service to various forts". In his covering letter, dated 20 March 1900, forwarding Kekewich's Despatch to the Secretary of State for War, Lord Roberts lauded Kekewich's achievements, concluding that: "I confidently recommend this officer to the favourable consideration of Her Majesty's Government."

Kekewich went on to be promoted to Brevet Colonel and, after successful operations during the war, was promoted Major-General in August 1902. Following ill-health and depressed, he sadly committed suicide in 1914.

Endnotes

1. Later Lt Col W A J O'Meara CMG, and a Barrister-at-Law of the Inner Temple. Long after Rhodes and Kekewich had died, he wrote a book *Kekewich in Kimberley*, published in 1926. Having been at the centre of things in Kimberley, he described the siege and exposed Rhodes' behaviour, something that had been well known in military circles but not released publicly.

2. An expression meaning that they communicated by telegraph, from the telegraph office, using telegraphists to relay their one-to-one personal conversation in Morse code. The telegraphists earned a very good reputation for not divulging confidential information so obtained. In those days, before the invention of the thermionic valve, which could amplify the complex waveform of the human voice, telephones could only work over short distances, which is why they were only used for local communications, such as the defence outposts at Kimberley for example. Some authors have not understood this meaning of 'talked', and erroneously refer to long distance telephone conversations.

3. O'Meara. *Kekewich in Kimberley.* p35.

4. The Postal Telegraph Department, *The Siege of Kimberley.* Africana Library, Kimberley.

5. *Royal Engineers Journal*, 1 December 1903, p 269. Report of 7th (Field) Company RE, No 1 Section (approx 50 men), who were assigned to the defence of Kimberley on 18 September, arriving there on the 21st. The RE sappers were taught how to operate the exchange.

6. The author's research led him, in Cape Town, to the Cape Archives and the Reports of the Postmaster General over this period. It was interesting to find that a Cape Government House of Assembly Select Committee reported in September 1900, eight months after the siege ended. It had sat for some months and was attended over many days by the Prime Minister himself, Merriman – a procedure that today appears rather unusual. It dealt with a long list of grievances by the civilian staff of the Post and Telegraph Services about what might generally be called terms and conditions of service – excessive hours, overtime, pay, absence, holidays, travel allowances, *etc*. One of the principal witnesses was Mr. J. E. Symons, head of the 'Telegraph Squad', whose name appears above. They no doubt gave good service, despite a little bit of 'blowing their own trumpets' recorded above, but did they not realise there was a war on?

7. O'Meara. *Kekewich in Kimberley*. p102. O'Meara's description differs from that found in *The Times History of the War* where it says: 'On the 4th [December] Kekewich signalled that he had 40 days provisions and plenty of water'. It goes on to amplify in a footnote: 'Electric searchlight and heliograph communication with Kimberley had begun on the evening of the fight at Enslin [25 November], but had not worked satisfactorily till the night before the Modder River battle.' As far as the heliograph is concerned, this is questionable, not least because of the time of day and the sunlight; if heliograph was established, why continue over the next month or so with much inferior searchlight? It seems more likely that the footnote refers only to searchlight, and that heliograph communications were as stated by O'Meara.

8. *The Life of a Regiment* (The History of the Gordon Highlanders) by Lieutenant Colonel A D Greenhill Gardyne, vol 3, p 107. Pub 1934. It was the regiment's 1st battalion at Enslin. The 2nd battalion, which had rushed to Natal from India just before the war started, was now besieged in Ladysmith. The two battalions were to reunite later, with great rejoicing, when the western and eastern armies met.

9. Article in *The Post Office Militant, 1899-1902*. PO archives.

10. De Putron's personal papers, including his Certificate of Signalling and the message sent by Elmhirst Rhodes to Cecil in Kimberley, both reproduced above, are held in the archives of the Royal Signals.

11. O'Meara. *Kekewich in Kimberley*. p 120.

Chapter 7

The Advance to Bloemfontein

Lord Roberts Arrives

Since the battle of Magersfontein on 11 December 1899, matters on the Western Front had settled into stalemate. Field-Marshal Lord Roberts, the new Commander-in-Chief, had landed at Cape Town on 10 January 1900. He had found telegrams waiting for him from General Buller in Natal, and replied to the effect that Buller was to carry on in Natal and do the best he could. Roberts himself concentrated on the situation in the Cape Colony, where the troops were now at least holding their own.

He decided that the Boers invading the eastern part of the Cape Colony were too sluggish to be any great threat, and that the potentially rebellious inhabitants there were too half-hearted to do anything, despite having had every chance during November and December when the British were weak in the area. As far as the Western Front was concerned, he recognised the need to relieve Kimberley. He also wanted to concentrate his forces ready to resume the original plan - the advance to Bloemfontein and then Pretoria. Reinforcements from England were still arriving, and so he decided to remain on the defensive and consolidate before making any new advance. He undertook considerable troop redeployment in Cape Colony in late January. Troops assembled between Orange River and Modder River, the area where the Telegraph Division was operating. He left Cape Town secretly on 6 February with Major-General Lord Kitchener, arriving two days later at the once pretty Modder River Station, now despoiled by ten weeks of increasing military occupation, where the troops awaited new leadership and decisions.

The Telegraph Division

Since their arrival the limited resources of the Telegraph Division on the Western Front had been kept very busy, initially with the advance from De Aar to the Modder River in the second-half of November, then the detachment of Mackworth's section to Colesberg on 29 November, whence it returned in February just as the advance to Bloemfontein resumed. As well as maintaining communications between De Aar and Modder River, there had been several sorties with reconnaissance forces along telegraph spurs to the west, undertaken by Lieutenants Moir and Henrici. This was followed by the deployment of Jelf's section to Natal on 23 December, where it remained. Telegraph traffic levels remained high throughout all this period.

Godfrey-Faussett's diary records that on 20 January Lieutenant-Colonel Hippisley, Captain Fowler, and he discussed and agreed the arrangements for handing over the permanent lines and future responsibilities. Hippisley, the Director of Army Telegraphs, was still based in Cape Town, where he was responsible for staffwork involving liaison, strategic planning, and obtaining the necessary resources, leaving Captain Godfrey-Faussett responsible for providing telegraph communications in direct support of Lord Roberts' operations and in executive command of the sections. The permanent line party under Captain Fowler and Lieutenant Hepper (see chapter 3), with the role to repair, maintain, and operate the permanent telegraph lines of the existing system, took over most of the responsibilities for the line and telegraph offices from Orange River to Modder River, as well as Naauport to Rensburg. The successful use of the permanent lines was vital, as they provided greater capacity and were far superior in robustness and transmission quality than the tactical airlines or field cable constructed by the sections under Godfrey-Faussett.

As part of the preparations for the advance to Bloemfontein, the telegraph sections concentrated at Enslin. The two sections from Modder River, under Lieutenants Moir and Henrici, now relieved of their permanent line duties, went there on 9 February. Lieutenant Mackworth's section returned from the Colesberg operation and joined them there a few days later. Altogether they now consisted of four officers, 69 NCOs and men, 63 native boys, 290 mules, 32 ponies, and 32 wagons of various sorts, 100 miles of airline, and 70 miles of cable. The native boys were mostly used in looking after the animals and in the airline detachments to do the hammer and jumper work for erecting poles, thus freeing the soldiers for more skilled work. Ten of the wagons carried additional stores to replenish the stores used as lines were constructed.

The general method of working was that, as they advanced, the leading troops would be kept in communication by field cable laid on the ground quickly from a cable cart immediately behind them, using

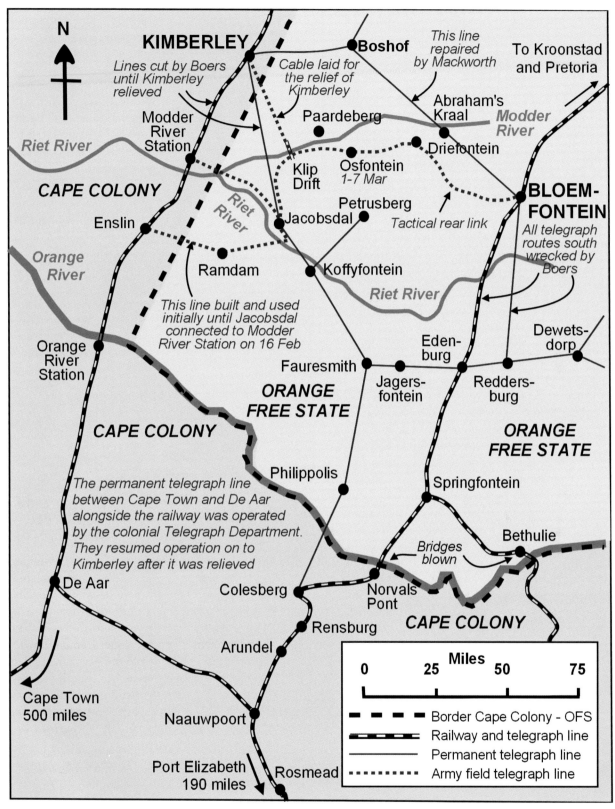

a 'vibrator' to communicate. The vibrator was a sensitive instrument used by the telegraphist, and could operate considerable distances over lines of poor quality; its disadvantage was that it caused interference on telegraph lines connected to other instruments, so was best used on a one-to-one link. The cable cart would be followed more slowly by an airline construction party building a line of better quality, able to use instruments that could handle a greater volume of telegraph traffic. The airline party would recover the field cable as they advanced, to be used again later. In this way communication was maintained throughout. The field telegraph lines shown on the map on page 69 do not distinguish between cable and airline; as described, they were mostly cable initially, then reconstructed as airline. The line parties were provided with escorts while they went about their work of construction and repair in territory not yet under full British control.

The Advance begins

Field-Marshal Lord Roberts' force for the advance comprised an army HQ and three divisions (the 6th Division under Lieutenant-General Kelly-Kenny, the 7th Division under Lieutenant-General Tucker, and the Cavalry Division under Major-General French), altogether some 37,000 men. His plan was to advance eastwards, using the Cavalry Division and their complement of mounted infantry and horse-drawn artillery as a mobile force to range well ahead, capture the drifts across the Riet river, then make a dash northwards across the Modder river to retake Kimberley, while the rest of his force following in strength as fast as possible headed for Bloemfontein 100 miles away to the east. He intended somehow to deal with Cronje's force still at Magersfontein, but how that was to be done depended on Cronje's reaction. Absolute secrecy about the plan was maintained, and few officers knew what was afoot.

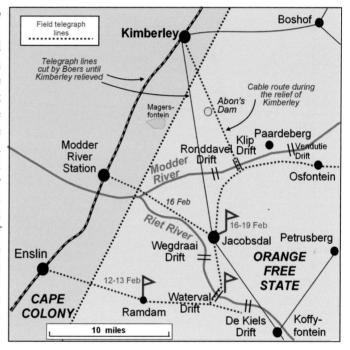

The advance out of the Cape Colony and into the Orange Free State started on Sunday 11 February and the 7th Division reached Ramdam that afternoon, its HQ being kept in communication with Enslin by cable. Airline was erected as soon as possible, but constructing nine miles of airline in intense heat in ground so hard that nearly every pole had to be dug in with pick and shovel was a purgatorial task. The Cavalry Division started from Modder River Station, leaving all their tents standing as a deception, and feinted south to Ramdam. The line of advance was considerably influenced by availability of water supply in that arid region, not just for the troops but for the large body of cavalry with thirsty horses, and Ramdam had wells. Roberts moved his HQ to Ramdam on the 12th, while the leading troops continued onwards.

Also on the 12th, starting at the early hour of 2:30 am, the Cavalry Division advanced to De Kiel's drift on the Riet River, dealing successfully with some Boer opposition on the way, and securing the crossing. With them during their advance was Lieutenant Moir and the cable cart detachment, keeping them in communication. A Boer force, under De Wet (a very able Boer commander, with a small force operating in the area), retired to a nearby hill and watched developments.

Two squadrons of the 10th Hussars patrolled forward, looking unsuccessfully for watering facilities ahead, and with orders to cut the permanent telegraph line between Koffyfontein and Jacobsdal, which they did. The permanent telegraph lines in this part of the Orange Free State, which were used by the Boer forces, are marked on the map on page 68. This telegraph route was part of the initial route built in 1876 to connect Kimberley with Cape Town after diamonds were discovered but before the railway was built. In this lightly populated area these telegraph circuits, which had become part of the Orange Free State permanent network, were simply hand-speed Morse operated between telegraph offices.

Relieving Kimberley

On 13 February French set his sights directly on Kimberley, the change of direction northwards from De Kiel's drift at this point disclosing his hitherto well-concealed intention. Firstly there were twenty miles of waterless sandy veld to the Modder River at Ronddavel drift, then the river to cross, and then another twenty miles to Kimberley. It was, so witnesses report, an impressive sight as three brigades of cavalry and one of mounted infantry, with supporting artillery, suddenly swung north towards Ronddavel drift and Klip drift on the Modder river. Behind them came their supplies and not least the cable cart under Lieutenant

Moir. Signallers today would be appalled at the thought of attempting to provide reliable rear link communications for a division by a single cable laid out of the back of a cart on to the ground for that distance across a semi-desert, vulnerable to damage, and with enemy patrolling around waiting to pounce on stragglers or isolated parties, but that was the only option then.

The advance of this force was watched from his vantage point by De Wet, who realised that the troops were heading, not for Koffyfontein as expected, but for the Modder river. His force was not strong enough to attack them, although he seems to have missed an opportunity by not attacking stragglers. Instead he sent a message by galloper to Cronje near Magersfontein, the permanent telegraph line between Koffyfontein, Jacobsdaal, and on to Magersfontein that would otherwise have been used having been cut by the cavalry patrol of the 10th Hussars the previous day.

While French was advancing to the Modder river, the 7th Division was following up behind him and attempting to cross the Riet River at De Kiel's drift. There was extreme congestion at De Kiel's as the large number of troops and their transport tried to cross the bottleneck, and the unseasoned 7th Division, newly arrived from England as reinforcements, found it difficult to cope with the heat, many dropping out. Kitchener, who Lord Roberts invariably used as his troubleshooter on such occasions, was sent forward to investigate. To relieve the pressure, Kitchener decided to use another nearby drift, Waterval drift, as the main crossing point. This occurred just as the telegraph section building the airline had reached De Kiel's, so five miles of airline had to be pulled down again before it was used, and new cable laid to Waterval drift where a telegraph office was opened at 3:00 pm on the 13th. This change of plan also broke the cable link laid by Moir behind French's advancing force, with the result that they were out of communication for a period, until reconnected after the line was constructed across Waterval drift.

Returning to French's advance to the Modder, some surprised Boers along the route skirmished lightly, but the cavalry advanced at a searing pace; horses dropped out, exhausted by the distance, the pace, the heat, the dust, and the lack of water. The bewildered Boers at the drifts bolted, caught by surprise as French's force galloped up, and a bridgehead was established across the Modder around Klip drift and Ronddavel drift. By 5:00 pm, as the stragglers crawled wearily into camp, French was able to report the successful accomplishment of his task, although some 500 horses had been lost during the day's advance. But things were now moving – and the cable cart was still with them! There is some doubt, however, about the state of communications. *The Times History of the War* states that "the cable cart had kept up with the force, but a veld fire which had sprung up behind had burnt the wire, and the message was eventually sent by despatch rider". [1] But in the various records of the telegraph sections there is no mention of the cable having been damaged by fire. Instead, it appears that Kitchener's change of plan, and consequently the change of routing of the airline to Waterval drift, meant that Moir's cable from De Kiel's drift was not connected as originally intended.

During the night of 13 February the 6th Division crossed the Riet River at Waterval drift and marched to Wegdraii drift, accompanied by cable laid by Lieutenant Mackworth's section. On the 14th the Division pushed on to Ronddavel drift and, using the cable laid by Mackworth's section, telegraph communication was re-established with the Cavalry Division.

Early on the 14th the Telegraph Division managed to get sole use of Waterval drift for half-an-hour, and the heavily-laden airline wagons managed to get across using double teams to pull them. The airline was constructed to Wegdraii by 1.00 pm, and onwards a further two-and-a-half miles towards Ronddavel drift by nightfall, ready for the next day.

By the 15th, other troops from Lord Robert's force had followed up behind French to secure his rear and he was ready for the final dash to Kimberley via Abon's Dam. The Boers held the ground just north of Klip drift and opened rifle fire prematurely, disclosing their positions. There was an artillery duel and the Boer guns were knocked out. General French boldly decided that the Boer positions were not strongly held and ordered a cavalry charge in brigade strength – probably one of the last such in British history. It was spectacular, and irresistible. The Boers scattered and fled, or were dealt with by the Lancers.

As he advanced, French passed close to Cronje's laager, which was crowded together in a depression some four miles to his left, waiting for the British to attack Magersfontein again. Had French been aware of this, and realised how demoralised the Boers were at the situation, he might have captured them there and then.

But he didn't, and that event was to be enacted some weeks later at Paardeberg. His objective was Kimberley. Beyond Abon's Dam, Kimberley came into view some ten miles away at 2:30 pm. Events had been moving quickly, and the besieged garrison did not know the current situation. Now that Kimberley was within sight, regimental signallers from French's leading brigade used a heliograph to communicate with them, but it took an hour to convince the Kimberley signallers that they were not being deceived by Boer heliographers and that the messages were from the British relieving force.

French's force closed on Kimberley and, after some minor skirmishing with Boers on the outskirts, they entered the town. The siege of Kimberley was over. General French rode into the town at 6:30 pm on 15 February.

For Moir and the cable cart detachment, following immediately behind them during the final twenty-five mile charge from Klip drift to Kimberley - the last lap of a sixty-mile cable laying expedition from Ramdam - the dust must have been dreadful. A high proportion of French's cavalry horses had died of thirst and exhaustion in the searing heat. That evening, using the field cable laid by Moir, General French was able to report the relief of the town by telegraph using that tenuous field cable. However, it was not to last long, as will shortly be described.

The cable cart - a sketch.

The History of the War in South Africa records that:

> French pushed on to Abon's Dam, which he reached at 11.45 am. A field telegraph cart which he had with him enabled General French to report his success to the Chief of the Staff [Kitchener], who at once telegraphed it on from Klip drift to Lord Roberts. ….. French's brigades ... met with little further opposition, and that evening he was able to send by telegram direct to Klip drift, and by flashlight [searchlight] *via* Lord Methuen's camp the following report to the Chief of Staff… [2]

Corporal Charlish, one of the telegraphists, describes in a letter how he handed the message to Lord Roberts at his headquarters, then near Wegdraai drift:

> I am now with Lord Roberts' staff. My duties are to take a set of duplex telegraph apparatus and fit up at each halting place. Of course this brings me in touch with Lord Roberts himself. It was at Wegdraai drift, five miles from Jacobsdal, that Lord Roberts received the telegram from Lord Kitchener [at the forward HQ at Klip drift] stating that General French had entered Kimberley and dined at the club. I had the pleasure of handing Lord Roberts the telegram personally. He read it aloud and remarked: 'Well, that's good.' I had several chats with Lord Roberts as he was often waiting near the [telegraph] tent for news. [3]

Behind French's thrust to Kimberley, the remainder of the army continued their advance south of the Modder river. Army HQ had advanced to Waterval drift on the 13th, and the airline was completed there by 3:00 pm. The infantry continued their advance to Wegdraai drift, and cable and airline were run there the next morning. Following the infantry, a cable was laid from Wegdraai to Ronddavel drift on the Modder, and there it was joined to Moir's cable to Kimberley, cutting out the portion between there and De Kiel's drift. Good communication was kept with General French to Kimberley. The cable to Ronddavel drift was replaced by airline on the 15th.

During the 14th and 15th De Wet's forces attacked and destroyed supply convoys in the area of Waterval drift, and the airline to Enslin which by this time was carrying the rear link traffic from the Army HQ was cut. As a result of this attack, the most effective of the Boer actions which were otherwise not more than minor skirmishes, there was a serious shortage of stores and the army had to exist on half-rations until the end of the month.

Paardeberg

After the relief of Kimberley on the 15th the elderly and unimaginative General Cronje, sitting passively in his laager near Magersfontein since the battle there some two months before, was bewildered by the sudden turn of events. His subordinate commanders were all urging retreat while it was still possible. His ineptitude over the next few days sealed their fate. He decided to retreat up the line of the Modder river towards Bloemfontein. His laager was no sort of military outfit; in Boer tradition it consisted of many wagons, and a lot of extraneous matter - including wives! When the time came to move, on the night of the 15th after French had swept past, the huge mob, some 5,000 strong, straggled in bright moonlight eastwards across the route taken by the British force, but all unobserved by the British! In doing so they mangled the cable connecting Roberts to Kimberley as they crossed, and cut great pieces out of it, some of which were found later in their defeated laager at Paardeberg.

Early the next morning a cloud of dust was seen, and the capture of a straggling wagon confirmed what had happened. It was well into the next morning before Methuen's force, still holding the British position near Magersfontein, discovered that the trenches in front of them were empty! Many of the Boers had escaped and gone their separate ways; not all followed Cronje's column, which by the 17th got as far as Wolvekraal, near Paardeberg.

With the cable now broken by the retreating Boers, it could not be used for communicating orders to General French in Kimberley in the operations that shortly ensued to head off Cronje as he tried to escape. Alternative means – written message sent by despatch rider, searchlight, and heliograph via Enslin - were all used instead. Apart from communication difficulties the operation was also hampered by the state of the cavalry's horses; many had died of exhaustion in the extreme heat compounded by lack of water, and few of those still alive were capable of going anywhere.

As a cavalryman, General French, good tactician though he was, gained an unenviable reputation in South Africa for not conserving his horses. In fact the British army as a whole was deficient in horsemastership in a country where the horse was the key to tactical mobility. The mostly urban British could not compete with the born-on-the-veld Boer fighting on his own territory, riding horses bred for the conditions. The Boers mostly rode Basutoland ponies, small and agile, able to graze from the veld. They well knew how to get the best out of them, and they each usually had two of them. The British rode large cavalry horses, unacclimatised to South Africa, used to being fed forage and unable to graze, overloaded with excess kit. Horse problems were to afflict the British throughout the war, and many of these unfortunate animals were to die during its course.

Although the cable to Kimberley was cut, the telegraph worked well on the 15th and 16th between Waterval, Wegdraai, Jacobsdaal and Klip drift, and Lord Roberts, from his HQ at Wegdraai and then at Jacobsdaal, was able to communicate with Kitchener at Klip drift about Cronje's escape and to coordinate the action to pursue him.

On 16 February, the day after the relief of Kimberley, the Army HQ moved to Jacobsdal, and a telegraph office opened working to Enslin. A direct airline from Jacobsdal to Modder River was also constructed on the same day, by Fowler from Modder River Station and Henrici from Jacobsdal, the two meeting half-way. This new route cut out the roundabout route via Enslin which had followed the initial advance, and all the line that had previously been laid from Enslin was abandoned. Lord Roberts stayed at Jacobsdal until the 19th.

On the 17th the 6th Division continued their advance, Mackworth's section laying cable behind them. Meanwhile the airline between Wegdraii and Ronddavel drifts had been cut in two places, and two linemen sent out to repair it had been captured. It became a favourite trick of Brother Boer to cut the line and then lie in wait for the repair party so, that lesson quickly learnt, escorts had to be provided. Mackworth's section constructed airline as far as Drieput's drift that day.

It was not until late on the 17th that Lord Roberts was fully aware of what had happened to Cronje. A confused sequence of events followed, which need not be described. General French and the cavalry, the most mobile part of the army but much depleted due to horse exhaustion, hot-footed it back from Kimberley to try and cut him off. Some six miles north of Paardeberg they came across a party of Boer heliographers trying to contact reinforcements; they captured them, and from them had it confirmed that the dust cloud they could see to the south-east was indeed Cronje's retreating remnants. Artillery was quickly and effectively brought into action. More British mounted infantry arrived, and Cronje was trapped on the north bank of the Modder River at a place near Paardeberg called Vendutie drift. Other Boer forces from the area attempted to break the cordon, but it held.

THE ADVANCE TO BLOEMFONTEIN

The telegraph was extended by cable to Sterkfontein, near Signal Hill (see map on page 75, obviously a British heliograph station), and when Lord Roberts moved his HQ forward to that area on the 19th, it was run right into his camp. On the next day the cable was replaced by airline. On the 22nd the airline was extended further forward to Koedoesrand, which remained the most forward point of the telegraph until 7th March.

A cable was run right into the Paardeberg battlefield, giving direct communication between Lord Roberts, who had become indisposed, at his HQ in Jacobsdaal and Lord Kitchener at his tactical HQ on top of Signal Hill. From there Kitchener was in heliograph communication with French on the battlefield, where he had been sent to ensure the necessary impetus to the operation. A telegram was sent from the indisposed Roberts, placing Kitchener in command.

Kitchener's tactical conduct of the subsequent battle has attracted criticism on account of the unnecessarily high British casualties it caused. It was a confused battle, stemming from inadequate staff arrangements and unclear operational orders at the outset, but that detail is not part of this narrative. Heliograph communications played a part in command and control but they were not well organised.

While the Boers were surrounded, and things became static for a while, the overworked linemen had a short period of rest, although as a consequence of De Wet's attack on the supply convoy, food was scanty, water bad, and forage for the animals was in short supply. The Boers, upstream in their besieged laager, were polluting the river; the stench in the laager being appalling, they were also disposing of dead animal carcasses killed by British shellfire, polluting the river further. The thirsty British soldier – it was high summer here, and very hot – perhaps had neither the knowledge, nor the resources, nor the time, to boil his water before drinking it. The cause of imminent health problems for the Army was being incubated here, as we shall see later.

To withstand the siege, the Boers entrenched deeply into the ground, making themselves almost impervious to artillery, but they had no escape route. Cronje, the author of his own defeat, surrendered to Lord Roberts on 27 February, the same day that Buller's troops in Natal broke through the Boer lines on the Tugela river and, more ignominious for them, the anniversary of Majuba Day (when in 1881 they had secured a significant victory over the British in the First Anglo-Boer war). Over 4,000 Boers were captured at Paardeberg some 2,600 from the Western Transvaal and 1,400 Free Staters. They were all, including Cronje and his wife, sent to a prisoner-of-war camp on the south Atlantic island of Saint Helena.

Within a two-week period Kimberley had been relieved, a large Boer force had been defeated at Paardeberg, and Ladysmith had been relieved. It was the turning point in the war - although it was going to drag on in other ways for another two years.

After the battle, on 1 March, the Army HQ was moved a short distance upstream from the insanitary conditions at Paardeberg to Osfontein. Up to this time the HQ telegraph office had been worked with double-current duplex apparatus. Jacobsdal and Klip drift were kept open as linesmen's offices. The traffic on this line to the Army HQ was very heavy, so it was decided to get a quadruplex apparatus from Modder River, enabling greater traffic capacity, and this arrived on 26 February. (Duplex meant that it was possible to transmit messages down a wire in opposite directions simultaneously, thus doubling the capacity of the line. Quadruplex meant that four messages could be sent simultaneously, two in each direction along the same line.) A convoy containing equipment for a further thirty miles of airline came up, and more airline was built to Koodoesrand drift in preparation for a further advance.

The linemen may have been having a well-earned rest during this pause but the telegraph operators weren't - there was an enormous surge in traffic. Godfrey-Faussett in his diary records how:

> The operators had a hard time, as nearly every public body in England thought it necessary to telegraph their congratulations to Lord Roberts. Wheatstone would have been invaluable but the only set available was in use elsewhere and the line was worked duplex day and night without cessation. [4]

The telegraph lines from Britain and elsewhere down to that remote, previously unheard of place on the Modder River, could not cope with the level of traffic. It is said that duplex circuits have little value because the traffic usually flows heaviest in one direction, not compensated by an equal flow in the other direction, and this situation was a good example. Whether a Wheatstone would have helped is a matter of conjecture; it generally required a higher quality of line than was possible with tactical airline such as was at Osfontein; that there wasn't one available is probably academic. No mention is made in any of the reports why this sort of non-operational traffic was not filtered out at De Aar or Modder River Station and delivered to the Army HQ by despatch rider, rather than cluttering the forward area operational communications system with its limited capacity. There was, however, an officer present in the telegraph office at the Army HQ to try and sort out message priorities for outward traffic.

After the advance to Paardeberg and the ensuing battle the army needed time to resupply and consolidate. This time was also used to repair and strengthen the telegraph lines. Lieutenant Moir took an airline detachment and repaired the permanent line from Jacobsdal to Kimberley (see map page 69). With the permanent lines south from Kimberley now repaired and operational from 20 February, this enabled people and equipment to be moved from Modder River Station to Kimberley, where accommodation and facilities were much better. Modder River Station, after some three months heavy influx of military, reduced considerably in importance. Moir remained in this rear area maintaining the lines, on which breakdowns were frequent. Most of these were due to convoys, which in that featureless area found the telegraph lines a useful navigation guide, but in the process unhelpfully knocked them down. On one occasion nothing was found of two miles of line but mangled pieces of wire and scraps of poles. On 3 March a convoy brought in more line equipment to the HQ, needed for the further advance.

Tactical Communications at Paardeberg

That the tactical conduct of the battle of Paardeberg was found wanting was largely due to command and staff work but also in part to failure in tactical communications. Like Spion Kop, it was a battle involving a large number of troops, and the inexperience of senior commanders in conducting a battle of this size with modern weapons against a modern enemy was apparent. *The Times History of the War* commented at the time:

> For much of it the fault lay in defective peace training; for still more in the failure to realise the enormous difficulties modern battle conditions have put in the way of intercommunication, and to provide for them. The difficulty of securing co-operation between the Highlanders on the right bank and the rest of MacDonald's brigade, the hours wasted by Hannay and Stephenson facing east, Smith-Dorrien's inactivity in the north, the unreported approach of De Wet, are all instances of the results of defective intercommunication. Paardeberg, like Spion Kop, indicates that for the battles of the future, signallers, orderlies, and heliographs, will be required on a far larger scale than hitherto, and that the field telegraph and telephone will play a part undreamt of before. [5]

After the battle Lord Roberts wrote to Lord Lansdowne, the Defence Minister in London. He criticised the British Army transport and supply system, blaming Lord Wolseley's reforms (there was no love lost between those two) which had downgraded the Quartermaster-General's Department, although his own reforms of the transport system in South Africa had also attracted much criticism. He also criticised army signalling, saying:

> Our signalling arrangements are very faulty, and the necessity for their being capable of expansion in times of war was evidently overlooked at Aldershot [Does he mean the Army HQ or the Army Signalling School, both at Aldershot?]. This country is admirably adapted for heliograph and flag signalling, but we cannot take full advantage of it for want of sufficient number of heliographs and properly trained men. Signalling is essentially a duty which belongs to the QMG's Department. [6]

While one may not agree with his proposed solution, the criticism clearly indicates that all was not well with signalling within the field formations. The difficulties of visual signalling in the immediate area of the Modder River, due to its flat and featureless ground, which might perhaps have coloured Lord Roberts' report a little, but onwards past Jacobdal the topography changes and there are numerous kopjes that could have provided the necessary facilities.

But in general it seems that little had changed since the time of the Zulu War twenty years earlier when Major Hamilton, OC 'C' Telegraph Troop, the Director of Military Telegraphs and by default also Director of Signalling (because there was no effective regimental system), had reported in similar terms – lack of heliographs and poor training of regimental signallers. Although the Army Signal School at Aldershot had been set up and training regimental signallers for some twenty-five years, much was still lacking. The nub of the matter probably lay with the attitude and lack of commitment of most regimental officers who had not experienced a war of this kind. Signalling was, in their eyes, neither important nor interesting, and they would have been very happy to see it shovelled off to the QMG's Department. It did not in most cases assume any importance in their careers in the Victorian army. There were of course exceptions, and some regiments had efficient signallers and interested signal officers. Also, in England, where conditions for its use were not favourable, the motivation to teach and use the heliograph was not high..

There was another fault with the organisation. Each regiment had a small scale of signallers – a section of about ten to fifteen – but when it came to fighting battles at Army, Division, or Brigade formation level, there was no permanent organisation. When the need arose, a divisional signalling officer had to be appointed from somewhere, and he had to prise the signallers from the regiments and organise them into some sort of ad hoc team for the task in hand. Once the immediate need had passed, the team dispersed and reverted to their regiments. As the war progressed the message was taken on board and all aspects of regimental signalling improved considerably, but for the moment it was defective.

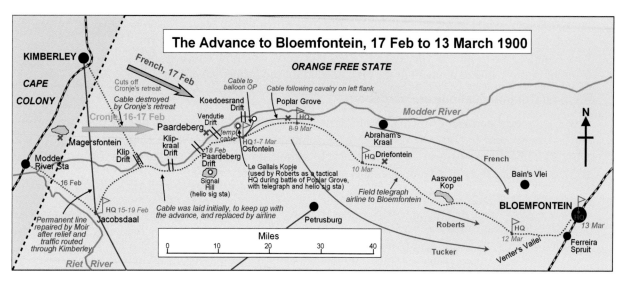

On to Bloemfontein

After their period of consolidation and resupply, the advance continued from Osfontein on 7 March with an attack on the Boer position at Poplar Grove, starting soon after daylight - to the surprise of the Boers! President Kruger had just arrived nearby and was with De Wet, discussing plans for an address to rally the faltering burghers when the first artillery shells arrived amongst the Boer positions. The aged Kruger was hastily sent away from the battlefield. The Boers beat a fast retreat from their positions, and the battle of Poplar Grove was over by midday with relatively few British casualties. Lord Roberts, now recovered from his short illness, viewed it from Le Gallais' Kopje, halfway between Osfontein and Poplar Grove, where he had arrived before daybreak. The cable cart was with him throughout the advance, providing rearward communications. The Kopje was also set up as a heliograph station (signs at last of some coordinated planning, as there had been at Colesberg), and from his elevated viewpoint Roberts was able to signal to his divisional commanders by heliograph, but there are no details of the visual signalling operation. A cable cart under Mackworth accompanied the cavalry on the flank, running out twenty-six miles of cable from Osfontein. Another cable cart under Henrici provided communication with the observation balloon and the force on the left flank. The telegraph office at Osfontein was moved up to Poplar Grove later that same day, effected quickly by using the cable that had been laid for the balloon, airline replacing the cable by early the next morning.

The next day eight miles of airline were constructed towards Driefontein under protection of an escort, and Mackworth ran cable over twenty miles with the cavalry to Abraham's kraal

Orders were issued on 9 March for the next phase of the advance to Bloemfontein. Robert's plan had to depend on ease of movement and water supply. Also, he had learnt that the Boers expected him to advance by the main road, which would take him to Bain's Vlei (see map), where they were preparing a defensive position. Roberts had reason to avoid a fight there, which would only hold him up; he wanted to surprise the Boers and advance quickly to Bloemfontein because he was intent on capturing trains and rolling stock, which he needed for his further advance along the railway line to Pretoria, before the Boers could withdraw

THE ADVANCE TO BLOEMFONTEIN

them to the north. With these things in mind, he divided his force into three columns. The left column under General French was to continue towards Bain's Vlei to keep the Boers guessing, but turn south towards the central column before reaching there; Roberts himself would command the centre column and head for Driefontein and Aasvogel Kop (vulture hill); and the right column under General Tucker would head south for Petrusberg and then east to Bloemfontein. Bloemfontein would then be attacked from the south-west by the three columns converging at Venter's Vallei. Before converging the columns would be some ten miles apart and advancing as rapidly as possible.

The advance resumed at 6:00 am on 10 March. Lord Roberts was aware that the Boers were in a position at Abraham's kraal, on French's route close to the Modder River, but he was unaware that they also occupied a position some six miles further south on his own intended route near Driefontein. There, Colonel Martyr, commanding a Mounted Infantry Brigade, was advancing with some of his troops on an unusual mission. He was acting as mounted escort to the telegraph section which Robert's, confident that the enemy had fled from this area, had sent on to construct airline twelve miles ahead of the main body. Proceeding to Driefontein, the section suddenly came under fire from Boer patrols on a nearby ridge. The escort managed to dislodge the Boers and secured the south-east corner of the ridge. The Boers were found to be in greater strength over a wider front than had previously been thought, and during the day, after reinforcements arrived, were eventually cleared from the area in what was known as the battle of Driefontein. Construction of the airline was able to continue.

The Capture of Bloemfontein

After Driefontein, the Boers retreated to positions that had been prepared for them about five miles from the town in a semi-circle around its western frontage. French's column was reunited with Roberts on the 11th at Aasvogel Kop and they made a dash for Brand Kop (Fire Hill), a prominent cluster of hills directly overlooking Bloemfontein, about four miles to the south-west. On the way there it was discovered that several trains had gone south to bring back the Boer commandos from Norval's Pont, and a detachment was sent to cut the railway line and prevent it returning. Brand Kop was found to be defended but, with the bit between their teeth, it was rushed by a small but determined force and in the gathering gloom the dispirited Boers panicked and fled. That night a Royal Engineers demolition party, guided by a local British man, rode right round the east of Bloemfontein and at 4:30 am on 12 March blew up a culvert carrying the railway line north of the town and cut the telegraph wires, also fighting off a Boer patrol. This occurred none too soon, for some trains including one with President Steyn aboard had already got away earlier in the night, but eleven trains and a hundred trucks were secured, just as Roberts had intended, and were soon to be put to use by the British.

Using a captured prisoner as his emissary, Lord Roberts had the previous day sent a proclamation to Bloemfontein to the effect that if his entry were unopposed, all inhabitants would be protected and he would refrain from artillery shelling, adding that any resistance would incur loss of life and damage. It was to prove enough. Bloemfontein was a pretty little town with many English inhabitants, whose political views were moderate, and most of whom had opposed the policy of allying with the Transvaal and dragging them, they thought, into an unnecessary war. Their view prevailed, and the Boers withdrew, although a number of spies remained in the town.

On 13 March, while they were waiting for the formal surrender, Captain Godfrey-Faussett was with the headquarters staff on a kopje some five miles to the south of the town and observing developments. Apparently unconcerned or unaware of the situation, a cable cart under Sergeant Longley could be seen trekking steadily into the town on its own. Perhaps Sergeant Longley had decided for himself that it was all over and he just wanted to get the cable laid. Godfrey-Faussett eventually galloped after Longley and his cable cart, but Sergeant Longley was right - there was no resistance forthcoming from the town, and together with a few press men, they were the first to enter.

The formal surrender soon followed. The Mayor rode out and presented Lord Roberts with the keys of the town as a token gesture, and Lord Roberts rode into the town shortly after 1:00 pm - a few hours after Sergeant Longley.

The entry was quite dramatic from the point of view of the Telegraph Division. They entered Bloemfontein, and rode through the streets to the post office – which they found in full swing, as though it were a normal working day. Godfrey-Fausset's diary reads as follows:

THE ADVANCE TO BLOEMFONTEIN

> Entered Bloemfontein at 11.15 and seized station and PO telegraph offices, and stopped all work. Cable between Ferreira and Venters Vallei unfortunately dis [disconnected, ie unserviceable for some reason] till 3.5 [?], and airline went dis behind at 2.45, so was unable to get Lord Roberts message off. A bitter disappointment. Got Bloemfontein office ready for work. Found no instruments there but morse inkers. (7)

Another description of the capture of the Bloemfontein telegraph office comes from an entry in *The Telegraph Chronicle*, entitled 'Relieving the Postmaster-General', and written by a sapper:

> Our captain walked into the General Post Office on the morning Bloemfontein surrendered, pulled all the wires off the test box, and ordered all civilians off the premises. Presently an old 'toff' came up and said, "I'm the Honourable Mr Falck, Postmaster-General of the Orange Free State". Our captain said, "I'm Captain Godfrey-Faussett, of the Royal Engineers, and I command you to leave the building at once." And he did, too. (8)

Mr Falck, along with many others, was later re-instated.

The same sapper gave an insight into counter work at the front. He described it thus:

> Rudyard Kipling was in the office this morning sending a 'gram to his wife at Cape Town. There are no coppers in circulation here and 'Rud' had 1s. 1d. to pay for his message. I offered to toss him 'two bob or nothing' which 'Rud' did, and lost.

Unfortunately, as Godfrey Faussett's diary extract describes, at about midday, just as Lord Roberts wanted to report the capture of the town by telegram via Kimberley (which it will be recalled was where the 125-mile length of airline now went after the re-routing at Jacobsdal to the repaired permanent line), the line had been cut. Godfrey-Faussett was extremely frustrated, and instead a mounted despatch rider was sent off to Kimberley, 100 miles away, with the news. But at 6:00 pm the next day, the 14th, some thirty hours later, the operator who was sitting watching the useless instrument suddenly obtained signals and Lord Roberts telegram was immediately sent, beating the despatch rider by two or three hours. Professional pride was restored.

What had happened was that Lieutenant Moir and his detachment, having been left in the rear to maintain the line, was coming forward to join them at Bloemfontein, together with the infantry detachments protecting them, on the express order of Lord Roberts who feared for their safety as, in the rapid advance from Paardeberg, he had not thoroughly cleared the area of Boers. Moir was testing the line as he came, had discovered the cut, and repaired it as he came through. After this the airline worked for ten days without any maintenance, but it was a tenuous rear link for so large a force in an area not fully cleared of enemy, and something more reliable was urgently needed - to be described in the next chapter.

Corporal Charles Shaw, a telegraphist from Liverpool, wrote from Bloemfontein on 27 March describing how the telegraphists had operated during the the advance to Bloemfontein. Readers today will recognise what is probably the origin of that well-known signals procedure for 'stepping up' and 'changing command'.

> The Army Post Office Corps is a complete failure and is to be thoroughly reorganised. Rather different to the praise Lord Roberts gave the Telegraphists for their work during the march [to Bloemfontein]. He told the Captain [Godfrey-Faussett] to tell us he was highly pleased and astonished at the way the telegraphs sprang up and were worked. We were working DX [duplex] to Kimberley all the time, and as we had to march sometimes as much as 16 miles a day and every inch of line had to be put up it meant a lot of work. The way we managed it was to split the head-quarter [telegraph] staff into two. I had charge of one half and another Corporal took the other half. I would go on to where the Field Marshal intended to stop and join through to the station I had just left, who would exchange signals to see the new piece of line was good, and would immediately join through to Kimberley, dismantle his office, and come on, and pass our station, go on to the new one, exchange signals with me; then I dismantled my station and passed him again. So you see we were always working through, and one station did not stop work till the other was ready to start from the front. We were very busy, especially after some of the fighting, when they handed in lists of casualties and despatches, some of the messages being from 1,500 to 3,000 words each; but we got over it all and reached Bloemfontein at last, where we have the Post Office. There are about fifty telegraphists here, so there is plenty to do the work now. Rather different to when we were on the march when there were only twenty-three of us. I expect we will leave here before you get this. Pretoria this time! ….. (9)

It was the end of a successful period of intense operations, but the war was far from over and, without a break, new problems had to be addressed.

Endnotes

1. *The Times History of the War in South Africa*, Vol III, p 387 footnote.
2. *The History of the War in South Africa*, Vol II, p 36.
3. *The Telegraph Chronicle*, 13 April 1900.
4. Diary, Captain Godfrey-Faussett, Royal Signals archives.
5. *The Times History of the War in South Africa*, Vol III, pp 451-2.
6. *The Boer War*, Carver, p 91.
7. Diary, Captain Godfrey-Faussett, Royal Signals archives.
8. *The Telegraph Chronicle*, 25 May 1900.
9. *The Telegraph Chronicle*, 25 May 1900.

Chapter 8

At Bloemfontein

A Friendly Reception

The troops were surprised at the friendly reception, but Bloemfontein was essentially a British town. They found streets brightly decorated with British flags, and cheering crowds. After forty-six years and two days the flag of Britain was flying once more over the capital of the old Orange River Sovereignty, so hastily handed over in 1854 under the terms of the Bloemfontein Convention in one of the many poor political decisions made by British governments in the history of South Africa.

The Orange Free State south of Bloemfontein seemed happy to surrender, and relatively small numbers of British troops were able to take over the countryside to the south. Many government administrators resumed their posts under the British – including the dismissed Postmaster-General, the Honourable Mr Falck! This general spirit of apparent cooperation around Bloemfontein misled Lord Roberts into believing that the whole of the Orange Free State would react similarly. An amnesty was offered, and some obsolete guns were handed in as gestures. But out in the veld, their natural habitat, Boers were regrouping.

In a few brief weeks the course of the war had been changed. Lord Roberts was keen to maintain the pressure on a retreating and demoralised enemy, but the army needed to pause for resupply and reorganisation. Events conspired to make this pause rather longer than he would have wished, and there was to be a period of seven weeks at Bloemfontein before the army could resume its advance to Pretoria in early May.

Much to be Done

At Bloemfontein the immediate problem was logistics. The army had arrived there with but five days of food in hand, and hardly any forage for the animals. Food supplies had quickly been requisitioned on arrival but these would allow only a short period of relief, and there was need for reinforcements, remounts, medical equipment, stores of all sorts, and the build up of a reserve of supplies before a further advance could be sustained. All of this depended on the lines of communication. Also, major health problems afflicted the army at Bloemfontein.

Rail traffic with the south was cut due to the demolition of the railway bridges over the Orange River at Bethulie and Norvals Pont. The nearest port, East London, was 400 miles away, Port Elizabeth was 450 miles, Cape Town, the principal base, was 650 miles, and the railways from these ports to Bloemfontein converged into a single line at Springfontein, ninety miles south of Bloemfontein. Fortunately there were experienced and able officers in charge of the logistics and the supply chain in Lord Roberts' HQ, and a most efficient railway operating team of the Royal Engineers. But none of them could operate without communications. As a major contribution to the logistic operation, efficient telegraph communications were needed urgently.

Restoring the Telegraph Communications

To support the army as it sorted out these problems, a number of issues had to be addressed quickly by the Telegraph Division. Paramount was restoring and operating permanent telegraph communications, for which high levels of traffic capacity were going to be needed. Then there was work in support of tactical operations to clear Boers from pockets of resistance to the south and east of Bloemfontein. Finally, preparations had to be made for the continuing operation - the advance to Pretoria.

It required a coordinated effort between the Telegraph Division operating under Captain Edmund Godfrey-Faussett in Bloemfontein, the railway communications operating under their Superintendent of Telegraphs, Lieutenant Graham Manifold, and the permanent line party under Captain John Fowler who was presently restoring permanent lines in the Cape Colony in conjunction with the colonial Telegraph Department. To assist in this and other staffwork the Director of Army Telegraphs, Lieutenant-Colonel Richard Hippisley, moved from Cape Town to Bloemfontein on Sunday 18 March.

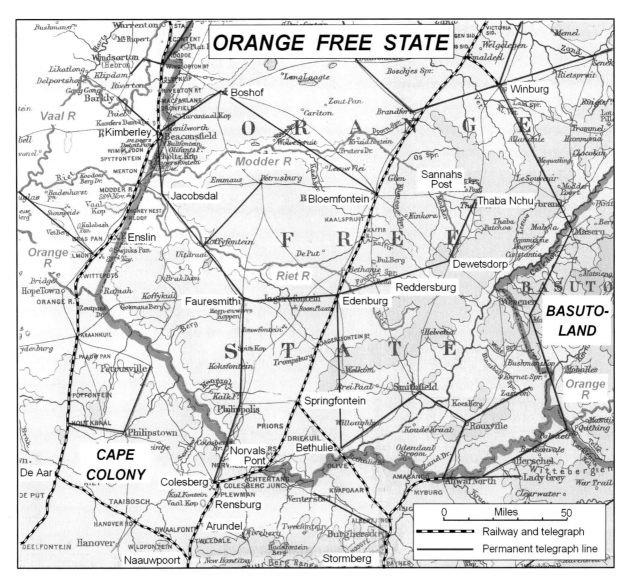

It is an appropriate moment to describe the railway telegraphs, a vital part of the extensive railway network that was going to be crucial to the army operations in the two Boer republics. These railway lines were to be run by the Royal Engineers. In October 1899, at the start of the war, Major Percy Girouard RE had been appointed Director of Railways. He had held a similar position in the Sudan in the campaign led by Lord Kitchener there in 1896-98, when the construction and operation of a desert railway had become necessary. During that appointment Girouard had noted the competence (and also his ability to handle the bullying Kitchener) of the young Director of Telegraphs, Lieutenant Graham Manifold. Now, with telegraph an essential part of Girouard's responsibilities, he secured the services of the capable and experienced young Manifold. After the Sudan campaign Manifold had returned to England in early 1899, got married, and been posted to the 2nd Division of the Telegraph Battalion. After but a few months he was sent to South Africa, where he was now responsible for creating, staffing, and working the railway telegraph system for about 1,000 miles of captured railway lines in the two Boer republics, much of it in a demolished condition. Despite the difficulties it was to be achieved successfully, demonstrating the reputation he had achieved as a junior officer for initiative, drive, and organising capacity. The course of his work under the command of Girouard involved much liaison with the Telegraph Division.

The first priority after capturing Bloemfontein was to improve the Army's rear link communication. It will be recalled that after the relief of Kimberley the permanent line there, running northwards beside the railway through De Aar, Orange River Station, Modder River Station, and on to Kimberley was quickly restored and the Kimberley Telegraph Office re-opened on 20 February with three lines southwards. Responsibility for the permanent line communications on that side, all within Cape Colony and now free of enemy apart from a few subversives, had been handed back to the Cape Colony Telegraph Department.

AT BLOEMFONTEIN

The rear link from Bloemfontein was temporarily being provided by means of the unguarded 125-mile tactical airline to Kimberley, constructed by the Telegraph Division along the route of their earlier advance south of the Modder River to Bloemfontein, and it had to be replaced quickly by something more robust and with greater traffic capacity. South of Bloemfontein, both in the Orange Free State and onwards into those parts of the Cape Colony previously occupied by the Boers, the permanent lines had been wrecked to varying degrees and would take some time to repair, so the first possibility for improved communications appeared to lie in using the permanent line from Bloemfontein to Kimberley via Boshof (see map, page 80).

The Line to Boshof

In the process of taking over the telegraph office in Bloemfontein on 13 March it was discovered, by reading messages and the office diary, that the Boers had been using the telegraph line between Bloemfontein and Boshof until the last minute, but the line was no longer working and had been cut.

On 14 March, the day after Bloemfontein was occupied, Lieutenant Mackworth with a cable cart mended a break in the Boshof line near Bloemfontein, but it was still cut further on towards Boshof, over sixty miles away. Harry Mackworth and Sergeant Cadwell volunteered to ride out, find the fault, and repair it. It was a dangerous ride as the country through which the line passed had not been cleared of Boers, although the British were by then in occupation of Boshof itself. The Telegraph Battalion Historical Record relates that three years earlier, in 1897, the then Corporal Cadwell won the regimental championship for mounted sports - so he was clearly the right man for this job! Having received Lord Roberts' sanction for this venture - Roberts was well aware of the danger, but the need for communications was urgent - they started about midday; no mounted escort was available. At about sunset they came across a party of Boers near Abraham's Kraal, and managed to remain hidden until dark. One of their horses had become too exhausted to continue, but under cover of darkness they succeeded in capturing one of the Boers' horses and rode on to Boshof, where they arrived at about 10:00 am next day having found the fault and repaired the line on the way. After his earlier efforts at Colesberg, the young Mackworth was having quite an exciting war.

As a result of their efforts, and those of Lieutenant Macfie in Kimberley who had repaired the line from Kimberley to Boshof, good communications between Bloemfontein and Kimberley were possible via Boshof from 15 March, and all military traffic was cleared by the 16th, as well as some private work. Godfrey-Faussett's diary notes: "took private work – great rush." The tactical airline to Kimberley, despite no attention for three days, was also by some miracle still working.

Mackworth returned to Bloemfontein on Tuesday 20 March with an escort, bringing with him a Wheatstone automatic which had been sent up from De Aar. It was going to be essential in clearing the volume of traffic being generated, and it was working between Bloemfontein and Kimberley on 21 March.

Sapper Stimson, one of the Post Office Telegraphists who formed part of the advanced Wheatstone detachment and had been at Modder River, then later Kimberley, describes in a letter how they brought the Wheatstone to Lord Roberts at Bloemfontein:

> We are making great progress with the campaign, as you have read in the *Daily Mail* I expect. We started from Modder River to join Roberts' column, but were unable to catch it up. After getting half-way to Jacobsdal we had to return because the roads were so heavy, and try from Kimberley. When we reached there we were hanging on for a convoy, and were eventually stopped until Roberts had reached Bloemfontein. We spent a very nice week in Kimberley. …

> We at last got orders to start, and go via Boshof, and there to pick up an escort; but on arrival found none, and had to wait for one from Kimberley. We started in the afternoon and bivouacked at night, the distance being thirty-two miles. It was a pure bit of South African veldt life. We soon got the pot singing and had a decent meal on the bank of a pond. … It was a splendid night, clear moonlight and perfect stillness. I thought of the buzz at TS [Central Telegraph Office, London] at the time. We started at daybreak and reached Boshof about 4 p.m. It is a nice little place, very much like an English village. We camped on the market green with four companies of the Yorkshire Light Infantry, who were garrisoning the place. All the women and children had cleared out. They were told that we should kill all the old women and make the young ones slaves – these are the lies which have been spread about. It was terrible to see so many people in black. The Yorkshires gave an open air concert at night, and a very good one it was. All the loyal inhabitants were invited and we passed a very enjoyable evening.

> Our escort arrived on Sunday night [18 March] and we started off at daybreak the next morning. We only had twenty Mounted Infantry, not much protection. We managed to get some ducks and fowls, and any amount

of vegetables. We could have got enough furniture to stock an eight-roomed house. All the [Boer] inhabitants had fled as soon as they heard our people were marching on Bloemfontein. It was a very wet evening and night; we were lucky to be able to sleep in a coach house at Ortle's farm, where Kruger and Steyn tried to rally their forces.

We started off again at daybreak and had to walk some distance on account of the roads being so heavy. We were to get to Bloemfontein as soon as possible, for we had the Wheatstone apparatus with us and it was greatly needed, as they had about a day and a half's [message traffic] delay. The next night we took possession of an old post house for our halt and had a little excitement. We had the pot on the fire cooking some chickens and ducks and all manner of vegetables when suddenly we heard the report of a rifle. We immediately paddled off for our carbines and thought we were in for it, but it turned out that our sentry had challenged some men in a cart and they would not reply so he fired at them. Then they answered and said they had a pass. We had a very restless night as there was so much livestock in the house – three of us cleared out during the night to sleep outside. I am still scratching.

We started off again early, and passed an uninhabited farm with plenty of vegetables and tomatoes of which we bagged a plentiful supply to take to Bloemfontein. It is a grand life to commandeer what you want without opposition.

Bloemfontein was reached about noon, and is a very decent place. There are some very good buildings. It is very English – English being spoken by everyone, and most of the shopkeepers are English. Our troops had no opposition in taking it, the mayor going out to meet Roberts. We are using the Post Office which is a fine-built place, but the sanitation of the town in general is very poor. Fancy, in a large building like this, no water!

We are doing eight hours on and eight off – a bit thick, don't you think? Don't know how long it is going to last. Am pleased to say I am in good health myself, but we have had a good deal of sickness. We left three behind at Modder in hospital with fever, and lost Hawkins (TSF) at Orange River of enteric. Poor little chap! He was going to get married when he returned. Such are the fates! [1]

The Boshof line was again cut by the Boers on 22 March. The Boers were known to be laagered under the line in the area of Abraham's kraal, just north of Driefontein and thirty miles from Bloemfontein, but it was impracticable at that time for Lord Roberts to send a force to dislodge them, so the line again could not be used. But it had served well during that critical period, and by that time lines south from Bloemfontein had been repaired so that alternative arrangements could be made.

Health Problems

On arrival at Bloemfontein there was an epidemic of typhoid and enteric fever, the prime cause being the insanitary conditions and contaminated water at Paardeberg mentioned earlier, and now made worse by the concentration of troops living closely together at Bloemfontein. At that time it was only a small town that could not accommodate the troops, so they were concentrated in tents and without proper sanitation. Many died there of disease due to clean water supply problems and hygiene compounded by unsatisfactory hospital facilities and understaffed nursing support. In fact the Boer War, by the time it ended, had claimed more British lives from disease (12,911) than from enemy action (7,091) and was the last war where this adverse ratio was to occur. It was not just hygiene discipline. Drinking water was difficult to obtain in the quantities needed and supplies often had to be transported some distance, an ongoing supply problem. While tested and approved at source, it was often contaminated during transport in an unhygienic design of water cart. The design of the army water bottle – the type used subsequently in both world wars, with a narrow neck and cork, that could not be cleaned properly – was also unhygienic.

As at Ladysmith, the medical arrangements at Bloemfontein were not satisfactory, somewhat reminiscent of the scandals of the Crimea, and there were nothing like enough nurses. This was well recognised by Lord Roberts and the deficiencies in hospitals and medical organisation and resources was to lead to the formation of the Queen Alexandra's Imperial Nursing Service, later the Queen Alexandra's Royal Army Nursing Corps. When he returned to England from South Africa, and became Commander-in-Chief, Lord Roberts was in communication with the new Queen, Alexandra (Victoria having died in 1901), and made a number of proposals about the need for such a service, including that she be its patron. The suggestions fell on sympathetic ears, and out of this the Army's new nursing service was formed, but it is fair to say that it stemmed from Lord Robert's experience at Bloemfontein.

AT BLOEMFONTEIN

Restoring the Permanent Lines South of Bloemfontein

The obvious long term solution, now that the Army had moved eastwards to the other railway line, was to use the permanent telegraph line south from Bloemfontein, much of it having been wrecked on both sides of the Orange River. To this end, Captain Fowler and his permanent line party, after repairing the lines between Modder River and Kimberley, had gone round to Naauwpoort and from there restored the permanent lines within Cape Colony up to the Free State border at Norvals Pont and Bethulie, country which had been in Boer occupation. Both bridges over the Orange River at these places were destroyed and it was not known which would be repaired first but it was clearly going to be extremely urgent to get Bloemfontein into reliable telegraphic communication as soon as possible. Considerable work was needed to accomplish this, the distance from Bloemfontein to the Cape Colony border at Norvals Pont being about 125 miles.

By 15 March, two days after Bloemfontein had been occupied, the line had been repaired through Rensburg, Colesberg, and the minor intermediate stations up to Norvals Pont, the line detachments following the British troops in that area as they advanced northwards into the Free State. On the Bethulie side, some thirty miles further east, following up behind the advancing troops of Generals Knox and Gatacre, the line was repaired from Steynsburg and Stormberg through Burgersdorp as far as Bethulie Bridge by 13 March, the day that Bloemfontein was occupied, and Bethulie Station and Springfontein in the Orange Free State by 16 March.

Meanwhile, from Bloemfontein itself, the line southwards towards the two railway bridges was inspected and repaired by Lieutenant Henrici. There were three wires beside the railway but two of these were required for the operation of trains, leaving only one for army telegraph communications. On 14 March, the day after Bloemfontein was captured and the same day that Mackworth was on his way on horseback to Boshof to try and repair that line, Henrici departed by rail on an engine with a small party heading for Norvals Pont. By that evening, the line was repaired to Edenburg, and this also brought into communication the line via Fauresmith, Philippolis, and Colesberg, which had only been lightly damaged and was quickly repaired. Henrici continued on by rail to Bethulie and that line was through via Springfontein by the evening of 16 March so direct communication with the Cape Colony then became possible and Godfrey-Faussett managed to get in communication with Fowler. On the next day, 17 March, the line through Norvals Pont to Naauwpoort was opened, giving a second outlet to the Cape Colony system.

It was desirable to open a third line route, to provide alternative routing in case of damage and to increase available capacity. Captain Godfrey-Faussett, with a small detachment, a scotch cart, and an escort, set off from Bloemfontein on 19 March, travelling partly by train and partly marching, with the intention of opening up the alternative route south which went from Bloemfontein by a wire beside the road through Reddersburg to Edenburg, where it divided. One route went from Edenburg to Bethulie, and was found to be damaged. The other route went to Fauresmith, and from Fauresmith through Philippolis to Colesberg. The line to Reddersburg was intact, and the telegraph office there was opened. The line on to Edenburg was also intact. As this area was still occupied by Boers it was not safe to inspect the line, but a native lineman managed to get through and repaired it, thus giving a second route from Bloemfontein to Bethulie. The Edenburg, Fauresmith, Philippolis, Colesberg line was working and was set up as a through circuit from Bloemfontein to Colesberg. This deprived Fauresmith, Jagersfontein, and Philippolis of any telegraph, so Lieutenant Henrici was sent to recover thirty miles of the now disused tactical airline from Jacobsdal to Bloemfontein and it was re-erected as an auxiliary airline alongside the permanent route to serve Edenburg, Fauresmith, and Jagersfontein; this was completed by 2 April.

So to summarise that rather involved sequence of events, there were now three circuits using permanent lines south from Bloemfontein for use by the Army Telegraphs: the line alongside the railway to Norvals Pont and Naauwpoort, which was regarded as the primary line; another through Reddersburg, Edenburg, and Fauresmith to Colesberg; and a third through Reddersburg, Edenburg, and Bethulie to Burghersdorp. In addition, the railway telegraphs were working. By some very well-organised and highly active work, the Telegraph Division had achieved all this within a week of capturing Bloemfontein.

Kimberley was eventually connected via Naauwpoort and De Aar. Because the Cape Colony Telegraph Department could not deal with the handover of traffic at either Naauwpoort or De Aar for some reason, presumably staffing problems and the high volume of traffic, Kimberley became the handover point between the Army Telegraph and the Colonial Administration – a rather roundabout route to send it back down to Cape Town! This circuit, to which the Wheatstone was transferred after the Boshof line was finally cut on the 22nd March, became the main rear link route from Bloemfontein for many weeks and handled a considerable volume of traffic.

Godfrey-Faussett's diary notes that:

> 24th March: Began working 8 to 2 [8.00am to 2.00pm] and 6 to 10 [6.00pm to 10.00pm] duplex to NOM [telegraph address of Naupoort Military], and 2 to 6 and 10 to 8 (all night) Wheatstone to KB [Kimberley] via Naauwpoort and De Aar. Got Bethulie SC [single current] and sent all Eastern work [*ie* to Port Elizabeth, Natal, *etc*] that way. Timed Wheatstone simplex at 55 messages per hour for 10 hours, and duplex key speed at 35 [messages] an hour for 4 hours. Heard of arrival of 3 new sections.

> 25th March: Took no private work at all [because military traffic took priority]. Good steady days work, and got clear by midnight. Henrici back [from recovering 30 miles of the tactical airline to Kimberley that was no longer required]. Timed Wheatstone simplex at 110 messages an hour for 3 hours.

It is worth reminding the reader that these permanent telegraph lines were not just used for normal communications purposes; some of the telegraph lines were separately used by the railways for the operation of their trains, and more will be said later about railway communications and the technical and resource conflicts between the two different user requirements. Until these permanent telegraph lines were operational there were neither communications facilities nor railway services. The permanent line party, although not at the sharp end with the Army HQ, carried out a vital job with great efficiency, assisted as ever within the Cape Colony by the Cape Telegraph Department with their local knowledge, men, and materiel.

The Telegraph Office at Bloemfontein.

The telegraphists in the Bloemfontein telegraph office were kept very busy. Sapper A. W. York, who was later to be captured at Heilbron on 22 May, wrote to Colour Sergeant Kemp on 20 April:

> The great trek of Lord Roberts' column has not come off yet, consequently I am still in the town office here, where my section had orders to report on 7th inst, the staff already here being very short-handed, while the delay was two or three days on military work, and more on private telegrams, a state of affairs which was soon changed, though even now there is far more delay than would be allowed at home. This is due to the fact that

many wires are cut, and our only outlet is a well-guarded single wire to Kimberley, which is running slip day and night. We moved our tents to the garden of a commandeered house in order to be near the office, as the hours are very long, viz., midnight to 7 a.m., 2 p.m. to midnight one day, and 7 a.m. to 2 p.m. next day. Our neighbours were Scotch people, who were very glad the British occupied the town, and their little children prattle merrily to us over the garden wall. ..... We have fine walks to the surrounding camps on our mornings off, and usually end by ascending a kopje, on many of which the R.E.'s may be seen at work building fortifications, while at the top there are artillery camps, and the heliograph is busy by day, giving place to the signal lamps at night. (2)

Operations in the Area of Thaba Nchu and Dewetsdorp

It was said earlier that Boers in the veld were regrouping. Operations took place in the areas of Thaba Nchu and Dewetsdorp (the small town, or 'dorp', was named after General De Wet's father) to try and clear the country south-east of Bloemfontein. Some 5,500 Boers had managed to escape from positions they had occupied around the area of Colesberg and Bethulie, and from there they scattered north and east, remaining a threat to Lord Robert's impending advance northwards through the Free State and into the Transvaal.

On 18 March General French and the Cavalry Division left Bloemfontein to clear the country to the east as far as Thaba Nchu, reinforcing a small detachment which had been sent on the 15th to guard the strategically important waterworks supplying Bloemfontein at Sannahs Post. Telegraph communications were provided by Lieutenant Moir and his section and a number of temporary telegraph offices were opened along the route. The cable was repeatedly cut by guerrillas and communications were only maintained by the efforts of many linemen sent out from both ends.

General French's column returned to Bloemfontein, but on 31 March part of it was badly surprised by an ambush at Sannahs Post when attacked by General De Wet. As a result, 159 British were killed or wounded, 421 were taken prisoner, and seven artillery guns and eighty-three wagons loaded with stores fell into De Wet's hands. Sergeant Shergold and his line detachment of Corporal Williams, Sapper Hay, and Driver Preston, who were reeling up the cable as the column was withdrawing towards Bloemfontein, were captured and taken to Pretoria. The cable cart was subsequently found in pieces at Winburg, and the instruments at Harrismith, near the Natal border! The battle at Sannahs Post had been a severe 'wake up' call to over-complacency about the Boers apparent disinclination to continue to fight in the Orange Free State – and much more was to follow. It was also a reminder about the vulnerability of unescorted line parties.

General De Wet.

Meanwhile, Lieutenant Mackworth with a small party had been assisting the 3rd Division (Major-General Gatacre) in similar clearance operations around Reddersburg since 23 March. Reddersburg was held by the Boers until 4 April, and the towns of Dewetsdorp and Wepener, on the Basutoland border, were still in Boer hands. Mackworth being needed back at Bloemfontein for preparations for the advance northwards, he and his detachment were replaced on 13 April by Moir and his section after their efforts at Thaba Nchu. The permanent line to Dewetsdorp was still mostly intact, and Moir kept the Division in contact from its various halting places along the route to Dewetsdorp, by cable to the nearest point on the permanent line. On 25 April the Boers evacuated Dewetsdorp, the British troops marched in, the telegraph office was opened, a break in the line was repaired, and they were in communication with Bloemfontein. On the 26th they continued the clearance operation northwards to Rietput on the line to Thaba Nchu, where a temporary telegraph office was established, and on the 27th they reached Thaba Nchu, where by good fortune the permanent line was found to be in working order, giving access to the Army HQ at Bloemfontein by the route through Dewetsdorp and Edenburg. At that point we shall leave Moir and his section, for they were soon to be busy again providing communications for the force that was about to advance northwards protecting the right flank of Lord Roberts' main column from Boer attack from the east. That will be described shortly.

What might have struck the reader after these descriptions of events is how the resources of the Telegraph Division were scattered over a wide area, in small groups, operating independently on a variety of tasks. It called for strong leadership and capability at junior officer and senior NCO level. That they achieved what they did in this period of time is noteworthy.

AT BLOEMFONTEIN

Replenishing Stores

In the period 29 March to 4 April Captain Godfrey-Faussett and a small administrative team went by train to replenish stores from the depot at De Aar, the place they had set off from towards the Modder River the previous November – a long time ago, it must have seemed to them. In his absence Lieutenant Moir, back from Thaba Nchu, looked after the camp at Bloemfontein which he also moved to a new site just south of the town. Godfrey-Faussett and his party got back on 4 April after a long and slow railway journey via Norvals Pont with three wagons full of stores and some additional native drivers.

Reorganisation

As well as stocking up with stores, some reorganisation was needed to prepare for the advance to Pretoria. This was going to be done by a number of columns advancing on different and well-separated routes, all needing to be kept in communication. Experience had shown the necessity for a cable section able to run long cable lines without borrowing from and so disorganising the remaining sections. At the same time there was a need to increase the numbers in an airline section to enable two working detachments instead of one, as the rate of construction with only one detachment was insufficient to keep pace with a day's march.

Additional men were found by getting volunteers from other RE Field Companies – ten men from the 6th Company, six from the 26th Company, and six from the 38th Company. Lieutenant Mackworth was placed in charge of a new cable section consisting of four detachments, each with its own mounted men and a clerk (as telegraphists were than called), and each detachment under a mounted sergeant (Sergeants Blackburne, Cadwell, Glue, and Longley). Lieutenants Moir and Henrici were re-equipped as enhanced airline sections. All stores were sorted and loaded accordingly, and in the period 12 to 16 April the new sections were practised in their drills. On 19 April fifty-six mules were drawn, branded with their identity, and arranged into teams. This formed part of a wider reorganisation needed for the advance to Pretoria.

Three More Telegraph Sections Arrive

On 23 March three more sections under Captain H. B. H. Wright (who had been left in Aldershot in charge of the rear elements of the Telegraph Battalion when Major Hawkins and the original section went to Natal in September) had arrived at Cape Town as reinforcements from England. From Cape Town they went initially to De Aar to be equipped and mobilised. On 18 April the Director of Army Telegraphs, Lieutenant-Colonel Hippisley, went to De Aar to brief them on the forthcoming operations. Wright and Lieutenant L. A. Sherrard with two sections were sent to Kimberley, and Lieutenant O. T. O'K. Webber's section went to Bloemfontein, arriving on the 24th.

The Reorganised Telegraph Division

Also by now the section which had been locked up in Ladysmith, originally the first to arrive in South Africa, had been released and, after a period of recuperation and re-equipment, was fit for further operations. With the recent reinforcement of three sections there were eight sections now available and they were reorganised and redesignated to support the forthcoming advance as follows:

In Bloemfontein:

Lieutenant-Colonel R. L. Hippisley, Director of Army Telegraphs.

Captain E. G. Godfrey-Faussett in command, and Lieutenant Bannerman:

'C' Section. Lieutenant H. L. Mackworth, with four cable detachments.

'D' Section. Lieutenant J. P. Moir, with two airline detachments.

'E' Section. Lieutenant E. O. Henrici, with two airline detachments.

'F' Section. Lieutenant T. O'K. Webber, recently arrived, with two airline detachments.

In addition there was the permanent line party under Captain J. S. Fowler and Lieutenant A. W. Hepper, now assisted by the cooperative former employees of the Orange Free State civil telegraph administration. Captain C. de W. Crookshank joined on 4 May, and was deputed to Fowler, mainly to assist in financial matters. Their combined task was to maintain and operate the permanent telegraph system in the Orange Free State as it was taken over by the advancing force.

In Kimberley:

Captain H. B. H. Wright, a recent reinforcement, was assigned to General Hunter's column. Lieutenant D. S. MacInnes, previously staff officer to Lieutenant-Colonel Kekewich in Kimberley and now assigned to telegraph duties with Lord Methuen's column. MacInnes had previously served with the Telegraph Battalion, having taken part in the 1896 Ashanti Expedition, so was well experienced in his new role.

'G' Section. Lieutenant W. C. Macfie.

'H' Section. Lieutenant L. A. Sherrard, recently arrived.

In Natal:

Major W. F. Hawkins, Captain R. B. H. Boys, and Captain J. N. C. Kennedy.

'A' Section. Lieutenant A. B. R. Hildebrand.

'B' Section. Lieutenant E. V. Turner, commanding the section which took part in the relief of Ladysmith, previously under Lieutenant Jelf.

Again, it is perhaps an appropriate moment to reflect on the high level of responsibility that was devolved to relatively low ranks in the Telegraph Division. Godfrey-Faussett, for example, as a Captain, was in daily contact with the Army Commander, a Field-Marshal, and responsible for providing him with his communications. Lieutenants were responsible to Generals for keeping their widely separated Divisions in communication. And Sergeants had many responsible tasks. The Boer War and its extensive communications requirements, despite what today might seem to be the limited technology of the time, were to be the turning point for developing a better resourced and better structured army communications organisation for the future.

The regular troops provided the framework, but the part played by what might generally be called the 'reserve army' was crucial. It is interesting to look back over the history of how this reserve of telegraphists developed – not from any great strategic military planning decisions but from a political decision to 'nationalise' the private telegraph companies in 1869, and the consequent shortage of skilled civilian manpower leading to the army providing manpower. It is fortunate that at that time education and literacy had improved, most certainly amongst those who served as telegraph reservists, to the extent that they could undertake great responsibility throughout their ranks.

Endnotes

1. *Telegraph Chronicle*, 11 May 1900. Sapper Alfred Hawkins, aged 22, who died of enteric fever at Orange River on 13 March 1900, was the first casualty from amongst the Post Office reservists. TSF was the name of his office – London Overseas Telegrams. Further detail about Hawkins, including a photograph, will be found in *Khaki Letters from My Colleagues in South Africa*, p54.

2. *Khaki Letters from My Colleagues in South Africa*, p184.

Chapter 9

The Advance to Pretoria

The Plan

The advance through the Orange River Colony was to be far from the easy task it had at first appeared after Bloemfontein capitulated. The earlier cooperative attitude by the largely British inhabitants of the town was in complete contrast to the attitude of the Boers remaining undefeated out in the veld. Their President, Steyn, had escaped from Bloemfontein shortly before it was captured and continued to lead them with great determination. The eastern area of the Colony had in the past month become very active with resurgent Boer commandos, led principally by General Christiaan De Wet. They had shown themselves to be highly mobile, and capable of effective 'hit and run' raids - and they communicated amongst themselves, as Gideon Scheepers had planned, by heliograph. [1] The railway line to Pretoria and hundreds of miles of telegraph lines through often remote areas were easy targets. Consequently the Boer Commandos tied down a disproportionately high number of British troops defending these vital means of communication.

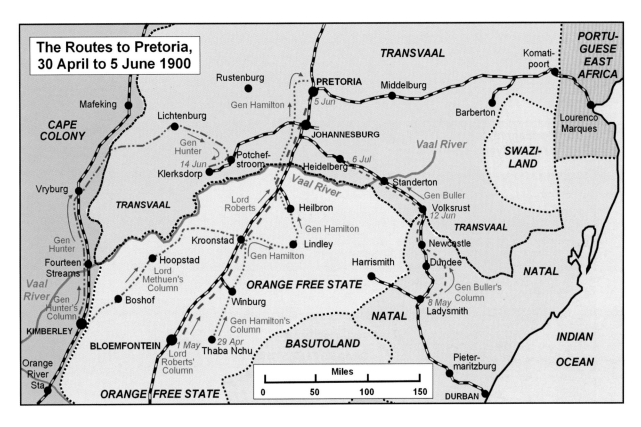

Lord Roberts' plan was to advance to Pretoria on a wide front in five column as shown in the map above.

- The main force, about 25,000 strong, under Roberts' own command, to advance from Bloemfontein along the line of the railway. They were to be joined after about a week by General French's Cavalry Division which had been held up while refitting after carrying out operations in the eastern part of the Colony. To this column was assigned Captain Godfrey-Faussett with 'C' and 'F' Sections under Lieutenants Mackworth and Webber.

- General Ian Hamilton's force, some 14,000 men, to move northwards from Thaba Nchu, parallel to the main column, to join the main force at Kroonstad. The main purpose of this column was to protect Roberts' right flank (ie from the East, where the Boer commandos were most active) and fill the large gap between himself and Buller in Natal. To this column was assigned 'D' and 'E' Sections, under Lieutenants Moir and Henrici, with Moir in overall charge.

THE ADVANCE TO PRETORIA

- To the west, Lieutenant-General Lord Methuen's Division, 10,000 men, still in the area of Kimberley, to advance through Boshof and Hoopstadt, and then to be ready either to cross the Vaal river and merge with Hunter on his left, or to close in on Lord Roberts' main thrust to his right, the decision to depend on progress. In the event, he closed in on Lord Roberts at Kroonstad, and remained in the Orange River Colony to deal with the Boer commandos there. To this column was assigned Lieutenants D. S. MacInnes (having completed his duties with Kekewich in Kimberley), and Lieutenant L. A. Sherrard, newly arrived with 'H' Section.

- On the extreme west, General Sir Archibald Hunter with his Division, 10,000 men, starting also from Kimberley but to advance via Fourteen Streams (near Warrenton) and Vryberg. To this was assigned Captain Wright as ADAT with 'G' Section under Lieutenant Macfie.

- The Natal force, now 45,000 men under the command of General Sir Redvers Buller, comprising three Divisions, was originally intended to clear Natal and then join up with Roberts in the Orange River Colony but, due to their slow progress, advanced from northern Natal along the line of the railway and into the Transvaal through Volksrust and Standerton. With this column were those remaining after the relief of Ladysmith - Major Hawkins, Captain Kennedy, Captain Boys, and 'A' and 'B' Sections under Lieutenants Hildebrand and Turner (Jelf's replacement).

In total it was some 100,000 men who advanced to Pretoria. Additionally there were over 50,000 who were to be left at various points, to secure the country behind the advancing troops and on its flanks, to guard the line of communication, or to act as reserves. Other troops and militia battalions guarded the railway and into the Cape Colony. All of these were to be kept in communication by the Orange River Colony permanent telegraph system under Captain J. S. Fowler and Lieutenant A. W. Hepper.

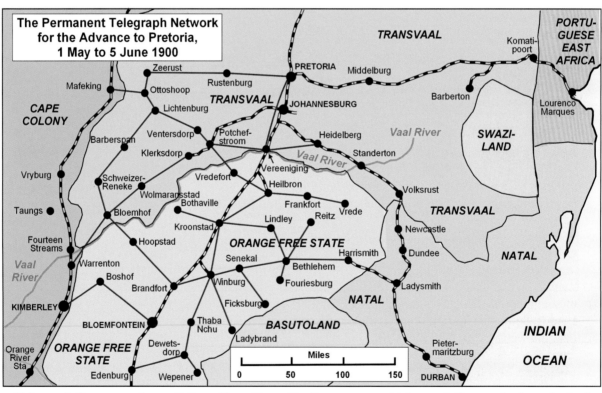

Telegraph lines run along all the railway lines. On the permanent telegraph lines only the principal routes and offices are shown.

The Communications Plan

Clearly, over these distances, both width of frontage (over 300 miles) and distance to Pretoria (285 miles from Bloemfontein), the telegraph communications plan was based around use of the existing permanent telegraph network in the Boer republics; it was recognised that they were extremely vulnerable to Boer disruption and possibly the interception of telegraph traffic, but there was no alternative.

THE ADVANCE TO PRETORIA

The movement of the various columns was to be controlled by Lord Roberts, by means of the telegraph, as he advanced up the main route of the railway from Bloemfontein to Johannesburg and Pretoria. The map on page 89, which covers the same area as the map on page 88, shows the principal permanent telegraph lines serving this huge area. Note the scale of the map to appreciate the size of the task.

It was to be the most ambitious plan for military telegraph that had so far been undertaken by the British army (and probably any others), organised and executed by the Telegraph Division, with a supporting cast of some civilian assistance in the Cape Colony and the Orange River Colony (where many of the Post Office staff had voluntarily returned), and a native labour force (there were about 900 native labourers in support of the telegraph operations). Without all those years of close co-operation with the Post Office in Britain, as a consequence of which the military telegraphists fully understood the working of a civil system, coupled with a system of drawing their trained reservists from the Post Office in wartime, this communications plan could never have succeeded.

When Methuen's column reached Kroonstad it stayed in the Orange River Colony to conduct operations against the Boer commandos, and his telegraph resources, 'H' Section with MacInnes and Sherrard, became available to Fowler. The balance of their work shifted away from field work to repair, maintenance, and working of the permanent telegraph lines. To them was added the loyal members of the Orange River Colony Telegraph Department who had returned to work, but they were mostly confined to telegraph offices.

Lord Roberts' Advance from Bloemfontein

Lord Roberts and his headquarters set off from Bloemfontein on 3 May 1900. They were accompanied by Captain Godfrey-Faussett, one cable section under Lieutenant Mackworth, one airline section under Lieutenant Webber, and the Telegraph Division HQ with its telegraphists. The damage done by the retreating Boers was extensive, and taxed the resources of the linemen to repair it. Remembering their efforts with the tactical lines in the time around Modder River and on to Bloemfontein, their repair of the permanent system up to Bloemfontein after it was captured, and now their efforts during the advance to Pretoria and subsequently their efforts at keeping damaged lines open throughout the breadth of the whole country despite considerable personal risk, the well-known determination of army linemen must have stemmed from these days in South Africa.

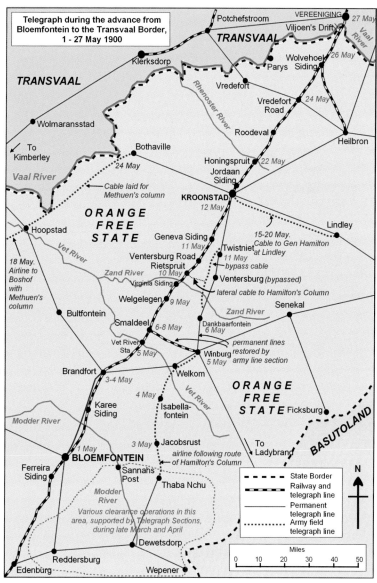

As Lord Roberts' central column advanced the routine was that, after an early start, the day's march was usually completed by 1:00 pm, and by the time the camp had been established the telegraph office was open and ready for traffic. Godfrey-Faussett described it:

Lord Roberts controlled the movement [of all the columns] in considerable detail. This threw a large quantity of traffic on the central column, and for this a Wheatstone receiver was used with great success. While the column was on the move the base office at Bloemfontein punched up the messages as they came in from the rear and as soon as the headquarters bivouaced these were sent through, leaving the line clear for work from the front. At headquarters the large mass of received work was written up during the night, and the messages were delivered early in the morning. This night work put a heavy strain on the operators, but arrangements were made to carry them on the march in wagons, when they were able to get some sleep. It was found necessary to tell off 3 wagons for the headquarters office, which marched and bivouaced separately under a senior N.C.O. [2]

A cable cart, displaying a large white and blue flag so that it was given priority of passage at the congested river crossings, always accompanied Lord Roberts, and he was able to send and receive messages during temporary halts at any time of the day. Being in the lead, the crew of the cable cart also temporarily mended any breaks in the line that they came across, if necessary by a cable 'patch'.

The method of working that developed as the advance progressed was described in detail by one of the officers:

..... The cable-cart also tapped in at the end of the day's march until the other instruments were ready, and also when these instruments were dismantled in the morning until headquarters moved off. This buzzer [vibrating sounder] worked always to an advanced transmitting office, which had to be kept in front of Railhead, the reason being that the buzzer will not work over sufficiently long distances to buzz all the way to Bloemfontein; therefore, the advanced transmitting office had to take all the buzzer work from the front, and transmit it with a D.C. [double current] key over the through wire to Bloemfontein. As soon as the headquarter office was ready to start, the intermediate office was instructed to go out of circuit, and front worked D.C. duplex direct to Bloemfontein. We also repaired a second wire as we advanced. This was used as a S.C. [single current] circuit between the advanced office and Railhead, and when headquarter was open it was also in circuit; the object of this circuit was supply, a huge supply park being always formed at Railhead and advancing with it; and the reason why this advanced transmitting office had to be ahead of Railhead was that the wire had to be given up to the railway as it got up; so if Railhead got ahead there was no communication with the supply park. When headquarters came into circuit this wire also afforded a convenient means of working off all the less important messages to the advanced office, which, in turn, could get rid of them to Bloemfontein during the day, which was the slack time when the through wire was divided.

As regards the instruments, we started with two G.S. wagons, one fitted up with a D.C.X. office, the other Wheatstone. However, the Wheatstone was not ready when we started, and the transmitter was left behind, but caught us up at Kroonstad; however, we had a spring receiver. No doubt D.C. is the thing for all long important lines, as it enabled us to work through the most amazingly full earths by the somewhat violent expedient of increasing battery power until you got a current through at the other end; it was all one could do, as the work had to be disposed of one way or another.

..... The office was in charge of Sergt. Ancell, and he and the clerks had a precious hard time of it on the march, as they had to work all night at high pressure, and get what sleep they could by day on top of a jolting G.S. wagon.

..... We were able nearly always to keep ahead of the infantry, and knowing the blocks that would occur at drifts used to get off before other people started; reveille, 4 a.m., march generally about 5.45, just as there was light enough. The difficulty was to get the [native] boys to inspan the mules in the dark, and precious cold it was at starting, too. We haven't yet got a hold of the country north of Kroonstad. It is in practically the same state as two months ago; there are about 5 miles either side of the railway, where it is fairly safe to travel by day, but by night nothing is safe outside the outposts of the posts along the line, and brother Boer comes down nightly, cuts the wires, and blows up the line. They also have wiped out [captured] the whole of the original lot of linemen left between here and Viljoen's Drift, with the exception Corpl. Houghton, who was here, and had gone out after a fault when this place was held up in June; however, he lost all his kit. Corpl. MacSeveney was left as inspector at Kroonstad, and he disappeared early. I don't know what has become of all the men, as only one of them, Sapper Seymour, is reported by the U.S. Consul-General to be at Nooitgedacht [a POW camp in the Transvaal]. [3]

Lord Roberts reached Brandfort on the evening of 3 May, and the telegraph office opened at 6:00 pm, then Vet River on 5 May where there was a minor battle in progress, and the telegraph office was opened in a platelayer's hut beside the railway line. Smaldeel Siding was reached on 7 May, where there were problems with the line, which ended up in communicating with Bloemfontein via Winburg by vibrator, despite the fact that there was a gap of several yards in the as yet unrepaired permanent line, and thence General

Hamilton's rear link through Thaba Nchu, Dewetsdorp, and Reddersburg – a roundabout route (see map on page 90). In situations like this the extremely sensitive vibrator was the only instrument that would work.

Between Rietspruit and Geneva Siding the line was extensively damaged, taking much time and effort to repair. From Rietspruit, Lieutenant Mackworth ran a lateral cable to the cable laid for General Hamilton's column which was now closing on the main column. Unfortunately this cable was much damaged by troops of Hamilton's own column as they advanced, but it was repaired.

On 11 May they reached Geneva Siding, Godfrey-Faussett's diary on that date noting cryptically that "Glue tapping for Chief " - in other words, Sergeant Glue, the NCO with the cable cart, had tapped into the permanent line, and was using his Morse key and vibrator to send and receive messages for Lord Roberts while waiting for the telegraph office to be set up and opened for communications.

On 12 May Lord Roberts marched into Kroonstad, the column having marched 130 miles in ten days, and the telegraph office in the town was taken over. By this time the Brandfort-Winburg-Kroonstad permanent line had been repaired by Lieutenant Henrici's section with General Hamilton, so there were two circuits immediately available, working duplex, from Kroonstad to Bloemfontein.

Lord Roberts stayed at Kroonstad until 22 May, to enable repair of the railway and telegraph lines, and thus permit supplies to catch up. The railway line, and with it the telegraph line, had been demolished in seventeen different places between Bloemfontein and Kroonstad; every bridge had been destroyed. The main column was kept supplied, but the units of the protecting flank column were more difficult to reach and were rarely on full rations. Also there were water supply problems at Kroonstad, and some soldiers, insufficiently recovered from the medical problems at Bloemfontein, were taken ill again. It became noticeable that whenever there was a halt for a few days, sickness became rife, whereas the hard conditions of the march in fresh and uncontaminated countryside produced little illness; hygiene discipline was not strictly enough enforced.

Meanwhile the Boers were in a state of confusion and lacking in any coherent plan. They squandered the opportunity to shore up their defence offered by Lord Roberts' pause at Kroonstad - something which they had enforced by their destruction of the railway. Despite their undoubted individual skills at things that mattered at the tactical level – shooting, fieldcraft, horsemanship, individual initiative, and local knowledge - they were an unprofessional army when it came to planning, organisation, and discipline.

On 13 May the two sections under Mackworth and Webber moved forward to Jordaan Siding, camped, and opened a telegraph office, the Boers having made themselves scarce. Moir's section, having completed their task for General Hamilton, also moved into Kroonstad.

On 15 May Lieutenant Mackworth's section ran a fifty-mile cable to Lindley for General Hamilton. This cable was damaged by veld fires and its insulation stripped - but it still worked. The troops left Lindley on 20 May and any of the cable worth recovering was reeled up. That same day Mackworth also sent a cable detachment under Sergeant Longley with General Hamilton to Heilbron.

The Advance from Thaba Nchu to Kroonstad

During this same period, to the east of the main column, and acting as its flank guard while they passed through the area where General De Wet had become active, Hamilton's force had left Thaba Nchu on 30 April. Lieutenant Moir with 'D' Section was initially tasked with providing the telegraph communications to keep Hamilton's column in touch with the Army HQ as Lord Roberts advanced northwards up the main railway and telegraph line from Bloemfontein. This was to be an airline from Thaba Nchu to Winburg, to be constructed as the column advanced, and from Thaba Nchu via the permanent line from there through Dewetsdorp, Reddersberg, and Bloemfontein . As a glance at the map will show, it was a long line route for what was actually a much shorter distance between the two HQs.

On 1 May, Moir's section was joined by Lieutenant Henrici with 'E' Section from Bloemfontein, bringing fifty miles of airline to replenish Moir's almost exhausted supply. Today this simple statement might pass without further thought. It is worth remembering that fifty miles of airline weighed some three tons in line alone, excluding the poles - probably something like five tons in all – and this should be placed in the context of animal transport, cargo capacity, the need to feed, water, and husband the animals, the distances involved, the poor roads, the few sources of water in the area, and the precarious supply of rations. It was not then as easy as it may sound today.

On 3 May the two sections constructed over sixteen miles of airline to Verkeerde Vlei, three miles north of Isabellafontein, and on 4 May it was extended to Welkom. On 5 May the column arrived at Winburg, and the telegraph line was completed into the town. They had built the sixty-five miles of airline from Thaba Nchu to Winburg in five days. General Hamilton mentioned the good work of the detachment in his despatch to Lord Roberts. Moir remained at Winburg until 8 May, his section repairing the permanent line from there to Brandfort and to Smaldeel and thus enabling a much more direct line route to the main column (see map), before he returned on 8 May to rejoin the HQ of the Telegraph Division with the Army HQ, then at Smaldeel.

Henrici's section remained with Hamilton's column. As the column continued its advance they restored the permanent line from Winburg to Kroonstad. By 9 May they had reached Bloemplaats, just south of the Zand River, and on 10 May Boschkop, just north of the Zand River. But from there General Hamilton's route veered to the left, closing in on Lord Roberts' line of march, and the permanent line to Kroonstad, some miles to his right, could no longer be used. As Henrici's field cable route became dangerously extended, Lieutenant Mackworth's section was detached from the main column and laid a cross cable directly from the railway line to Boschkop, enabling Henrici to reel up the cable to Bloemplaats behind him. General Hamilton reached the railway line four miles south of Kroonstad on 12 May, where a telegraph office was opened for him on the main telegraph line at Kroonspruit. On 13 May Lieutenant Henrici and his section rejoined the Telegraph Division HQ at Jordaan Siding in Kroonstad. General Hamilton, now on the main route with Lord Roberts, had been kept in communication for the duration of his flanking march.

It gives some idea of the level of activity, the diversity of their tasks, and the independence of their operations when it is realised that the bivouac at Jordaan Siding on 13 May held four of the Telegraph Division's sections together in one place for the first time since 16 November 1899, some six months previously, when they had been concentrated at De Aar before the original advance towards Modder River had begun. "And a most imposing spectacle we made in camp", said one of the officers, "six ranks of wagons drawn up by sections in parallel lines. In front 'A' Section, then the three airline, then the mule transport, and in rear the ox wagons."

Into the Transvaal

On 22 May the advance of the main column resumed and they reached Honingspruit, where the telegraph office was set up in a coach house, the lines working well. Lieutenant Webber was sent to Heilbron along the route of the permanent line to repair the line and then open an office for a cavalry Brigade which had been ordered to go there.

The next day, 23 May, they reached Rhenoster River, just north of Roodeval, where Godfrey-Faussett heard that plans for sending the cavalry Brigade to Heilbron (Webber's destination) had changed, and they were going elsewhere. Typical perhaps of the sort of confusion mentioned earlier about poor staffwork and lack of coordination, he sent a rider to bring Webber back. On 24 May the column reached Vredefort Road, where there were problems with the lines and communications were poor. On 25 May they were at Grootvlei, just south of Wolvehoek Siding. There, the rider sent for Webber returned to say he had been unable to find him. On 26 May, by which time the headquarters was at Taaibosch, north of Wolvehoek, Webber's native boy returned to say that Webber and his detachment of three had been captured by Boers, who had resumed possession of Heilbron, and they had been taken to Pretoria.

While the main column was in this area the leading forces, cavalry commanded by General French, crossed the Vaal River at Parys on 24 May. The next day General Hamilton's column crossed the Vaal ahead of the main column. Both these columns kept pressure on the Boers, allowing them no time to prepare an organised defence. It is unfortunate that no records have been found of the signalling operations conducted by the signalling officers within these columns in this highly mobile phase, but the level of coordination and control that was achieved indicates that they must have been effective. General French had a good reputation for ensuring that his signalling communications were efficient. The hilly countryside across the Vaal, a change from the mostly flat and monotonous Orange River Colony through which they had advanced so far, lent itself to visual signalling.

On 27 May, twenty-four days after setting off from Bloemfontein, the main column crossed the Vaal River at Viljoen's Drift and entered the Transvaal, spending the night at Vereeniging. The advance from Bloemfontein through the Orange River Colony had, on the days that they marched, averaged sixteen miles a day, although the combat troops had done much more. The eighty-five miles since Kroonstad had been unopposed. The Boers were now in full flight, and due to confusion in their command structure - always a weak point - were for the moment incapable of organised resistance on any scale.

Vereeniging was an important junction in the permanent telegraph network, with lines west to Kimberley, north to Johannesburg, and east to Natal. The actual line arrangements around these more populated areas were much more complex than the simple telegraph routes shown in the sketch map suggest There were also telephone systems and exchanges - things that did not exist in the more remote country areas. (4) Fortunately

Crossing the Vaal river at Viljoen's Drift. Roberts and Kitchener riding together at the head of the group.

in Vereeniging both the telegraph office and the exchange for local telephones were undamaged and good communications were quickly established. Over 500 messages were received there that night. The lines were tapped in the hope of intercepting Boer communications, but nothing of any value was obtained.

After the night at Vereeniging the march of the main column continued, along the line of the railway. On 28 May they spent the night at Klip River Station, and next day they reached Germiston where they spent two nights. The columns of Generals French and Hamilton to the west, and another to the east, encircled Johannesburg. Some excellent individual actions were fought by these columns as they closed the noose. Lord Roberts demanded the surrender of the town, which its Commandant, Dr Krause, agreed to. On 31 May Lord Roberts entered the town, marching to Orange Grove on its northern side. The town looked forlorn and delapidated, but the mines were left intact. The *Vierkleur* (the Transvaal flag, literally four colours) was hauled down, and the Union Jack replaced it to the sound of 'Three Cheers for Queen Victoria'. During the three days of consolidation at Johannesburg the town's telegraph office was taken over and reorganised on military lines.

By now the line of communication was tenuous in the extreme. Between Johannesburg and Bloemfontein there were 263 miles of railway and telegraph line to guard. In fact the Transvaal railway lines were not damaged; by some local arrangement they belonged to the Dutch railway officials, who had shown great enthusiasm in destroying the Orange River Colony railway lines but refrained from damaging their own. Many urged Lord Roberts to stay in Johannesburg and consolidate his line of communication through the Orange River Colony before pressing on, but he boldly decided to strike out for Pretoria, giving the Boers no time to regroup. There were also some 5,000 English prisoners in Pretoria (amongst them a number of the Telegraph Division), who might be removed elsewhere.

The advance continued to Pretoria on 3 June, and a minor engagement was fought on the southern approaches to the town. Telegraph communication was maintained with Lord Roberts throughout. On 5 June Pretoria surrendered. Roberts, with accompanying cable cart, stopped outside the town while the detail of the surrender was negotiated. The terms agreed, he continued to the town's railway station and was quickly in communication with Bloemfontein.

Soon afterwards the Pretoria telegraph office was taken over, the Director of Army Telegraphs, Lieutenant-Colonel Hippisley, quickly moving there from Bloemfontein. The officer prisoners having released themselves from confinement, Lieutenant Webber, captured at Heilbron, reported himself for further duty. Sergeant Shergold and his detachment, captured at Sannahs Post, were also released and rejoined soon afterwards.

In Pretoria Lord Roberts and his staff were accommodated in the area still known as Sunnyside, about one mile from the principal telegraph office, so a cable was laid and a branch office was opened in the headquarters. All telegraph wires out of Pretoria except those needed and used by the British were cut, beating the Boers at their own game, to prevent leakage of information. Lieutenant Mackworth went down the railway line to Irene, repairing the railway telegraphs, while Lieutenant Moir repaired the permanent lines to Johannesburg. There was an efficient telephone system in the town, still working. Military users were connected to it, and all others were disconnected, again for security reasons. There was also a telephone trunk line to Johannesburg, interconnecting the exchanges in both places, and this was to prove useful.

Communications south through the Orange River Colony were continually being cut by the Boers, despite the best efforts of many maintenance parties and their escorts. There was also concern that the lines were being tapped by the Boers to extract information, although little hard evidence of this emerged - they were, after all, not over-endowed with telegraphists capable of reading the messages or commanders capable of reacting to any intelligence obtained - but as a precaution extensive use was made of cipher.

The anonymous officer who wrote so freely after the advance was over, quoted above, gives us more of his views on various interesting matters. All linemen, he said, should be able to ride.

> ...a man who can't move about faster than his own legs can carry him is simply an encumbrance. In the O.R.C. we are putting them all up on country ponies. ... you don't want a man with a cavalry seat; all you want is one who can get about the country rapidly; let him stick on the best way he can, he will soon get accustomed to it.

He continued with some strong views on equipment:

> The cable-cart is another thing. I have had nothing to do with cable work, but the present cart is absolutely unsuitable, owing to the shifting balance and short length of cable it carries. I believe those that have had most experience are in favour of a timber wagon. Up to now we have been accustomed to regard the drums as part of the cart; this is entirely wrong, they are part of the cable, and when empty are dumped on the ground at the end of each length of cable. Each drum should, therefore, be of uniform pattern, the reason being that when laying long cables you cannot return to pick them up. When laying forwards you must also reel up forwards or you will lose means of communication; so you must have an empty cart at your starting point to reel up, and if no drums are shed on the way you must waste a lot of room with a duplicate set of drums. Probably the pair of cable-wagons, with their transport, will have to carry about 50 miles of cable between them. Out here we have been reduced to using the packing-drums on which the manufacturers supply the cable for this reason.

> The air line wagon is much too much of a gimcrack affair, they all fall to pieces. It is a question whether the air line equipment is suitable; certainly it is quite unsuitable out here, as you can't jump holes more than 10 inches deep, so it falls down directly you are out of sight; more particularly as a rubbing-post is an unobtainable luxury to cattle out here; also, if it is to follow a main line of advance, it should be able to carry two wires, one for the through work, the other for communication between the posts of the line of communication. Then the wire is of far too high resistance. It will be necessary to find wire which, while weighing no more, is of lower resistance; there are objections to copper or bronze, which led to the adoption of zinc. Perhaps aluminium may get over the difficulty; it is now much cheaper than copper, and bulk for bulk its conductance is 50 per cent that of copper, while weight for weight it is 50 per cent greater.

> Then take the instruments. No doubt D.C. [Double Current] working is the thing for long lines; Quad[ruplex] working has turned out the most satisfactory way of dealing with the traffic on main through wires, with Wheatstone in reserve. The Wheatstone duplex comes in when the wire fails. During the time the circuit is down the staff can be employed punching up slip, which is rattled off directly wire comes right. The Wheatstone is far too often out of order for regular work, and wastes a lot of staff who might be more usefully employed. It is found, in consequence, that Quad deals with the work faster, with fewer staff, than Wheatstone can. All the P.O. instruments so far have turned out satisfactory ; but for single current working it is a question whether the present baseboard is suitable. ... The experience on the omnibus circuit between the posts on lines of communication would seem to show that a direct sounder, with the same figure of merit [sensitivity] as the military inker, would be preferable. The inker is in many ways an admirable instrument encumbered by a lot of unnecessary machinery. The Siemens' inker is in universal use in this country, and has many advantages over our inker, as it has top and bottom contacts, so can be joined up in pairs as a repeater; it also can be connected up to work D.X.[duplex] It has the disadvantage that it absorbs an enormous amount of power. If some of the advantages of the two could be combined, and the clockwork abolished, you might produce a highly useful military instrument. Although I believe we have never made use of a repeater here, such an arrangement would have been highly useful on our inter-post circuit, and would have saved a lot of transmission.

> ..... I hope this campaign will have sealed the death warrant of the field Leclanché; as a production combining all the disadvantages of the dry cell and the ordinary Leclanché, with none of their advantages you would have to go far to equal it.

Another detail which requires organizing is that of stationery. It is difficult to realize the amazing amount of work dealt with and the stationery consumed in consequence. The number of messages dealt with by the clearing-house will be something to make one stand aghast; it already amounts to several millions. Perhaps when one has seen a few of the military offices at work one might realize it. Most of them have enough to keep them steadily at it for 16 hours out of the 24, and the average message runs to perhaps 50 words; up to 200 are not at all uncommon, and I have frequently seen ones of from 500 to 700 words, so stationery is swallowed up like water going through a sieve. We have long run out of army forms, but fortunately have captured a fair supply; but the printing works at Cape Town are now hard put to it to keep pace with the demand. Three patterns are now being printed for us - the Cape post office form, the Imperial military telegraph form, and an imitation of the A.T. [Army Telegraph] form - all of which differ. But it is difficult to keep offices supplied, and they are often reduced to using any old scraps of paper. Probably some very much improved and extended form of dividing up into unit boxes will meet the difficulty. However, it will have to be gone into. And there are heaps of other things, all more or less important. (5)

Although in parts perhaps a little technical, these detailed criticisms illustrate some of the practical problems being encountered. It is refreshing that somebody was prepared to air them; most contemporary diaries avoid criticisms that might appear politically incorrect or career-damaging to the author.

The telegraph running along the railway originally carried three wires, one used for the work between Bloemfontein and Pretoria, and the other two for railway work, these being used as a through wire and a station-to-station wire. When the guard posts to protect the line were set up, at about seven-mile intervals and not colocated with the railway stations, it became necessary to add a fourth wire so that they could intercommunicate. Inkers, some military pattern and some captured or commandeered, were used on this line

With the capture of Johannesburg and Pretoria, the army was now responsible for the operation of the entire telegraph networks in the two Boer Republics. Their capability for such a task was to draw heavily on the experience they had gained in the preceding thirty years working alongside the Post Office in England and the fact that so many Post Office reservists now formed a substantial proportion of their total strength. This changed situation was going to lead to a reorganisation, to be described later. Meanwhile, what of the other columns that had set off to Pretoria, from Kimberley in the west and from Natal?

The Western Advance

While Roberts was advancing to Pretoria, the two western columns were advancing from Kimberley and Boshof, General Hunter's column following the more northerly route through Vryburg and Lichtenburg, and Lord Methuen's column following the route through Hoopstad to Kroonstad (see map). As mentioned earlier, communications for these columns were provided by Captain Wright and Lieutenant Macfie with Hunter, and Lieutenants MacInnes and Sherrard with Methuen. Macfie and MacInnes had, since early March, generally been responsible for repairing damaged lines around Kimberley, Boshof, and Warrenton. In particular,

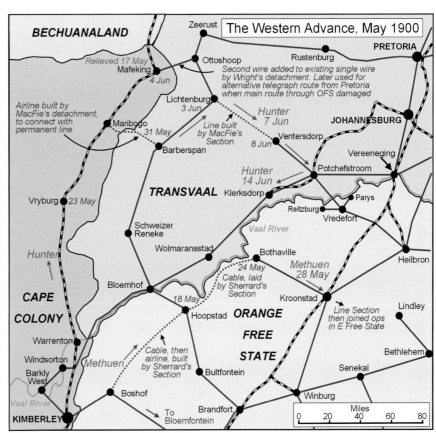

THE ADVANCE TO PRETORIA

MacInnes had constructed a second line between Kimberley and Boshof, so that when the area was cleared and the line from Bloemfontein to Boshof restored, the main line could be used for through traffic between Kimberley and Bloemfontein, and the new wire could be used for local traffic. Wright and Sherrard, who had arrived in South Africa with other reinforcements on 23 March, joined from De Aar with a section in late April after various delays awaiting mules.

Wright and Macfie joined General Hunter at Warrenton on 2 May, and opened an office there. The advance started, and Macfie crossed the Vaal river with Hunter's force on 4 May. On 6 May an observation balloon ascended to act in support of an attack on a Boer position at Fourteen Streams, near Warrenton, and it was connected by telephone. By 23 May they reached Vryburg, communications having been maintained variously by cable, airline, or repair of the permanent line as appropriate.

Hunter continued along the railway about halfway to Mafeking (which had been relieved on 17 May by a specially selected column) communications being maintained by using one of the permanent wires along the line of the railway. Small local systems of outpost telephones had been set up at Warrenton, Vryburg, and Mafeking during the advance. Hunter then turned into the Transvaal, heading for Lichtenburg and beyond, remaining in communication by means of an airline built by Macfie's section across to Barberspan, between Lichtenburg and Schweizer Reneke, and then continuing by using the permanent line to Lichtenburg. The line to Barberspan was completed on 31 May, and Lichtenburg was reached on 3 June.

Meanwhile other elements of the section continued on to Mafeking, repairing the railway telegraph line as they went, reaching Mafeking on 4 June, and bringing the town back into telegraphic communication with the south. On arrival there, Captain Wright discovered that the lines to Zeerust and Lichtenburg were in working order, and a second wire was run between Mafeking and Zeerust. On numerous future occasions, when guerrilla activity had damaged the lines south from Pretoria to Bloemfontein, telegraph communication between those two places was maintained by this long and rather remote diversion through Rustenburg, Mafeking, and Kimberley using this second wire that Wright's detachment had added. The more direct permanent line between Johannesburg and Kimberley was never reliable, suffering continuous disruption by Boers in the area of Wolmaransstad, country not yet traversed by British troops. Communication was maintained as Hunter advanced, Macfie's section building a line to cover the gap between Lichtenburg and Ventersdorp, the last eight miles being bare wire laid on the ground.

Turning to Lord Methuen's column, their advance through the western Orange River Colony towns of Hoopstad and Bothaville started from Boshof on 14 May. There was no permanent line between Boshof and Hoopstad. They reached Hoopstad on 17 May, being kept in communication by cable, replaced as soon as possible by airline. The airline to Hoopstad, sixty-eight miles, was completed on 18 May. At Hoopstad it was found that the line through Bultfontein to Brandfort was intact and working, but Bultfontein was still in the hands of the Boers, who were apparently unaware of the British presence. It was decided to let sleeping dogs lie, and not use the line. For the continuing advance to Bothaville Lieutenant Sherrard laid a field cable on a route near the Vaal river, reaching Bothaville on 24 May.

From Bothaville the column now headed for Kroonstad, a permanent line connecting these two places, although damaged and not working. By day the column was kept in communication by cable; by night the permanent line was repaired and used while the Boers slept. They entered Kroonstad on 28 May, Lord Roberts' column having passed through some two weeks earlier.

When Methuen's column reached Kroonstad it stayed in the Orange River Colony to conduct operations against the Boer commandos, and its telegraph resources, 'H' Section with MacInnes and Sherrard, became available to Fowler. The balance of their work shifted overwhelmingly away from field work to repair, maintenance, and working of the permanent telegraph lines.

The Advance from Natal

The relief of Ladysmith on 28 February 1900 had seen the Boers beat a hasty retreat from the area. Over the next few weeks various rearrangement of the telegraph and telephone lines around Ladysmith and the surrounding area were made. The large relieving force there, some 50,000 men, now had no independent strategic purpose other than to defend Natal. After two months of largely unexplained inactivity, the force under Sir Redvers Buller began its advance on 7 May, entering the Transvaal at Volksrust on 13 June, over a week after Pretoria had been occupied (see map on page 89). Although a certain amount of half-hearted Boer resistance was encountered it was seen by many as a rather leisurely advance by 'Sitting Bull', as

Buller became known, whose camp comforts, even for the animals, was legendary, and in contrast to Lord Roberts' driving force. There had been only some minor skirmishing on the way, but such operations as there were had been successfully executed. Some 15,000 troops were left in Natal to defend the colony. By 23 June Buller's HQ had reached Standerton. [6]

The telegraph communications for Buller's advance were provided by the original section sent to Ladysmith at the start of the war, now relieved and back in action, and the reinforcement section originally under Lieutenant Jelf, their composition shown at the start of the chapter. The main route to Newcastle consisted of a combination of newly built airline to Helpmakaar, where the permanent line through Greytown to Pietermaritzburg was repaired and connected, the work carried out largely by Lieutenant Hildebrand's section. Other elements of the force advanced along the line of the railway to the north, and that permanent line was also repaired and put in working order. Newcastle was reached and opened on 18 May. Numerous airline spurs were built from the permanent line to various camps and headquarters as the advance progressed. On 28 May Captain Boys began to open the permanent line from Newcastle to Utrecht to support flanking operations into the Transvaal. An airline along a separate route to Utrecht was also built by Lieutenant Hildebrand's section. The various columns, as they moved, were always kept in communication by similar use of airline and permanent line. A turning movement around Laing's Nek, in the direction of Botha's Pass, was supported by cable and airline. The country was rough, stony, and hilly, making the construction of airline difficult and slow. When this line was later dismantled the line party was attacked by Boers, and part of it was abandoned.

The main force advanced to Standerton, which was reached on 23 June after a four-day march. The railway telegraph line was repaired during the advance. The permanent line along the road to Standerton was found to be in working order, although afterwards constant interruptions were caused by Boers damaging the lines. On 30 June the advance continued to Heidelberg, the line being repaired behind the troops as they advanced and offices opened. Just beyond Heidelberg the line parties met a repair party coming from Johannesburg, to complete the junction of the lines to Johannesburg and Pretoria. Damage to the line by Boers continued, and parties had to repair it many times.

But by now Pretoria had been captured, and the situation changed. The defeated Boer forces, such as remained, were retreating eastwards from Pretoria to Komatipoort. Buller's slow-moving force was now going to be diverted northwards to Belfast, to participate in operations against the retreating Boer remnants, to be described in the next chapter.

Endnotes

1. The heliograph actually used by De Wet is displayed in the museum of the Royal Signals.

2. Extract from a post-war report by Captain Godfrey-Faussett. A copy is held in the Royal Signals archives.

3. From an account written by 'An Officer' at Vredefort Weg, Orange River Colony, 10 August 1900, and published in the *Royal Engineers Journal*, 1 November 1900.

4. The detail will be found in such documents as the *History of the Telegraph Operations during the War in South Africa, 1899-1902*, or the *Sketch Map of South African Telegraphs 1896*. Copies of both are in the archives of the Royal Signals.

5. *Royal Engineers Journal*, 1 November 1900.

6. It had been Lord Roberts' intention that the Natal troops should swing through into the Orange River Colony at an early opportunity, to converge with his main force on Pretoria, rather than fight its way northwards through the difficult country of north Natal and the hills of the eastern Transvaal. Although the telegraph communications were in operation during March and April, and numerous messages were passed, there was less than consensus and full cooperation between Buller and Roberts and things did not work out as Roberts intended. Historians have dwelt on the subject of Buller's lack of cooperation and 'drive' at this stage.

Chapter 10

Two Years of Guerrilla War

The War continues

Despite the optimistic strategic plan drawn up in England before the war (see chapter 1 - 'Home for Christmas!') the eventual capture of Pretoria on 5 June 1900, and the formal ceremony that took place in the town a few days later, was far from being the end of the war. Boer commandos, De Wet prominent, were still at large in the open veld of the Orange River Colony (as the Orange Free State had been re-titled after its annexation on 24 May 1901). Mostly untouched by Roberts' advance to Pretoria, they continued to attack and disrupt the line of communication back to Bloemfontein. Substantial forces had to be left there to deal with them. The Transvaalers, under Generals Botha and De la Rey, although demoralised and in disarray, withdrew from the area of Pretoria into the Eastern Transvaal, generally along the line of the railway leading to the border with Portuguese East Africa at Komatipoort and in the hills to the north. Unfortunately Roberts had not cut off their obvious line of retreat quickly enough, either just north of Pretoria as part of the operation to capture the town, or further east along the railway line, where the Boers established a temporary headquarters at Balmoral. The ongoing operations now needed to deal with these two separate problems, in the Orange River Colony and the Eastern Transvaal.

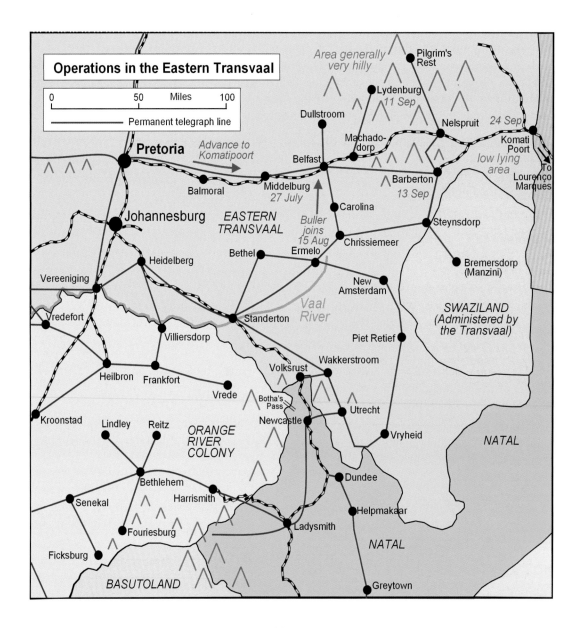

The Orange River Colony

During the advance through the Orange River Colony to Pretoria, when four line sections had been deployed initially (a further two were with the western columns in the area of Kimberley, and two were in Natal), the procedure had been that these sections undertook temporary repair of the permanent line and carried out field work for the detached columns, to enable communications by the quickest means possible. Following in their rear were line parties under Captain Fowler making permanent repairs and attending to the continual damage to the line being inflicted by Boer raiding parties. When the main body advanced into the Transvaal, Fowler's men became responsible for the permanent line from the Orange River to the Transvaal border. Also involved was Lieutenant Graham Manifold, responsible for the operation of the railway telegraph system. Many small guard posts were deployed along the railway line for security purposes, also needing to be in communication, and the work of Fowler's men became more than they could cope with. To help them, two of the field sections ('E' and 'F') were detached from the main body and assigned to permanent work, leaving only two sections available for the ongoing field work in the Transvaal, one airline section and one cable section. These were 'C' and 'D' Sections, under those redoubtable two Lieutenants, James Moir and Harry Mackworth.

Boers caught cutting the telegraph line.

Keeping open the rearward line of communication through the Orange River Colony, a distance of nearly 200 miles from Bloemfontein to the Transvaal border, most of it passing through desolate and lightly populated farming country, and then another eighty miles to Pretoria, was a difficult task. The railway and telegraph lines were soft targets, and were frequently damaged. Between Johannesburg and Vereeniging the line was cut, as was the alternative route through Heidelberg. To provide yet another alternative route, Lieutenant Henrici had to build an additional wire on the railway telegraph line, the forty-mile line being completed in two days.

When the telegraph lines were disrupted communication between Bloemfontein and Pretoria was maintained by using alternative routes. One was through Potchefstroom, Klerksdorp, Wolmaransstad, Fourteen Streams, and Kimberley, although due to Boer activity in uncleared and unoccupied country around Wolmaransstad this was very unreliable. Another route was through Rustenburg, Zeerust and Mafeking, making use of the second wire built by Captain Wright's detachment between those two places. The third alternative was through Heidelberg, Standerton, Newcastle, Pietermaritzburg, King Williams Town, and De Aar. This last and very circuitous route, about 700 miles long, was worked with Wheatstone simplex - some achievement. None of these diversionary routes was an efficient way to pass the high volume of traffic, and they were all subject to interruption. It was an anxious time for those responsible for the communications. Moreover, the railway line and its telegraph was essential, and that could not be diverted.

Lord Roberts was determined to stop these attacks, made by Boer commandos with the connivance of local farmers. On 14 June, from Pretoria, he telegraphed to Kitchener, who had been left behind to command operations in the area:

> We must put a stop to these raids on our railway and telegraph lines, and the best way will be to let the inhabitants understand that they cannot be continued with impunity. Methuen's troops are now available [having completed their advance to Kroonstad] and a commencement should be made tomorrow morning by burning De Wet's farm, which is only three or four miles from the Rhenoster Railway Bridge. He, like all Free Staters fighting against us, is a rebel and must be treated as such. Let it be known all over the country that in the event of any damage being done to the railway or telegraph the nearest farm will be burnt to the ground. A few example only will be necessary, and let us begin with De Wet's farm. Tell Methuen he must arrange for the Heilbron telegraph line to Kroonstad be kept open as well as the one by the railway.

This marked the beginning of the policy of farm burning, which was controversial and disliked by many on the British side. It led in turn to a 'scorched earth' policy and later the emotively named 'concentration camps' in which the displaced Boer women and children were interned while their men rampaged in the veld. The overcrowded camps were badly administered and unhygienic, leading to many deaths, and were also to be the cause of much controversy, not to mention lingering anti-British resentment for many years after the war finished. [1]

All telegraph communications in the Orange River Colony, both permanent line and field operations, were now under the control of Captain Fowler. He had Lieutenant Hepper and his permanent line party, as before, but he had been heavily reinforced. As already mentioned he had 'E' and 'F' Sections, left behind as the main column went on to the Transvaal. He also had Lieutenants MacInnes and Sherrard with 'H' Section, previously with Lord Methuen's advance to Kroonstad, and Lieutenant Macfie with 'G' Section, previously with General Hunter on their march to Klerksdorp, all now after their various travels re-assigned either to permanent line work or the field operations to capture the Boers. As fast as permanent lines were repaired, they were cut again by Boers. There were not the troops available to act as escorts, either to permanent line repair parties working in isolated territory or to cable laying detachments working in support of the mobile operation, all able to be observed by Boer commandos and their many local sympathisers, making the risk of capture high. Numerous linemen had already been captured.

During July a major operation was launched against the Boer commandos, who were mostly in the eastern half of the Orange River Colony, a more fertile and hilly area than the monotonous karoo in which the war had been conducted so far. The operation, conducted by columns under Generals Lord Methuen, Clements, Hunter, Paget, and Rundle, lasted about a month and culminated in the entrapment and surrender of some 5,000 Boers at the Brandwater Basin, a natural circle of hills with few ways out, centred on Fouriesburg (see map on page 89).

The telegraph communications to support this fast-moving and complicated operation leading to the Brandwater Basin endeavoured to keep the column commanders in touch with each other and rearwards by the combined use of the existing permanent lines connecting Lindley, Reitz, Bethlehem, Senekal, and Fouriesburg (see map on page 99) and the rapid laying of much field cable to connect into the permanent lines. Captain Fowler was responsible for the communication arrangements, and Captain Webber and Lieutenant Elkington, recently arrived in South Africa with reinforcements, as well as Lieutenants MacInnes and Sherrard with their sections, carried out all the line work.

A detailed description of the line work in support of this mobile operation involving five columns is not possible. Communications were not perfect, but the main columns were kept in contact with each other and rearwards to Lord Roberts at his headquarters for a high proportion of the time, and as the noose finally tightened around the Brandwater Basin the four column commanders there were able to intercommunicate. A telegraph office was opened at Fouriesburg on 27 July, connected to the permanent line. On 30 July the Boers capitulated at a farm called Verliesfontein, the location marked today by a large sign, 'Surrender Hill' (as it became known), on the road between Fouriesburg and Harrismith, and just a few miles from the Basutoland border. The field telegraph office dealt with a large amount of traffic concerning arrangements for the disposal of the many prisoners and their surrendered weapons.

A determined group, however, including President Steyn and General De Wet, managed to wriggle away at the eleventh hour under cover of darkness through an unguarded escape route, Slabbert's Nek. They were pursued over 200 miles into the western Transvaal yet, characteristic of many similar chases that ensued around the vast tracts of South Africa, they still managed to evade their hunters who were dogged by poor mobility and a certain amount of ineptitude.

By August much greater control of the main permanent telegraph lines had been achieved. Linemen were stationed about every seven miles between Vereeniging and Kroonstad, and working parties about every fifteen miles, including those attached to breakdown trains and armoured trains.

Reorganisation of the Army Telegraphs

In July, while the operations in the Orange River Colony were under way, it was widely anticipated that the war would soon end; the extent of the forthcoming guerrilla activity was greatly underestimated. In the towns, especially the larger ones, the conventional war was effectively over; the gold mining industry, an important source of revenue, was needed to sustain the economy, and most townspeople were eager to resume their business whatever was still going on in the veld. The country's permanent communications were now needed for both civil and military purposes, but in the Boer republics, with some staff either unwilling to cooperate or untrustworthy in such a role, who was to run them?

The answer, of course, was the Army Telegraph Division. It was just as well that as a result of decisions taken thirty years previously, when after much procrastination the army Telegraph Troop had been formed (for a role in Europe, where they were never deployed), and the Postal Telegraph Companies were formed soon afterwards (for reasons associated with the nationalisation of the Post Office and consequent labour shortage rather than any military purpose), the Telegraph Division had accumulated a wide experience of Post Office communications. Now, supported by its many Post Office reservists, it was well versed in running civil communications and able to take on the task in South Africa.

Thus the scope and nature of the telegraph operations changed. Field work was diminishing and the entire operation and management of a permanent system serving civilian and military needs was now the predominant requirement. Further reorganisation of the Telegraph Division was necessary. At its head, as Director of Army Telegraphs (DAT), and with many years experience working with the British Post Office while serving in the Postal Telegraph Companies, remained Lieutenant-Colonel Richard Hippisley, now based in Pretoria. Local Lieutenant-Colonel Reginald Curtis, who arrived in Pretoria on 19 July, was appointed as his Assistant Director of Army Telegraphs (ADAT). Curtis had been commander of the detachment of the Telegraph Battalion sent to Ashanti in 1895-96. Initially deployed to the operation in the Eastern Transvaal, Curtis's appointment did not last long, for he was sent to be Chief Staff Officer of the South African Constabulary in November 1900. In Pretoria a stores organisation and centralised clearing house,

The Telegraph Office in Johannesburg.

dealing with money matters, was set up, amalgamating others previously in Bloemfontein and Natal. These centralised departments were overseen by Lieutenant Bannerman.

Other matters were delegated on a regional basis. In the Transvaal, Captain Godfrey-Faussett became the Director of Transvaal Telegraphs, with an organisation under him dealing with engineering construction and maintenance work, a surveyors branch, a personnel branch dealing with the civilian staff re-employed, and a stores branch which also controlled the tradesman - carpenters, instrument mechanics, and so on. (The telegraph system around Johannesburg had employed many of British origin - the so-called Uitlanders, whose presence had been a cause of friction with the Boers. Many had fled to the two British Colonies when the war started, and most now returned to take up their former employment.) As well as telegraph there was also, in Pretoria and Johannesburg and other large towns, a quite sizeable telephone network. The Transvaal was subdivided into a separate Western District, where Captain Wright was placed in charge, with similar responsibilities.

Probably the most important role of all was that of District Inspector, of whom in the Transvaal there were ten. These were Sergeants, each assigned to an area, who were responsible for all details of the telegraph service in their area - traffic, telegraph offices, control of operating and engineering staff, repair and maintenance of lines and instruments, planning construction work, cash accounts, payment of wages, and discipline. It was a great responsibility, with a wide purview, and the non-commissioned officers rose to the challenge magnificently.

Organisational arrangements were similar in the Orange River Colony, where a high proportion of civilian Post Office employees were of British origin. Captain Fowler, who had stayed there during the advance into the Transvaal, was appointed Director of Telegraphs.

The original organisation of the 1st Division of the Telegraph Battalion, as constituted before the start of the war, with four sections, was reformed, albeit with some changes of personalities. They were assigned to field work. Major Hawkins, in command of the 1st Division at the start of the war, and in Natal during the siege of Ladysmith and subsequently, reassumed his command of what were now three airline sections and one cable section. These four sections were going to be used in support of the army's widespread operations which continued against Boer commandos conducting guerrilla warfare. [2]

Laying cable by bicycle.

The Advance into the Eastern Transvaal

While the diversion of attention to the Orange River Colony was in its closing stage, and still waiting for Buller to catch up from Natal, operations in the Transvaal resumed. The British advanced into the eastern Transvaal along the line of the railway, reaching Middelburg on 27 July after various minor battles and skirmishes. They were joined shortly afterwards by Buller's column which, from the area of Standerton, had struck northwards through Ermelo. The detail of the operations need not concern this narrative, but the result was that on 24 September, having fanned out to capture towns on the flanks such as Lydenburg and Barberton, the British reached Komatipoort, on the border with Mozambique, which was found abandoned. Many Boers had defected, surrendered their arms, and returned to their farms, but a hard core remained and evaporated into the surrounding countryside, mostly in the hills to the north. Hounded by the British, the aged President Kruger had fled to Lourenço Marques on 11 September, and later caught a Dutch naval ship to Europe. [3]

Lieutenant-Colonel Reginald Curtis, in his new appointment as Assistant Director of Army Telegraphs, took charge of the telegraph arrangements in support of these operations in the Eastern Transvaal. Lord Roberts was in command, and operated from a special train, accompanied by his own army telegraphist, able by tapping the line to put him in communication whenever the train stopped. Buller's column joined on 15 August, brought into communication, after a heliograph message indicated his position to be at Twyfelaar, by a cable run out from the permanent line at Wonderfontein, to the west of Belfast, by Lieutenant Mackworth's section. With Buller's column came welcome reinforcements of the Telegraph Battalion from Natal, supplementing the depleted resources brought about by the ongoing work in the Orange River Colony. The telegraph work during the advance to Komatipoort was heavy, and a through line between Pretoria and Belfast, where the Headquarters was located from 25 August, was established with a Wheatstone. The section between Middelburg and Belfast was worked on a separate local line, and temporary lines were run to wherever the headquarters of Generals Buller and French were located, respectively to the north and south of the railway line. The line resources under Captains Boys and Crookshank and Lieutenants Mackworth and Moir, were again kept very busy, their work consisting of repair of permanent lines and lots of cable-laying to accompany the numerous detached columns that scoured the country to left and right of the main route. Offices were opened along the railway line by Captain Crookshank as the force advanced to Komatipoort. From there the line through Portuguese territory to Lourenço Marques was found to be in good working order. [4]

The nature of these operations, with mobile columns deployed on both flanks, led generally to greater integration of communications, something first seen at Colesberg during General French's operations there nine months earlier. A good example occurred in the hilly country to the north of the railway line, around Lydenburg, where the permanent telegraph line ended. East of Lydenburg, towards Sabie, on the route to Pilgrims Rest, one of the last skirmishes of this phase of the war took place. A cable was run from the termination of the permanent line at Lydenburg for about ten miles to a hill top dominating the area, one of a group of four known as 'The Devil's Knuckles', and a heliograph signalling station giving coverage for miles around was located there to control the operations. Nowadays this wonderfully scenic area (in what is now called Mpumalanga) is known as the 'Long Tom Pass', named after the long-range Creusot gun used by the Boers as their principal artillery piece. Soon afterwards their last remaining gun, used during the battle there, was captured; they fought the continuing guerrilla war with no artillery.

Partying at Pretoria

Happily, it was not all work. On 28 September 1900 the Telegraph Division held a 'smoking concert' in Pretoria, mainly to establish bonhomie with the civilian telegraphists they now worked with. It was later reported in the Telegraph Chronicle.

> On Friday evening, September 28th, the members of the Telegraph Division at present in the Transvaal capital invited their civilian telegraph friends to a smoking concert in the new carpenter's shop of the telegraph stores department. The proceedings from start to finish were immensely successful. ..... the room was most artistically decorated with bunting, flags, and flowers, and illuminated by thirty electric lamps. .... So confident was our enterprising District Engineer, Sergeant W. Ancell, of a successful evening that he erected an invisible telephone with a sounding funnel in the rafters of the roof immediately in front of the stage, and made other arrangements which enabled not only those who were on duty at the Pretoria telegraph office but also friends in town, and even the staff at the Johannesburg office, to hear all that went on most distinctly.
>
> Our very esteemed and genial Sergeant Major, C. J. Aplin, took the chair at 6.30 p.m., and in the course of a brief introductory speech said his experience that evening was very different to that of 1881. Then when he was in Pretoria, the [1st Anglo-Boer] war was over. Now peace had not been declared, and they were actually in an enemy's country.
>
> A long and most interesting programme of songs and sketches, interspersed with speeches, was then gone through ..... In the midst of the concert Colonel Hippisley and other officers entered the room and were received with great enthusiasm and musical honours. Their health was proposed by Mr. Sprawson, an ex-R.E., who remained in the Transvaal after the 1881 war, and was for some years Director of the Pretoria Office, and is now surveyor of the Transvaal Telegraphs.
>
> Colonel Hippisley, in reply, thanked the non-commissioned officers and men for their cordial welcome. He said all had done good service throughout the campaign .....and then referred to the manner in which Lord Roberts's travelling headquarters telegraph staff under the able management of Sergeant Ancell had performed their duties from Enslin to Bloemfontein, and thence Pretoria. [5]

Hippisley continued by recounting the good work done by many of them, and the congratulations he had received on their behalf by many senior officers. Captain Godfrey-Faussett also spoke, in similar vein. Sergeant Ancell proposed a toast to "The Visitors" (ie the civilian telegraphists). He referred to his prior misgivings as to the reception they would be accorded by the Pretoria telegraphists, and he had "never been more agreeably surprised in his life, for he found these men to be the most brilliant lot of operators he had ever met. Most of them had once been in the home service, and a few of them were old personal friends".

The civilian telegraphists both in Pretoria and Johannesburg were of British origin who, attracted by the opportunities of a young and expanding country, came to work and live in the Transvaal. They formed part of the Uitlander community, which provided most of the skilled labour in those two principal towns, but had been at odds with 'Krugerism'. For example, they were not allowed to become enfranchised, Kruger afraid that their increasing numbers would overcome his Afrikaner voting majority. Their loyalty strained, they were placed in a difficult position. Some remained at their posts during the war, but were only too happy when Pretoria was taken over by the British, to whom they immediately transferred their allegiance. Their services were essential to the insufficiently staffed Telegraph Division, and they integrated well.

Change of Command

Pretoria had fallen, Kruger had fled, the Boer army, such as it was, no longer existed, and the South African Republic had been annexed on 1 September, reverting once more in name to the Transvaal. Many, including Lord Roberts, now fondly imagined the war as good as over. On that basis he handed over command to Lord Kitchener, and returned home in December after an eventful year in South Africa. General Sir Redvers Buller also departed, given an emotional send off by his troops, despite his tactical failures at Colenso and the Upper Tugela in Natal.

A despatch from General Buller, describing operations between 3 March and 18 May 1900, refers to "the excellent work done by the Telegraph Section, under Major Hawkins and Captain Kennedy". It will be recalled that Major Hawkins was the Director of Telegraphs who was locked up in Ladysmith with the section there when the town was put under siege. A further extract from the same despatch is interesting, referring to an incident when an unescorted working party constructing a telegraph line were attacked by a Boer raiding party but defended themselves and drove them off:

> I have to add that on the morning of 13th June I sent back the Telegraph Detachment under an escort of 150 men of Thorneycroft's Mounted Infantry. ... They were attacked by superior forces south of Gans Vlei, whom they drove off, and the wagons were brought back safely via Botha's Pass and Schain's Hoogte, with the loss of about 7 miles of their line, which they were unable to pick up.

And yet more plaudits, as others get mentioned in another of General Buller's despatches, dated 9 November, 1900:

> Major W. F Hawkins, Royal Engineers: As Director of Army Telegraphs with the Force, he has given the greatest satisfaction. With a thorough knowledge of technique and unwearying perseverance, the Telegraph service has been maintained in the highest state of efficiency.
>
> Captain R.H.H. Boys, Royal Engineers: Has ably assisted Major Hawkins.
>
> There may be others I should mention, as Major Hawkins has been too fully employed to render me a report; but I cannot omit a reference to the late Lieutenant Jelf R.E. A young officer of singular talent and promise, he lost his life from devotion to his duties.
>
> All officers, non-commissioned officers and men of the Telegraph Department have done exceptionally well. The only fault I have had to find with them has been that they have been sometimes too anxious to keep their line up, and have incurred undue risk.

The Guerrilla War

But it was all premature, for the Boer hard core were determined, still dedicated to their cause (independence), and in the veld where they roamed were supported by the local population. In the Transvaal, following the lead of the Orange River Colony, they too turned to guerrilla warfare - a method of operation ideal for the situation that had been reached, and in many ways much better suited to their national character and capabilities. They organised themselves into commandos under competent leadership - principally Generals Koos De la Rey and Louis Botha in the Transvaal and President Steyn and General Christiaan De Wet in the Orange River Colony. The success of guerrilla operations grew, and they spread also into the northern part of the Cape Colony. [6]

In the ensuing hide-and-seek operations, they tied down a disproportionately large number of British troops who were required to hunt the elusive enemy, mostly unsuccessfully, and protect their defences and lines of communications. It was to prove a costly and frustrating period, as the Boers intended it to be.

It is not necessary to describe in what would be repetitive detail the field communications that supported all the many scattered operations of the guerrilla war. As far as the field telegraph was concerned, they were conducted as already mentioned by the reconstituted Telegraph Division. Their first task was to send a section under Lieutenant Hildebrand on clearance operations lasting several months in the area north of Pretoria. Other similar operations, with sections commanded by Lieutenants Macfie and Mackworth, took place to the west of Pretoria and then back in the Eastern Transvaal. Throughout the remainder of the war there were many operations of this type across the breadth of the two former republics. Their general format was to use the permanent line as far as possible, then use cable to reach the various HQs, and open offices as and when required. Signalling was also better integrated into the overall communications arrangements, the heliograph being essential to this fast-moving, long-range type of warfare. Under the new organisation, the Assistant Director of Army Telegraphs in the two former Boer republics (Godfrey-Faussett in the Transvaal and Fowler in the Orange River Colony) became responsible for arranging the temporary telegraph lines needed, and integrating them into the permanent network, for the so-called 'moveable columns' that hunted Boer commandos. As Boer commando operations spread to the Cape Colony, Captain Boys later became ADAT there, and assumed a similar responsibility. The brevity of this description should not detract from the reality; for well over a year it was continuing hard and tedious work, but the telegraph operations themselves followed a similar pattern throughout, and need not be described. [7]

Visual Signalling Improves

After their setbacks in the first six months of the war the British army had learnt vital lessons and improved in all respects. The most important progress was in mobility and the development of mounted infantry.

After obtaining horses, and equestrian training, worthwhile numbers could now ride long distances and undertake raids on Boer commandos.

Also, field commanders at all levels had come to recognise the need and utility of tactical communications, and they knew they neglected it at their peril. General French was perhaps the first, as he had demonstrated at Colesberg, but most commanders were now converted. The 'Aldershot syndrome' - if that connotation may be given to the artificiality of a pre-arranged exercise around a limited area with a well-defined duration (a few hours), a precise finishing time and place, using runners or gallopers to deliver generally unimportant messages over short distances, all symptomatic of the late-Victorian army - was overtaken, after many hard lessons, by a more professional attitude to tactical intelligence gathering and command and control, the key to which at that level was the heliograph.

Limited though the means may have been in those days, they were now being used in a much more intelligent way. Basic visual signalling skills were better, attained if nothing else by constant practice. The use of the heliograph, after the faltering start at the beginning of the war, in a country where the topography and sunny climate suited it admirably, improved enormously. The changed nature of the guerrilla operations, now a less conventional and more mobile form of warfare, with small columns rather than large formations, overcame the inherently flawed organisational deficiency of signalling at formation level without established formation signalling units. For all these reasons the latter stages of the Boer war became the heyday of the heliograph in the British Army.

Moreover, communications planning was better integrated. Visual signalling and the electric telegraph was increasingly coordinated, cable being laid by line sections, usually led by NCOs, from the permanent line to the signalling station or some other convenient point. This was not without risk, for the line parties were usually unescorted, and their cables were frequently cut. Sergeant Cadwell, mentioned earlier in the description of the repair of the line from Bloemfontein to Kimberley, was one of those linemen captured along with his detachment in remote country near Ermelo when laying cable in support of operations being carried out by General French.

Spandau Kop - once a British heliograph station.

A photograph of the rather striking Spandau Kop, its shape typical of the regional geology, and not unlike Coles Kop, pictured in Chapter 4. It is located just outside the historic town of Graaff Reinet (see map at page 3), in what is now the Eastern Cape.

During the guerrilla war British troops, principally a battalion of the Coldstream Guards, formed the garrison at Graaff Reinet (the town is just to the left of the picture above) to deal with Boer commando incursions into that part of the Cape Colony. They used Spandau Kop as a heliograph signalling station to communicate with their roving mounted patrols. The cliff near the summit was a useful natural defence feature for such a vulnerable target, and the signallers had to be hauled to the top by a rope and pulley arrangement. The heliograph station could cover a large area and communicate with the patrols, passing messages to and from the garrison in the town below. There were numerous other heliograph stations performing a similar purpose

It was also at Graaff Reinet that one of the Boer commandos, Gideon Scheepers, who had commanded the Orange Free State signalling unit (see Chapter 1) before becoming one of the guerrilla leaders operating in the Cape Colony, was executed by firing squad.

Blockhouse Lines

After many unsuccessful hunts for Boer commandos, Lord Kitchener decided early in 1901 to introduce a network of blockhouse lines, firstly along railways and then across vast tracts of the open veld, both to protect the lines of communications and to restrict Boer movement as part of the 'scorched earth' policy. The important railway lines between Pretoria and Norvals Pont, Pretoria and Portuguese East Africa, and the line to Natal, were all protected by blockhouses. Then others were built in lines across the veld. When the war finished, at the end of May 1902, there were 3,700 miles of blockhouse lines.

The blockhouses were of various designs. Some, of quite substantial strength, had been built earlier in the war to guard particularly vulnerable locations - the blockhouse guarding the strategically important railway bridge over the Modder River, still standing, and pictured on page 30, is a good example. The expanded blockhouse network, introduced subsequently, consisted of much simpler and quickly erected affairs, typically made of a circular corrugated iron inner and outer shell, infilled with stone shingle to provide bullet-proof protection (the Boers had by then lost all their artillery, so nothing more was needed), and a roof to keep out the rain. They were of a standard design, developed by Major Rice of the RE, so that the parts could be prepared centrally and then distributed wherever needed for easy do-it-yourself assembly. Between the blockhouses was a barbed wire fence, fitted with an insulated wire; any pressure on the wire, or cutting it, would ring a bell in the nearest blockhouse. These prefabricated edifices, which in hot weather had the characteristics of a good oven, were built at intervals of about 1,000 yards over many hundreds of miles, each in sight of its neighbour. They were manned by legions of very bored soldiery, seven to each blockhouse. Unimaginative as they might seem, and initially ridiculed by the Boers, they went some way towards restricting freedom of movement for the marauding commandos, as they themselves later acknowledged. Numerous 'drives' within the blockhouse areas hounded and harried the increasingly desperate Boers, bringing them to their last legs, and ultimately to the conference table at Vereeniging. (8)

The standard 'Rice' blockhouse.

The blockhouse lines, however, became a new and additional communications requirement. The magnitude of this task involved the use of 9,361 miles of wire (4,413 in the Transvaal, 2,343 in the Orange River Colony, 2,513 in Cape Colony, and 92 in Natal), and the provision of 1,945 telephones. Some of this laborious work was undertaken by the various civil telegraph departments, but the majority was undertaken by the army. Two wires were provided along each blockhouse line, one for a telegraph circuit between the main command posts some distance apart, and the other for telephones in each blockhouse about 1,000 yards apart. The magneto telephones, needed in large quantities, were procured mostly from the Swedish company Ericsson, who must have been delighted at this sudden upsurge in business. The telephones were connected in series into a section of about ten blockhouses. The occupants were left to use the telephones themselves. "It was feared at first that, in the absence of a little preliminary instruction in the working of telephones, there would be some trouble, but, though a few stupidities did occur, they were not frequent, and the men in the blockhouses learnt to appreciate the value of what was their only communication with the outside world and were careful", reported Lieutenant-Colonel Hippisley. (9)

As part of this work, the opportunity was taken to construct a new and relatively secure alternative trunk line from Pretoria to the Cape along the blockhouse line through Klerksdorp to Mafeking, this being extended from Mafeking to Kimberley by the Cape Colony Telegraph Department. This new route avoided the line from Klerksdorp through Wolmaransstad, an area still giving trouble. It also met the heavy traffic demand being experienced; military traffic was heavy, and there was increasing demand from the rapidly growing commercial community of Johannesburg who simply wanted to get on with their business. As it happened, the war finished only a few weeks after this line was brought into service.

The Telephone

The Victorian army staff officer had a rather Luddite aversion to using the telephone, despite the fact that it had been invented some thirty-five years before. Whatever they may have thought about it, telephones and local telephone exchanges (as they were then called) were introduced widely, something that the DAT, Lieutenant-Colonel Hippisley, had strong views about - "in no case was the pernicious system followed of employing clerks to transmit messages." (He meant that telegraphists were not to speak on behalf of officers who would not deign to use a telephone. Times have changed!) [10]

As already mentioned, because the human voice could not yet be amplified, telephones could only be used over a limited distance, but they had great utility in the present situation. Their use in local defences has already been described, as in the defence of the besieged towns. Now the scope was widened, and they were used by the army for operational and general administrative purposes in about seventy towns throughout the Boer republics and the British colonies where the army was deployed, as well as the blockhouse lines. Altogether there were 4,105 miles of wire connecting these various local town military telephone systems.

Endnotes

1. The matter was famously taken up by Emily Hobhouse, and became the subject of much critical debate in the British Parliament. Most books about the war describe the topic in some detail.

2. Hawkins is not mentioned further in the *History of the Telegraph Operations*. It appears that he returned to England and assumed command of a Field Company RE, returned to South Africa, and became responsible for building blockhouses. He retired in the rank of Colonel in 1907.

3. Kruger died in July 1904 in a small town in Switzerland called Clarens. In December that year his body was brought to Pretoria for a state funeral, when he was accorded appropriate honours by both Boers and British. The small town of Clarens, near Fouriesburg in what is now the Free State, was named in memory of him in 1912.

4. This is a summary of the main work which was actually very intensive. Detail is given in *The History of the Telegraph Operations during the War in South Africa*, pp 28-32 and diagram 12.

5. *The Telegraph Chronicle*, 7 December 1900.

6. Interesting descriptions of Boer guerrilla operations will be found in these books: *The Three Years War* by Christiaan De Wet; *Commando* by Deneys Reitz (who later fought for Britain in the First World War, and commanded a Scottish regiment in France); *Jan Christian Smuts* by J. C. Smuts (his son); *De la Rey, Lion of the West* by Johannes Meintjes.

7. Detail is given in *The History of the Telegraph Operations during the War in South Africa*, pp 32-46.

8. More about blockhouses will be found in *The Times History of the War*, vol 5, chapter XIV.

9. *The History of the Telegraph Operations during the War in South Africa*, p 49.

10. Ibid. pp 49-50.

Chapter 11

The War Ends

Stalemate

As the war dragged on frustration increased on both sides, as well as back in England where attitudes had changed considerably since the patriotic outburst at the start of the war. The Boers were prolonging things, with occasional success on their part - on 7 March 1902, they captured General Lord Methuen during an action at Tweebosch, Western Transvaal - but their numbers were being whittled away and they were slowly getting worn out by what was now a war of attrition. Meanwhile, Kitchener, although he commanded some 250,000 men, could only tactically deploy less than a third of them to hound the remaining guerrillas. The rest of his army were either sick, wounded, or employed on static tasks such as patrolling the railways, protecting towns and isolated garrisons, and manning blockhouses, most of which was very tedious. In Britain attention focussed on the cost of this apparently interminable and increasingly controversial war - the cost over £200 million so far, and with arguments about concentration camps, what had been achieved? Everybody, apart from a few fanatical Boer leaders, wanted closure. But on what terms?

The Treaty of Vereeniging

There had been a previous attempt to negotiate a deal, between Kitchener and Botha at Middelburg in February 1901, but it failed. In early April 1902, as the South African winter approached, Kitchener again tried for peace. He allowed the commandos free passage so that they could meet and confer. There is the story of the young officer at a British outpost who thought he had won everlasting military glory by 'capturing' the much-hunted De Wet, only to be disappointed by his presentation of the magic travel pass! The Boers who passed through the outposts were given unstinting hospitality by their well-supplied foes.

So it was that the senior members of the two Boer governments, Transvaal and Orange Free State, met at Klerksdorp in early April to discuss their policy. Schalk Burger had replaced Kruger as the Acting President of the Transvaal. The generals reported on their respective situations. The real hardliner amongst them was the intransigent President Steyn of the Orange Free State - if the English refused to grant independence, the war must continue. After some days of deliberation, proposals for a 'Treaty of Friendship and Peace' were agreed by the Boer leaders. On 12 April a small delegation brought them to Kitchener in his house in Pretoria where representatives from both sides met. The two Boer republics were willing to agree to numerous issues that were previously contentious, but both presidents stubbornly refused to give up their independence - an issue that brought deadlock to the preliminary negotiation.

Matters were then discussed separately by British and Boers. There were differences of opinion amongst the leaders on both sides. Just as there were differences between the Orange Free State and the Transvaal representatives, there was divergence on the British side between Milner and Kitchener about the terms that should be demanded. As had been the case at Middelburg a year earlier, Milner wanted unconditional surrender but Kitchener was prepared to be more emollient and persuasive by offering terms which included compensation. Milner's view was that everything was subordinate to founding the new colonies immutably on British supremacy, the Boers must be clearly and unmistakably conquered, and he was prepared to fight on until the guerrilla war eventually fizzled out. Kitchener was more pragmatic. The Boers had been continually underestimated, the war wasn't going to fizzle out - it would be an ongoing irritant unless brought to a definite end. And he did not wish to haggle on detail - the Boers were onside and talking. He wanted to induce them to a settlement by generous conditions.

After the first meeting a telegram to London from Kitchener got the reply from the British government that it "would not entertain any proposals based on the continued independence of the former republics, which have been formally annexed to the British Crown". The British position was unyielding. They met the Boer leaders again on 17 April and the reply was read out. Major Birdwood, one of Kitchener's staff officers, attended the meeting and recorded in his diary: "Milner nearly spoilt the whole show by getting on his high stool. K[itchener] said it made his blood run cold for fear he should ruin everything." The meeting broke up on 18 April, the matter still unresolved. Nevertheless, Kitchener agreed that the Boer leaders could consult their people. The venue for that meeting - Vereeniging, on the railway line at the border between the two countries.

THE WAR ENDS

General Smuts, one of the brains of the Boer outfit, previously the State Attorney of the South African Republic, and a trained lawyer, was recalled from a remote part in the north-west of the Cape Colony, where he was leading a guerrilla commando as well as trying unsuccessfully to stir up a rebellion by Cape Afrikaners. Given free passage, he arrived in the Transvaal for a meeting with Botha. He was accompanied by his aide, Deneys Reitz, son of the Secretary of State for the Transvaal and a former President of the Orange Free State. Deneys Reitz, who with the rest of his commando had been living off the unscorched land in the Cape, was appalled by what he saw when he returned to the former Boer republics. In his book *Commando* he describes the scene:

> Here about 300 men were assembled. They were delegates from every Commando in the Eastern Transvaal, come to elect representatives to the Peace Congress to be held at Vereeniging, and nothing could have proved more clearly how nearly the Boer cause was spent than these starving, ragged men, clad in skins or sacking, their bodies covered in sores from lack of salt and food, and their appearance was a great shock to us, who came from the better-conditioned forces in the Cape. Their spirit was undaunted, but they had reached the limit of physical endurance, and we realised that, if these haggard, emaciated men were the pick of the Transvaal commandos, then the war must be irretrievably lost. [1]

On 15 May 1902 sixty delegates representing areas throughout the Transvaal and Orange Free State met the ten or so leaders in a tented camp set up by the British at Vereeniging, and proceedings began. The representatives reported on the situation in their districts, covering matters such as numbers and state of people, cattle and horses, stocks of food and grain, and the attitude of their 'kaffirs' - on whom many Boers were now dependent for food, but who were themselves suffering and becoming increasingly hostile. There were some proponents of 'fight till we die', and negotiating concessions were suggested, such as handing over Swaziland and the goldfields - to which Secretary of State Reitz replied: "What benefit have they ever done us? Did the money they brought ever do us any good? No, it did us harm! It was the gold which caused the war." It was recognised there was going to be no help from Europe and no Cape Afrikaner rebellion, as once thought. After many emotional contributions, Generals Botha and De la Rey from the Transvaal gave powerful, well-reasoned, sensible speeches, both concluding that it was futile to fight on. De Wet from the Free State also gave a powerful speech - but did not agree! Landdrost (Magistrate) Bosman of Wakkerstroom boldly hit the nail on the head by saying that the reason for not giving in was "obstinacy" - that part of the Boer psyche which hindered progress in their countries. It is strange that the Transvaal, which initiated the war, was ready to give in, but the Orange Free State, which had joined them as allies, wanted to fight on.

The meeting continued for three days, until 17 May. Eventually, edging forward fractionally, a new set of peace proposals were offered, but the nettle was not grasped - the fundamental question of independence was again avoided. These new proposals were taken to Pretoria where there was a meeting between Lords Kitchener and Milner and the Boer National Representatives - Generals Botha, De la Rey, de Wet, Smuts, and Judge Hertzog. The imperious and insensitive Milner continued in his 'unconditional surrender' mode, while the Boers could not bring themselves to admit loss of independence; that really was the only stumbling block.

So it was handed over to a sub-committee of the lawyers - Hertzog, Smuts, and Richard Solomon, Milner's legal adviser - and that saved the day. Kitchener, who was described by Smuts' son in his biography of his father as "a kindly understanding man [not a universal view!] who appreciated the feelings of a vanquished foe", managed to have a private tête-à-tête with Smuts and sowed a few fertile seeds. As they walked outside, Kitchener said: "I can only give it to you as my opinion ... that in two year's time a Liberal government will be in power, and ... it will grant you a constitution for South Africa".

Two days later the sub-committee had drawn up a draft settlement. Sensitive in terminology, it was drawn up as a treaty, not a surrender; this form of words made the terms more acceptable to the proud Boer spirit. For example, in the preamble, the Boer signatories were described as representing the governments of the South African Republic and the Orange Free State - to which the British, having annexed them as colonies, turned a blind eye. The first clause stated that "the burgher forces shall lay down their arms ... and shall refrain from any further opposition to His Majesty King Edward VII, whom they acknowledge as their lawful sovereign" - in other words, they were surrendering and weren't going to be independent, but the offensive words weren't mentioned. Subsequent clauses, without further explanation, refer to the Transvaal Colony and the Orange River Colony - which the Boers had never previously recognised but to which they

THE WAR ENDS

did not now object! And in substance the rest of it was the terms the Boers had rejected at Middelburg, more than a year previously. There were some additions about treatment of Natal and Cape Colony rebels, financial matters, and enfranchisement of natives. But in essence, the sacrifices of the last year had gained the Boers nothing.

The sub-committee went back to the plenary session. Botha fastened on to the financial aspects. The sensitive issue of independence was overtaken by money. There were arguments about all sorts of financial affairs and compensation. Eventually, after lengthy haggling, Milner settled for one million pounds: "Would that meet the Boer view?" "No", said Botha. With great largesse, Kitchener weighed in: "Would two or three million be sufficient then?" Not as daft as it sounded - he knew what the war was costing the British taxpayer (£1.25 million per week) and the difficulties of maintaining a large army in the country. Moreover, he wanted out, to take up his next appointment as Commander-in-Chief in India. The Boers quickly agreed £3 million. It was then agreed that the draft terms be telegraphed to London. If approved by London, the Boers must consult with their people and then answer 'Yes' or 'No', no further negotiation.

For a week the telegraph wires between South Africa and London hummed to capacity, every detail to be minutely examined and amendments agreed at both ends. On the 28th the delegates met again at Kitchener's house in Pretoria. Milner laid down the terms, reaffirmed it was to be 'Yes' or 'No', and the answer in three days time, by midnight Saturday 31 May. The Boer delegates returned to their folk in the big marquee at Vereeniging. They had done all that they could. The last act of the drama was about to unfold.

There were two days of further Boer haggling amongst themselves, raking over a multitude of Boer grievances - the refusal to negotiate on independence, the devastation the British had caused to their republics, the concentration camps, the hostility of the natives, the threats to their property being made to those who continued fighting, the overpowering forces against them, and so on. The day of reckoning arrived, 31 May 1902. Then the unthinkable happened. To his honour, and yielding to strong personal appeals from Botha and De la Rey to use his influence in favour of a unanimous vote, De Wet, acting President of the Orange Free State (the intransigent Steyn had resigned in protest and was no longer party to the negotiation), proposed an adjournment, gathered the thirty Free Staters in his tent, and persuaded all but a few irreconcilables to accept the position he had most reluctantly come to, that they were beaten and that they should now in the interest of unanimity accept the terms. In the afternoon the vote was held. The motion was carried by fifty-four votes to six. The commissioners returned to Pretoria by special train only one hour before the expiry of the deadline, and signed the Treaty. The war was over. [2]

Events after Vereeniging

This strange war between the incompatible white races in South Africa, which had smouldered for years, had finally been ignited in 1899 by the chemistry of conflicting personalities: Chamberlain, the British Colonial Secretary, imperialist, but never travelled out of Europe; Milner, intelligent, able administrator, but arrogant and inflexible; Rhodes, the arch-imperialist, now dead; Kruger, President of the South African Republic, arch-Boer, obstinate and uncouth, now fled. Sensible diplomacy on both sides had been lacking at the outset. The war had lasted two-and-a-half years and achieved little or nothing of any long-term worth.

The Treaty of Vereeniging brought about a new frame of mind on both sides and saw the reconstruction and rehabilitation of the former Boer republics, and the development of more efficient organisation and administration in which Afrikaners participated to a great extent. As Kitchener had predicted, a Liberal government came to power in Britain under Campbell Bannerman, and in January 1906 the Transvaal and Orange River Colonies were granted self-government (the difference between 'independence' and 'self-government' being somewhat obscure). In 1910, only eight years after the Treaty of Vereeniging, the Afrikaners, by sheer weight of numbers, had gained political control of the Union of South Africa, and were largely left to run it themselves. Botha was appointed Prime Minister of the new country, and Smuts was given three key ministries. The Afrikaners united to form a new pan-South African Afrikaner party. Such apparent harmony and cooperation amongst them soon evaporated as the inbred mentality of a hard core of Boers took over. In 1913 squabbling broke out between the politicians of the Transvaal and the Orange Free State, there was a miners strike, followed by a railway strike in 1914, all dealt with rather forcefully by Smuts. The 'Old Boers' set up a new National Party to fight the duo of Botha and Smuts, and the Maritz anti-government rebellion erupted; it was successfully defeated by government forces under Smuts and Botha.

THE WAR ENDS

By then the First World War had started and overtook South Africa's internal conflicts. Twelve years after the Boer War had ended, and thanks to the more cooperative elements, South Africa was fighting on the British side. During the First World War, Smuts formed the South African Defence Force, and in 1915 he and Louis Botha led the South African army into German South-West Africa (Namibia) and conquered it. In 1916 General Smuts commanded the British Army expedition in German East Africa (Tanganyika), and was a member of the British War Cabinet from 1917 to 1919. In the Second World War he became a field-marshal in the British Army and served in the Imperial War Cabinet under Winston Churchill. He served as Prime Minister of the Union of South Africa from 1919 until 1924 and from 1939 until 1948, and had a reputation as an international statesman. It was all rather different from the young man who led a commando force on operations against the British during the Boer War!

These last few paragraphs have perhaps been a diversion from the main narrative, but have been included to show what a misguided situation existed over the unnecessary Boer War and how post-war pacification changed the scene entirely.

Military Lessons and Reorganisation

Politics apart, the British army had learnt many hard lessons, and over the next decade reforms were put in place. If nothing else, the results achieved by the underestimated Boers had served the purpose of modernising the moribund organisation, tactics, and administration of the late-Victorian army before the First World War. The lessons of the Boer war also impinged on modernising military communications.

The men of the Telegraph Battalion who set off from England at the start of the Boer War can have had no idea of the extent of the work they were about to undertake - its duration, the numbers that became involved as they expanded rapidly with more regular soldiers, many reservists from the British Post Office, and a large local labour force to meet unprecedented demand. And nobody can have foreseen the diversity of their work, both in the field telegraph operations and the permanent communications network, much of the latter only summarised in this narrative. All of it was on an unprecedented form and scale, and very different to their first operational venture into the same area just over twenty years earlier when in 1879 the first established army telegraph unit, 'C' Telegraph Troop, just over 200 strong, went on their first operation to the Zulu War.

It was recognised after the war that the organisation of army communications needed to be reviewed, taking into account permanent telegraph, field telegraph, and tactical signalling requirements, the basic technology still unchanged, but wireless now coming on the scene as it developed further. The separation of signalling between regimental level and higher formation level was now seen as a barrier to a coherent communications system. The use of telegraph below divisional headquarter level had shown its worth on numerous occasions, firstly with General French at Colesberg (chapter 4) and latterly on numerous occasions during the guerrilla war. The telephone, once it had been accepted by the staff officer, was found to have been most useful in static situations and needed to be properly absorbed into the planning of communications. Most importantly of all, communications needed to form an integral part of a better structured General Staff organisation.

A War Office committee appointed in 1906 addressed these and many other issues. They recommended centralised direction of communications planning under a Director of Telegraphs and Signalling, and the formation of a Signal Service to provide communications for all headquarters down to Brigade level, to work under the General Staff. Meanwhile a Royal Commission under the Secretary of State for War, Lord Haldane, was looking at overall army structure. In the light of that, a further committee in 1911 looked at signalling structure for the army, and made three suggestions: either, to form a separate Signal Corps to provide communications for all branches of the army; or, to form an all-arms Signal Service, combining RE and regimental signallers; or, to form a Signal Service restricted to RE only. The first two recommendations were rejected for financial and administrative reasons. The third recommendation, to form a RE Signal Service was adopted, and it was formed in 1912.

While these changes were being discussed numerous organisational changes to expand the army's telegraph resources had taken place. In 1905 the Telegraph Battalion was disestablished and replaced by separate Telegraph Companies, including Wireless Telegraph Companies. Over the next few years further Telegraph Companies with different roles were raised to expand capability and fit in with the overall army organisation that was being developed. The two separate training establishments, electric telegraph under the RE at Chatham and regimental signalling at Aldershot, were combined. There were many changes as a consequence of the experience of the Boer War, the changing army organisation, and advances in technology, notably wireless. The detail need not be elaborated here; it is summarised in *The History of the Royal Corps of Signals* [3]

* * * * * * * * * * * * * *

Epilogue

Medals and Awards

Two British campaign medals were awarded during the war, the Queen's South Africa medal and the King's South Africa medal, reflecting the fact that Queen Victoria died in January 1901, to be succeeded by King Edward VII. These two medals are shown below.

The Queen's and King's South Africa medals.

All soldiers who participated in the war qualified for the Queen's South Africa medal. Twenty-six bars, for victorious battles, were awarded with this medal. The production of these medals caused controversy because they were initially issued with the dates 1899-1900 on them. When it was clear the war was going to last longer than expected the modifications had to be made at great expense. The Queen's death in January 1901 caused further argument when the new King, Edward VII, insisted that a special medal be struck with his head on it. The reverse of the King's South Africa medal is the same as the reverse of the Queen's medal - Britannia holding a wreath. The King's medal was awarded to all who served in the war on or after 1 Jan 1902. Only two bars were issued for the King's medal - South Africa 1901 and South Africa 1902.

In addition to the campaign medals the officers of the Telegraph Battalion received recognition for their services. Many were Mentioned in Despatches, some of them several times. The notable awards were these:

- **Lieutenant-Colonel R. L. Hippisley,** Director of Army Telegraphs, was appointed CB.

- **Major W. F. Hawkins,** Director of Army Telegraphs, Natal Field Force was awarded CMG.

- **Captain J. S. Fowler DSO**, Director of Orange River Colony Telegraphs, was appointed to the Brevet rank of Major.

- **Captain E. G. Godfrey-Faussett**, Director of Transvaal Telegraphs was appointed to the Brevet rank of Major.

But most notably were the awards to the two Lieutenants who had been so active throughout the war, near the front line in its early stages and then in the field during the guerrilla phase - **Lieutenants H. L. Mackworth** and **J. P. Moir.** They were both awarded the DSO, "in recognition of services during the operations in South Africa", a high award for such junior officers. They both earned eight clasps to their campaign medals.

THE WAR ENDS

Another notable award, although he has not been mentioned much in this narrative, was to Lieutenant Graham Manifold. As explained in chapter 8, he had been responsible for the railway telegraphs, an essential part of the vital railway system. For his work Manifold was also Mentioned in Despatches and awarded the DSO, as well as the Queen's South Africa medal with three clasps. Unfortunately, after eighteen months in South Africa, he contracted severe dysentery and was invalided back to England in August 1901.

The subsequent lives of these officers who had shown such ability in the difficult circumstances that they faced is summarised below. Their later achievemnts reflects the capability they had shown in the Boer War.

Colonel R. L. Hippisley CB

Richard Hippisley was promoted Brevet Colonel on 14 July 1902, and on return to England he was employed as Secretary to the Telegraphs Committee. In August 1904 he was appointed CRE and later Chief Engineer, Scottish Command. In August 1908 he went on half-pay (pending a further appointment) and retired on 2 July 1910. During the First World War he was re-employed as Deputy Director, Army Signals, Colonel Richard Lionel Hippisley died on 7 December 1936 at the age of 83.

Lieutenant-General Sir John S. Fowler KCB KCMG DSO, Colonel Commandant Royal Corps of Signals

John Fowler returned to Aldershot in October 1902, and in the following January took up the appointment of Staff Officer to the CRE in Ireland (he was of Irish extraction). He remained in Dublin until March 1905, when he was appointed as DAA&QMG of the 2nd Division in Aldershot, and served there for four years. On returning to regimental duty he took up the command of the 2nd Air Line Company at Limerick, one of the newly formed signal units, and held this appointment until the end of 1910, when he received the brevet of Lieutenant-Colonel and was selected by Major-General Sir William Robertson, the Commandant, to be an Instructor at the Staff College. He was promoted substantive Lieutenant-Colonel at the end of 1911. In April 1913, he was transferred to be Commandant of the Army Signal Schools at Aldershot and Bulford. At the start of the First World War he was appointed Director of Army Signals to the British Expeditionary Force and was at GHQ in that capacity throughout the war- an appointment in which he excelled during a period of considerable expansion of signal units.

In 1915 he was created CB, and in January 1916 he received the brevet of Colonel and was promoted Major-General a year later; he was made KCMG in 1918. His services were Mentioned in Despatches eight times and he also received, besides the war medals, the American Distinguished Service Medal, the French Legion of Honour 3rd Class, and the Order of St. Vladimir, 4th Class with swords.

He served in the Far East from 1921 to 1925, firstly as GOC Singapore, then Hong Kong and China. He was promoted Lieutenant-General in February 1926, the year in which he was made KCB.

When the Royal Corps of Signals was formed, he became its first Colonel Commandant in 1923, an appointment he held until 1934. He retired from the Army in March 1928, and died at Harrogate on 20 September 1939 aged 75.

Brigadier-General E. G. Godfrey-Faussett CB CMG

After the war he returned to the 2nd Division of the Telegraph Battalion at Aldershot, followed by staff appointments in London. In 1910 he returned to Aldershot in command of 'A' Signal Company, this new unit being part of the re-organization and expansion of the Signal Service described above.

1914 saw him in France in the first week of war as OC GHQ Signals. When the First Army was formed he became its Deputy Director of Army Signals, an appointment he also held in the Fifth Army in 1917. He was created CMG in 1915, CB in 1919, obtained his Brevet Colonelcy, and was six times Mentioned in Despatches. After the war he became Commandant of the Signal Service Training Centre, firstly at Bedford and in 1920 at Maresfield Park in Sussex until his retirement in 1922. During that period, in 1920, he saw the birth of the Royal Corps of Signals. As the responsibilities and technical requirements of Signals had expanded during the war, he had argued forcefully about the need for the formation of a separate Corps, and the new Corps owed much to him for its establishment.

He retired in May 1922 in the rank of Brigadier-General but was soon involved in other activities, amongst them the Boy Scout movement in which he played a leading role for many years. Edmund Godfrey-Faussett died on 29 May 1942 at his home at Hadlow Down, Sussex, at the age of 73.

Colonel Sir H. L. Mackworth CMG DSO

Harry Mackworth later served in Somaliland 1903-04 (at Jidballi, where he won the African General Service medal with two bars), Sudan, Australia (in 1913, where he was Director of Army Signals, and also met and married his wife, Leonie), Gallipoli, Palestine and Egypt. During this period he was appointed to: the Order of the Osmanieh; the Order of the White Eagle, by the king of Serbia, and was appointed CMG in 1918. He transferred into the Royal Signals on their formation in 1920, and rose to the rank of Colonel before retiring in 1927 to Studland, Dorset. He succeeded to the baronetcy in 1948. Sir Harry Llewellyn Mackworth CMG DSO died in 1952, aged 74. For a family history, see *Burkes Peerage and Baronetage*.

The medals of Colonel Sir Harry Mackworth CMG DSO, displayed in the Royal Signals museum.

Lieutenant-Colonel J. P. Moir DSO

After the war James Moir was employed in south Nigeria from 1905 to 1908. He was promoted to Major in February 1912,and then served as Director of Posts and Telegraphs, Egyptian Army, from 1912 to 1917. He was awarded the Order of the Nile, 3rd Class, and was given the Brevet of Lieutenant-Colonel in January 1917. He continued to serve, with postings in south Nigeria and the Sudan, before retiring in 1929. He died in the King Edward Convalescent Home for Officers in Osborne House, Isle of Wight, in 1934, aged 61.

Major-General M. G. E. Bowman-Manifold KBE CB CMG DSO, Colonel Commandant RE

Having in 1901 acquired the additional surname Bowman, Graham Bowman-Manifold returned to work in the Post Office within the 2nd Division of the Telegraph Battalion, taking some years to recover fully from the illness he had contracted in South Africa. He then continued his career in various regimental and staff appointments in England, became an interpreter in Russian, and on the outbreak of war in 1914 went with the British Expeditionary Force to France in charge of Signals, 1st Corps, under Sir Douglas Haig. In 1915, and as Brevet Lieutenant-Colonel, he was appointed Director of Army Signals to the Mediterranean Expeditionary Force in the rank of Colonel, later promoted to Brigadier. This was followed by a similar appointment to the Egyptian Expeditionary Force which in 1918 , under Sir Edward Allenby, invaded Jerusalem and advanced up to Aleppo, involving the planning and operation of communications over a very large area. For all his various work during the war he was Mentioned in Despatches eight times, and awarded the CB, CMG, and KBE. In August 1921 he was promoted to Major-General, and retired in 1924, taking up civilian employment. He died in 1940, aged 69.

Endnotes

1. *Commando.* pp316-7. See bibliography.

2. For a full description of the intricacies of the negotiations see *The Boer War* by Pakenham, Chapter 42, and *The Times History of the War,* vol 5, Chapter XXI.

3. *The Royal Corps of Signals - A History of Its Antecedents and Development* by Maj Gen R. F. Nalder, London, Royal Signals Institution, 1958, pp 50-53.

Bibliography

The war produced a profusion of books, mostly published a few years after the war, but consequently now out of print and difficult to obtain. They are, however, usually available in specialist libraries.

Predominant amongst them are:

Amery, L. S. (Gen Editor). *The Times History of the War in South Africa 1899-1902.* 6 volumes. Pub London, Sampson Low, Marston, 1907.

> An immense work of great detail written in the years immediately after the war, with contributions from many authors, but impaired by Amery's partisan support for the now discredited Lord Milner, whose inflexibility, arrogance, and high-handedness largely instigated this controversial war, as Pakenham's 1979 book (see below), using extensive archive material, reveals. Vol 6, pp 354-363 contains a post-war report on the army telegraphs.

Maurice, Maj Gen Sir Frederick. *The History of the War in South Africa 1899-1902.* Compiled by the direction of His Majesty's Government. 4 volumes. Pub London, 1906.

> Written shortly after the war, another immense work of great detail, concentrating on the military course of the war. Diplomatic activity to heal the scars of the war were under way, and the authors of this 'Official' history were under government orders to avoid all political and other controversial matters, so it is a sanitised but detailed account of operations only.

Wilson, H. W. *With the Flag to Pretoria: A History of the Boer War, 1899-1900,* Pub London, 1901.

> An interesting general history of the first part of the war.

Doyle, A. C. *The Great Boer War*, Pub London, Smith, Elder & Co, 1902.

> A general description by the author better known for his (Conan Doyle) Sherlock Holmes novels.

* * * * *

A short list of other books that the reader might consider are:

Pakenham, Thomas. *The Boer War.* Pub London, Weidenfeld & Nicolson, 1979. ISBN 0-349-10466-2.

> This well-written masterpiece is the one to read first. Very well researched using much archive material not available in the immediate aftermath of the war, it gives a good general description of the war, and the controversial political background, as well as revealing the animosity that existed between many of the prominent personalities, Boer and British, military and political. However, it says next to nothing about communications in the war! Easily obtainable, and available in paperback.

Hippisley, Lt Col R. L. *History of the Telegraph Operations during the War in South Africa, 1899-1902.* Pub for HMSO by Eyre and Spottiswoode, London, 1903.

> The author of this book was the Director of Army Telegraphs throughout the war, and it contains considerable detail of the telegraph operations. Copies are held in the Royal Signals archives and the RE library.

Kemp, R. E. *Khaki Letters from My Colleagues in South Africa.* Reprint published by the Naval and Military Press in association with The National Army Museum, London.

> A reprint of a book which was originally published soon after the war. It contains letters giving details of their personal experiences from the many reserve army Post Office telegraphists who were sent from England to reinforce the army's Telegraph Battalion, compiled by Colour Sergeant R. E. Kemp of the 24th Middlesex (P.O.) Rifle Volunteers. It gives an excellent soldier's insight to the war, as seen by the telegraphists on the ground as events happened.

Riall, Nicholas. *Boer War. The Letters, Diaries and Photographs of Malcolm Riall.* Pub London, Brasseys, 2000.

> Never a book that was going to hit the best-seller lists, but its particular interest in this context is that Malcolm Riall was a regimental signalling officer with his battalion (The West Yorkshire Regiment, commanded by Walter Kitchener, Lord Kitchener's brother) in the Boer War, and also an amateur photographer. He carefully recorded events, both in his diary and with his camera. His grandson, Nicholas, the author, recovered this material from a family attic and wrote the book. A useful insight into regimental signalling in the war - an area which otherwise lacks much recorded detail.

Carver, Field Marshal Lord. *The Boer War.* Published Sidgwick and Jackson, 1999.

> Written in conjunction with the National Army Museum, it gives a concise history of the war using material held by the Museum, but does not look at the wider scene.

De Wet, C. R. *The Three Year's War.* Published 1902, Constable's Indian and Colonial Library.

> The elusive De Wet's account of his very active part in the war, particularly as a Boer commando leader.

Reitz, D. *Commando.* Published 1929, Faber & Faber.

> Another Boer's story of the war. The author, Deneys Reitz, son of a senior Boer politician and a Smuts protégé, refused to sign up to British rule after the war and went to Madagascar. He returned to South Africa after a few years, became pro-British, volunteered to fight for Britain in the First World War and commanded a Scottish infantry regiment in France. He became involved in South African politics, and ended up as South African High Commissioner in London in 1943. A fascinating story.

19th Century Telegraphy - a Glossary of Terms

Airline	Generally a military term. Uninsulated wire suspended on poles and insulators, using an earth return circuit. The insulators prevented electrical 'leakage' to earth. Airline gave a better quality electrical circuit than cable (see below), enabling better instruments to be used. It was less prone to damage by one's own troops, but was slower to build and an easy target for disruption by enemy troops, the latter very frequent in hostile territory.
Baseboard	A board on which were mounted the telegraphist's instruments - Morse key, sounder, relay, and galvanometer. They were permanently connected, so that the only other connections needed were to line and to battery for which there were clearly marked terminals.
Cable	An insulated and flexible single wire made up of twisted strands of copper and steel to combine electrical conductivity and strength. It could be wound on reels and laid directly on the ground from a horse- or mule-drawn cable cart, thus was fast to lay and could be rewound, but, lying on the ground, it was prone to damage, and more expensive than wire. Various cable carts were designed to carry and lay the cable. It was common practice to lay a cable quickly behind the advancing HQ, to be replaced by airline later.
Cipher	A cipher is a rearrangement or replacement of the characters of a message, almost always with the aim of secrecy. In the 19th century, cipher was either by the 'cipher wheel' or, more securely, by use of cipher books held by the originator and recipient of the message.
Code	From *codex*, Latin for a book. The word 'code' became used rather loosely to embrace two different concepts – as a signalling alphabet (eg Morse code), and as an abbreviation to economise on time of transmission or message length (eg SOS). A code is not a security device – see cipher.
	The idea of a signalling alphabet had been used by earlier inventors such as Gauss, Weber, Steinheil, and others, but the alphabet that came into use with practical telegraph systems in the mid 1830s was the Morse alphabet, which of course is universally called the Morse code - although it was actually invented by Morse's partner, Alfred Vail. (Like many of their inventions, Vail did the work and Morse took the credit!) The original Morse code was designed for the English language, so that the most-used letters were the shortest code (eg the most frequent letter is 'e' is simply a 'dot', whereas an infrequent letter such as 'q' is 'dash-dash-dot-dash'); it was later modified to become the International Morse code, to take account of other languages. The Morse code, originally designed for electric telegraph, was also used for visual signalling by heliograph, flag, or lamp, and later for hand-operated wireless telegraphy.
	Operating codes were developed to speed transmission, and these were definitely codes not alphabets. Typical is the 'Q' code, being abbreviations for commonly used operating procedures or instructions. They eventually became rationalised by international agreement. Special military operating codes were also used.
	As telegraphy developed from the end of the 19th century new 5-bit and 7-bit machine codes were developed, such as Baudot, Murray, ASCII, etc but these were not in military use in the 19th century. Morse remained the code for manual transmission throughout the 20th century although this method has has by now ceased to exist.
Commutator	A rotating device that acts as a directional switch.

Diplex	Sending two messages at the *same time* in the *same direction* (contrast with duplex). Achieved by using two relays at the receiving station: one a polarised relay which which works when the *direction* of the current is changed, whatever its strength; and the other a non-polarised relay which works whatever the direction of the current but only when the *strength* of the current is increased. Combining diplex with duplex produces quadruplex – see below.
Direct working	Method of connection of a telegraph circuit so that the current that works the sounder at the receiving station is drawn from the battery at the sending station. Depending on the detailed method of connection, direct working may be either 'intermittent' or 'continuous'. Intermittent saves on battery power, drawing current only when the key is pressed down. Continuous working maintains a current in the circuit when messages are not being transmitted, thus deflecting the galvanometer needle. The principal advantages are that it enables any break in the line to be observed immediately, and a battery is not needed at every station. The disadvantages are that the battery becomes exhausted more quickly, stations without batteries located on the far side of any fault that may occur are unable to work, and the single current key is not suitable for this type of working. Direct working was avoided if possible.
Down station	Stations in a telegraph circuit that are subordinate to the Up station. See 'Up station'.
Double current	A technique which improved the accuracy of Morse transmission over long lines, especially submarine cables. The principle was to transmit the current one way down the line during the spaces between the dot and dash signals and to reverse the direction of the current when either the dot or the dash were transmitted. This reduced errors by causing a very positive change in the position of a relay-controlled switch.
Double current key	Type of morse key needed for double current working.
Duplex	A technique for transmitting messages down a wire in opposite directions simultaneously, thus doubling the capacity of the line (although in practice at any one time the weight of traffic was usually in one direction). Two methods were used: the differential relay developed by Stearns from about 1873, and a bridge method developed by Muirhead in 1875, the latter in response to the rather different electrical characteristics of submarine cable.
Earth return	Field telegraphy always used a single wire and an 'earth return' circuit. This technique was discovered early in the development of telegraphy by Steinheil. It saved on the length of wire needed, and thus the weight of line - an important factor in field telegraphy. in days when there wers such transport difficulties. The earth's resistance is very low, but it depends on a good earth connection (achieved by driving a metal spike into the ground) and this was often difficult in dry, stony ground such as in many parts of Africa.
Electromagnet	A vital discovery in the early development of telegraph, accredited to Stearns in 1825, although its full potential was not realised until years later. The electromagnet is a vital part of the relay, acting as a switch when it detected current changes, which is the basis for actuating receiving equipment and the telegraph repeater.
Field sounder	Instrument used by the Telegraph Battalion, consisting of a sounder, galvanometer and Morse key, grouped on a board with terminals for line, earth, and battery.
Galvanic electricity	Today we would call it direct current (dc), and all telegraphy to the end of the 19th century depended on dc principles. It is the electricity that flows from a power source such as the original Voltaic cell, or as it would be called today, a battery.

	This form of electricity, in contrast to earlier known forms such as static electricity, was not understood until Ohm's Law was developed.
Galvanometer	A needle, magnetised by a coil, designed to indicate if a current is flowing, its direction, and to some extent its strength. It became a sensitive receiving device used in early needle telegraph systems. Later it was used mainly for test purposes, the instrument being known as the galvanometer, single and duplex, or in those days commonly the 'S and D galvo'.
Inker	See Recorder.
Leakage	Loss of current to earth along a line due to leakage caused typically by poor insulation of the line at the supporting poles, and made worse by damp or wet conditions. Overcome by the use of relays (see below).
Leclanché cell	Primary cell most commonly used in telegraphy in the late 19th and early 20th century. Consisted of plates of zinc and carbon immersed in ammonium chloride, delivering about 1.5 volts. Could be made in 'dry' form, therefore portable and much used as the power source for field telegraphy. Deteriorates in storage, especially at higher temperatures.
Local circuit	A low resistance circuit set up at the end of a long line, typically in a telegraph office, with its own battery, so that a weak received signal can be boosted by the local battery to operate a receiving instrument such as a sounder (which on its own would not be sensitive enough).
Morse key	The morse key used by a telegraphist to transmit messages. A very good telegraphist could achieve speeds of 35 wpm (words per minute), but the norm was about 20 wpm. See also 'single current key' and 'double current key'.
Polarised (and non-polarised)	Term applied to a telegraph instrument which depends for its operation on the correct direction of flow of the current. Eg a relay, which depends on the operation of an electromagnet. Conversely a non-polarised instrument does not depend on how it is connected. Eg a sounder. (See also Up station.)
Quadruplex	A further development of duplex (see above), enabling four messages to be sent simultaneously, two in each direction along the same wire. Developed by Edison in 1874. Very useful when traffic levels were high. Requires careful setting up of instruments and experienced telegraphists.
Recorder	A recorder (often called an 'Inker', and in America it was called a 'register') is a telegraph receiving instrument which will make a permanent record of the received message, thus not needing a skilled operator to write the message down as it is received. A paper tape is driven by clockwork and as 'dots' and 'dashes' of electric current are received, an armature is magnetised, causing a lever to dip the paper tape into an ink trough. When the current ceases, a spring action lifts the moving tape out of the ink. It is more sensitive than a sounder, so could be used on longer lines. However, the inked tape subsequently has to be transcribed on to a message form. With good operators it was much quicker to read by ear from a sounder and write the message immediately.
Register	The American term for a Recorder (see above).
Relay	A sensitive instrument, powered by a local battery, and depending on the operation of an electromagnet, used on long lines either as a repeater (to relay the signal onwards down the line), or, in conjunction with a local circuit and a sounder, to receive the message.
Repeater	An instrument used to retransmit signals down a further section of line (see also 'translation'). In field telegraph it was usually confined to single current working.

	Repeaters were able to work on double current circuits, and duplex, but these were generally too complex for field telegraph.
Separator	A simple instrument consisting of a combination of an electro-magnet and two capacitors (they were called 'condensers' at the time). They enable one line to be used for two separate circuits - one with vibrators and one an ordinary Morse circuit without interference, and thus double the working capacity of the line.
Single current	The current flows in a single direction along the wire – either it is there and detected by a sensitive device such as a galvanometer, or it is not present. See also 'double current'.
Single current key	A type of Morse key (see above) for use with single current working.
Sounder	An aural receiving device used with Morse code transmissions. Using his Morse key, the sending telegraphist produced Morse code 'dots' and 'dashes' which turned the electric current on and off for short or long periods. At the receiving station this was detected by the sounder. When the current was 'on' it would magnetise an electro-magnet, which would attract a lever causing it to strike a bracket, making a 'click'. When the current was 'off' the lever would return to its rest position, and in doing so it would make a different sound - a 'clack'. Thus the intervals between 'clicks' and 'clacks' indicated whether a short 'dot' had been sent, or a longer 'dash' had been sent. To operate effectively the sounder needed careful adjustments by a well-trained telegraphist, and it needed a current of at least 55 milliamperes. It was preferred to earlier needle telegraphs as it was faster, one operator writing down the message as it was received, rather than one observing the needle and another writing the message. The sounder was the basic Morse receiving instrument used by army telegraphists.
Tapping a line	The same as teeing-in (see below), but more generally meant to indicate 'listening in', to send a message or get information, possibly 'sigint'.
Teeing-in	Connecting to an existing telegraph circuit, usually an airline, somewhere along its length, so that telegraph contact could be made with one of the telegraph offices along the line, using a morse key and typically a vibrator. This method was often used in war, typically by a senior commander and his accompanying telegraphist, to communicate with HQ. First recorded use was in India during the Mutiny in 1857, when General Roberts tee'd-in to the civil telegraph to receive and transmit.
Translation	The process of receiving a weak signal at the end of a long section of line and, without operator intervention, retransmitting it on the next section of line. This is typically achieved by a relay at the receiving station operating a local circuit and another relay, with local battery, to retransmit the signal down the next section. The process requires the addition of a translating sounder so that the operator at the repeater station can hear the signal to ensure satisfactory operation and make any adjustments to the relays. The instrument is known as a 'translator' or 'repeater'.
Up station	In any telegraph circuit one station, usually the most important, is known as the 'Up' station. The remaining telegraph stations on the line are known as the 'Down' stations. The line from the 'Up' to the 'Down' stations is known as the 'Down' line, and in the opposite direction the 'Up' line. The object of this arrangement is that the current always flows in an agreed direction and therefore the batteries and instruments on the line are connected correctly (see 'polarised').
Vibrator	More properly, a vibrating sounder. A robust and sensitive receiving device, making it suitable for military use, having the same mechanism as a electric bell but without the bell's dome, so that we might today call it a 'buzzer'. It derived its sensitivity from the use of a telephone as the receiving instrument. The first

	vibrating sounder was invented by Captain Phillip Cardew, Royal Engineers, one of a number of inventions of his.
Wheatstone automatic	An instrument that enabled much higher speeds than were obtainable by manual methods (eg over 200wpm, about ten times as fast as an average manual telegraphist). Thus its traffic capacity was high, and it was relatively secure as in practice it could not be be intercepted at that time. Messages could be prepared off-line on punched tape by a number of operators. It was a complicated instrument requiring particularly trained telegraphists, not robust enough for general military service, so not issued to the army, but it was used in the Boer War by Post Office telegraphists in permanent offices over permanent lines when the civil system was taken over.
Wire	The uninsulated wire used in airline, as distinct from cable. At the end of the 19th century military field telegraphs used stranded galvanised iron wire less than 5mm thick, but for long lines No 14 hardened copper wire was used, with better conductivity but difficult to manipulate and prone to breaking. The early measurement for a wire's diameter was BWG (Birmingham Wire Gauge), replaced later by SWG (Standard Wire Gauge).

A Glossary of Afrikaans Words

An explanation of names used in the narrative or found on maps

Bad	A spring, bath.
Berg	A mountain.
Biltong	Dried meat.
Boer	Literally a farmer; a generic term for South Africans of Dutch origin.
Brandwacht	An outpost, or picket; literally beacon or camp fire.
Bult	A ridge; literally a hump.
Burg	A town; literally a borough.
Burgher	A male inhabitant or citizen of one of the Boer Republics who possessed full political rights (in contrast to Uitlanders).
Bush	Country covered in a varying degree with trees and undergrowth.
Bushveld	Generally used in the Transvaal in reference to the low veld, in contrast to the high veld of the south and east.
Commandant	Senior officer of a commando; a commander.
Commando	A Boer military force of any size, usually the fighting force of one district.
Corporal	Assistant to a Veld-Cornet.
Donga	A ditch or cutting made by the action of water draining, often dry.
Dopper	A Boer Calvinist religious sect, and to some extent political.
Dorp	A village.
Draai	A bend.
Drift	A ford.
Fontein	A spring; literally a fountain.
Hoek	A re-entrant in a range of hills, literally corner; also used for a blind valley (no exit at the other end).
Hout	Wood.
Inspan	To attach transport animals of any kind to their vehicles; to get ready to march, to harness-up.
Kloof	Ravine, a gorge; literally a cleft.
Kop	A hill; literally head.
Kopje	A small hill.
Kraal	Native village, or collection of huts; an enclosure for cattle.
Kranz, or krans	Cliff.
Krijgsraad	War council.
Laager	Camp, bivouac.
Landdrost	Boer magistrate.
Nek	A pass or saddle between two hills of any height.

Pan	A pond, full or empty; a saucer-like depression, usually dry in winter.
Plaats	House or farm. The term is equivalent to an English estate.
Pont	A ferry-boat or pontoon, worked by ropes or chains.
Poort	A gap, breaking a range of hills; literally gate.
Rand	Ridge or edge, *eg* the edge of a plateau.(eg 'Witwatersrand' is white water ridge)
Schanz	Stone entrenchment or breastwork.
Sloot or sluit	Open watercourse; an artificial ditch or gutter.
Span	A team of animals, usually oxen.
Spruit	A stream or small river, sometimes dry.
Stad	Town.
Stoep	A platform in front of a house; a verandah.
Uitlander	Literally 'Outlander'. Immigrants who worked in the Transvaal.
Uitspan	To detach animals from their waggons; to halt; to unharness.
Vallei	Valley.
Vecht-general	Fighting General, as opposed to the Administrative General.
Veld	The open country, as opposed to the town; literally field.
Veld-cornet	The senior officer of a ward or sub-district.
Vlei	A small lake or pond.
Volksraad	Parliament; People's Council.
Weg	Way, road.
Winkel	Shop or store.
Zarp	A member of the Transvaal Police (Zuid-Afrikaansche Republiek Politie).
Zwart or swart	Black.

Be recognised for your professionalism

With you now and for the rest of your career.

Professional registration provides recognition of your military skills and experience and may mean you are eligible for up to £3,000 once achieved*.

Become professionally registered with the IET.

We are licensed by the Engineering Council to award CEng, IEng, EngTech and ICTTech. With IET membership discounts available for technicians and annual fee reimbursement by the MOD, there is no better time to apply.

Find out more by registering for our latest webinar

What you will learn
- An overview of the IET
- Membership benefits
- Professional Registration Categories
- Overview of the UK SPEC
- The benefits of Professional Registration
- The Professional Registration Application process
- The guidance and support we offer

Register here:
theiet.org/royal-signals

Heather Brophy, MOD Development Manager
heatherbrophy@theiet.org

*More information can be found by asking your Trade or Branch Sponsor about the Engineering Professional Registration Award (EPRA).
The Institution of Engineering and Technology is registered as a Charity in England and Wales (No. 211014) and Scotland (No. SC038698).
The Institution of Engineering and Technology, Michael Faraday House, Six Hills Way, Stevenage, Hertfordshire SG1 2AY, United Kingdom.

© Crown copyright 2019

A STRATEGIC PARTNER TO THE BRITISH ARMY

IN SHAPING A MULTI-DOMAIN WARFARE NETWORK

Elbit Systems UK is committed to offering advanced and proven technologies to the British Armed Forces, delivered locally through our UK subsidiaries addressing requirements for UK sovereignty and freedom of action.

elbitsystems-uk.com

in Elbit Systems   @ElbitSystemsLtd

Promoting friendship and comradeship across the World to the serving and retired Corps family.

AIMS

Promoting friendship and comradeship across the world to the serving and retired Corps family.

One of our aims is to foster friendship among all members of our Corps family. We are especially keen to provide more for younger members and appreciate any constructive ideas you may have to help us appeal to this group.

ELIGBILITY

All serving and retired members of the Corps, Regular and Re-serve, and past members of the ATS and WRAC attached to the Corps are eligible to become Life Members of the Association.

Associate Membership is open to spouses, Widows/Widowers and anyone who promotes the Aims of the RSA and becomes an active member of a Branch.

BRANCHES

The Association has over 60 branches and a number of affiliated groups spread throughout the UK and across the world. These normally meet on a monthly basis.

The Association also has virtual branches who share a common interest, or are based on now historic units who meet on an ad-hoc basis. RHQ branch is for members who are do not want to join any of the branches above.

EVENTS

The RSA has various events throughout the year from Corps Weekend to Founders day and the Cenotaph.

Branches hold their own events throughout the year, more information is available on the website or through your branch.

CONTACT US

ROYAL SIGNALS ASSOCIATION SECRETARY - MRS AMY THORPE

Email: RSA@royalsignals.org or amy.thorpe108@mod.gov.uk

Office Tel: 01258 482090

Mobile: 07542 167963

Website: www.royalsignals.org/royal-signals-association

Facebook: www.facebook.com/groups/royalsignalsassociation